Mechanochemistry: From Functional Solids to Single Molecules

Montréal, Canada

21–23 May 2014

FARADAY DISCUSSIONS

Volume 170, 2014

ROYAL SOCIETY OF CHEMISTRY

The Faraday Division of the Royal Society of Chemistry, previously the Faraday Society, founded in 1903 to promote the study of sciences lying between Chemistry, Physics and Biology.

Faraday Discussions (Print ISSN 1359-6640, Electronic ISSN 1364-5498) is published 8 times a year by the Royal Society of Chemistry, Thomas Graham House, Science Park, Milton Road, Cambridge, UK CB4 0WF.

Volume 170 ISBN-13: 978-1-78262-170-6

2014 annual subscription price: print+electronic £827, US $1542; electronic only £785, US $1465. Customers in Canada will be subject to a surcharge to cover GST. Customers in the EU subscribing to the electronic version only will be charged VAT. All orders, with cheques made payable to the Royal Society of Chemistry, should be sent to RSC Order Department, Royal Society of Chemistry, Thomas Graham House, Science Park, Milton Road, Cambridge, CB4 0WF, UK.
Tel +44 (0) 1223 432398; E-mail www.orders@rsc.org

If you take an institutional subscription to any RSC journal you are entitled to free, site-wide web access to that journal. You can arrange access via Internet Protocol (IP) address at www.rsc.org/ip. Customers should make payments by cheque in sterling payable on a UK clearing bank or in US dollars payable on a US clearing bank.

Printed in the UK

Faraday Discussions documents a long-established series of Faraday Discussion meetings which provide a unique international forum for the exchange of views and newly acquired results in developing areas of physical chemistry,
biophysical chemistry and chemical physics.

Mechanochemistry

Faraday Discussions

www.rsc.org/faraday_d

A General Discussion on Mechanochemistry was held in Montréal, Canada on the 21st, 22nd and 23rd of May 2014.

CONTENTS

ISSN 1359-6640; ISBN 978-1-78262-170-6

Cover
Image courtesy of Dr Richard Blair, University of Central Florida, and Hope College, Michigan, USA.

INTRODUCTORY LECTURE

PAPERS AND DISCUSSIONS

SPONSOR

Retsch®

Solutions in Milling & Sieving

CONCLUDING REMARKS

ADDITIONAL INFORMATION

Faraday Discussions

PAPER

Introductory Lecture: Mechanochemistry, a versatile synthesis strategy for new materials

William Jones* and Mark D. Eddleston

Received 28th August 2014, Accepted 2nd September 2014
DOI: 10.1039/c4fd00162a

Mechanochemistry deals with reactions induced by the input of mechanical energy – for example by impacts within a vibratory ball mill. The technique has a long history with significant contributions from Ostwald, Carey Lea and, notably, Faraday. Mechanochemistry has subsequently seen application in a variety of areas of materials science including mechanical alloying in metallurgy, the synthesis of complex organic molecules and, more recently, the discovery and development of new solid forms of active pharmaceutical ingredients. This paper overviews the broad areas of application of mechanochemistry, some key features which make it a particularly attractive approach to materials synthesis and some mechanistic aspects highlighted within the literature. A significant part, however, will focus on recent applications in the area of pharmaceuticals and its important role in exploring the rich variety of solid forms available for small, drug-like, molecules.

1. Introduction

Mechanochemistry is a long-established method for the synthesis of solids and molecules. Its chemical importance was recorded by Faraday, who in 1820 demonstrated the reduction of AgCl to Ag by grinding in a mortar and pestle a mixture of AgCl and Zn.[1] Significant progress followed, especially through the work of Carey Lea (1823–1897),[2] and the position of mechanochemistry within the framework of chemistry as a whole was established by Ostwald who noted the equivalency of electrochemistry, thermochemistry, photochemistry and mechanochemistry.[3–5]

Long held as an efficient process for the synthesis of metallic alloys, a recent extensive review by James *et al.*[6] has highlighted several of the emergent areas now being actively pursued with regard to the application of mechanical energy in driving various transformations – indeed a working definition from that review is "mechanochemistry refers to reactions, normally of solids, induced by the impact of mechanical energy such as grinding in ball mills". A formal definition by IUPAC states "a mechano-chemical reaction is a chemical reaction that is induced by

Department of Chemistry, University of Cambridge, Lensfield Road, Cambridge, CB2 1EW, UK. E-mail: wj10@cam.ac.uk

mechanical energy".[7] Although a narrow definition of mechanochemistry may lead to a suggestion that the making and breaking of covalent bonds would be a prerequisite for a process to fall under the definition of "mechanochemistry",[8] following the appreciation of the importance of "non-covalent bonds" in supramolecular chemistry it seems reasonable to include changes in crystal packing to be a mechanochemical event. Links to mechanochemical activation by ultrasound have also been proposed.[9,10]

This paper will briefly summarise recent developments in mechanochemistry, particularly in its application in pharmaceutical materials science and its role in the development of new solid drug dosage forms, to complement earlier literature which has covered extensively metals and alloys and inorganic (ionic) systems. As we will describe later, the amount of energy which can be imparted to a system under mechanical activation can be significant – certainly sufficient to break chemical bonds[6] – but can also be reduced when working with systems such as soft molecular crystals in order to minimise chemical activation while still enabling processes such as the conversion of a crystalline solid to an amorphous product,[11] or the readjustment of molecules within a lattice. Such readjustments may lead to the introduction of lattice imperfections,[12] to polymorphic transformations[13] or to the intimate mixing of two separate crystalline molecular solids to create a crystalline multicomponent product indistinct from that which would be obtained by a conventional solution crystallisation.[14] Particularly advantageous in mechanochemical transformations is the avoidance of the need for large volumes of solvent (as required, for example, in the recrystallization of poorly soluble molecules), and hence a strong "Green Chemistry" aspect to mechanochemistry is recognised.[15,16] In terms of commercial interest, the review by James *et al.* describes a search of patent applications for the terms "mechanochemical" and "mechanochemistry" which revealed a significant increase in usage in the patent literature from 1980 onwards.[6]

2. Areas of application and development

The areas in which mechanochemistry has been applied with success are varied and include: catalysts,[17] nanoparticles,[18–20] organic synthesis,[21–27] MOFs,[28] oxides,[29] biomaterials,[30,31] supramolecular chemistry,[32–34] process engineering,[35] dyes and pigments,[36] fluorophores,[37] exfoliation of graphene nanosheets[38] and modification of fullerenes (see Fig. 1).[39] Numerous further examples are given in the review by James *et al.*[6] and several other reviews on the application of mechanochemistry are available.[28,40–47] In particular, the contributions of V. V. Boldyrev and his colleagues to both experimental and theoretical aspects of mechanochemistry have been important to the subject.[3,9,48,49]

From a synthetic chemistry viewpoint an important observation is that substances can react by a significantly different pathway in the solid state than they would in solution, leading to new products or to much higher selectivities.[50,51] This ability to generate a much broader product-spectrum, as we will demonstrate later, is also particularly important in areas other than chemical transformations *e.g.* in crystal polymorph conversions – and it is to be noted that the addition of very small amounts of liquid may have a significant impact on the outcome of a solid–solid reaction, even though quantities are much less than a conventional solvent-based reaction might require – see later.

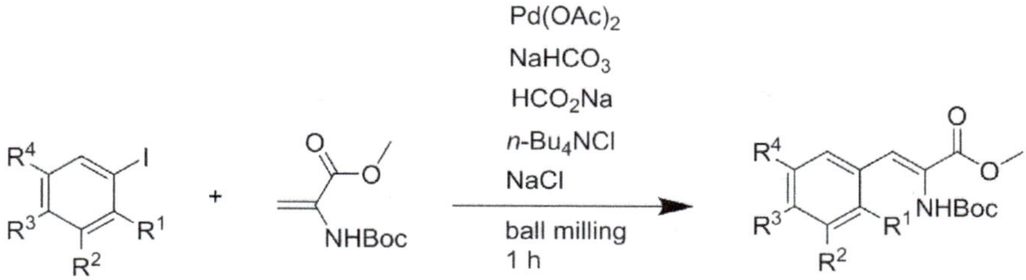

Fig. 1 General scheme for the Heck reaction under ball milling conditions. Reproduced from ref. 6.

A key area of active research remains the understanding of the processes by which mechanically induced reactions occur.[52] Given the variety of materials of interest, ranging from metals and ceramics to pharmaceuticals, the specific details of how a process occurs will likely be varied and complex.[3,53,54] Also to be distinguished is the time-scale over which an event is studied. During the few milliseconds following the impact of two solids at high velocity, significant local consequences will ensue. Hot spots, magma–plasma regions, rapid induction of defects (point, linear and planar) and then propagation of these defects through the crystal are some of the concepts which have evolved. The distribution of imparted energy will also be different in those solids with high rigidity and few pathways for energy dissipation (so that the effects at the specific point of impact will be great) than it is in those systems where structural movement (*e.g.* slippage of crystal planes) may allow the energy (and its consequences) to be distributed over a much larger distance.[3] During this early stage, the making and breaking of chemical bonds would seem possible. Following the initial short-term consequences, there will also be the longer term events associated with the final energy dissipation and relaxation of the system. The slow crystallisation of an intermediate amorphous phase may be such an example (see Fig. 2 and 3).[11,52,55,56] It is possible, then, to envisage that an individual particle, for example within a milled pharmaceutical material, may be completely amorphous initially, but over time

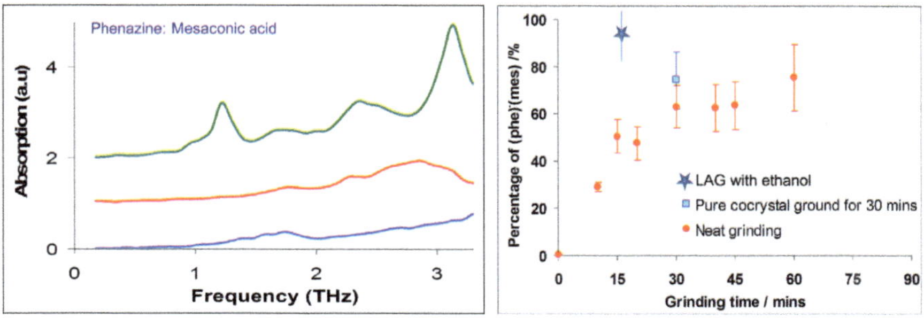

Fig. 2 (a) Terahertz spectra obtained from (top to bottom): a 1 : 1 mesaconic acid–phenazine cocrystal; mesaconic acid; phenazine. (b) Analysis (*ex situ*) of the extent of formation of the cocrystal from a mixture of phenazine and mesaconic acid, as a function of neat grinding time, as measured using the characteristic cocrystal band at 1.1 THz (red circles). The blue star, by comparison, shows close to 100% conversion in 15 minutes as a result of the presence of liquid in a LAG experiment. The blue square represents the percentage of a cocrystal sample that remained crystalline during dry grinding.[56]

may change to consist of domains of crystalline solid embedded within an amorphous shell (or more likely *vice versa* where, for example, humidity from the surrounding atmosphere results in each amorphous particle crystallising from the outside). This model might suggest that a description of a product as 25% amorphous does not mean that 25% of all the particles are amorphous and 75% are not – but that each individual particle within the product consists of domains of crystalline and amorphous in a 3 : 1 ratio.

A key factor for potential commercial exploitation in many instances would seem to be difficulties associated with scale-up. The progression from gram-scale, lab-based synthesis to kilogram-scale or higher outputs is an issue that has been addressed in many industrial applications[3] and more recently in the area of pharmaceuticals, where twin extrusion approaches are employed (Fig. 4).[57,58] Furthermore, it is noteworthy that frequently the generation of a new material (*e.g.* in the case of molecular crystals) at the gram level can allow, through subsequent seeding of a supersaturated solution, production *via* a conventional solvent-based crystallisation route.[59]

It is also necessary to recognise the need for methods of structurally characterising the products of mechanochemical processes. When applied to chemical conversions which may occur, molecular identity can be established by conventional analytical methods (*e.g.* solution NMR and mass spectrometry). For solid–solid transformations, where the identity of crystalline intermediates or final states is required, major improvements in crystal structure determination from powder X-ray diffraction data (frequently coupled with solid-state NMR) have been pivotal.[60–64] In addition, the structural analysis of individual crystallites that result from mechanochemical processes is possible through utilisation of the electron diffraction facility of the transmission electron microscope.[65–67]

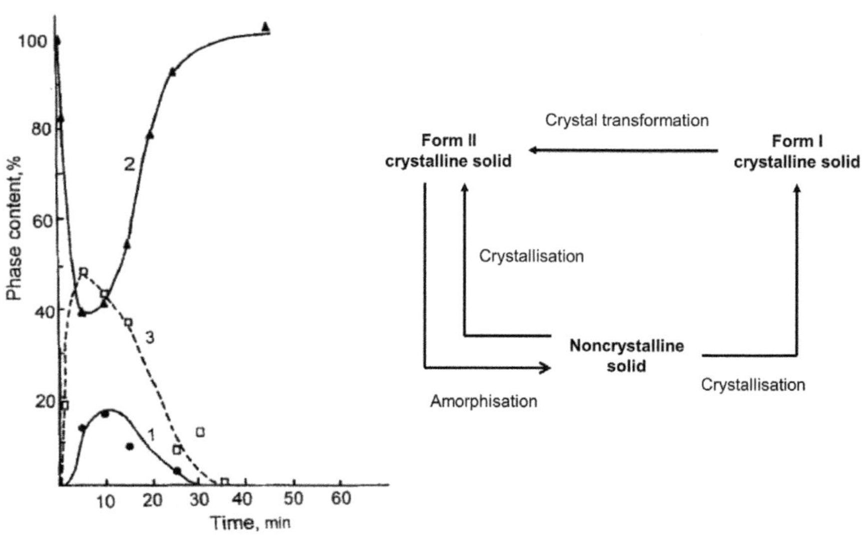

Fig. 3 (a) Schematic of the different pathways occurring during the mechanochemical treatment of sulphathiaziole. Mechanical treatment resulted in the partial amorphisation of the drug and the interconversion of Forms I and II. (b) Time profile for the phase interconversion for sulphathiazole: 1 – indicating generation of Form I; 2 – initial loss and subsequent regeneration of Form III; 3 – the existence of a transient amorphous phase. Figure adapted from ref. 55.

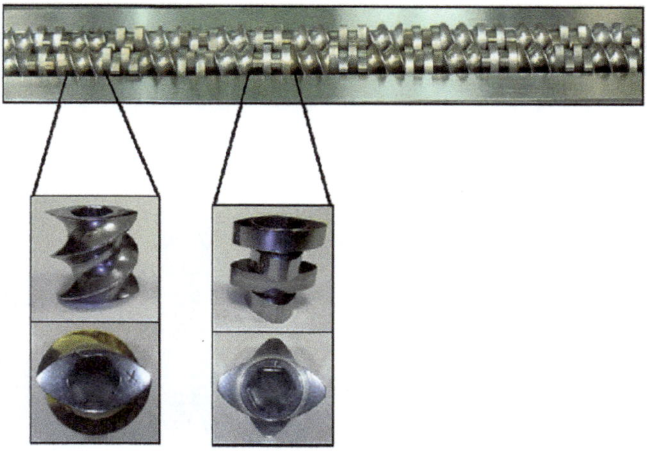

Fig. 4 Typical twin extrusion configuration used for the formation of cocrystals. The conveying and mixing zones are shown. Typical output rate is approximately 0.5 g min^{-1}. PXRD suggests comparable properties for the product to that obtained by conventional solution growth. The possibility of simultaneously blending the cocrystal product within a polymer matrix is an important additional advantage of the method. Reproduced from ref. 58.

3. Pharmaceutical materials

Within various areas of process engineering (including food science, chemicals and pharmaceuticals) particle size reduction is an important step.[35,68] Approaches to size reduction include compression (*e.g.* rollers) or impact (*e.g.* impact mills and ball mills). All methods involve transfer of varying amounts of energy to the

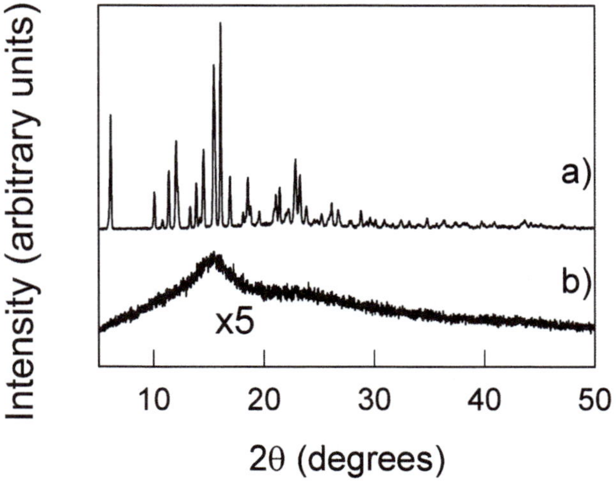

Fig. 5 Demonstration of the amorphisation of trehelose as a result of milling at room temperature within a high energy planetary micro-mill. PXRD patterns (a) before and (b) after milling. Adapted from ref. 11.

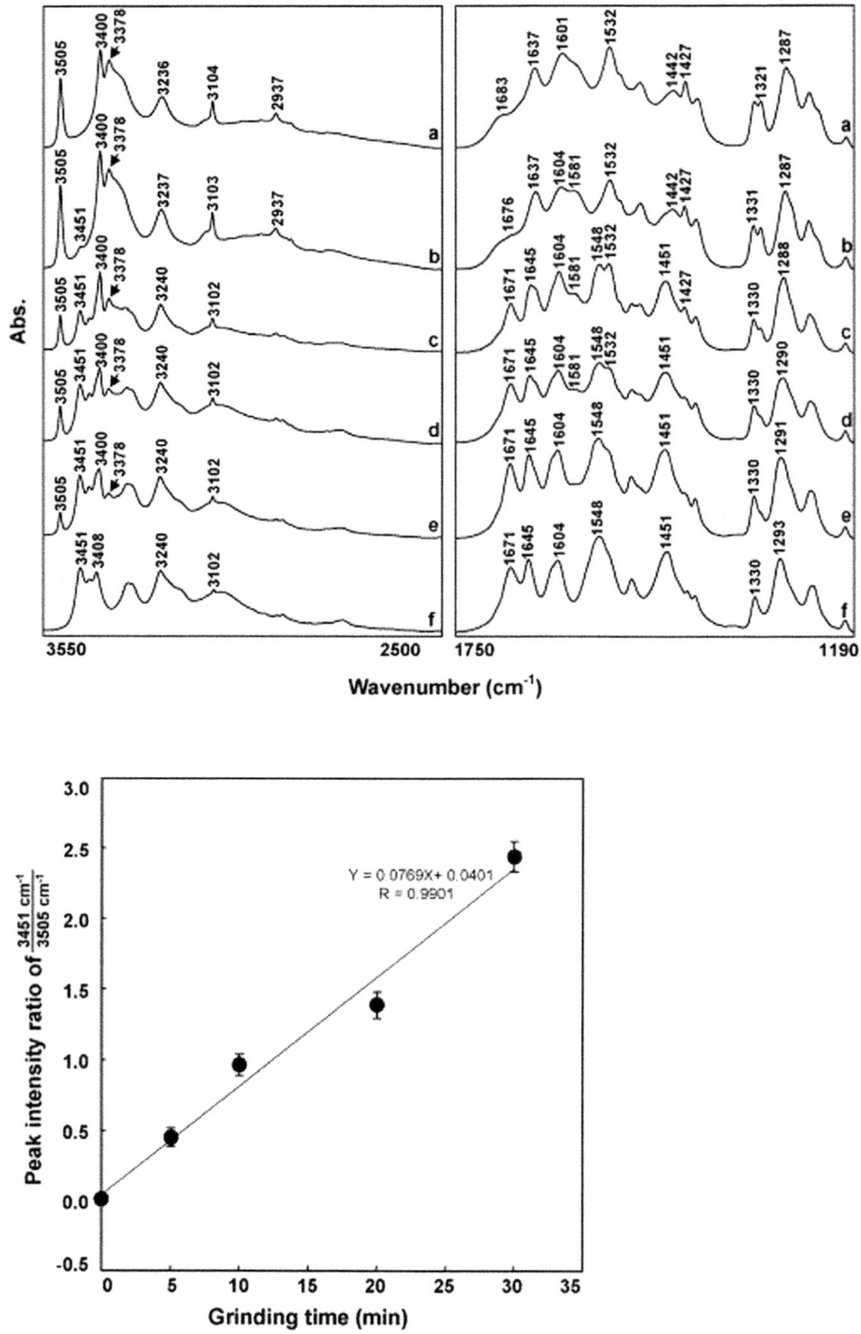

Fig. 6 (a) Grinding-time dependence of the FTIR spectra for famotidine showing from top to bottom the emergence of the absorption at 3451 cm^{-1} which is used to monitor the formation of Form A of the material. (b) Plot of peak intensity ratio (I_{3451}/I_{3505}) for the spectra above indicating a possible zero order rate law for the conversion. Reproduced from ref. 78 with permission from Elsevier.

particle(s), and as a result other consequences can accompany the size reduction including the introduction of defects (such as dislocations, point defects and stacking faults)[12] and in particular amorphisation[69] or polymorphic transformations.[70]

The work of Descamps and others with regard to pharmaceutical materials has clearly demonstrated the critical role of the glass transition temperature, T_g, in determining whether or not a persistent amorphous phase will be generated.[11,71–73] The combination of increased milling intensity and the absence of any plasticisers, which might lower T_g (either as liquid or vapour, *e.g.* water, from the surrounding atmosphere), will encourage the formation of an amorphous product (Fig. 5). The distinction between amorphous and crystalline products is particularly important for pharmaceutical applications given the inherent increased transient solubility (and hence improved bioavailability) associated with amorphous compared to crystalline solid forms.[69,74–76]

Frequently, the amorphisation of a material can be a desirable outcome, especially as a means of increasing the transient solubility of the material or of eliminating compression issues with the crystalline equivalent.[68] This attribute is frequently used in the pharmaceutical industry, and the complex nature of the resulting solids has been extensively studied (especially with regard to the instability of the amorphous form with respect to the crystalline state).[74]

Polymorphic transformations from the stable to a metastable form (again with increased apparent solubility) can also be a useful consequence of the application of mechanical energy (Fig. 6).[13,77–79] Jet-milling, a particle-size reduction method frequently used within the pharmaceutical industry where particle–particle collisions occur within a carrier gas (typically air), has, for example, been found to induce polymorph interconversions along with the size reductions.

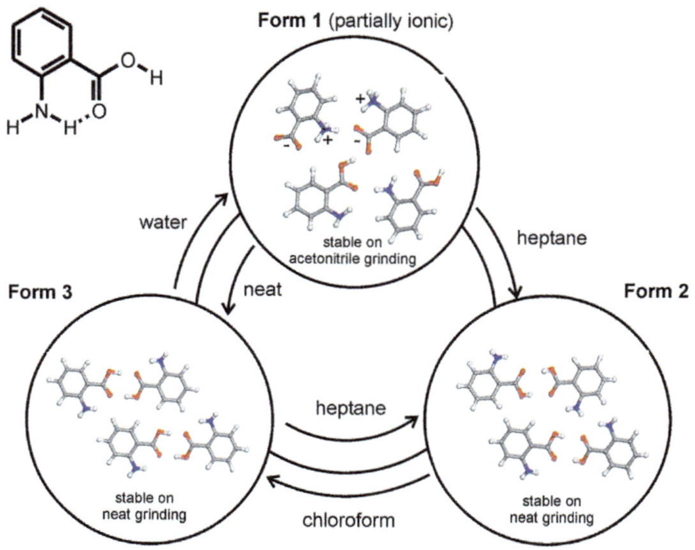

Fig. 7 Schematic of the observed transformations between the three polymorphic forms of anthranilic acid with the pathways depending on the conditions under which mechanochemical treatment was performed.

With grinding experiments, an approach for directing the polymorphic outcome through addition of a small amount of liquid was described by Trask *et al.* for the cases of anthranilic acid and succinic acid (see Fig. 7).[80] For anthranilic acid, interconversion between the three known polymorphic forms occurred depending upon the choice of the added liquid.[81] In the case of succinic acid, while the β-polymorph appeared stable to neat grinding, addition of a few drops of a low polarity liquid (*e.g.* hexane) caused conversion to the α-polymorph. More polar liquids (*e.g.* water, acetonitrile or methanol) appeared not to induce the transformation. The actual mechanism of the solid–solid transformation is an area of ongoing interest, especially where the change appears to be crystal–crystal in nature.[82]

An additional reason for interest in mechanochemical processes comes from the fact that the majority of drugs (80–90%) are formulated as tablets, which are seen as a reliable and general way of delivering medicines, and that the tableting process itself requires the application of significant applied pressures (see Fig. 8).[55,83–90] Importantly, tablets contain other components (excipients), which are added to enhance performance *e.g.* lubricants, dissolving agents and so on. When pressure is applied during compression, there is a risk that surface–surface reactions between the components and drug, such as that between dicalcium phosphate dihydrate and aspirin, will lead to enhanced instability.[91]

More recent interest has focused on the formation of cocrystals (see Fig. 9) by milling or grinding.[92–94] One of the earliest examples of a cocrystal obtained by grinding is that of quinhydrone reported in the work of Wohler published in 1844.[95] The advantage for this system compared to solution crystallization was shown by Patil *et al.*,[96] who demonstrated that simple grinding of mixtures was significantly superior to attempted solution growth, where self oxidation–reduction lowered the efficiency of complex formation. Subsequently, numerous examples of host–complex formation by solid–solid mixing were reported by Toda and co-workers.[50] Work by Etter and colleagues also demonstrated how solid–solid grinding could be used to produce hydrogen bonded cocrystals of adenine and thymine derivatives, where a marked selectivity amongst various base pairs was observed.[14] Hollingsworth *et al.*[51] used grinding as a means of creating cocrystals of dinitriles and urea, as part of a wide study of inclusion crystals, in cases where solution growth was not possible. Pedireddi *et al.* demonstrated that in

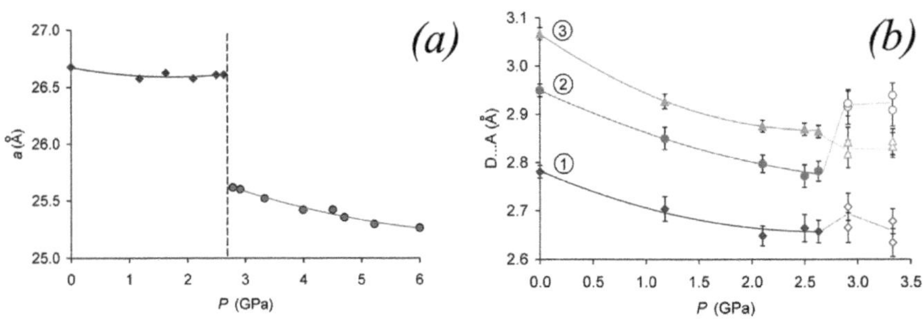

Fig. 8 (a) Hydrostatic pressure induced transition for chlorpropamide in a saturated ethanol solution. The phase change results in a first order change at approximately 3.5 GPa. The change is illustrated for the a-axis. (b) Changes in non-hydrogen atom positions as the transition pressure is reached. Adapted from ref. 46.

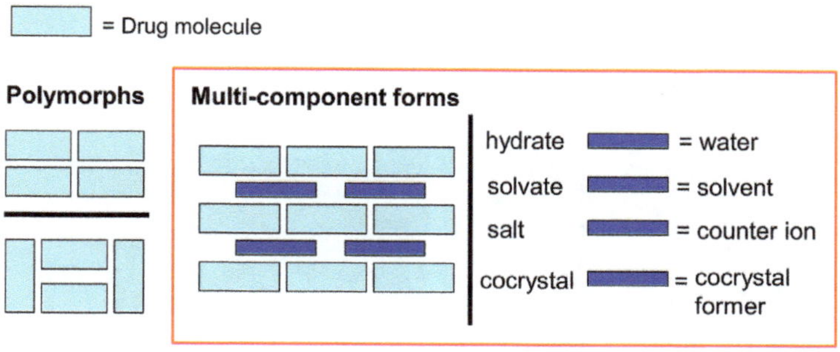

Fig. 9 Schematic of the characteristics of polymorphs and multicomponent systems. Cocrystals represent a sub-set of the multicomponent group, frequently that characterised by the requirement that the components be neutral and solid at room temperature.

some cases a third component is necessary in order for a cocrystallisation reaction to take place[97] and Kuroda and her colleagues studied the influence of the addition of a third component on the optical properties of the resulting cocrystals.[98]

Frequently, the solid state properties of a pure drug can present processing or application problems *e.g.* particle flow, thermal stability, stability to moisture or ability to be compacted,[68] and, increasingly, poor solubility is an issue.[99,100] The challenge with a badly behaving drug, therefore, is to be able to control the properties of the molecule in the solid state without recourse to chemical modification by covalent chemistry. The use of multicomponent systems, especially pharmaceutical cocrystals, and more recently ionic cocrystals,[101,102] has been demonstrated to be an effective strategy in such instances.[103–112] In terms of property improvement with pharmaceutical cocrystals, various examples are known including: enhancing stability against hydration of an anhydrate form (see

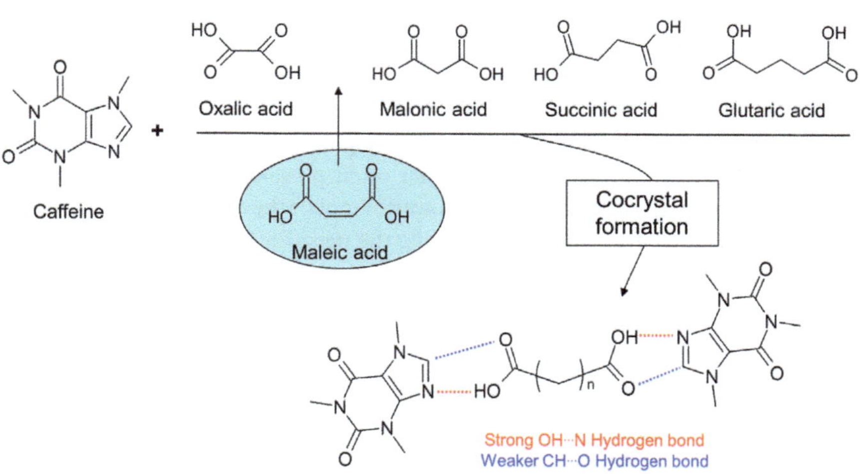

Fig. 10 Supramolecular synthon approach used in the design of cocrystals of caffeine with various dicarboxylic acids. This approach lead to a variety of solid forms of caffeine which could be tested for stability under conditions of high humidity. Mechanochemistry was the technique used to both screen for cocrystal formation and also prepare powders for humidity assessment.

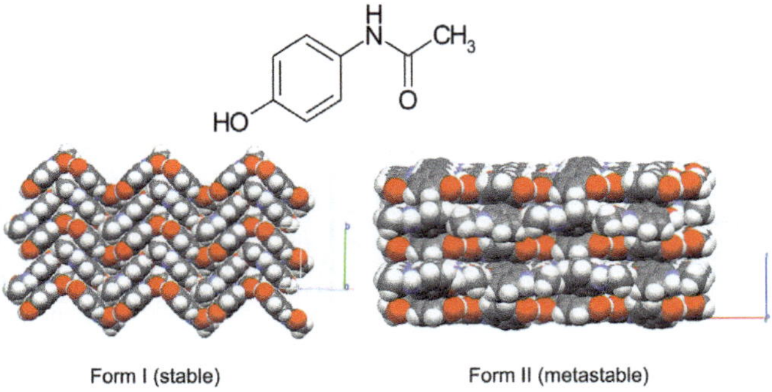

Fig. 11 Packing diagrams for Form I and Form II paracetamol. Whilst Form I is the stable phase it does not compress well during tablet formation. The metastable Form II has a layered structure more appropriate for compression, but readily transforms to Form I.

Fig. 10),[113–116] developing forms with improved compression properties (for tablet formation) (see Fig. 11 and 12),[117,118] significantly improving the bioavailability of poorly soluble drugs and controlled release strategies *e.g.* for insulin.[119] Trask has reported on the potential intellectual property implications of pharmaceutical cocrystals,[120] and recent interest has also focused on regulatory guidelines.[121] In non-pharmaceutical applications, benefits of cocrystallisation have been shown for fluorescent materials[37,122] and explosives.[123,124]

An early indication of the potential to form pharmaceutical cocrystals by grinding was provided by Caira *et al.* in 1995 who used grinding to prepare a series of crystalline products containing sulfonamide (similar to those obtained from solution) in a facile manner.[125] They also reported detailed solid state kinetic data based on powder X-ray measurements and were able to show a distinct selectivity between reactants during grinding.

An ongoing challenge, however, is selecting possible coformers which might cocrystallise with the API rather than remain as a separate and distinct crystalline phase. Whilst various strategies have been developed to predict the likelihood of cocrystal formation,[126–129] experimental screening for viable coformers remains vital (Fig. 13).[130–132] While solution growth remains the ideal approach – especially with regard to scale-up in later stages of development – significant differences in solubility between the drug and coformer often presents difficulties.[115,116,133–135] Mechanochemistry has emerged as a powerful tool in the experimental screening area, and has the advantage of minimising the need for large volumes of solvent (Fig. 14). It has also emerged as a powerful way of searching for different

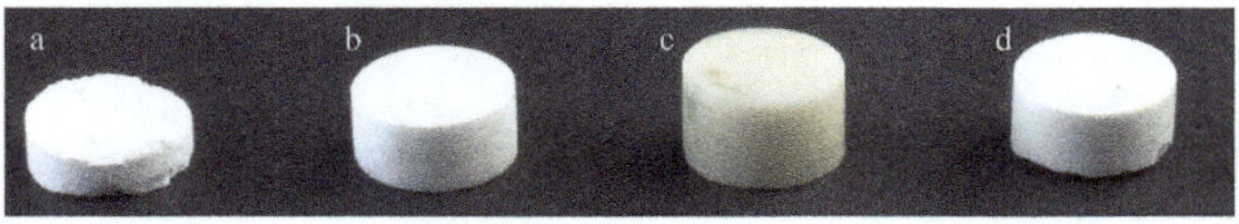

Fig. 12 Micrographs of tablets made from (a) paracetamol Form I and (b to d) cocrystals of paracetamol with theophylline, naphthalene and oxalic acid respectively. Tablet (a) is poorly formed as a result of the inappropriate packing of the molecules in the lattice, whereas the cocrystal structures are amenable to deformation during tableting.[118]

polymorphic forms of cocrystals[135,136] as well as for various stoichiometric compositions *e.g.* 1 : 1, 1 : 2 and 2 : 1.[59] Furthermore, mechanochemistry has been used to identify chiral agents suitable for converting a racemate into two distinct diastereomeric cocrystal phases[137] as well as for coformer exchange.[138]

An extension to simple (*i.e.* neat or dry) grinding of mixtures was reported by Shan *et al.* who added a few drops of liquid to the reaction mixture (liquid assisted grinding, LAG) (see Fig. 15).[139] Initial attention focused on LAG as a means of accelerating cocrystallisation kinetics, although it was subsequently shown to provide a means of increasing product diversity. For example, the outcome of mechanochemical formation of cocrystals has been compared with that from using other approaches to cocrystallsation. Trask *et al.* reported three possible

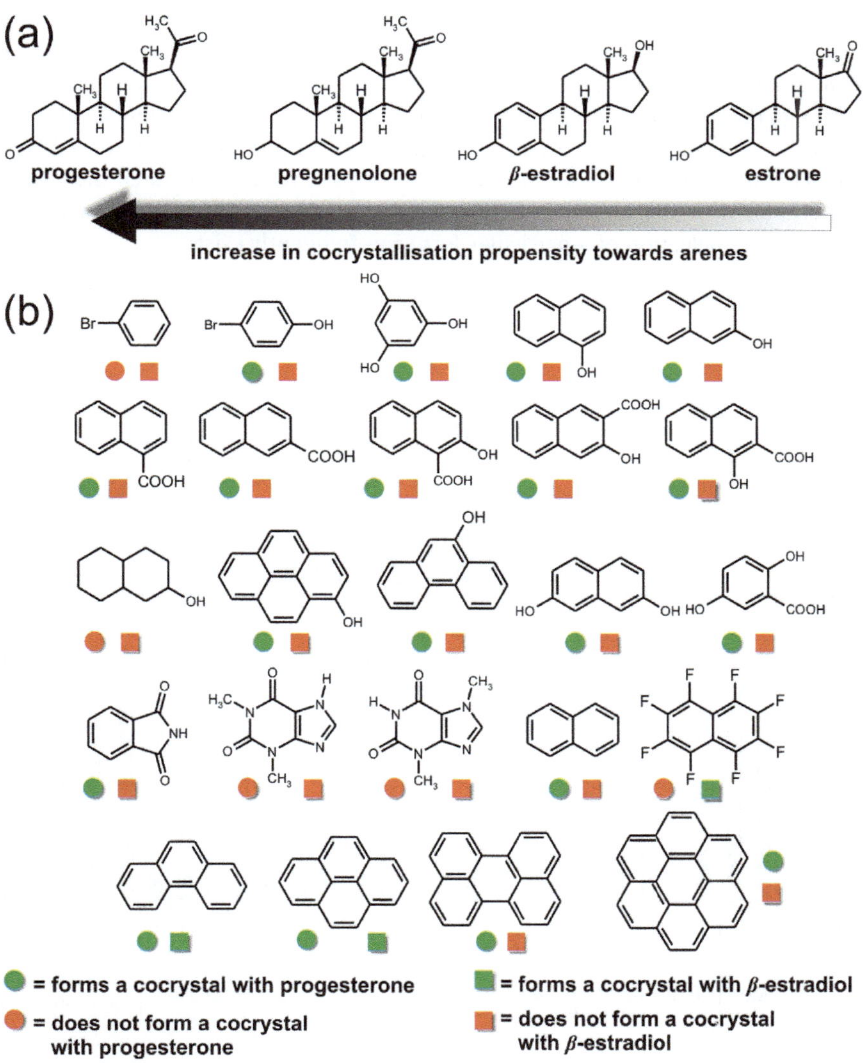

Fig. 13 The use of mechanochemistry to explore the various possible combinations of coformers for a family of steroid drugs. The various outcomes are delineated by the accompanying legend. See ref. 161 for details.

outcomes in the case of cocrystals with caffeine: (i) solution growth and milling produced identical products; (ii) a different stoichiometric product could be obtained by milling compared to solution; and (iii) different polymorphic forms were possible.[59] Numerous recent publications have demonstrated that LAG is likely to be the most efficient way of screening for cocrystal formation.[140–144] The key advantage of LAG over solution based approaches was interpreted by Davey,[145,146] Childs[147] and others[148] on the basis of solution phase chemistry (Fig. 16). Other developments, such as ion- and liquid-assisted grinding (ILAG)[149] and vapour digestion of solid mixtures,[150,151] have also been reported.

Further insight has been given by the recent work of Bučar *et al.* for the case of caffeine and benzoic acid.[152] No cocrystals for this pair of molecules were known despite a strong expectation based on studies of analogous systems[153] and despite extensive experimental efforts to generate such a species over many years. A systematic study using a large range of cocrystallisation methods including neat grinding, liquid assisted grinding, melt crystallization and various solution methods was employed (Fig. 17). All failed to produce the elusive cocrystal even though theoretical work had clearly confirmed a cocrystal was thermodynamically feasible. During grinding experiments, however, the addition of a few seed crystals of an isostructural fluorobenzoic acid cocrystal readily generated the elusive cocrystal, with the inference being that, as in solution crystallisation, a high barrier to nucleation of a particular form can frequently exist.

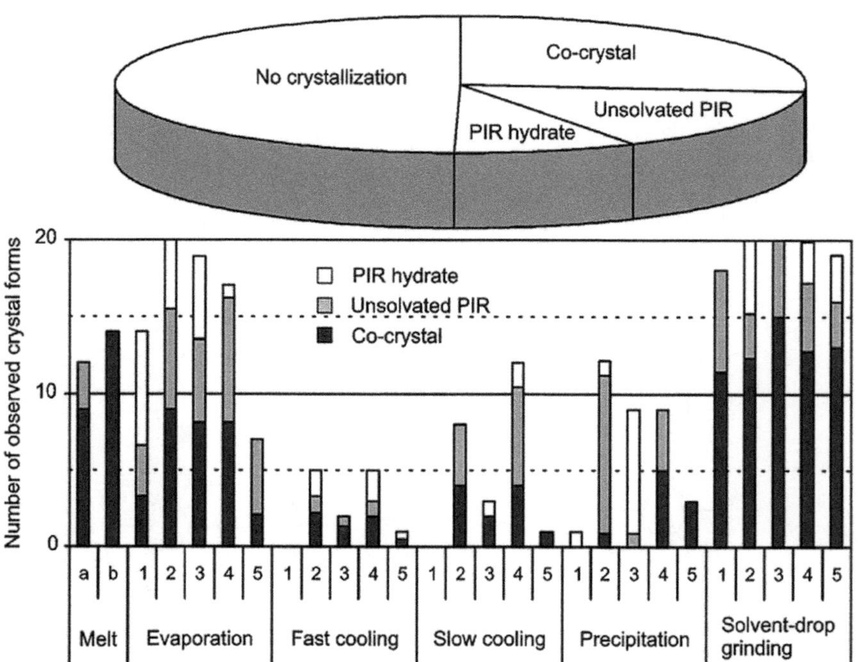

Fig. 14 Overview of the results obtained by various cocrystallisation attempts for piroxicam. Solvent-drop grinding (LAG) is shown in this case to be the most efficient and straightforward screening method. The numbers 1–5 indicate the various solvents used in the experiments. Reproduced with permission from ref. 142.

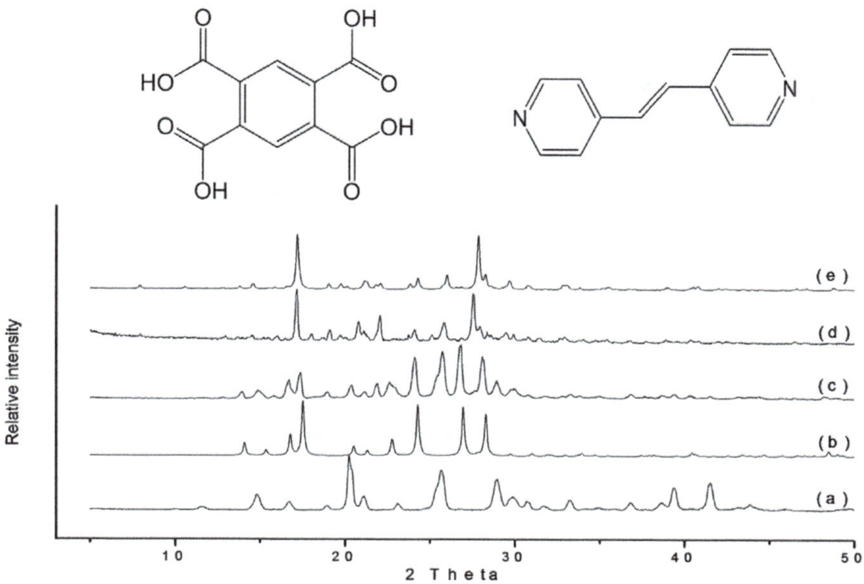

Fig. 15 Demonstration of the enhanced kinetics of cocrystal formation upon addition of a few drops of methanol to the grinding jar. PXRD patterns for (a) tetracarboxylic acid, (b) bipyridyl, (c) outcome of neat grinding for 60 minutes (as the sum of the reactants), (d) the product obtained when a few drops of methanol are added and (e) the simulated pattern for the cocrystal.

Accompanying the recent developments in mechanochemistry has been the development of techniques to follow the course of solid–solid reactions using both *ex situ* and *in situ* approaches.[56,122,154–158] *Ex situ* analysis has frequently suggested that there may be distinct intermediates generated (Fig. 18).[154,159] The application of synchrotron X-ray diffraction for *in situ* studies has indeed confirmed the presence of crystalline intermediates and a possible role of amorphous phases.[157] To better understand the processes occurring during

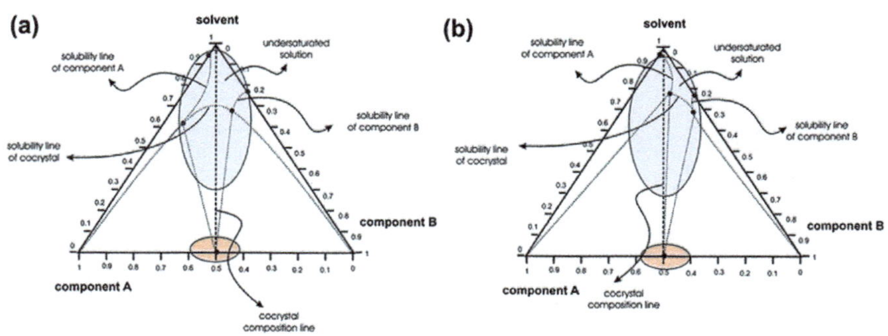

Fig. 16 Typical three-component phase diagrams contrasting (a) congruent and (b) incongruent solubility profiles indicating the difficulties which will arise during conventional evaporative solution growth crystallisation. The fact that LAG experiments, with minimal amounts of solvent present, places the system close to the cocrystal composition illustrates the ability of LAG to form cocrystals when solids of very different solubility are investigated. Reproduced from ref. 34.

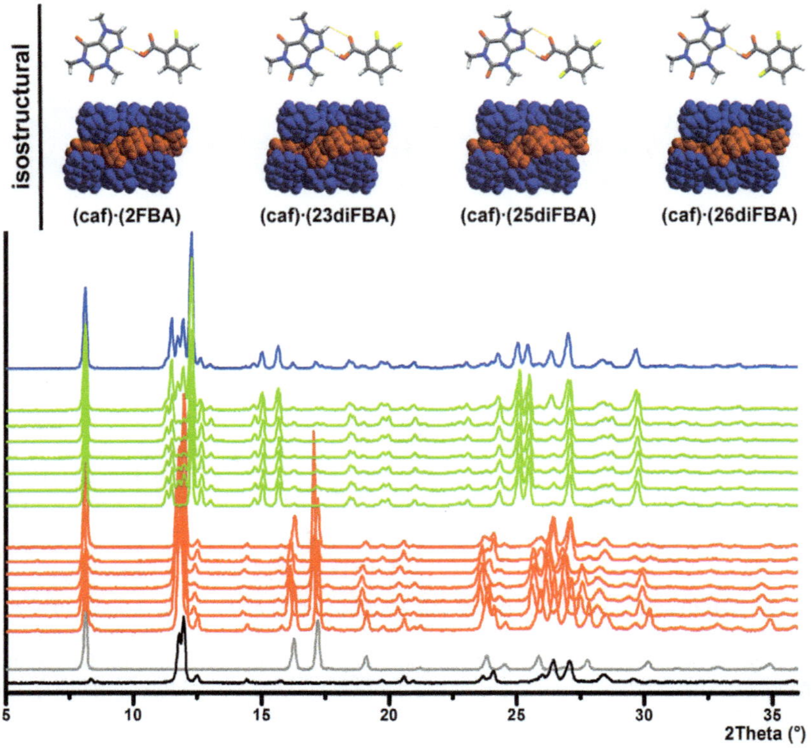

Fig. 17 Top. Isostructural cocrystals of caffeine with various fluorobenzoic acids which have similar packing to that predicted for the likely structure of the caffeine–benzoic acid cocrystal. Bottom. PXRD traces of (a) product without deliberate addition of seeds, but prepared in an environment where the caffeine–benzoic acid cocrystal had previously been obtained, (b) products of various seeding reactions, (c) products of various unseeded reactions, (d) benzoic acid and (e) caffeine. The main points to note include the similarity of the patterns in (a) and (b) and that those in (c) are a simple combination of the starting materials in (d) and (e). See ref. 152 for more details.

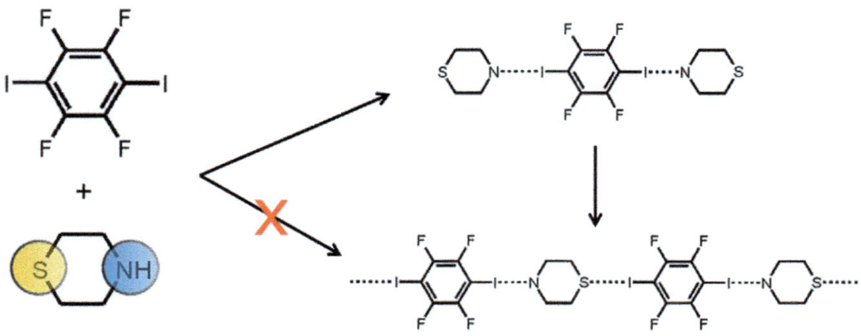

Fig. 18 The discrete trimer units and extended polymer chains (based on halogen bond interactions) believed to be formed during the LAG reaction. When the strength of the halogen bond is weak the product is limited to the trimer unit.

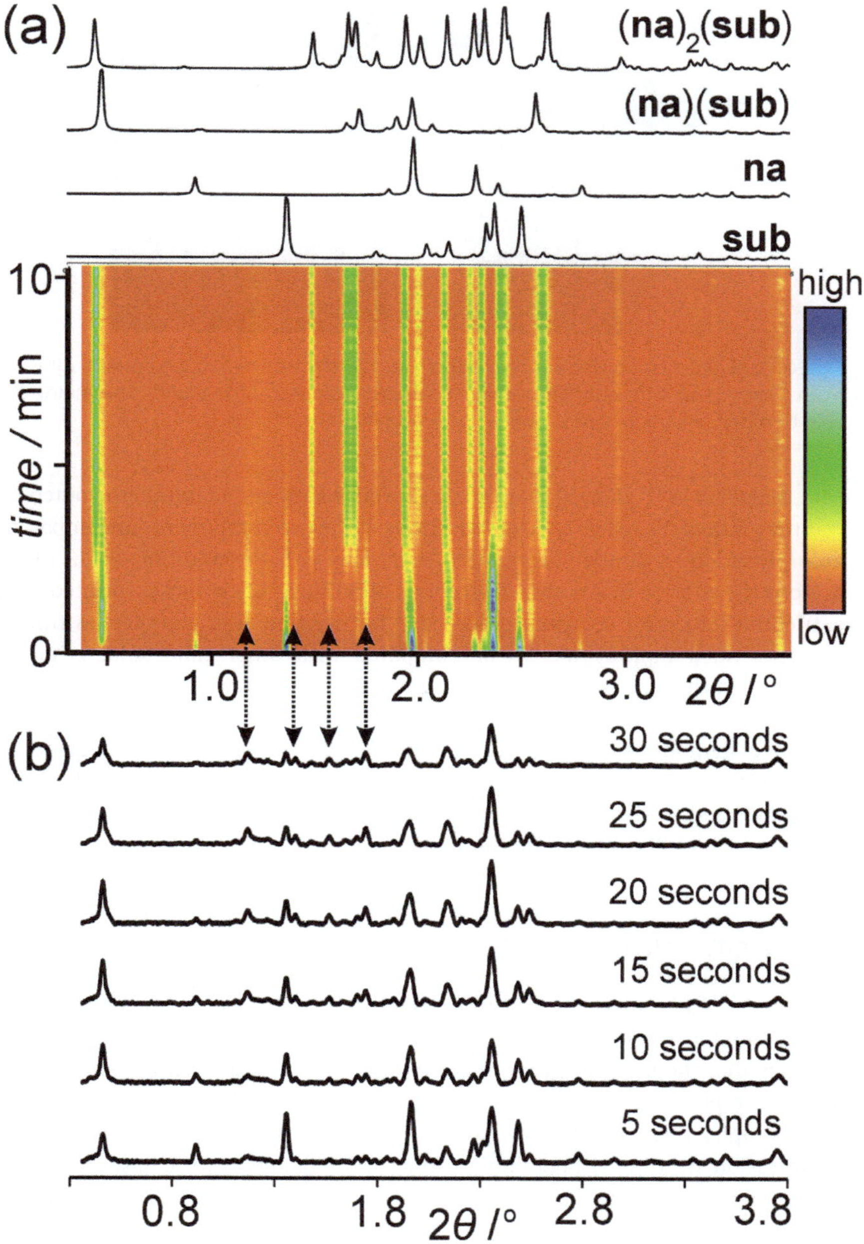

Fig. 19 *In situ* synchrotron data for the milling of suberic acid and nicotinamide in a 2 : 1 ratio. (a) The PXRD patterns of the phases expected to be formed during milling along with the data obtained as a function of time. The inset shows the particular region at 0.4 to 0.5 degrees two-theta showing the initial formation of the 1 : 1 cocrystal followed by the 2 : 1 product. (b) The specific reflections indicated by arrows indicate the intermediate phase which is not detected by conventional *ex situ* studies. See ref. 157 for further details.

milling, several recent efforts have concerned developing methods to follow *in situ* the course of the reaction. Noteworthy has been the use of synchrotron X-ray diffraction. In a study of the milling of a mixture of carbamazepine and succinic

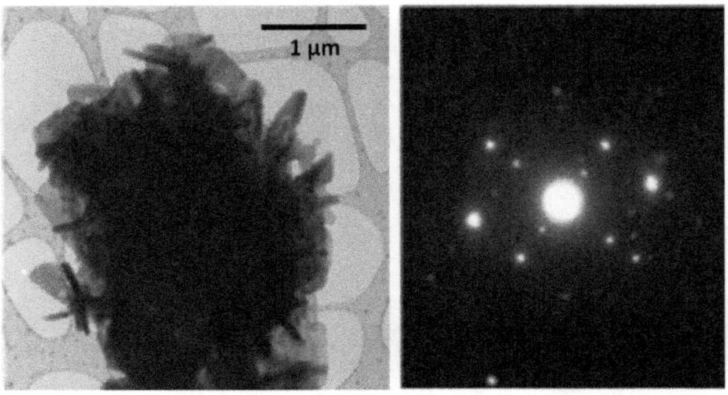

Fig. 20 (a) Transmission electron micrograph and (b) electron diffraction pattern obtained from a sample of a theophylline:L-malic acid cocrystal prepared by LAG. The method is able to monitor particle size as well as polymorphic form.[160]

acid, Halasz *et al.*[157] showed that, in the absence of any added liquid, the outcome was the gradual disappearance of reflections of the two components and apparent amorphisation. A similar experiment, but now with the addition of 50 μL of acetonitrile showed that within 10 seconds of milling, reflections associated with a triclinic (cbz)(sac) cocrystal emerged. The reaction was complete within 4 minutes. Such rapidity was to be compared with other solid–solid grinding systems (*e.g.* the reaction of ZnO with imidazoles) where reaction timescales were much longer. Similar *in situ* studies were conducted for suberic acid and nicotinamide demonstrating that formation of a 2 : 1 cocrystal during grinding proceeds *via* a 1 : 1 cocrystal intermediate (see Fig. 19). Friščić and his colleagues have also used *in situ* methods, based on the changes in fluorescence[122] accompanying cocrystallisation, to monitor directly events within the grinding jar.

4. Concluding remarks

In the case of pharmaceutical solids, recent work using atomic force microscopy (AFM), terahertz spectroscopy and transmission electron microscopy (TEM) suggests that these techniques will be important for characterising mechanically treated samples and understanding changes which occur during such processing. For example, there are particle–particle effects – especially important during compression in a tablet containing various components. In the case of aspirin (acetylsalicylic acid) and dicalcium phosphate, when the excipient was attached to the tip of the AFM cantilever and different pressures and contact times under various humidity conditions there was evidence for enhanced reactivity induced by surface contact with excipient.[91] Terahertz spectroscopy has emerged as a sensitive method of monitoring changes in the solid form during mechanical activation,[56,156] whereas TEM has been shown to be a powerful technique for characterising the size and shape of particulates generated mechanochemically, as well as determining their crystal form (see Fig. 20). TEM can also be used to identify defects in pharmaceutical crystals (see Fig. 21), and will enable a greater understanding of how defects introduced into crystals during mechanical processing influence material properties – for example a link between defects in theophylline crystals and crystal fracture has been established.[160]

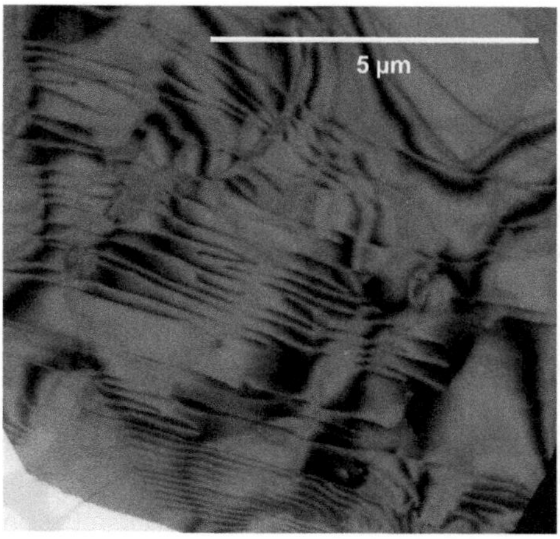

Fig. 21 Extensive network of dislocations in a sample of theophylline as revealed by transmission electron microscopy. In general, such dislocations can result from the mechanical treatment of crystals.

Mechanochemistry, as mentioned at the outset, provides many opportunities for developing environmentally-friendly processes (especially the avoidance of large volumes of solvent). It also provides routes to materials not readily obtained by other methods. It has a long history and no doubt will continue to demonstrate useful applications whilst at the same time generate considerable activity in rationalising the mechanistic aspects behind the many types of transformations observed.

References

1 L. Takacs, The mechanochemical reduction of AgCl with metals revisiting an experiment of M. Faraday, *J. Therm. Anal. Calorim.*, 2007, **90**, 81–84.

2 L. M. Takacs, Carey Lea, the father of mechanochemistry, *Bull. Hist. Chem.*, 2003, **28**, 26–34.

3 V. V. Boldyrev, Mechanochemistry and mechanical activation of solids, *Russ. Chem. Rev.*, 2006, **75**, 177–189.

4 P. Baláž, *Mechanochemistry in Nanoscience and Minerals Engineering*, Springer-Verlag, Berlin, Germany, 2008.

5 L. Takacs, The historical development of mechanochemistry, *Chem. Soc. Rev.*, 2013, **42**, 7649–7659.

6 S. L. James, C. J. Adams, C. Bolm, D. Braga, P. Collier, T. Friščić, F. Grepioni, K. D. M. Harris, G. Hyett, W. Jones, *et al.*, Mechanochemistry: opportunities for new and cleaner synthesis, *Chem. Soc. Rev.*, 2012, **41**, 413–447.

7 A. D. McNaught and A. Wilkinson, *IUPAC Compendium of Chemical Terminology*, Blackwell Scientific Publications, Oxford, 1997.

8 G. Kaupp, Mechanochemistry: the varied applications of mechanical bond-breaking, *CrystEngComm*, 2009, **11**, 388–403.

9 V. V. Boldyrev, Mechanochemistry and sonochemistry, *Ultrason. Sonochem.*, 1995, **2**, S143–S145.

10 G. Cravotto, E. C. Gaudino and P. Cintas, On the mechanochemical activation by ultrasound, *Chem. Soc. Rev.*, 2013, **42**, 7521–7534.

11 M. Descamps, J. F. Willart, E. Dudognon and V. Caron, Transformation of pharmaceutical compounds upon milling and comilling: The role of T_g, *J. Pharm. Sci.*, 2007, **96**, 1398–1407.

12 R. Hüttenrauch, S. Fricke and P. Zielke, Mechanical activation of pharmaceutical systems, *Pharm. Res.*, 1985, **02**, 302–306.

13 H. Li, J. G. Stowell, X. He, K. R. Morris and S. R. Byrn, Investigations on solid–solid phase transformation of 5-methyl-2-[(4-methyl-2-nitrophenyl)amino]-3-thiophenecarbonitrile, *J. Pharm. Sci.*, 2007, **96**, 1079–1089.

14 M. C. Etter, S. M. Reutzel and C. G. Choo, Self-organization of adenine and thymine in the solid state, *J. Am. Chem. Soc.*, 1993, **115**, 4411–4412.

15 G. W. V. Cave, C. L. Raston and J. L. Scott, Recent advances in solventless organic reactions: towards benign synthesis with remarkable versatility, *Chem. Commun.*, 2001, 2159–2169.

16 A. S. Cannon and J. C. Warner, Noncovalent derivatization: green chemistry applications of crystal engineering, *Cryst. Growth Des.*, 2002, **2**, 255–257.

17 K. Ralphs, C. Hardacre and S. L. James, Application of heterogeneous catalysts prepared by mechanochemical synthesis, *Chem. Soc. Rev.*, 2013, **42**, 7701–7718.

18 F. K. Urakaev and V. V. Boldyrev, Theoretical estimation of conditions for the preparation of nanoscale systems in mechanochemical reactors, *Russ. J. Phys. Chem.*, 2005, **79**, 563–572.

19 V. K. Smolyakov, O. V. Lapshin and V. V. Boldyrev, Mechanochemical synthesis of nanosize products in heterogeneous systems: macroscopic kinetics, *Int. J. Self-Propag. High-Temp. Synth.*, 2008, **17**, 20–29.

20 P. Baláž, M. Achimovičová, M. Baláž, P. Billik, Z. Cherkezova-Zheleva, J. M. Criado, F. Delogu, E. Dutková, E. Gaffet, F. J. Gotor, *et al.*, Hallmarks of mechanochemistry: from nanoparticles to technology, *Chem. Soc. Rev.*, 2013, **42**, 7571–7637.

21 M. C. Etter, G. M. Frankenbach and J. Bernstein, Solid-state nucleophilic aromatic substitution reaction of a carboxylic acid cocrystal, *Tetrahedron Lett.*, 1989, **30**, 3617–3620.

22 M. A. Mikhailenko, T. P. Shakhtshneider and V. V. Boldyrev, On the mechanism of mechanochemical synthesis of phthalylsulphathiazole, *J. Mater. Sci.*, 2004, **39**, 5435–5439.

23 R. Trotzki, M. M. Hoffmann and B. Ondruschka, Studies on the solvent-free and waste-free Knoevenagel condensation, *Green Chem.*, 2008, **10**, 767–772.

24 A. Bruckmann, A. Krebs and C. Bolm, Organocatalytic reactions: effects of ball milling, microwave and ultrasound irradiation, *Green Chem.*, 2008, **10**, 1131–1141.

25 V. Štrukil, M. D. Igrc, L. Fábián, M. Eckert-Maksic, S. L. Childs, D. G. Reid, M. J. Duer, I. Halasz, C. Mottillo and T. Friščić, A model for a solvent-free synthetic organic research laboratory: click-mechanosynthesis and structural characterization of thioureas without bulk solvents, *Green Chem.*, 2012, **14**, 2462–2473.

26 J. Stojaković, B. S. Farris and L. R. MacGillivray, Vortex grinding for mechanochemistry: application for automated supramolecular catalysis and preparation of a metal–organic framework, *Chem. Commun.*, 2012, **48**, 7958–7960.

27 G.-W. Wang, Mechanochemical organic synthesis, *Chem. Soc. Rev.*, 2013, **42**, 7668–7700.

28 T. Friščić, Supramolecular concepts and new techniques in mechanochemistry: cocrystals, cages, rotaxanes, open metal–organic frameworks, *Chem. Soc. Rev.*, 2012, **41**, 3493–3510.
29 V. Šepelák, A. Düvel, M. Wilkening, K.-D. Becker and P. Heitjans, Mechanochemical reactions and syntheses of oxides, *Chem. Soc. Rev.*, 2013, **42**, 7507–7520.
30 S. M. Hick, C. Griebel, D. T. Restrepo, J. H. Truitt, E. J. Buker, C. Bylda and R. G. Blair, Mechanocatalysis for biomass-derived chemicals and fuels, *Green Chem.*, 2010, **12**, 468–474.
31 T. Kleine, J. Buendia and C. Bolm, Mechanochemical degradation of lignin and wood by solvent-free grinding in a reactive medium, *Green Chem.*, 2013, **15**, 160–166.
32 D. Braga, D. D'Addario, L. Maini, M. Polito, S. Giaffreda, K. Rubini, F. Grepioni. Applications of Crystal Engineering Strategies in Solvent-free Reactions: Toward a Supramolecular Green Chemistry, in *Frontiers in Crystal Engineering*, ed. E. R. T. Tiekink and J. J. Vittal, John Wiley & Sons, Ltd, Chichester, UK, 2006.
33 A. M. Belenguer, T. Friščić, G. M. Day and J. K. M. Sanders, Solid-state dynamic combinatorial chemistry: reversibility and thermodynamic product selection in covalent mechanosynthesis, *Chem. Sci.*, 2011, **2**, 696–700.
34 A. Delori, T. Friščić and W. Jones, The role of mechanochemistry and supramolecular design in the development of pharmaceutical materials, *CrystEngComm*, 2012, **14**, 2350–2362.
35 C. F. Burmeister and A. Kwade, Process engineering with planetary ball mills, *Chem. Soc. Rev.*, 2013, **42**, 7660–7667.
36 D.-K. Bučar, S. Filip, M. Arhangelskis, G. O. Lloyd and W. Jones, Advantages of mechanochemical cocrystallisation in the solid-state chemistry of pigments: colour-tuned fluorescein cocrystals, *CrystEngComm*, 2013, **15**, 6289–6291.
37 D. Yan, A. Delori, G. O. Lloyd, T. Friščić, G. M. Day, W. Jones, J. Lu, M. Wei, D. G. Evans and X. Duan, A cocrystal strategy to tune the luminescent properties of stilbene-type organic solid-state materials, *Angew. Chem., Int. Ed.*, 2011, **50**, 12483–12486.
38 N. G. Shang, P. Papakonstantinou, S. Sharma, G. Lubarsky, M. Li, D. W. McNeill, A. J. Quinn, W. Zhou and R. Blackley, Controllable selective exfoliation of high-quality graphene nanosheets and nanodots by ionic liquid assisted grinding, *Chem. Commun.*, 2012, **48**, 1877–1879.
39 S.-E. Zhu, F. Li and G.-W. Wang, Mechanochemistry of fullerenes and related materials, *Chem. Soc. Rev.*, 2013, **42**, 7535–7570.
40 N. Blagden, D. J. Berry, A. Parkin, H. Javed, A. Ibrahim, P. T. Gavan, L. L. De Matos and C. C. Seaton, Current directions in co-crystal growth, *New J. Chem.*, 2008, **32**, 1659–1672.
41 P. Baláž and E. Dutková, Fine milling in applied mechanochemistry, *Miner. Eng.*, 2009, **22**, 681–694.
42 X. Guo, D. Xiang, G. Duan and P. Mou, A review of mechanochemistry applications in waste management, *Waste Manage.*, 2010, **30**, 4–10.
43 A. L. Black, J. M. Lenhardt and S. L. Craig, From molecular mechanochemistry to stress-responsive materials, *J. Mater. Chem.*, 2011, **21**, 1655–1663.

44 E. V. Boldyreva, Mechanochemistry of organic solids: where are we now?, *VI International Conference on Mechanochemistry and Mechanical Alloying (INCOME 2008)*, NML Jamshedpur, India, 2011, pp. 17–30.

45 A. Nasser and U. Mingelgrin, Mechanochemistry: A review of surface reactions and environmental applications, *Appl. Clay Sci.*, 2012, **67–68**, 141–150.

46 E. Boldyreva, Mechanochemistry of inorganic and organic systems: what is similar, what is different?, *Chem. Soc. Rev.*, 2013, **42**, 7719–7738.

47 G. Majano, L. Borchardt, S. Mitchell, V. Valtchev and J. Perez-Ramirez, Rediscovering zeolite mechanochemistry – A pathway beyond current synthesis and modification boundaries, *Microporous Mesoporous Mater.*, 2014, **194**, 106–114.

48 V. V. Boldyrev and K. Tkáčová, Mechanochemistry of solids: past, present, and prospects, *J. Mater. Synth. Process.*, 2000, **8**, 121–132.

49 V. V. Boldyrev, Mechanochemical modification and synthesis of drugs, *J. Mater. Sci.*, 2004, **39**, 5117–5120.

50 F. Toda, K. Tanaka and A. Sekikawa, Host–guest complex formation by a solid–solid reaction, *J. Chem. Soc., Chem. Commun.*, 1987, 279–280.

51 M. D. Hollingsworth, B. D. Santarsiero, H. Oumar-Mahamat and C. J. Nichols, New series of 1 : 1 layered complexes of α,ω-dinitriles and urea, *Chem. Mater.*, 1991, **3**, 23–25.

52 T. Friščić and W. Jones, Recent advances in understanding the mechanism of cocrystal formation *via* grinding, *Cryst. Growth Des.*, 2009, **9**, 1621–1637.

53 V. V. Boldyrev, Hydrothermal reactions under mechanochemical action, *Powder Technol.*, 2002, **122**, 247–254.

54 A. A. L. Michalchuk, I. A. Tumanov and E. V. Boldyreva, Complexities of mechanochemistry: elucidation of processes occurring in mechanical activators *via* implementation of a simple organic system, *CrystEngComm*, 2013, **15**, 6403–6412.

55 T. P. Shakhtshneider, Phase transformations and stabilization of metastable states of molecular crystals under mechanical activation, *Solid State Ionics*, 1997, **101–103**, 851–856.

56 K. L. Nguyen, T. Friščić, G. M. Day, L. F. Gladden and W. Jones, Terahertz time-domain spectroscopy and the quantitative monitoring of mechanochemical cocrystal formation, *Nat. Mater.*, 2007, **6**, 206–209.

57 C. Medina, D. Daurio, K. Nagapudi and F. Alvarez-Nunez, Manufacture of pharmaceutical co-crystals using twin screw extrusion: a solvent-less and scalable process, *J. Pharm. Sci.*, 2010, **99**, 1693–1696.

58 D. Daurio, K. Nagapudi, L. Li, P. Quan and F. Alvarez-Nunez, Application of twin screw extrusion to the manufacture of cocrystals: scale-up of AMG 517–sorbic acid cocrystal production, *Faraday Discuss.*, 2014, DOI: 10.1039/C3FD00153A.

59 A. V. Trask, J. van de Streek, W. D. S. Motherwell and W. Jones, Achieving polymorphic and S. diversity in cocrystal formation: Importance of solid-state grinding, powder X-ray structure determination, and seeding, *Cryst. Growth Des.*, 2005, **5**, 2233–2241.

60 E. Y. Cheung, S. J. Kitchin, K. D. M. Harris, Y. Imai, N. Tajima and R. Kuroda, Direct Structure determination of a multicomponent molecular crystal prepared by a solid-state grinding procedure, *J. Am. Chem. Soc.*, 2003, **125**, 14658–14659.

61 S. Karki, L. Fábián, T. Friščić and W. Jones, Powder X-ray diffraction as an emerging method to structurally characterize organic solids, *Org. Lett.*, 2007, **9**, 3133–3136.

62 E. Salager, G. M. Day, R. S. Stein, C. J. Pickard, B. Elena and L. Emsley, Powder crystallography by combined crystal structure prediction and high-resolution 1H solid-state NMR spectroscopy, *J. Am. Chem. Soc.*, 2010, **132**, 2564–2566.

63 K. D. M. Harris, Powder diffraction crystallography of molecular solids, *Top. Curr. Chem.*, 2012, **315**, 133–178.

64 M. Arhangelskis, G. O. Lloyd and W. Jones, Mechanochemical synthesis of pyrazine:dicarboxylic acid cocrystals and a study of dissociation by quantitative phase analysis, *CrystEngComm*, 2012, **14**, 5203–5208.

65 U. Kolb, T. E. Gorelik, E. Mugnaioli and A. Stewart, Structural characterization of organics using manual and automated electron diffraction, *Polym. Rev.*, 2010, **50**, 385–409.

66 M. D. Eddleston, K. E. Hejczyk, E. G. Bithell, G. M. Day and W. Jones, Polymorph identification and crystal structure determination by a combined crystal structure prediction and transmission electron microscopy approach, *Chem.–Eur. J.*, 2013, **19**, 7874–7882.

67 M. D. Eddleston, K. E. Hejczyk, E. G. Bithell, G. M. Day and W. Jones, Determination of the crystal structure of a new polymorph of theophylline, *Chem.–Eur. J.*, 2013, **19**, 7883–7888.

68 Y. Qiu, Y. Chen and G. G. Z. Zhang, *Developing solid oral dosage forms: Pharmaceutical theory and practise*, Academic Press, USA, 2009, p. 943.

69 N. S. Trasi and S. R. Byrn, Mechanically induced amorphization of drugs: A study of the thermal behavior of cryomilled compounds, *AAPS PharmSciTech*, 2012, **13**, 772–784.

70 G. Bruni, V. Berbenni, F. Sartor, C. Milanese, A. Girella, D. Franchi and A. Marini, Quantification methods of amorphous/crystalline fractions in high-energy ball milled pharmaceutical products, *J. Therm. Anal. Calorim.*, 2012, **108**, 235–241.

71 S. Desprez and M. Descamps, Transformations of glassy indomethacin induced by ball-milling, *J. Non-Cryst. Solids*, 2006, **352**, 4480–4485.

72 J. F. Willart, N. Descamps, V. Caron, F. Capet, F. Danede and M. Descamps, Formation of lactose–mannitol molecular alloys by solid state vitrification, *Solid State Commun.*, 2006, **138**, 194–199.

73 V. Caron, J. F. Willart, F. Danede and M. Descamps, The implication of the glass transition in the formation of trehalose/mannitol molecular alloys by ball milling, *Solid State Commun.*, 2007, **144**, 288–292.

74 B. C. Hancock and M. Parks, What is the true solubility advantage for amorphous pharmaceuticals?, *Pharm. Res.*, 2000, **17**, 397–404.

75 S. B. Murdande, M. J. Pikal, R. M. Shanker and R. H. Bogner, Solubility advantage of amorphous pharmaceuticals: I. A thermodynamic analysis, *J. Pharm. Sci.*, 2010, **99**, 1254–1264.

76 A. Newman, G. Knipp and G. Zografi, Assessing the performance of amorphous solid dispersions, *J. Pharm. Sci.*, 2012, **101**, 1355–1377.

77 M. Pudipeddi and A. T. M. Serajuddin, Trends in solubility of polymorphs, *J. Pharm. Sci.*, 2005, **94**, 929–939.

78 S.-Y. Lin, W.-T. Cheng and S.-L. Wang, Thermodynamic and kinetic characterization of polymorphic transformation of famotidine during grinding, *Int. J. Pharm.*, 2006, **318**, 86–91.
79 P. L. D. Wildfong, K. R. Morris, C. A. Anderson and S. M. Short, Demonstration of a shear-based solid-state phase transformation in a small molecular organic system: chlorpropamide, *J. Pharm. Sci.*, 2007, **96**, 1100–1113.
80 A. V. Trask, N. Shan, W. D. S. Motherwell, W. Jones, S. Feng, R. B. H. Tan and K. J. Carpenter, Selective polymorph transformation *via* solvent-drop grinding, *Chem. Commun.*, 2005, 880–882.
81 N. Madusanka, Solid state transformations in single-component and multi-component pharmaceutical crystals, M.Phil thesis, University of Cambridge, UK, 2012.
82 I. Halasz, Single-crystal-to-single-crystal reactivity: Gray, rather than black or white, *Cryst. Growth Des.*, 2010, **10**, 2817–2823.
83 H. G. Brittain, Effects of mechanical processing on phase composition, *J. Pharm. Sci.*, 2002, **91**, 1573–1580.
84 E. V. Boldyreva, V. Dmitriev and B. C. Hancock, Effect of pressure up to 5.5 GPa on dry powder samples of chlorpropamide form-A, *Int. J. Pharm.*, 2006, **327**, 51–57.
85 I. D. H. Oswald, I. Chataigner, S. Elphick, F. P. A. Fabbiani, A. R. Lennie, J. Maddaluno, W. G. Marshall, T. J. Prior, C. R. Pulham and R. I. Smith, Putting pressure on elusive polymorphs and solvates, *CrystEngComm*, 2009, **11**, 359–366.
86 M. D. Moore, A. M. Steinbach, I. S. Buckner and P. L. D. Wildfong, A structural investigation into the compaction behavior of pharmaceutical composites using powder X-ray diffraction and total scattering analysis, *Pharm. Res.*, 2009, **26**, 2429–2437.
87 I. S. Buckner, R. A. Friedman and D. E. Wurster, Using compression calorimetry to characterize powder compaction behavior of pharmaceutical materials, *J. Pharm. Sci.*, 2010, **99**, 861–870.
88 V. Mazel, C. Delplace, V. Busignies, V. Faivre, P. Tchoreloff and N. Yagoubi, Polymorphic transformation of anhydrous caffeine under compression and grinding: a re-evaluation, *Drug Dev. Ind. Pharm.*, 2011, **37**, 832–840.
89 R. Roopwani and I. S. Buckner, Understanding deformation mechanisms during powder compaction using principal component analysis of compression data, *Int. J. Pharm.*, 2011, **418**, 227–234.
90 N. P. A. Christensen, C. Cornett and J. Rantanen, Role of excipients on solid-state properties of piroxicam during processing, *J. Pharm. Sci.*, 2012, **101**, 1202–1211.
91 A. M. Cassidy, C. E. Gardner, T. Auffret, B. Aldous and W. Jones, Decoupling the effects of surface chemistry and humidity on solid-state hydrolysis of aspirin in the presence of dicalcium phosphate dihydrate, *J. Pharm. Sci.*, 2012, **101**, 1496–1507.
92 D. Braga and F. Grepioni, Solventless reactions: Reactions between or within molecular crystals, *Angew. Chem., Int. Ed.*, 2004, **43**, 4002–4011.
93 A. V. Trask and W. Jones, Crystal engineering of organic cocrystals by the solid-state grinding approach, *Top. Curr. Chem.*, 2005, **254**, 41–70.
94 D. Braga, L. Maini and F. Grepioni, Mechanochemical preparation of co-crystals, *Chem. Soc. Rev.*, 2013, **42**, 7638–7648.

95 F. Wöhler, Untersuchungen über das Chinon, *Justus Liebigs Ann. Chem.*, 1844, **51**, 145–163.
96 A. O. Patil, D. Y. Curtin and I. C. Paul, Solid-state formation of quinhydrones from their components. Use of solid–solid reactions to prepare compounds not accessible from solution, *J. Am. Chem. Soc.*, 1984, **106**, 348–353.
97 V. R. Pedireddi, W. Jones, A. P. Chorlton and R. Docherty, Creation of crystalline supramolecular arrays: a comparison of co-crystal formation from solution and by solid-state grinding, *Chem. Commun.*, 1996, 987–988.
98 R. Kuroda, Y. Imai and N. Tajima, Generation of a co-crystal phasewith novel coloristic properties *via* solid state grinding procedures, *Chem. Commun.*, 2002, 2848–2849.
99 N. Blagden, M. de Matas, P. T. Gavan and P. York, Crystal engineering of active pharmaceutical ingredients to improve solubility and dissolution rates, *Adv. Drug Delivery Rev.*, 2007, **59**, 617–630.
100 S. J. Bethune, N. Schultheiss and J.-O. Henck, Improving the poor aqueous solubility of nutraceutical compound pterostilbene through cocrystal formation, *Cryst. Growth Des.*, 2011, **11**, 2817–2823.
101 D. Braga, F. Grepioni, L. Maini, S. Prosperi, R. Gobetto and M. R. Chierotti, From unexpected reactions to a new family of ionic co-crystals: the case of barbituric acid with alkali bromides and cesium iodide, *Chem. Commun.*, 2010, **46**, 7715–7717.
102 D. Braga, F. Grepioni, G. I. Lampronti, L. Maini and A. Turrina, Ionic co-crystals of organic molecules with metal halides: A new prospect in the solid formulation of active pharmaceutical ingredients, *Cryst. Growth Des.*, 2011, **11**, 5621–5627.
103 O. Almarsson and M. J. Zaworotko, Crystal engineering of the composition of pharmaceutical phases. Do pharmaceutical co-crystals represent a new path to improved medicines?, *Chem. Commun.*, 2004, 1889–1896.
104 G. P. Stahly, Diversity in single- and multiple-component crystals. The search for and prevalence of polymorphs and cocrystals, *Cryst. Growth Des.*, 2007, **7**, 1007–1026.
105 N. Shan and M. J. Zaworotko, The role of cocrystals in pharmaceutical science, *Drug Discovery Today*, 2008, **13**, 440–446.
106 N. A. Meanwell, The emerging utility of co-crystals in drug discovery and development, *Annu. Rep. Med. Chem.*, 2008, **43**, 373–404.
107 G. P. Stahly, A Survey of Cocrystals Reported Prior to 2000, *Cryst. Growth Des.*, 2009, **9**, 4212–4229.
108 T. Friščić and W. Jones, Benefits of cocrystallization in pharmaceutical materials science: an update, *J. Pharm. Pharmacol.*, 2010, **62**, 1547–1559.
109 N. Qiao, M. Li, W. Schlindwein, N. Malek, A. Davies and G. Trappitt, Pharmaceutical cocrystals: An overview, *Int. J. Pharm.*, 2011, **419**, 1–11.
110 H. G. Brittain, Pharmaceutical cocrystals: The coming wave of new drug substances, *J. Pharm. Sci.*, 2013, **102**, 311–317.
111 R. Thakuria, A. Delori, W. Jones, M. P. Lipert, L. Roy and N. Rodríguez-Hornedo, Pharmaceutical cocrystals and poorly soluble drugs, *Int. J. Pharm.*, 2013, **453**, 101–125.
112 J. W. Steed, The role of co-crystals in pharmaceutical design, *Trends Pharmacol. Sci.*, 2013, **34**, 185–193.

113 A. V. Trask, W. D. S. Motherwell and W. Jones, Pharmaceutical cocrystallization: Engineering a remedy for caffeine hydration, *Cryst. Growth Des.*, 2005, **5**, 1013–1021.
114 A. V. Trask, W. D. S. Motherwell and W. Jones, Physical stability enhancement of theophylline *via* cocrystallization, *Int. J. Pharm.*, 2006, **320**, 114–123.
115 M. D. Eddleston, R. Thakuria, B. J. Aldous and W. Jones, An investigation of the causes of cocrystal dissociation at high humidity, *J. Pharm. Sci.*, 2014, **103**, 2859–2864.
116 M. D. Eddleston, N. Madusanka and W. Jones, Cocrystal dissociation in the presence of water: A general approach for identifying stable cocrystal forms, *J. Pharm. Sci.*, 2014, **103**, 2865–2870.
117 C. C. Sun and H. Hou, Improving mechanical properties of caffeine and methyl gallate crystals by cocrystallization, *Cryst. Growth Des.*, 2008, **8**, 1575–1579.
118 S. Karki, T. Friščić, L. Fábián, P. R. Laity, G. M. Day and W. Jones, Improving mechanical properties of crystalline solids by cocrystal formation: new compressible forms of paracetamol, *Adv. Mater.*, 2009, **21**, 3905–3909.
119 M. L. Brader, M. Sukumar, A. H. Pekar, D. S. McClellan, R. E. Chance, D. B. Flora, A. L. Cox, L. Irwin and S. R. Myers, Hybrid insulin cocrystals for controlled release delivery, *Nat. Biotechnol.*, 2002, **20**, 800–804.
120 A. V. Trask, An overview of pharmaceutical cocrystals as intellectual property, *Mol. Pharmaceutics*, 2007, **4**, 301–309.
121 S. Aitipamula, R. Banerjee, A. K. Bansal, K. Biradha, M. L. Cheney, A. R. Choudhury, G. R. Desiraju, A. G. Dikundwar, R. Dubey, N. Duggirala, *et al.*, Polymorphs, salts, and cocrystals: What's in a name?, *Cryst. Growth Des.*, 2012, **12**, 2147–2152.
122 M. Frenette, G. Cosa and T. Friščić, Characterisation of organic solid forms and real-time *in situ* monitoring of their transformations using solid-state fluorescence, *CrystEngComm*, 2013, **15**, 5100–5106.
123 K. B. Landenberger and A. J. Matzger, Cocrystal engineering of a prototype energetic material: supramolecular chemistry of 2,4,6-trinitrotoluene, *Cryst. Growth Des.*, 2010, **10**, 5341–5347.
124 J. P. Shen, X. H. Duan, Q. P. Luo, Y. Zhou, Q. Bao, Y. J. Ma and C. H. Pei, Preparation and characterization of a novel cocrystal explosive, *Cryst. Growth Des.*, 2011, **11**, 1759–1765.
125 M. R. Caira, L. R. Nassimbeni and A. F. Wildervanck, Selective formation of hydrogen bonded cocrystals between a sulfonamide and aromatic carboxylic acids in the solid state, *J. Chem. Soc., Perkin Trans. 2*, 1995, 2213–2216.
126 L. Fábián, Cambridge Structural Database Analysis of Molecular Complementarity in Cocrystals, *Cryst. Growth Des.*, 2009, **9**, 1436–1443.
127 D. Musumeci, C. A. Hunter, R. Prohens, S. Scuderi and J. F. McCabe, Virtual cocrystal screening, *Chem. Sci.*, 2011, **2**, 883–890.
128 G. Springuel, B. Norberg, K. Robeyns, J. Wouters and T. Leyssens, Advances in pharmaceutical co-crystal screening: effective co-crystal screening through structural resemblance, *Cryst. Growth Des.*, 2012, **12**, 475–484.
129 P. A. Wood, N. Feeder, M. Furlow, P. T. A. Galek, C. R. Groom and E. Pidcock, Knowledge-based approaches to co-crystal design, *CrystEngComm*, 2014, **16**, 5839–5848.

130 V. André, D. Braga, F. Grepioni and M. T. Duarte, Crystal forms of the antibiotic 4-aminosalicylic acid: solvates and molecular salts with dioxane, morpholine, and piperazine, *Cryst. Growth Des.*, 2009, **9**, 5108–5116.

131 S. R. Bysouth, J. A. Bis and D. Igo, Cocrystallization *via* planetary milling: Enhancing throughput of solid-state screening methods, *Int. J. Pharm.*, 2011, **411**, 169–171.

132 H. Abourahma, J. M. Urban, N. Morozowich and B. Chan, Examining the robustness of a theophylline cocrystal during grinding with additives, *CrystEngComm*, 2012, **14**, 6163–6169.

133 E. Gagnière, D. Mangin, F. Puel, A. Rivoire, O. Monnier, E. Garcia and J. P. Klein, Formation of co-crystals: kinetic and thermodynamic aspects, *J. Cryst. Growth*, 2009, **311**, 2689–2695.

134 E. Gagnière, D. Mangin, S. Veesler and F. Puel, Co-crystallization in solution and scale-up issues, *RSC Drug Discovery Ser.*, 2012, **16**, 188–211.

135 M. D. Eddleston, S. Sivachelvam and W. Jones, Screening for polymorphs of cocrystals: a case study, *CrystEngComm*, 2013, **15**, 175–181.

136 N. Madusanka, M. D. Eddleston, M. Arhangelskis and W. Jones, Polymorphs, hydrates and solvates of a co-crystal of caffeine with anthranilic acid, *Acta Crystallogr., Sect. B: Struct. Sci., Cryst.Eng. Mater.*, 2014, **70**, 72–80.

137 M. D. Eddleston, M. Arhangelskis, T. Friščić and W. Jones, Solid state grinding as a tool to aid enantiomeric resolution by cocrystallisation, *Chem. Commun.*, 2012, **48**, 11340–11342.

138 D. Braga, F. Grepioni and G. I. Lampronti, Supramolecular metathesis: co-former exchange in co-crystals of pyrazine with (*R*,*R*)-, (*S*,*S*)-, (*R*,*S*)- and (*S*,*S*/*R*,*R*)-tartaric acid, *CrystEngComm*, 2011, **13**, 3122–3124.

139 N. Shan, F. Toda and W. Jones, Mechanochemistry and co-crystal formation: effect of solvent on reaction kinetics, *Chem. Commun.*, 2002, 2372–2373.

140 S. A. Myz, T. P. Shakhtshneider, K. Fucke, A. P. Fedotov, E. V. Boldyreva, V. V. Boldyrev and N. I. Kuleshova, Synthesis of co-crystals of meloxicam with carboxylic acids by grinding, *Mendeleev Commun.*, 2009, **19**, 272–274.

141 C. B. Aakeröy, A. B. Grommet and J. Desper, Co-crystal screening of diclofenac, *Pharmaceutics*, 2011, **3**, 601–614.

142 K. Fucke, S. A. Myz, T. P. Shakhtshneider, E. V. Boldyreva and U. J. Griesser, How good are the crystallisation methods for co-crystals? A comparative study of piroxicam, *New J. Chem.*, 2012, **36**, 1969–1977.

143 K. Yamamoto, S. Tsutsumi and Y. Ikeda, Establishment of cocrystal cocktail grinding method for rational screening of pharmaceutical cocrystals, *Int. J. Pharm.*, 2012, **437**, 162–171.

144 H.-L. Lin, T.-K. Wu and S.-Y. Lin, Screening and characterization of cocrystal formation of metaxalone with short-chain dicarboxylic acids induced by solvent-assisted grinding approach, *Thermochim. Acta*, 2014, **575**, 313–321.

145 R. A. Chiarella, R. J. Davey and M. L. Peterson, Making co-crystals – The utility of ternary phase diagrams, *Cryst. Growth Des.*, 2007, **7**, 1223–1226.

146 K. Chadwick, R. Davey, G. Sadiq, W. Cross and R. Pritchard, The utility of a ternary phase diagram in the discovery of new co-crystal forms, *CrystEngComm*, 2009, **11**, 412–414.

147 S. L. Childs, N. Rodríguez-Hornedo, L. S. Reddy, A. Jayasankar, C. Maheshwari, L. McCausland, R. Shipplett and B. C. Stahly, Screening strategies based on solubility and solution composition generate

pharmaceutically acceptable cocrystals of carbamazepine, *CrystEngComm*, 2008, **10**, 856–864.

148 T. Friščić, S. L. Childs, S. A. A. Rizvi and W. Jones, The role of solvent in mechanochemical and sonochemical cocrystal formation: a solubility-based approach for predicting cocrystallisation outcome, *CrystEngComm*, 2009, **11**, 418–426.

149 T. Friščić, D. G. Reid, I. Halasz, R. S. Stein, R. E. Dinnebier and M. J. Duer, Ion- and liquid-assisted grinding: improved mechanochemical synthesis of metal–organic frameworks reveals salt inclusion and anion templating, *Angew. Chem., Int. Ed.*, 2010, **49**, 712–715.

150 D. Braga and F. Grepioni, Making crystals from crystals: a green route to crystal engineering and polymorphism, *Chem. Commun.*, 2005, 3635–3645.

151 D. Braga, S. L. Giaffreda, F. Grepioni, M. R. Chierotti, R. Gobetto, G. Palladino and M. Polito, Solvent effect in a "solvent free" reaction, *CrystEngComm*, 2007, **9**, 879–881.

152 D.-K. Bučar, G. M. Day, I. Halasz, G. G. Z. Zhang, J. R. G. Sander, D. G. Reid, L. R. MacGillivray, M. J. Duer and W. Jones, The curious case of (caffeine)·(benzoic acid): how heteronuclear seeding allowed the formation of an elusive cocrystal, *Chem. Sci.*, 2013, **4**, 4417–4425.

153 S. Heiden, L. Troebs, K.-J. Wenzel and F. Emmerling, Mechanochemical synthesis and structural characterisation of a theophylline-benzoic acid cocrystal (1 : 1), *CrystEngComm*, 2012, **14**, 5128–5129.

154 D. Cinčić, T. Friščić and W. Jones, A stepwise mechanism for the mechanochemical synthesis of halogen-bonded cocrystal architectures, *J. Am. Chem. Soc.*, 2008, **130**, 7524–7525.

155 I. A. Tumanov, A. F. Achkasov, E. V. Boldyreva and V. V. Boldyrev, Following the products of mechanochemical synthesis step by step, *CrystEngComm*, 2011, **13**, 2213–2216.

156 Y. Du, Y. Xia, H. Zhang and Z. Hong, Using terahertz time-domain spectroscopical technique to monitor cocrystal formation between piracetam and 2,5-dihydroxybenzoic acid, *Spectrochim. Acta, Part A*, 2013, **111**, 192–195.

157 I. Halasz, A. Puškarić, S. A. J. Kimber, P. J. Beldon, A. M. Belenguer, F. Adams, V. Honkimäki, R. E. Dinnebier, B. Patel, W. Jones, *et al.*, Real-Time *In Situ* Powder X-ray diffraction monitoring of mechanochemical synthesis of pharmaceutical cocrystals, *Angew. Chem., Int. Ed.*, 2013, **52**, 11538–11541.

158 T. Friščić, I. Halasz, P. J. Beldon, A. M. Belenguer, F. Adams, S. A. J. Kimber, V. Honkimäki and R. E. Dinnebier, Real-time and *in situ* monitoring of mechanochemical milling reactions, *Nat. Chem.*, 2012, **5**, 66–73.

159 S. Karki, T. Friščić and W. Jones, Control and interconversion of cocrystal stoichiometry in grinding: Stepwise mechanism for the formation of a hydrogen-bonded cocrystal, *CrystEngComm*, 2009, **11**, 470–481.

160 M. D. Eddleston, E. G. Bithell and W. Jones, Transmission electron microscopy of pharmaceutical materials, *J. Pharm. Sci.*, 2010, **99**, 4072–4083.

161 T. Friščić, R. W. Lancaster, L. Fábián and P. G. Karamertzanis, Tunable recognition of the steroid α-face by adjacent π-electron density, *Proc. Natl. Acad. Sci. U. S. A.*, 2010, **107**, 13216–13221.

Faraday Discussions

PAPER

Liquid-assisted vortex grinding supports the single-step solid-state construction of a [2.2]paracyclophane

Jelena Stojaković, Brian S. Farris and Leonard R. MacGillivray*

Received 3rd February 2014, Accepted 11th March 2014
DOI: 10.1039/c4fd00006d

We employ vortex grinding to generate a [2.2]paracyclophane in the organic solid state. The vortex grinding is aided with the addition of a small amount of liquid phase to facilitate topochemical [2 + 2] photodimerizations that generate the cyclophane stereoselectively and in up to quantitative yield. The use of the liquid phase increases the scope of the vortex method.

1 Introduction

Mechanochemistry can be defined as chemical synthesis induced by external mechanical energy, such as grinding two solids using a mortar-and-pestle.[1] Historically, mechanochemistry has been more of a secondary approach to organic synthetic chemistry as compared to mainstream liquid-phase synthesis.[2] However, recent years have witnessed a rapid development in the use of mechanochemistry to generate organic molecules,[3] as well as supramolecular assemblies and frameworks.[4] A reason behind such increased development is related to the fact that mechanochemistry can facilitate reactions that result in covalent and/or noncovalent bonds that afford products difficult or even impossible to achieve *via* the liquid or solution phase.[5] Mechanochemistry can also offer more efficient, practical and inherently safer approaches to synthesis.[6] To employ fully mechanochemistry to generate molecular and supramolecular products of increasing complexity, as well as products connected to more mainstream chemistry, however, it will be necessary to develop a deeper understanding of the underlying mechanisms, principles, and scopes of the approach. With this in mind, we present here an application of a recently reported form of mechanochemistry known as vortex grinding[7] to facilitate the formation of the [2.2]paracyclophane **1** (Scheme 1). Specifically, we show that while dry vortex grinding alone fails to generate **1** in substantial amounts, the use of a small amount of liquid phase enables vortex grinding to generate the [2.2]paracyclophane stereoselectively and in up to quantitative yield. We expect the integration of the liquid

Department of Chemistry, University of Iowa, Iowa City, IA, USA. E-mail: len-macgillivray@iowa.edu

constituent to allow vortex grinding to be applied more generally to synthesize molecules of increasing complexity in organic solids.

Vortex grinding

Automated techniques to perform mechanochemical reactions and transformations (*e.g.* ball mills) as replacements to mortar-and-pestle grinding are attractive not only to alleviate manual work but to achieve reproducibility and minimize cross-contamination.[8] In this context, we have reported a method to facilitate topochemical [2 + 2] photodimerizations in the solid state that is based on vortex grinding.[7] In our research, we employ small ditopic molecules as templates (*e.g.* resorcinol, 1,8-naphthalenedicarboxylic acid, Rebek's imide) to assemble olefins into prerequisite geometries for photodimerizations in organic solids.[9,10] The templates assemble the olefins *via* hydrogen bonds within two-component solids in the form of co-crystals. Moreover, we have recently demonstrated that the vigorous shaking and vibrations provided by a simple and inexpensive vortex mixer can be used akin to mortar-and-pestle grinding to generate photoreactive co-crystals involving templates. The co-crystals are generated in small, transparent sample vials that are loaded with two metal balls or BBs (Fig. 1). The relatively small size of the vortex apparatus allows the hydrogen-bonded co-crystals to be generated within standard laboratory-sized photochemical reactors such that the grinding and application of ultraviolet (UV) light are applied simultaneously. To date, the method has only been applied for a photodimerization of the monoolefin *trans*-4,4′-bis(4-pyridyl)ethylene) that reacted to form *rctt*-tetrakis(4-pyridyl)cyclobutane. The monocyclobutane formed under completely dry, or neat, conditions stereoselectively and in quantitative yield.

Given that ditopic templates can assemble higher olefins (*e.g.* dienes) that react to form higher cyclobutanes as photoproducts (*e.g.* cyclophanes, ladderanes),[11,12] we were led to consider whether vortex grinding can be applied to generate **1**. Our selection was guided by the ready availability of both the template (4-benzyl-resorcinol or **4-benz-res**) and precursor diene (*p*-di[2-(4-pyridyl)ethenyl] benzene or **1,4-bpeb**), as well as a mechanistic knowledge of the photoreaction producing an intermediate monocyclobutane[13] when 2(**4-benz-res**)·2(**1,4-bpeb**) reacts to give **1**. From our studies, we demonstrate here a critical role of the use of

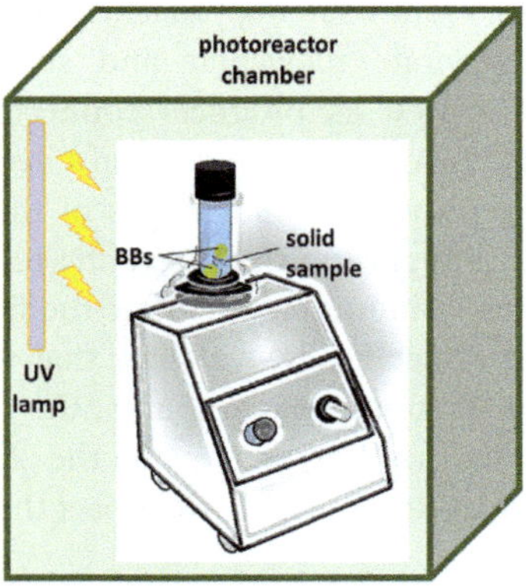

Fig. 1 Schematic vortex grinding apparatus with application of UV radiation.

a small amount of liquid phase to support the formation of **1** using the vortex method. It is well known that small amounts of liquid can be used to generate hydrogen-bonded co-crystals using mortar-and-pestle grinding.[14] Compared to dry grinding, such liquid-assisted grinding (LAG) can be faster, offer better yields, and generate solids of relatively high crystallinities. It has been suggested that the use of the liquid phase in LAG serves to enhance molecular diffusion without a dependence on solvent effects so as to provide essentially an ideal environment to explore molecular recognition and self-assembly.[15] Here, we employ liquid-assisted vortex grinding (LAVG) to generate **1** in a single step with outcomes that parallel the more conventional two-step solution growth approach (Fig 2).

2 Experimental

Materials

1,4-bpeb was prepared as reported.[16] **4-benz-res** was purchased from TCI Tokyo Chemical Industry Co., LTD. and was used as received. Methanol was purchased from Sigma-Aldrich and used as received. ^{1}H NMR spectra were collected using a Bruker 300 MHz spectrometer and DMSO-d$_6$ as solvent. Powder X-ray diffraction (XRD) data were collected on a Bruker D-5000 diffractometer equipped with a Bruker SOL-X energy-sensitive detector using Cu K$_\alpha$ radiation (λ = 1.54056 Å). Vortex grinding was performed using a VWR vortex mixer type Vortex Genie 2. All photoreactions were carried out in a photoreactor chamber equipped with a broadband, low-pressure UV lamp with a quartz Hg arc, Hanovia lamp, model number PC451.050 and ACE Glass Inc. Power supply: 230 V, 50 Hz, 450 W. Borosilicate glass vial of dimensions 21 × 70 mm were used as sample vials along with premium grade steel BBs (5 mm diameter) washed with soap, water, and acetone.

Manual grinding

In a typical grinding experiment, 50 mg of **1,4-bpeb** and 35 mg of **4-benz-res** (1 : 1 molar ratio) were placed in a mortar-and-pestle and ground for up to 1 h. The solid was analyzed using XRPD. LAG experiments were conducted in the same manner save for a single drop of methanol (50 µl) being added to the solid

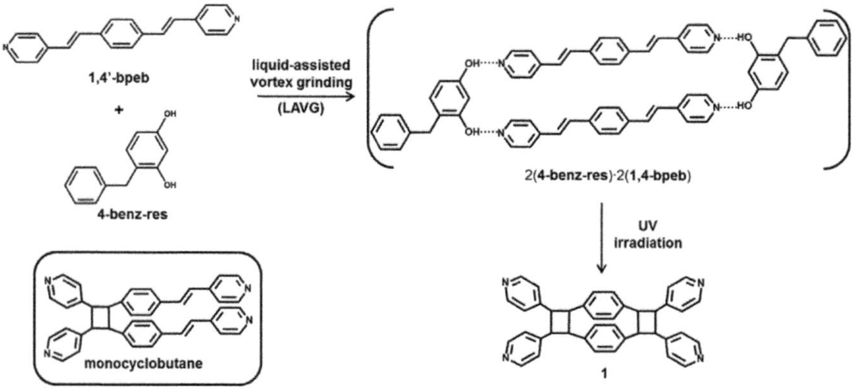

Fig. 2 Solid-state synthesis of **1** *via* LAVG combined with UV irradiation.

mixture. The resulting solid samples were then spread between glass plates and exposed to UV irradiation for a period of up to 120 h.

Vortex grinding

In a typical experiment, 50 mg of **1,4-bpeb** and 35 mg of **4-benz-res** (1 : 1 molar ratio) were placed in a capped sample vial with two metal BBs. The vial was mounted onto the vortex mixer and secured to a ring stand with a test tube holder. The samples were ground up to 1 h. The LAV vortex grinding was conducted in the same manner except for a single drop of methanol or nitromethane (50 μl) being added to the sample vial. To conduct simultaneously LAVG and UV irradiation, the vortex assembly was placed in the photochemical reactor, being approximately 15 cm from the UV lamp. Upon initiation of the vortex mixer, the photochamber was closed and sample was exposed to UV radiation for a period of 10 h.

3 Results and discussion

Initially, we subjected a mixture of **4-benz-res** and **1,4-bpeb** (1 : 1 molar ratio) to dry mortar-and-pestle grinding for a period of 1 h. A powder XRD analysis demonstrated partial formation of the co-crystal 2(**4-benz-res**)·2(**1,4-bpeb**) along with crystalline phases consistent with **4-benz-res** and **1,4-bpeb** (Fig. 3). Moreover, when the ground solid sample was exposed to broadband UV-radiation for a period of 60 h, the [2.2]paracyclophane **1** formed, along with intermediate monocycle (yield: 5%), in relatively low yield (yield: 35%) (Table 1). Neither **1** nor the monocyclized product formed in significantly greater amounts beyond 60 h of irradiation time. For comparison, **1** forms stereoselectively and in up to quantitative yield when the co-crystal 2(**4-benz-res**)·2(**1,4-bpeb**) that is grown from solution is subjected to UV light for a period of 20 h. An application of LAG using methanol was also unsuccessful in affording **1** in appreciable amounts. Specifically, a powder XRD diffractogram with peaks similar to the solid obtained *via* dry mortar-and-pestle grinding was obtained when **4-benz-res** and **1,4-bpeb** were ground for a period of 1 h in the presence of methanol (50 μL). UV-irradiation of

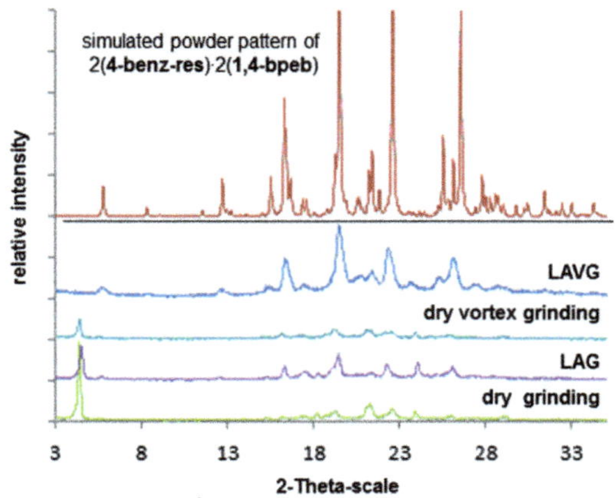

Fig. 3 X-Ray powder patterns of solid mixture of **4-benz-res** and **1,4-bpeb** after being subjected to different mechanochemical methods.

Table 1 Mechanochemistry involving **4-benz-res** to generate **1** from **1,4-bpeb**

Grinding conditions	Grinding time/h	UV-irradiation time/h	Yield of monocycle (%)	Yield of 1 (%)
Dry mortar-and-pestle	1	60	5	35
LAG-methanol	1	60	8	38
Dry vortex then UV	1	60	9	38
LAVG-methanol then UV	1	20	2	97

the solid for 60 h also generated a mixture of **1** (yield: 38%) and the monocyclized photoproduct (yield: 8%) in comparatively low yields.

We next turned to vortex grinding. In particular, when 50 mg of **4-benz-res** and 35 mg of **1,4-bpeb** (1 : 1 molar ratio) were placed in a transparent sample vial along with two metal BBs and subjected to vortex grinding for a period of 1 h, a powder XRD pattern similar to the pattern from the manual grinding experiments was obtained. Moreover, exposure of the solid to broadband UV-radiation for a period of 60 h also generated **1** (yield: 38%) and the monocyclized product (yield: 9%) in relatively low yields.

While the dry and LAG conditions involving mortar-and-pestle grinding, as well as dry vortex grinding, to form **1** quantitatively were unsuccessful, an application of vortex grinding in the presence of a small amount of methanol was successful.

In the experiment, a single drop of methanol (50 μL) was added to a solid mixture of **1,4-bpeb** and **4-benz-res** (1 : 1 molar ratio) in a capped sample vial containing the two BBs. The vial was then subjected to vortex grinding for a period of 1 h. Remarkably, in contrast to dry vortex grinding, the resulting powder XRD pattern matched the simulated pattern of the single-crystal X-ray structure of 2(**4-benz-res**)·2(**1,4-bpeb**) (Fig 3).[12] Exposure of the solid sample to UV-radiation for 20 h resulted in the stereoselective and near quantitative conversion of **1,4-bpeb** to **1**. The addition of the liquid phase, or application of liquid-assisted vortex grinding (LAVG), thus, served as a suitable mechanochemical method to generate **1** as compared to solution-grown 2(**4-benz-res**)·2(**1,4-bpeb**).

A successful application of LAVG was also achieved when the vortex grinding and application of UV light were performed simultaneously or in a single step. A mixture of **1,4-bpeb** and **4-benz-res** along with one drop of methanol were placed in a sample vial that was mounted on the vortex mixer. The assembly was then placed in the photochemical reaction chamber and the vial was exposed to UV-radiation under continuous vortex grinding. A ^{1}H NMR spectrum of the solid collected after a period of 4 h displayed peaks that correspond to the intermediate monocycle (yield: 16%) and **1** (yield: 60%). Moreover, a photoreaction to generate **1** stereoselectively and in near quantitative yield (yield: 97%) was complete in a period of approximately 10 h. The LAVG experiments were also successfully conducted using nitromethane as the liquid component. Thus, in addition to generating **1** in a single step, the photoreaction involving LAVG proceeded two times faster than the irradiated solution-grown co-crystals. We ascribe success of LAVG compared to LAG to form **1** to external stress exerted by the BBs in the vortex mixer, which may be comparable to a vibrating ball mill (*i.e.* 'shakers'). There may also be appreciable liquid vapour present in the vial during the grinding[17,18] that contributes to the formation of **1**. We note that exact reasons behind the dramatic effects of liquid on grinding, however, remain to be fully studied and understood.[19]

4 Conclusion

In this report, we have employed methods of mechanochemistry to generate the [2.2]paracyclophane **1** *via* topochemical [2+2] photodimerization in the organic solid state. We have demonstrated the ability of LAVG to afford access to **1** stereoselectively and in near quantitative yield. The results demonstrate the viability of LAVG as a practical approach to a complex organic molecule. We are now exploring applications of vortex grinding as a means to generate additional architecturally-rich molecules in solids.

References

1 S. L. James, C. J. Adams, C. Bolm, D. Braga, P. Collier, T. Friscic, F. Grepioni, K. D. M. Harris, G. Hyett, W. Jones, A. Krebs, J. Mack, L. Maini, A. G. Orpen, I. P. Parkin, W. C. Shearouse, J. W. Steed and D. C. Waddell, *Chem. Soc. Rev.*, 2012, **41**, 413–447.
2 L. Takacs, *Chem. Soc. Rev.*, 2013, **42**, 7649–7659.
3 A. Stolle, T. Szuppa, S. E. S. Leonhardt and B. Ondruschka, *Chem. Soc. Rev.*, 2011, **40**, 2317–2329.
4 T. Friscic, *Chem. Soc. Rev.*, 2012, **41**, 3493–3510.
5 T. Friscic, A. V. Trask, W. Jones and W. D. S. Motherwell, *Angew. Chem., Int. Ed.*, 2006, **45**, 7546–7550.
6 P. J. Walsh, H. Li and C. A. de Parrodi, *Chem. Rev.*, 2007, **107**, 2503–2545.
7 J. Stojakovic, B. S. Farris and L. R. MacGillivray, *Chem. Commun.*, 2012, **48**, 7958–7960.
8 A. L. Garay, A. Pichon and S. L. James, *Chem. Soc. Rev.*, 2007, **36**, 846–855.
9 L. R. MacGillivray, J. L. Reid and J. A. Ripmeester, *J. Am. Chem. Soc.*, 2000, **122**, 7817–7818.
10 L. R. MacGillivray, G. S. Papaefstathiou, T. Friscic, T. D. Hamilton, D. Bucar, Q. Chu, D. B. Varshney and I. G. Georgiev, *Acc. Chem. Res.*, 2008, **41**, 280–291.
11 X. Gao, T. Friscic and L. R. MacGillivray, *Angew. Chem., Int. Ed.*, 2004, **43**, 232–236.
12 T. Friscic and L. R. MacGillivray, *Chem. Commun.*, 2003, 1306–1307.
13 T. Friscic and L. R. MacGillivray, *Aust. J. Chem.*, 2006, **59**, 613–616.
14 M. B. J. Atkinson, D. Bucar, A. N. Sokolov, T. Friscic, C. N. Robinson, M. Y. Bilal, N. G. Sinada, A. Chevannes and L. R. MacGillivray, *Chem. Commun.*, 2008, 5713–5715.
15 T. Friscic and W. Jones, *Cryst. Growth Des.*, 2009, **9**, 1621–1637.
16 S. Z. Vatsadze, V. N. Nuriev, A. V. Chernikov and N. V. Zyk, *Russ. Chem. Bull.*, 2002, **51**, 1957–1958.
17 A. L. Garay, A. Pichon and S. L. James, *Chem. Soc. Rev.*, 2007, **36**, 846–855.
18 A. Pichon, A. Lazuen-Garay and S. L. James, *CrystEngComm*, 2006, **8**, 211–214.
19 I. A. Tumanov, A. F. Achkasov, E. V. Boldyreva and V. V. Boldyrev, *CrystEngComm*, 2011, **13**, 2213–2216.

Faraday Discussions

PAPER

Is the equilibrium composition of mechanochemical reactions predictable using computational chemistry?

Peter J. Bygrave, David H. Case and Graeme M. Day*

Received 29th December 2013, Accepted 15th January 2014
DOI: 10.1039/c3fd00162h

The ability of computational methods to predict the structures and energetics that determine the equilibrium of solid state mechanochemical reactions has been assessed. Two previously characterised base-catalysed metathesis reactions between aromatic disulfides are studied using crystal structure prediction methods and lattice energy calculations that combine molecular electronic structure methods with anisotropic atom–atom potentials. We find that lattice energy searches locate three of the six crystal structures as global minima on their respective crystal energy landscapes. The remaining structures are less successfully predicted, due to problems modelling relative conformational energies due to limitations of the density functional theory method for calculating intramolecular energies. Prediction of the overall reaction energies proves challenging for current methods, but the results show promise as a base on which to build more accurate and reliable approaches.

1 Introduction

The use of mechanochemical reaction conditions for synthesis has gained significant recent interest. These reactions are promoted by the input of mechanical energy and have been employed for the synthesis of a range of materials, such as metal–organic frameworks and organic molecular crystals,[1,2] whose construction involves the formation of reversible intermolecular or coordination bonds. Solid state reaction conditions have also been applied to the reversible formation of covalent bonds, providing solvent-free or, in the case of liquid assisted grinding, minimal solvent conditions for organic synthesis.

Recently, Belenguer *et al.*[3] demonstrated the thermodynamic equilibration of reversible covalent bond formation in mechanochemical base-catalysed disulfide metathesis reactions; the reactions involve the exchange of substituted aromatic groups between two symmetrical homodimeric disulfides, producing a non-symmetrical heterodimer (Scheme 1). Solution equilibration in the presence of

School of Chemistry, University of Southampton, Southampton, United Kingdom. E-mail: g.m.day@soton.ac.uk

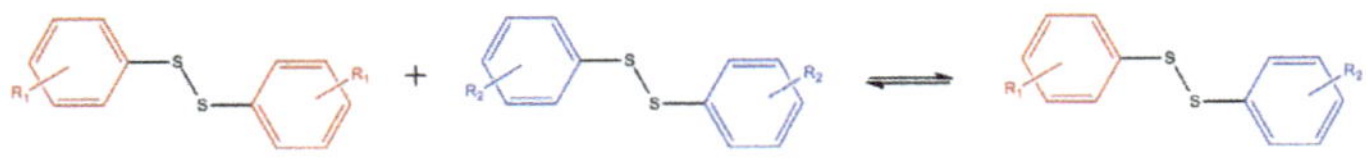

Scheme 1 The disulfide homodimer-to-heterodimer reaction scheme.

the base catalyst 1,8-diazabicyclo[5.4.0]undec-7-ene (**dbu**) of all reactions investigated leads to a statistical 1 : 1 : 2 mixture of the two homodimers and the heterodimer.

Ball mill grinding of the crystalline homodimer reactants with a small amount of **dbu** as catalyst leads fairly rapidly to a steady final composition. However, in contrast to the solution mixture, the final reaction mixture varies depending on the aromatic substituents. The solid-state reaction between bis(4-chlorophenyl) disulfide (**4Cl4Cl**) and bis(2-nitrophenyl)disulfide (**2NO$_2$2NO$_2$**) leads to a final mixture with a 98% mole fraction of the heterodimer (**4Cl2NO$_2$**). The analogous reaction between bis(4-methylphenyl)disulfide (**4Me4Me**) and **2NO$_2$2NO$_2$** produces only 20% of the heterodimer (**4Me2NO$_2$**), with 80% remaining as homodimers. The same final mixtures are obtained from different starting ratios of the homo- and heterodimers, indicating that the final mechanochemical reaction mixtures are a result of thermodynamic equilibration. The chemical diagrams of these systems can be found in Fig. 1.

The finding that the product mixture differs from that found in solution, and differs from reaction to reaction, demonstrates the importance of understanding the total energetics of the solid-state systems. Given that the reactants and products are crystalline, our hypothesis is that the equilibrium product distribution should be predictable from the lattice energies of the crystal structures of the molecules involved in the reaction. Here, we present an exploratory study of the challenges associated with computational prediction of the thermodynamic equilibrium of such solid state organic molecular reactions, taking the two previously characterised disulfide metathesis reactions described above as model systems.

We set the ultimate goal for computational chemistry in this area as the complete *ab initio* prediction of the outcome of mechanochemical reactions between molecular crystals. Such techniques would allow predictions that could be made without any experimental input regarding the structure or stability of the materials involved. While the combination of computational methods with experimental data is expected to be a powerful approach to understanding the energetics controlling these reactions, the ambition of completely *ab initio* predictions is necessary if computational work is to be used to screen possible reactions and decide which to pursue in the lab.

The computational studies performed in ref. 3 demonstrated that intermolecular interactions in the solid state have a key role in determining the lattice energy change across the reaction. Therefore, the crystal packing of reactant and product molecules must be known or predicted. Crystal structure prediction (CSP) has been a challenge for computational chemistry that has received significant attention, particularly for applications in polymorph screening of pharmaceutical molecules.[4,5]

The two main challenges involved in CSP, as applied to molecular crystals, are: (i) the dimensionality and complexity of the search space that must be explored to

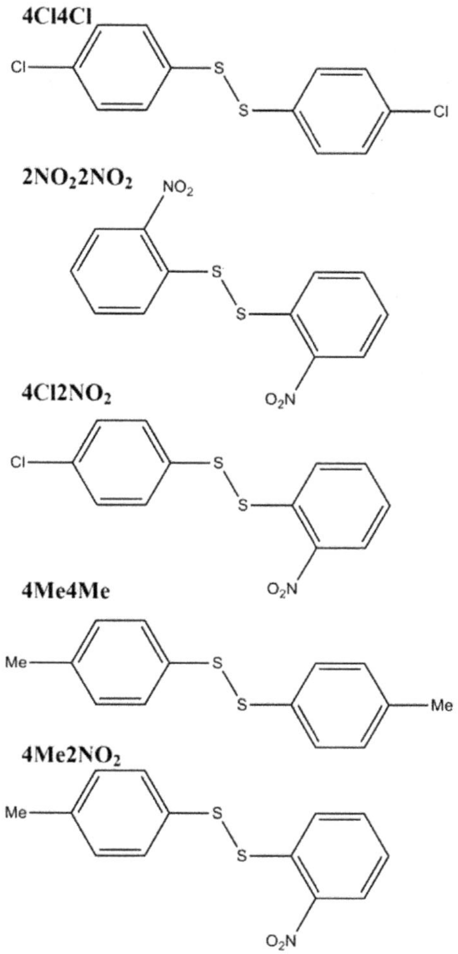

Fig. 1 Molecular diagrams of the disulfide homodimers and heterodimers.

identify all possible crystal packing configurations that might be adopted by the molecule; and (ii) the observation that, for most molecules, there exist a plurality of low energy possible crystal structures separated by small lattice energy differences, typically on the order of 1 kJ mol^{-1} or less. Therefore, practical methods for CSP must involve a rapid evaluation of the stability of a large number of trial structures at the early stages, followed by refinement of the most promising structures. The refinement techniques are able to assess relative energies with high accuracy but come with an increased computational cost and so can only be applied to a small number of structures. The development of CSP methods in the past few years has been promising, with reports of successful applications to a range of challenging, flexible organic molecules.[6–11]

In this study, we first apply CSP to the reactant and product molecules of both model reactions to assess the predictability of their structures. We then perform total energy calculations on their known crystal structures to evaluate whether, with knowledge of their structures, the thermodynamic preference for homodimer *versus* heterodimer is predictable.

Our expectation in starting this study is not that these reaction energies can be calculated using existing methods; this would involve both the correct prediction of crystal structures of the reactants and products, and the relative formation energies of the crystal structures. Instead, the study is to assess the challenges involved in these predictions and highlight the areas where further development of predictive computational methods is most urgently required.

Methods

Crystal structure prediction

CSP is approached by global exploration of the lattice energy surface to identify the most stable low energy minima,[12] corresponding to the most likely observable crystal structures. The computational methods therefore involve a search for promising trial structures, followed by their refinement and ranking by the calculated lattice energy.

Conformation searches

In this study, we separate the search for trial crystal structures into a conformational search for the stable geometries of the isolated molecule, followed by crystal structure generation, where the molecules are initially kept fixed at their idealised gas phase geometries.

The first step is a search for all energetically relevant gas phase conformations. Initially, conformations were generated using a force field description of intramolecular interactions; we employed the OPLS2005 force field,[13] using the low-mode conformational search,[14,15] as implemented in Macromodel.[16] The low-mode search explores conformational space by perturbing an initial molecular geometry along the molecule's low energy normal modes, followed by local energy minimisation. For each molecule, the search was continued until 10 000 energy minimisations had been performed and all unique conformers within 50 kJ mol^{-1} of the lowest energy conformers were kept. These conformers were subsequently re-optimised using density functional theory (DFT) to obtain more accurate molecular geometries for CSP, and for a more reliable ranking of the conformational energies. All DFT calculations were performed with Gaussian09 (rev. D01),[17] using the B3LYP functional,[18] 6-31G** basis set[19] and GD3BJ empirical correction[20] for dispersion interactions. The resulting conformers were clustered to remove duplicates and those within 30 kJ mol^{-1} of the lowest energy geometry for each molecule were kept as starting points for CSP. This search resulted in 2, 6, 2, 4, and 4 conformers being found for molecules **4Cl4Cl**, **2NO$_2$2NO$_2$**, **4Me4Me**, **4Cl2NO$_2$** and **4Me2NO$_2$**, respectively.

Crystal structure generation. Crystal structures were generated with each conformer of the five molecules studied in a set of the 25 most commonly observed space groups for organic molecular crystals[21] ($P2_1/c$, $P2_12_12_1$, $P\bar{1}$, $P2_1$, $Pbca$, $C2/c$, $Pna21$, Cc, $P1$, $Pbcn$, $P4_12_12$, $P4_32_12$, $P2_12_12$, Pc, $P3_1$, $P3_2$, $P4_1$, $P4_3$, $Fdd2$, $Pccn$, $P2/c$, $P6_1$, $P6_5$, $I4_1/a$, $R\bar{3}$) all with one molecule in the asymmetric unit. The searches were performed using our in-house Global Lattice Energy Explorer (GLEE) software,[22] by means of quasi-random sampling of the position and orientations of the molecule in the asymmetric unit, and of unit cell angles and dimensions. The dimensions of the molecule are estimated by projecting the atomic coordinates onto the (orthogonal) vectors which define the principal moments of inertia. These dimensions, in conjunction with the number of symmetry operations, yield a target volume and a set of bounds which are involved in preparing the trial unit cell lengths. Structures with overlapping molecules were rejected and the search was continued until 2000 physically realistic trial crystal structures were generated for each conformer in each space group, yielding 900 000 crystal structures in total.

Lattice energy minimisation. The trial crystal structures were lattice energy minimised in four stages, progressively increasing the accuracy while decreasing the number of structures considered. The first step involved rigid-molecule lattice energy minimisation using an empirically parameterised Buckingham interatomic potential, first with electrostatic interactions modelled by atomic partial charges. The W99 parameters[23–25] were used for the Buckingham potential, supplemented by sulfur parameters taken from Williams' modelling of elemental sulfur crystal structures[26] and an anisotropic chlorine repulsion–dispersion model derived non-empirically for chlorobenzene crystal structures.[27] Atomic partial charges were determined by fitting to the B3LYP/6-31G** calculated molecular electrostatic potential, using the CHELPG[28] scheme. Clustering and removal of duplicates at this stage was used to assess the completeness of the search and all unique structures within the lowest 15 kJ mol^{-1} for each conformer were re-optimised using the same method, but with electrostatics modelled using a more detailed atomic multipole model. The multipoles, up to hexadecapole on each atom, were derived from a distributed multipole analysis[29] of the B3LYP/6-31G** charge density.

The DMACRYS[30] crystal structure modelling software was used for all lattice energy calculations, with a 15 Å cutoff on repulsion–dispersion interactions, an Ewald summation for charge–charge, charge–dipole and dipole–dipole interactions, and all higher order electrostatics summed to a 15 Å cutoff. All clustering of crystal structures was performed using the COMPACK algorithm,[31] comparing interatomic distances within a cluster of 30 molecules.

At this stage, the total energy of each crystal structure was calculated as the sum of the DMACRYS intermolecular energy and the DFT energy of the relevant molecular conformation. For each molecule, structures within 20 kJ mol^{-1} of the global lowest total energy were retained for further structure refinement. This refinement allows for the distortion of the molecular geometry under the influence of intermolecular forces.

Structure refinement. We use the CrystalOptimizer program,[32] which has been developed to allow the optimisation of a specified set of intramolecular degrees of freedom. The forces result from combining intermolecular forces, from the interatomic potential calculated in DMACRYS, and intramolecular forces from electronic structure calculations in Gaussian09. The torsion angles around S–S–C–C, C–S–S–C, O–N–C–C and Cl–C–C–C bonds were chosen to be optimised directly in response to inter- and intramolecular forces. Rotation of the methyl groups on **4Me4Me** and **4Me2NO$_2$** had to be constrained at the orientation found in the original gas phase conformers to prevent convergence issues during the CrystalOptimizer optimisations. All remaining intramolecular degrees of freedom were optimised in the gas phase, but in response to changes in the torsions named above.

CrystalOptimizer calculations that are used during the CSP used the same atomic multipole-based model for intermolecular interactions as described above, with intramolecular forces and energies calculated at the B3LYP/6-31G** level of theory, with the GD3BJ empirical dispersion correction. Further CrystalOptimizer calculations were also performed on the observed crystal structures for the assessment of mechanochemical reaction energies. Here both B3LYP-GD3BJ/6-31G** and MP2/6-31G** electronic structure methods were independently used for both the molecular energy calculations and as a source of atomic multipoles. This allows us to assess the accuracy of the DFT approach and sensitivity to the description of intramolecular energies and charge density.

Final energy calculation. The influence of polarisation on the inter- and intramolecular contributions to the relative crystal energies was included in the calculation of total energies by performing a final single point molecular energy and DMACRYS lattice energy optimisation on the structures resulting from CrystalOptimizer (with no further refinement of the molecular geometry). This calculation takes the molecular energy and atomic multipoles from a molecular calculation performed within the polarising environment of a continuum dielectric, using the PCM model, as we have previously suggested for CSP.[33] The dielectric constant was set as $\varepsilon = 3.0$, as a value typical for the organic solid state.[34]

Results and discussion

Crystal structure prediction

We first assess the results of CSP on the five molecules by comparing the computer-generated structures to the known crystal structures of the reactants and products of both reactions. One crystal structure is known for each of **4Cl4Cl** (space group *Pbcn*, $Z' = 0.5$, Cambridge Structural Database code DCPHDS),[35] **4Me4Me** (space group $P2_1$, $Z' = 1$, CSD code IPIXUB),[3] **2NO$_2$2NO$_2$** (space group $P2_1/c$, $Z' = 1$, CSD code ODNPDS02)[36] and **4Me2NO$_2$** (space group $P2_1/c$, $Z' = 1$, CSD code FUQLEI).[37] The **4Cl4Cl** + **2NO$_2$2NO$_2$** reaction leads to a different polymorph of **4Cl2NO$_2$**, depending on whether the reaction is performed under neat grinding conditions or liquid assisted grinding (LAG), with a small amount of acetonitrile added to the grinding jar. The **4Cl2NO$_2$** *LAG* polymorph crystallises in the space group $P\bar{1}$ (CSD code FUQLIM),[37] while the **4Cl2NO$_2$** *Neat* polymorph crystallises in $P2_1/n$ (CSD code FUQLIM01).[3]

Molecular conformation

The approach used here to deal with conformational flexibility during CSP relies on the molecular geometry in the crystal structure being close enough to one of the optimised conformers of the isolated molecule. After geometry optimisation with molecular flexibility, one of the trial structures generated should lead to the observed crystal packing.

We find that all of the molecular geometries found in the observed crystal structures have a fairly good match among the generated gas-phase conformers. It is also useful to observe where each conformer lies on the energy landscape of the generated conformers. In all cases except **4Cl4Cl** and **4Me4Me**, the geometry adopted in the crystal structure corresponds to the global minimum energy conformer according the dispersion-corrected DFT calculations. For **4Cl4Cl** and

4Me4Me, the molecule adopts a conformation more than 5 kJ mol^{-1} above the most stable gas phase conformer (Fig. 2), highlighting the need to consider multiple conformers during CSP – in these structures, the energy penalty associated with adopting a higher energy conformation must be compensated by more favourable intermolecular interactions.

The overlays in Fig. 2 show the extent to which intermolecular forces due to crystal packing distort the molecules; the largest deviations from the nearest gas phase molecular geometry are seen in the orientation of the *para*-substituted ring for **4Me2NO$_2$** and the *Neat* polymorph of **4Cl2NO$_2$**, which are rotated 33.4° and 61.6° degrees away from the gas phase geometry in the crystal structure, respectively. We show in the following section that, for **4Cl2NO$_2$**, this leads to a failure of the structure generation to locate the structure of the *Neat* polymorph. The overlay for **4Cl4Cl** also shows a large RMSD$_1$, but this is due to small rotations of both rings, leading to a smaller change in the overall molecular shape.

Crystal structures

The multiple lattice energy minimisation steps described above are required for computational efficiency of the entire CSP procedure. Here we only show our analysis of the results after the final energy calculation, once the molecular geometry has been optimised within each crystal structure and polarisation of the

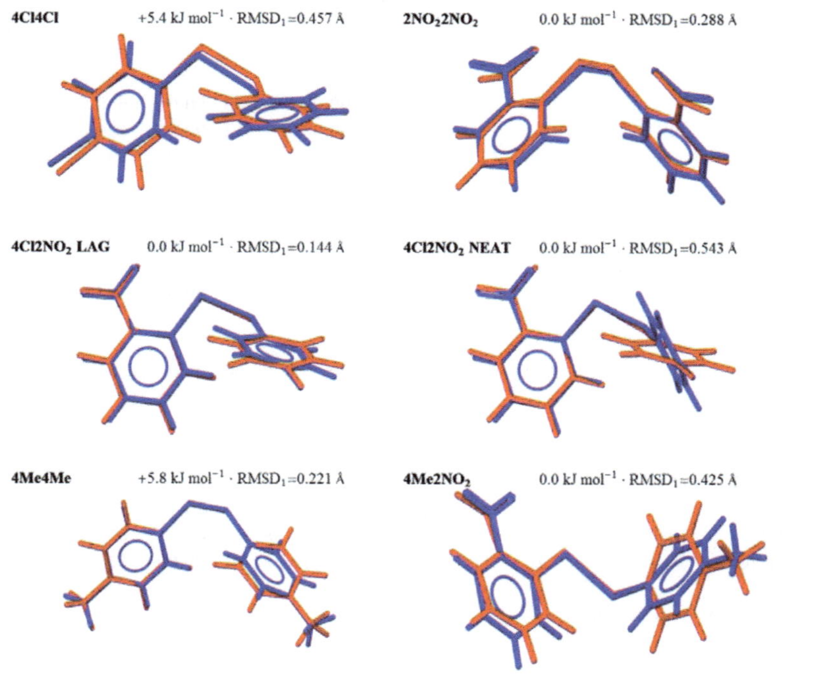

Fig. 2 Overlays of the molecular geometries in each experimentally determined crystal structure (blue) with the closest match among the computer-generated (B3LYP-GD3BJ/6-31G**) gas phase conformers (red). Also shown are the RMS deviations in atomic position (RMSD$_1$) and the calculated energy of the matching conformer relative to the lowest-energy calculated conformer (0.0 kJ mol^{-1} indicates that the global minimum conformer is adopted in the crystal structure).

electron density has been modelled. In analysing the results, we first concern ourselves with whether the experimentally determined crystal structures are found among the predictions and the quality of this match in terms of geometry. Secondly, we consider where this structure lies on the energy landscape of the prediction.

Crystal structure generation. We used the COMPACK algorithm[31] within Mercury[38] to search for matches between predicted and observed crystal structures, comparing the atomic positions in a cluster of 30 molecules taken from the structures being compared. If a match is found, we can measure an RMS deviation in atomic positions; which we refer to as $RMSD_{30}$. A successful match among the predictions was found for all systems except the *Neat* polymorph of **$4Cl2NO_2$**.

Table 1 Summary of the comparisons between observed (expt.) and predicted (pred.) crystal structure, along with optimised structures starting from the observed structures (expt.opt.). $RMDS_{30}$ is the RMS deviation in atomic positions between 30-molecule clusters taken from calculated and observed structures. For each CSP structure, the rank in the crystal energy landscape and energy difference to the global minimum is shown (in kJ mol^{-1})

Molecule	Rank (ΔE)	a/Å	b/Å	c/Å	α/°	β/°	γ/°	V/Å^3	$RMSD_{30}$/Å
$2NO_22NO_2$									
Expt.		7.103	22.925	7.777	90.00	95.54	90.00	1260.42	
Pred.	1 (0.0)	7.056	23.106	7.836	90.00	95.88	90.00	1275.05	0.190
Expt opt.		7.043	23.068	7.864	90.00	95.20	90.00	1272.21	0.215
4Me4Me									
Expt.		7.616	5.726	14.751	90.00	94.88	90.00	640.93	
Pred.	58 (6.3)	7.652	5.677	14.831	90.00	97.12	90.00	640.01	0.250
Expt opt.		7.606	5.662	14.846	90.00	94.40	90.00	637.39	0.129
$4Me2NO_2$									
Expt.		6.638	24.439	7.948	90.00	93.63	90.00	1286.82	
Pred.	1 (0.0)	6.891	23.847	7.610	90.00	93.30	90.00	1248.39	0.975
Expt opt.		6.898	23.988	7.525	90.00	93.39	90.00	1242.87	1.002
4Cl4Cl									
expt.		7.659	5.973	27.175	90.00	90.00	90.00	1252.08	
pred.	245 (13.8)	7.833	5.816	27.046	90.00	90.00	90.00	1233.54	0.268
expt opt.		7.685	5.863	27.460	90.20	90.00	90.00	1237.17	0.150
$4Cl2NO_2$ *LAG* polymorph									
Expt.		7.043	7.832	11.371	82.58	80.69	83.19	610.71	
Pred.	1 (0.0)	6.807	7.902	11.620	82.53	82.26	84.15	611.83	0.287
Expt opt.		6.776	7.901	11.645	82.81	82.03	86.00	611.71	0.357
$4Cl2NO_2$ *Neat* polymorph									
Expt.		13.456	7.103	15.707	90.00	124.00	90.00	1244.67	
Pred.[a]	18 (4.8)[a]	13.540	7.014	16.202	90.00	125.92	90.00	1271.40	0.370
Expt opt.		13.543	7.008	16.394	90.00	126.57	90.00	1249.33	0.464

[a] The *Neat* polymorph of **$4Cl2NO_2$** was not generated by the crystal structure search with the initial set of molecular conformers, but was generated in a search with the molecular geometry constrained to the observed conformation. The rank and relative energy reported for this polymorph refer to where this structure would be ranked within the original set of predicted structures.

Table 1 summarises the comparison between observed and predicted crystal structures and overlays of the best (**$2NO_22NO_2$**) and worst (**$4Me2NO_2$**) geometrical matches are shown in Fig. 3.

The one failure of the structure generation occurs for the crystal structure (*Neat* **$4Cl2NO_2$**) whose molecular geometry was farthest from one of the starting gas phase geometries. The approach of generating crystal structures starting with only gas phase molecular geometries has been successful for many conformationally flexible molecules.[7,8,39,40] However, this approach relies on the geometry adopted in the crystal being sufficiently close to one of the gas phase geometries that any molecular distortion due to crystal packing can be recovered during refinement of the trial structures. While previous CSP studies have shown that the crystal structure optimisation phase can allow for large changes in soft intramolecular degrees of freedom,[41] it is clear in this case that rotation of the *para*-substituted ring allows a crystal packing that is not available to the molecule in its gas phase geometry.

To satisfy ourselves that this polymorph would have been found had we started with the distorted molecular geometry, an additional crystal structure generation was performed using a molecular conformation where the S–S–C–C and C–S–S–C torsion angles were fixed to the values seen in the experimentally determined crystal structure. The intramolecular energy penalty for these constraints is only 2.3 kJ mol^{-1}, which must be balanced by improved crystal packing and intermolecular interactions. While the entire set of crystal structures from this additional prediction was not processed through the whole optimisation procedure, we found that the lowest energy structure after rigid-molecule lattice energy

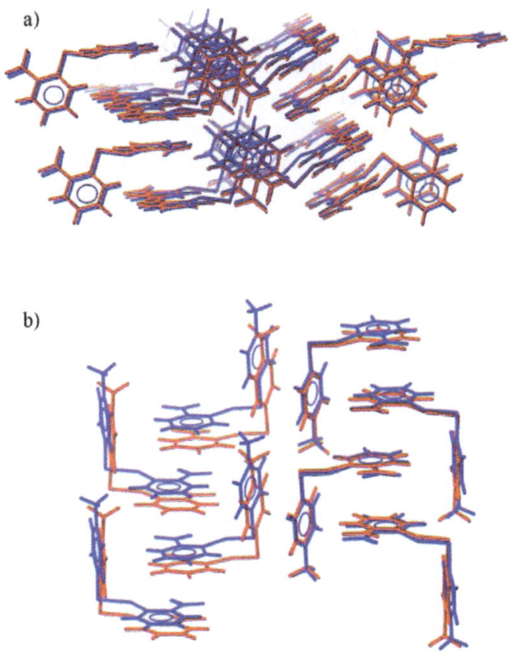

Fig. 3 Overlays of the observed (blue) and predicted (red) crystal structures of (a) $2NO_22NO_2$ and (b) $4Me2NO_2$.

minimisation with the atomic multipole electrostatic model matches the observed structure of the *Neat* polymorph. This confirms the cause of the failure as being due to the distortion of the molecular geometry; molecular flexibility has been introduced at too late a stage in the CSP methodology. We processed the matching structure through the CrystalOptimizer step to find its energy with respect to the rest of the predictions (Table 1).

Matches to the experimentally determined crystal structures were found for all remaining systems. The observed structures of **4Cl4Cl**, **4Me4Me**, **$2NO_22NO_2$** (Fig. 3a) and **$4Cl2NO_2$** are all reproduced very accurately, with an $RMSD_{30}$ of less than 0.3 Å and the worst lattice parameter being 3.4% different from experiment (most are reproduced to within 1% of observed values).

The structure of **$4Me2NO_2$** is reproduced less well (Table 1). The predicted and experimental structures differ by a 'slipping' of the planes of face-to-face stacked arene rings (Fig. 3b). The energy associated with this type of distortion is small and the calculated structure is likely to be very sensitive to small changes in the intermolecular potential. The difference might also be partly attributed to the lack of temperature in the calculated structures; slipping of these weak shear planes might accompany thermal expansion from the temperature-free predictions to the temperature of the observed structure.

Lattice energy ranking. The second, and equally important, assessment of the success of CSP is based on the energy ranking of the observed structure within the set of predicted structures. An assumption in the global lattice energy minimisation approach to CSP is that the observed crystal structures will be among the lowest-energy predicted possibilities. While this is often the case, it is important to note that there might be good reasons why the crystal structure with the lowest lattice energy might not be that observed in experiment. We ignore kinetic influences during crystal nucleation and growth on which crystal structure is formed. Furthermore, the influence of temperature on the relative stabilities is not included in the current method, and entropy differences between structures can sometimes be of the same order of magnitude as lattice energy differences.[42,43]

The calculated crystal energy landscapes are very crowded, showing that many nearly-isoenergetic crystal structures are available for each molecule (Fig. 4). The observed crystal structures of **$2NO_22NO_2$** (Fig. 4a), **$4Me2NO_2$** and the *LAG* polymorph of **$4Cl2NO_2$** all correspond to the global minimum structure from their respective CSP. While the *Neat* polymorph of **$4Cl2NO_2$** was not generated in the initial search, its energy would have ranked it 18th among the predicted crystal structures, 4.8 kJ mol^{-1} above the lowest energy crystal structure. For each of these molecules, the lowest energy gas phase conformer was the closest matching conformer to the molecular geometry in the crystal.

The CSP results are less successful for the two molecules **4Cl4Cl** and **4Me4Me**, whose observed crystal structure contains a higher energy conformer. The fine balance required between intramolecular and intermolecular energies is highlighted in the crystal energy landscape of **4Me4Me** (Fig. 4b); the observed crystal structure is the 4th lowest in energy among the crystal structures generated with that molecular conformation, just 1.6 kJ mol^{-1} above the best calculated packing of that conformation. The good in-conformation ranking, but relatively high global ranking, indicates errors with our chosen level of theory for treating intramolecular contributions to the total crystal energies. It seems likely that limitations of the DFT functional (B3LYP with empirical dispersion correction) or

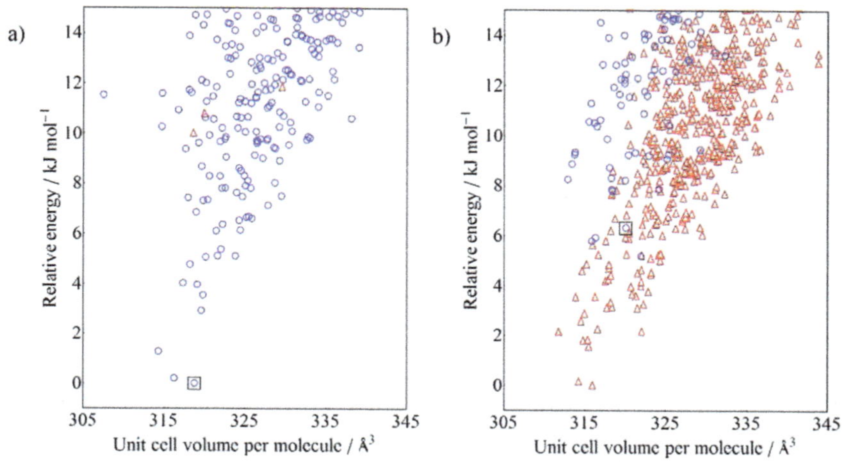

Fig. 4 Calculated crystal energy landscapes of (a) **$2NO_22NO_2$** and (b) **4Me4Me**. Each point corresponds to a distinct crystal structure, whose total energy is reported relative to the lowest energy predicted structure. The open blue circles are structures generated by the molecular conformer that is closest to the observed conformer, and open red triangles are from the other conformers. The prediction that matches the observed structure is indicated by an open square.

medium-sized basis set (6-31G**) affect this result. We see a similar distribution of the predicted crystal structures of **4Cl4Cl:** all of the lowest energy structures are generated from an alternative conformation to that found in the observed crystal structure. However, in this case, the true crystal structure is also poorly ranked within the structures generated with only the observed conformation, appearing at rank 23. This result causes most concern and is symptomatic of inaccuracies in both the intermolecular energy model and the method used for the intramolecular energies. In Fig. 5 we show the unit cells of the prediction which matches the observed structure, the lowest energy computer-generated packing of the observed conformer and the overall global energy minimum crystal structure. We note that the predicted and observed structures exhibit very different intermolecular interactions. The arene rings form T-shaped interactions in the experimentally observed packing (Fig. 5a), whereas the predicted structures with the lowest calculated energies have rings packed in a face-to-face arrangement (Fig. 5b and c). Furthermore, the observed crystal packing contains planes of close Cl---Cl intermolecular interactions, which are absent in the lowest energy predicted structures. We conclude that the intermolecular model potential does not describe these interactions accurately enough to rank correctly the predicted crystal structures. In particular, the chlorine atom parameters, which were added to the W99 intermolecular potential in an *ad hoc* manner, could be improved to provide a better balance between Cl---Cl interactions and Cl interactions with other atoms.

Reaction energies

From a synthetic point of view, the most important outcome is not the crystal structure, but the position of the equilibrium: does the solid-state reaction favour the heterodimers, homodimers, or are they finely balanced? In this work, we seek

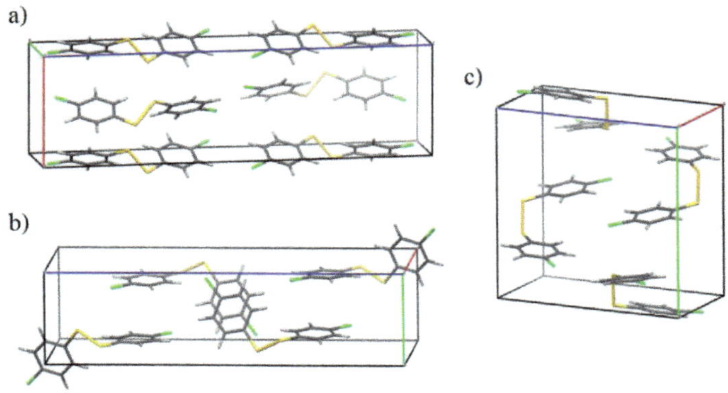

Fig. 5 Unit cells of (a) the predicted crystal structure that matches the observed structure, (b) the lowest energy computed structure of the observed conformer and (c) the global minimum predicted crystal structure of 4Cl4Cl.

to understand how well the computational methods, which have been developed for the application of CSP, perform for determining the relative stabilities of the homodimer and heterodimer crystal structures. We set our target as successfully predicting whether the equilibrium should favour the homodimers or heterodimer in each reaction. Until we can achieve this goal reliably, we make no attempt at predicting the precise equilibrium constant. For the two reactions under investigation, the experimental observations show clearly that, for the reaction of **4Cl4Cl** with **$2NO_22NO_2$**, heterodimers are favoured strongly enough that the grinding experiment leads to 98% yield of heterodimers. When **4Me4Me** is reacted with **$2NO_22NO_2$**, only 20% heterodimer is formed, demonstrating that the heterodimers are less stable than homodimers.

Fig. 6 shows the computed reaction energies using the same B3LYP-GD3BJ/6-31G** molecular calculations as used in the CSP studies, before and after applying the final PCM correction for polarisation contributions to the energies. All calculations are performed starting from the experimentally determined crystal structures, after applying the same optimisation procedure that was used in CSP (see Table 1 for the final structures). For comparison with DFT, we have performed the same CrystalOptimizer optimisation of the experimental structures with intramolecular energies and the intermolecular electrostatic model derived from MP2/6-31G** calculations. Because *Neat* and *LAG* grinding conditions lead to different polymorphs of **$4Cl2NO_2$**, we calculate the reaction energy separately for the two heterodimer polymorphs.

Overall, the results for the **4Cl4Cl** with **$2NO_22NO_2$** *LAG* reaction agree with experimental observations, irrespective of the details of the method used (DFT *vs.* MP2 molecular energies, with or without PCM estimate of polarisation contributions). This reaction is entirely dominated by the intramolecular energy change according to the DFT-based calculations, while the MP2 results give a description that is more evenly balanced between inter- and intramolecular contributions. Furthermore, it is important to note that the influence of crystal packing on the molecular geometry is crucial. At the geometries of the gas phase molecules, the B3LYP-GD3BJ/6-31G** molecular energies favour the heterodimers by only 0.9 kJ mol^{-1}. The packing forces in the individual crystal structures increase the

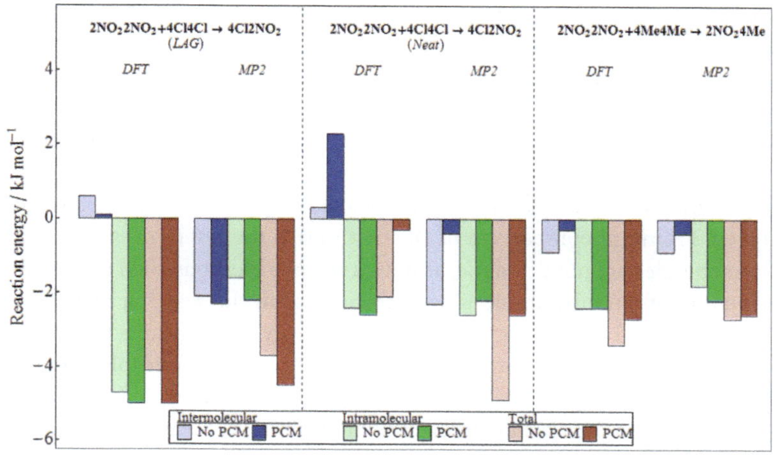

Fig. 6 Calculated reaction energies (in kJ mol^{-1}) for the reactions of **4Cl4Cl** + **$2NO_22NO_2$** to the *LAG* and *Neat* polymorphs, and the **4Me4Me** + **$2NO_22NO_2$** to **$4Me2NO_2$** reaction. Energies are shown both with and without the PCM corrected model described above. Contributions are separated into intermolecular, intramolecular and total energies and using both B3LYP-GD3BJ/6-31G** (DFT) and MP2/6-31G** (MP2).

intramolecular energy difference by an additional 3.7 kJ mol^{-1} by introducing more molecular strain in the homodimers than the heterodimer. Introducing the effects of polarisation using the PCM model enhances the relative stability of the heterodimers even further.

Although a different polymorph is produced under *Neat* grinding conditions, the heterodimer is still found to be favoured in the *Neat* reaction of **4Cl4Cl** with **$2NO_22NO_2$**. Again, the sign of the energy change agrees with experimental observations and, unsurprisingly, there is a similar intramolecular energy change as found for the *LAG* polymorph of **$4Cl2NO_2$**. However, unlike the *LAG* reaction, the total reaction energies are quite sensitive to the change from DFT to MP2. Interestingly, this sensitivity is found in the intermolecular contribution to the reaction energy, suggesting that these differences result from differences in the description of intermolecular electrostatics between DFT and MP2. One consistent finding between DFT and MP2 results is that the inclusion of polarisation into the calculation shifts the balance of intermolecular energies towards the homodimers. The intermolecular energy change cancels much of the favourable intramolecular energy change in the DFT results and allows the intramolecular energy to dominate the MP2 result. With polarisation included, the reaction energy drops to -0.3 kJ mol^{-1} in the DFT results (-2.6 kJ mol^{-1} with MP2).

Our results for the reaction of **4Me4Me** with **$2NO_22NO_2$** are less satisfactory. When DFT or MP2 is used to for the intramolecular contribution, our calculations would predict that this reaction also favours the heterodimers. Both the intra- and intermolecular energies are calculated to be lower in the heterodimer than the homodimers. This result contradicts what is observed in the grinding experiments. We note that this is the reaction for which the crystal structure of the heterodimer is poorly reproduced (Table 1 and Fig. 3b) and this is also true in the MP2 results ($RMSD_{30}$ = 1.340 Å). These errors in reproducing the heterodimer crystal structure may be a source of the inconsistency between the observed and predicted behaviour for this reaction.

Using DFT molecular energies and electrostatics, the three reactions each show quite a different balance between inter- and intramolecular energy changes, making these very challenging systems for computational prediction. It is a strength of the hybrid method used for the energy calculations here that the quality of the intermolecular and intramolecular energy model can be separately improved. From our comparison of DFT with MP2, the differences in the overall reaction energies are not dramatic, although the two methods yield quite different individual inter- and intramolecular contributions in the reaction of **4Cl4Cl** with **$2NO_22NO_2$**. Therefore, we suggest that the sensitivity of the results to the method used for molecular calculations is one area that needs further exploration.

With the small magnitude of the reaction energies, and the delicate balance of inter- and intramolecular energies, the inclusion of polarisation should be important in getting the energies correct. In the approach taken here, the polarisation of the molecules in their crystal structures has been modelled with a very approximate model. The crystalline environment is described by a structureless dielectric continuum with a single dielectric constant used to model all of the crystal structures. The model originates from the need to introduce polarisation in a computationally efficient manner for the large numbers of crystal structures considered in CSP studies.[33] The success of this polarisation model in CSP probably relies on fortuitous cancellation of error, because all of the crystal structures considered have the same chemical composition. It is inconclusive whether this method of including polarisation is useful in calculating reaction energies when the lattice energies are computed for different molecules. These types of reactions will be a challenging test for the development of more detailed polarisation models.

A relevant question is whether predictions of the thermodynamic balance between reactant and product crystal structures in solid state reactions can accommodate any error in the prediction of the crystal structures themselves. The magnitudes of the calculated reaction energies are similar to the calculated energy differences between the predicted crystal structures of each molecule. Therefore, at least for these reactions, mis-predictions of the crystal structure can lead to energetic errors that are significant. Certainly, the energy differences between the global minimum prediction and the observed structure for **4Cl4Cl** and **4Me4Me** are so large that using the global minimum predicted structure rather than the observed structure would have introduced substantial errors to the reaction energies.

Conclusions

The results of this assessment of the application of computational methods to solid-state aromatic disulfide metathesis reactions are a mix of successful results with some failures of the applied methodology that serve to highlight areas for development. We cannot conclude from these results that *current* theoretical methods are predictive of the solid state structures and energetics that govern the equilibrium in mechanochemical reactions. Nevertheless, there are positive results to build on and we find nothing to suggest that the goal of predictive computational methods in this area cannot be achieved.

The outcomes of the CSP part of this work are consistent with recent results for other moderately flexible organic molecules. Of the six target crystal structures, three are predicted as the global minimum structure on the molecule's crystal energy landscape; these very encouraging outcomes result from a successful conformational search, crystal structure generation and an accurate energy model for lattice energy minimisation. The structure prediction successes are tempered by the results obtained for the remaining three crystal structures, two being found at fairly high energies on the crystal energy landscape, and one of the known crystal structures being missed by the computational search. The three unsuccessful predictions are each due, at least in part, to problems related to conformational flexibility in the molecules, illustrating the two types of problems that molecular flexibility introduces in CSP. Failure to generate the monoclinic (*Neat* grinding) polymorph of **4Cl2NO$_2$** highlights the need for a more rigorous sampling of conformational space during the crystal structure search. The preference for unobserved conformations in the predicted crystal structures of **4Cl4Cl** and **4Me4Me** indicate the requirement for quantum chemical methods that provide more reliable relative intramolecular energies between conformers than the dispersion-corrected DFT that we have used here.

The later requirement is brought to the fore in our calculations of the overall reaction energies. The inter- *versus* intramolecular balance of the total lattice energy change from homodimer crystals to the heterodimer crystal varies between reactions and, especially for the reaction of **4Cl4Cl** with **2NO$_2$2NO$_2$**, is quite sensitive to the level of theory used for the molecular energy calculation. While the affordable computational cost of DFT is attractive, particularly when considering the large numbers of structures in CSP, the systematic improvability of wavefunction based methods may be required to give reaction energies in which we have confidence. The results also highlight the need for reliable structure prediction: the intermolecular interactions play a key role in determining the relative energies of the products and reactants, such that the crystal structure must be known to predict their equilibrium.

Overall, solid-state organic reactions are an attractive target for the application of computational methods, where we can build on the continuing development of quantum chemical methods used to study gas phase reactions and recent developments in predictive methods for molecular crystal structures. We believe that the results obtained in this study demonstrate that we have a solid base of methods on which to build.

Acknowledgements

P. J. B. and G. M. D. thank the EPSRC for financial support (grant EP/J01110X/1). D. H. C. and G. M. D. acknowledge the European Research Council for funding under grant ERC-StG-2012-307358-ANGLE. The authors acknowledge the use of the IRIDIS High Performance Computing Facility, and associated support services at the University of Southampton, in the completion of this work.

References

1 S. L. James, C. J. Adams, C. Bolm, D. Braga, P. Collier, T. Friscic, F. Grepioni, K. D. M. Harris, G. Hyett, W. Jones, A. Krebs, J. Mack, L. Maini, A. G. Orpen,

I. P. Parkin, W. C. Shearouse, J. W. Steed and D. C. Waddell, *Chem. Soc. Rev.*, 2012, **41**, 413–447.

2 T. Friscic, *Chem. Soc. Rev.*, 2012, **41**, 3493–3510.

3 A. M. Belenguer, T. Friscic, G. M. Day and J. K. M. Sanders, *Chem. Sci.*, 2011, **2**, 696–700.

4 S. L. Price, *Adv. Drug Delivery Rev.*, 2004, **56**, 301–319.

5 A. J. Cruz Cabeza, G. M. Day, W. D. S. Motherwell and W. Jones, *Cryst. Growth Des.*, 2007, **7**, 100–107.

6 G. M. Day and T. G. Cooper, *CrystEngComm*, 2010, **12**, 2443–2453.

7 A. V. Kazantsev, P. G. Karamertzanis, C. S. Adjiman, C. C. Pantelides, S. L. Price, P. T. A. Galek, G. M. Day and A. J. Cruz-Cabeza, *Int. J. Pharm.*, 2011, **418**, 168–178.

8 M. Baias, C. M. Widdifield, J.-N. Dumez, H. P. G. Thompson, T. G. Cooper, E. Salager, S. Bassil, R. S. Stein, A. Lesage, G. M. Day and L. Emsley, *Phys. Chem. Chem. Phys.*, 2013, **15**, 8069–8080.

9 R. M. Bhardwaj, L. S. Price, S. L. Price, S. M. Reutzel-Edens, G. J. Miller, I. D. H. Oswald, B. F. Johnston and A. J. Florence, *Cryst. Growth Des.*, 2013, **13**, 1602–1617.

10 H. Wu, M. Habgood, J. E. Parker, N. Reeves-McLaren, J. K. Cockcroft, M. Vickers, A. R. West and A. G. Jones, *CrystEngComm*, 2013, **15**, 1853–1859.

11 J. Kendrick, G. A. Stephenson, M. A. Neumann and F. J. J. Leusen, *Cryst. Growth Des.*, 2013, **13**, 581–589.

12 G. M. Day, *Crystallogr. Rev.*, 2011, **17**, 3–52.

13 J. L. Banks, H. S. Beard, Y. Cao, A. E. Cho, W. Damm, R. Farid, A. K. Felts, T. A. Halgren, D. T. Mainz, J. R. Maple, R. Murphy, D. M. Philipp, M. P. Repasky, L. Y. Zhang, B. J. Berne, R. A. Friesner, E. Gallicchio and R. M. Levy, *J. Comput. Chem.*, 2005, **26**, 1752–1780.

14 I. Kolossváry and W. C. Guida, *J. Am. Chem. Soc.*, 1996, **118**, 5011–5019.

15 I. Kolossváry and W. C. Guida, *J. Comput. Chem.*, 1999, **20**, 1671–1684.

16 *MacroModel, version 10.1*, Schrödinger, LLC, New York, NY, 2013.

17 *Gaussian 09 rev D.01*, M. J. T. Frisch G. W.; H. B. Schlegel; G. E. Scuseria; M. A. Robb; J. R. Cheeseman; G. Scalmani; V. Barone; B. Mennucci; G. A. Petersson; H. Nakatsuji; M. Caricato; X. Li; H. P. Hratchian; A. F. Izmaylov; J. Bloino; G. Zheng; J. L. Sonnenberg; M. Hada; M. Ehara; K. Toyota; R. Fukuda; J. Hasegawa; M. Ishida; T. Nakajima; Y. Honda; O. Kitao; H. Nakai; T. Vreven; J. A. Montgomery, Jr.; J. E. Peralta; F. Ogliaro; M. Bearpark; J. J. Heyd; E. Brothers; K. N. Kudin; V. N. Staroverov; R. Kobayashi; J. Normand; K. Raghavachari; A. Rendell; J. C. Burant; S. S. Iyengar; J. Tomasi; M. Cossi; N. Rega; N. J. Millam; M. Klene; J. E. Knox; J. B. Cross; V. Bakken; C. Adamo; J. Jaramillo; R. Gomperts; R. E. Stratmann; O. Yazyev; A. J. Austin; R. Cammi; C. Pomelli; J. W. Ochterski; R. L. Martin; K. Morokuma; V. G. Zakrzewski; G. A. Voth; P. Salvador; J. J. Dannenberg; S. Dapprich; A. D. Daniels; Ö. Farkas; J. B. Foresman; J. V. Ortiz; J. Cioslowski; D. J. Fox, Gaussian, Inc., Wallingford CT, 2009.

18 A. D. Becke, *J. Chem. Phys.*, 1993, **98**, 5648–5652.

19 J. S. Binkley, J. A. Pople and W. J. Hehre, *J. Am. Chem. Soc.*, 1980, **102**, 939–947.

20 S. Grimme, S. Ehrlich and L. Goerigk, *J. Comput. Chem.*, 2011, **32**, 1456–1465.

21 C. P. Brock and J. D. Dunitz, *Chem. Mater.*, 1994, **6**, 1118–1127.

22 G. M. Day, 2013.
23 D. E. Williams, *J. Mol. Struct.*, 1999, **485–486**, 321–347.
24 D. E. Williams, *J. Comput. Chem.*, 2001, **22**, 1154–1166.
25 D. E. Williams, *J. Comput. Chem.*, 2001, **22**, 1–20.
26 A. Abraha and D. E. Williams, *Inorg. Chem.*, 1999, **38**, 4224–4228.
27 G. M. Day and S. L. Price, *J. Am. Chem. Soc.*, 2003, **125**, 16434–16443.
28 C. M. Breneman and K. B. Wiberg, *J. Comput. Chem.*, 1990, **11**, 361–373.
29 A. J. Stone, *Chem. Phys. Lett.*, 1981, **83**, 233–239.
30 S. L. Price, M. Leslie, G. W. A. Welch, M. Habgood, L. S. Price, P. G. Karamertzanis and G. M. Day, *Phys. Chem. Chem. Phys.*, 2010, **12**, 8478–8490.
31 J. A. Chisholm and S. Motherwell, *J. Appl. Crystallogr.*, 2005, **38**, 228–231.
32 A. V. Kazantsev, P. G. Karamertzanis, C. C. Pantelides and C. S. Adjiman, in *Process Systems Engineering*, Wiley-VCH Verlag GmbH & Co. KGaA, 2011, pp. 1–42.
33 T. G. Cooper, K. E. Hejczyk, W. Jones and G. M. Day, *J. Chem. Theory Comput.*, 2008, **4**, 1795–1805.
34 *Handbook of Chemistry and Physics*, CRC, 1992.
35 J. K. Vandavasi, W.-P. Hu, C.-Y. Chen and J.-J. Wang, *Tetrahedron*, 2011, **67**, 8895–8901.
36 M. Song and C. Fan, *Acta Crystallogr., Sect. E: Struct. Rep. Online*, 2009, **65**, o2835.
37 C. Glidewell, J. N. Low and J. L. Wardell, *Acta Crystallogr., Sect. B: Struct. Sci.*, 2000, **56**, 893–905.
38 C. F. Macrae, I. J. Bruno, J. A. Chisholm, P. R. Edgington, P. McCabe, E. Pidcock, L. Rodriguez-Monge, R. Taylor, J. Van de Streek and P. A. Wood, *J. Appl. Crystallogr.*, 2008, 41.
39 D. E. Braun, M. Ardid-Candel, E. D'Oria, P. G. Karamertzanis, J.-B. Arlin, A. J. Florence, A. G. Jones and S. L. Price, *Cryst. Growth Des.*, 2011, **11**, 5659–5669.
40 M. Baias, J.-N. Dumez, P. H. Svensson, S. Schantz, G. M. Day and L. Emsley, *J. Am. Chem. Soc.*, 2013, **135**, 17501–17507.
41 G. M. Day, W. D. S. Motherwell and W. Jones, *Phys. Chem. Chem. Phys.*, 2007, **9**, 1693–1704.
42 B. P. van Eijck, *J. Comput. Chem.*, 2001, **22**, 816–826.
43 A. Gavezzotti and G. Filippini, *J. Am. Chem. Soc.*, 1995, **117**, 12299–12305.

Faraday Discussions

PAPER

The two-step mechanochemical synthesis of porphyrins†

Hannah Shy,[a] Paula Mackin,[a] Andrea S. Orvieto,[a] Deepa Gharbharan,[a] Geneva R. Peterson,[b] Nick Bampos[b] and Tamara D. Hamilton*[a]

Received 15th December 2013, Accepted 30th January 2014
DOI: 10.1039/c3fd00140g

Porphyrin synthesis under solvent-free conditions represents the "greening" of a traditional synthesis that normally requires large amounts of organic solvent, and has hindered the industrial-scale synthesis of this useful class of molecules. We have found that the four-fold acid-catalysed condensation of aldehyde and pyrrole to yield a tetra-substituted porphyrin is possible through mechanochemical techniques, without a solvent present. This represents one of the still-rare examples of carbon–carbon bond formation by mechanochemistry. Specifically, upon grinding equimolar amounts of pyrrole and benzaldehyde in the presence of an acid catalyst, cyclization takes place to give reduced porphyrin precursors (reversible), which upon oxidation form tetraphenylporphyrin (TPP). The approach has been found to be suitable for the synthesis of a variety of *meso*-tetrasubstituted porphyrins. Oxidation can occur either by using an oxidizing agent in solution, to give yields comparable to those published for traditional methods of porphyrin synthesis, or through mechanochemical means resulting in a two-step mechanochemical synthesis to give slightly lower yields that are still being optimized. We are also working on "green" methods of porphyrin isolation, including entrainment sublimation, which would hopefully further reduce the need for large amounts of organic solvent. These results hold promise for the development of mechanochemical synthetic protocols for porphyrins and related classes of compounds.

Introduction

Porphyrins and related structures play many roles in systems throughout nature, including light harvesting and metal transport. Of particular synthetic interest are the *meso*-tetrasubstituted porphyrins, since they can serve as a platform for a variety of important applications due to their shape and symmetry, and the

[a]*Department of Physical Sciences, Barry University, 11300 NE 2nd Ave., Miami Shores, FL 33161, USA. E-mail: thamilton@mail.barry.edu; Fax: +305-899-3479; Tel: +305-899-3431*

[b]*University Chemical Laboratory, University of Cambridge, Lensfield Rd., Cambridge, CB2 1EW, United Kingdom. E-mail: nb10013@cam.ac.uk; Fax: +44 1223 336017; Tel: +44 01223 336300*

† Electronic Supplementary Information (ESI) available: ^{1}H NMR spectrum of TPP. MALDI-TOF mass spectrum of TPP. EcoScale calculations tables. See DOI: 10.1039/c3fd00140g

possibility for substitution on the periphery which allows for the tailoring of optical, electronic and biochemical properties.[1] Despite the consistent high performance of the porphyrin systems in applications like catalysis, the industrial use of these species is barred because of the impracticality of the scale-up for their syntheses. Established synthetic approaches to *meso*-tetrasubstituted porphyrins involve large amounts of corrosive or toxic organic solvents or high temperature reaction conditions, often followed by extensive chromatographic purification.[2–6]

Mechanochemistry, where chemical reactions are promoted by grinding reagents together either using a mortar and pestle or automated ball mill, is an often-overlooked approach to molecular synthesis that is currently gaining ground due to the relative simplicity of the reaction conditions, the potential for the elimination of solvent and the possibility of generating products not attainable by other methods.[7] In addition, there are growing numbers of examples proving the effectiveness of mechanochemistry for the synthesis of several types of products including co-crystals and metal-organic frameworks, but the extent of the applicability of the approach for organic reactions, especially carbon–carbon bond-forming reactions is still emerging.[8] Recent reports have employed sequential mechanochemical steps to reach selected targets.[9]

The first synthesis of *meso*-tetraphenylporphyrin (TPP) was a sealed-tube anaerobic condensation in pyridine at 200 °C, resulting in a 5% yield.[2] *meso*-Tetrasubstituted porphyrins are now most commonly synthesized in solution using one of two methods. The Adler–Longo reaction produces a porphyrin in one step under aerobic conditions in refluxing acetic or propionic acid solvent (10–30% yield).[3] The Lindsey synthesis[4] requires an inert atmosphere and relatively high dilution in chlorinated solvents, and separates the process into two steps. Firstly, an equilibrium is established between the starting materials and the porphyrinogen macrocycle (six-electron, six hydrogen reduced porphyrin). This is followed by the addition of a gentle oxidizer such as 2,3-dichloro-5,6-dicyano-benzoquinone (DDQ) which, after stirring for several hours, generates the porphyrin (10–60% yield). The latter approach has led to higher yields than the former for several *meso*-tetrasubstituted porphyrins from aldehydes that are sensitive to the harsh conditions of the Adler–Longo synthesis, and also avoids purification problems caused by tar-like by-products for those porphyrins that do not crystallize at the end of the reaction. Two solventless syntheses of *meso*-tetrasubstitued porphyrins have been reported: one involves mixing the two reagents in the gas phase at >200 °C (10–23% yield)[5] and the other is a microwave synthesis on a silica gel support (1–10% yield).[6]

Here we present a mechanochemical synthesis of TPP (Scheme 1) involving the grinding of benzaldehyde and pyrrole in the presence of an acid catalyst to give a solid pink powder. This powder can then be oxidized either in air, in chloroform with an organic oxidizer, or in a second grinding step with an oxidizing agent to produce TPP as the product. We have also expanded the approach to include a series of *meso*-tetrasubstituted porphyrins. The use of mechanochemistry to bring about an acid-catalysed condensation between aldehydes and pyrrole followed by the oxidation of the intermediates to produce the porphyrin enables the elimination of both the solvent and high temperatures from the synthesis of these important compounds, and represents a significant reduction in environmental impact.

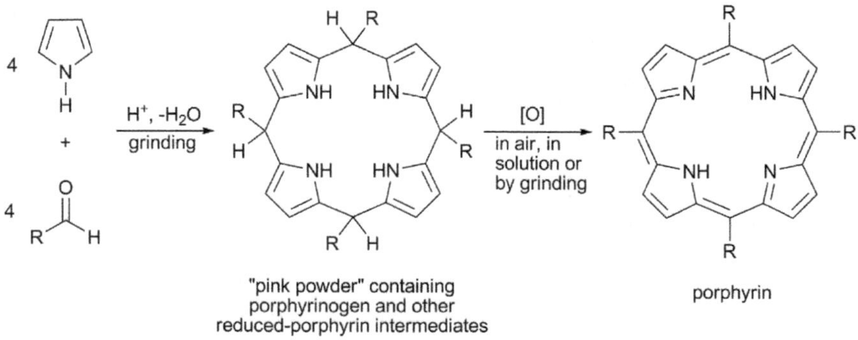

Scheme 1 The two-step synthesis of porphyrins involving mechanochemistry. For TPP, R = Ph.

Experimental

Materials and methods

All chemicals used were purchased from Sigma Aldrich and were used without further purification, unless otherwise noted. Benzaldehyde and pyrrole were purified by distillation once per week, or whenever discoloration appeared. Automated sample grinding was done using a Retsch Mixer Mill MM200. The electronic spectra were recorded on a Perkin Elmer Lambda 850 UV-Vis spectrophotometer, measured from 200–800 nm at 1 nm intervals. The samples were placed in quartz cuvettes with a 1 cm path length. ^{1}H NMR spectra were recorded on a Varian 300 MHz nuclear magnetic resonance spectrometer. The MALDI-TOF mass spectrum was collected on an ABI 4700 Proteomics Analyser.

Condensation by manual grinding

Manual grinding was done using an agate mortar and pestle. Pyrrole (0.259 g, 3.73 mmol) and toluenesulfonic acid (0.026 g, 0.151 mmol) were added to the mortar and ground together for approximately one minute. Benzaldehyde (0.380 g, 3.73 mmol) was added and the mixture was ground until a dry pink solid formed (~6 minutes).

The synthesis of *meso*-tetrasubstituted porphyrins - oxidation in chloroform

A 1 : 1 molar ratio of pyrrole (0.259 g, 3.73 mmol) and an aldehyde was combined with toluenesulfonic acid (0.026 g, 0.151 mmol) in a stainless steel grinding jar (10 mL volume) along with two stainless steel balls (5 mm diameter). The mixture was ground in a Retsch MM200 mill at a frequency of 25 Hz. The typical grinding time was 20 minutes until the formation of a dry, pink solid occurred. The pink solid was then dissolved in chloroform (*ca.* 50 mL). DDQ (2.54 g, 11.19 mmol) was added and the mixture was stirred for 2 hours. The mixture was then passed through a 1 cm plug of neutral alumina. Triethylamine (2–3 drops) was added to the resulting filtrate, which was then loaded onto a silica gel column with 9 : 1 ethylacetate : hexane as the mobile phase. Fractions were collected and assessed using thin-layer chromatography and visible spectroscopy. The fractions containing porphyrin were combined and the solvent was removed using a rotary evaporator. Tetraphenylporphyrin was obtained as a purple powder in 28% yield. ^{1}H NMR (300 MHz, $CDCl_3$): $\delta = 8.84$ (8H, s), 8.21 (8H, t), 7.76 (12H, d), 7.26 (chloroform), 1.60 (solvent residue), −2.79 (2H, s). UV-VIS ($CHCl_3$): $\lambda = 417$ (Soret band); $\lambda = 516, 549, 591, 645$ (Q-bands).

Mechanochemical oxidation

A pink solid was obtained through either manual or automated grinding of equimolar amounts of benzaldehyde and pyrrole in the presence of an acid, as described above. An oxidizer was then added in a 1 : 1.5 benzaldehyde : oxidizer molar ratio and the mixture was ground in a Retsch MM200 mill at a frequency of 25 Hz for varying times. In some experiments, a grinding or drying agent was also added. Each resulting dark purple solid was tested for the presence of TPP using visible spectroscopy.

Spectroscopic sample preparation to determine the yields from the mechanochemical oxidation

A stock solution of pure TPP was prepared by dissolving 30.0 mg TPP in 10.00 mL chloroform. This solution was serially diluted until a solution of concentration 3.00×10^{-6} g mL^{-1} was obtained. The diluted solution had a measured absorbance of 1.056 ($\lambda = 417$ nm). Solutions were then prepared for each oxidized sample in the same way, and the absorbance at 417 nm was measured and compared to that of the pure TPP solution, in order to obtain an approximate yield.

Results and Discussion

Condensation of benzaldehyde and pyrrole

The initial studies promptly showed that simple mortar-and-pestle grinding of two colourless liquids, benzaldehyde and pyrrole, in the presence of an acid catalyst yields first a gummy pink substance which, upon grinding for approximately six minutes becomes a uniform, dry, pink powder. No Soret band (~417 nm), which would be characteristic of TPP, is visible in the electronic spectrum of the freshly-ground powder (Figure 1), rather a broad peak at 500 nm is present. After exposure to air for two weeks however, a black powder is obtained. The electronic spectroscopy of this powder reveals a small Soret band at 417 nm, indicating the presence of TPP (3% yield of porphyrin in the mixture after 15 days). Thin-layer chromatography (TLC; 9 : 1 ethylacetate : hexanes) and MALDI-TOF mass spectrometry ($m/z = 615$, $[M + H]^+$) also confirm the presence of TPP in the black powder.

The two broad peaks (~500 nm and ~360 nm) in the UV-Vis spectrum of the freshly-ground pink powder must correspond to a mixture of reduced-porphyrin intermediates (Figure 2), which are able to be oxidised in air to form the porphyrin, since their absorbance decreases over time. Possibilities include dihydroporphyrins (phlorins or chlorins), tetrahydroporphyrins (bacteriochlorin, chlorin-phlorin, or porphomethene) and hexahydroporphyrin (porphyrinogen). All except the porphyrinogen have absorbance in the visible region. Those involving methine-bridge saturation (phlorin and porphomethene) are known to be the least stable to air oxidation.[10] Further studies are required to characterize

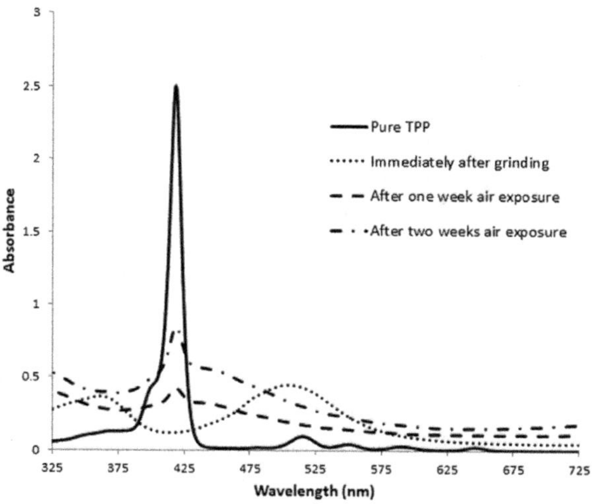

Fig. 1 The electronic spectrum of pure TPP ($CHCl_3$), freshly-ground reaction mixture, and the reaction mixture after having been exposed to air for one, and then two weeks. Pure TPP is shown at one-tenth the concentration of the other samples.

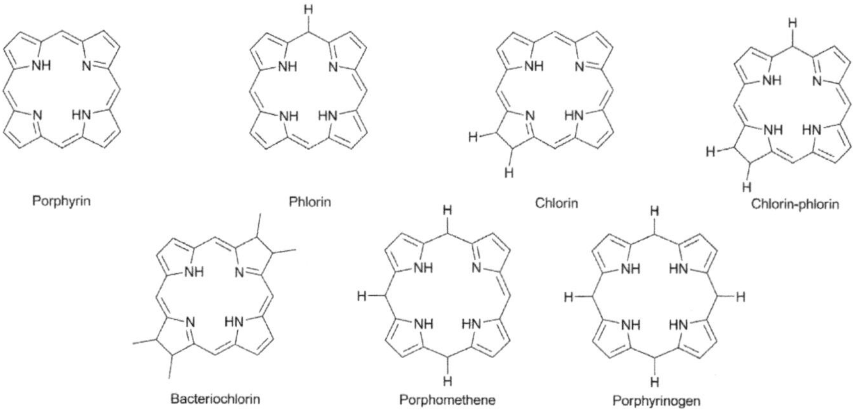

Fig. 2 Some reduced forms of porphyrins.

these products of the mechanochemical condensation, which would help give insight into the mechanism of this process.

The mechanochemical synthesis of several *meso*-tetrasubstituted porphyrins

Since the cyclocondensation reaction between an aldehyde and pyrrole is reversible,[4] it was hypothesized that letting the ground pink solid stand for several days while undergoing oxidation in air was allowing the cyclized product to revert to starting material before being oxidized to the porphyrin, resulting in low yields. Therefore, in order to evaluate the success of the mechanochemical cyclocondensation reaction, the pink solid was oxidized immediately by stirring for two hours in chloroform with DDQ, followed by column chromatography to isolate the porphyrin and determine the yield. Also, in order to have better control of the mechanochemical reaction conditions, the process was automated using a Retsch MM200 mixer mill. Coupled with these modifications, the scope of the synthesis was expanded to include a series of substituted benzaldehydes. The yields from these reactions are reported in Table 1. The addition of auxiliary grinding agents during the milling process such as magnesium sulfate (to remove water), silica or sodium chloride (to alleviate "stickyness") did not lead to an increase in the yield.

The results show that the mechanochemical condensation between an aldehyde and pyrrole to give the reduced porphyrin precursors is successful for several aromatic aldehydes, producing yields similar to those reported for the solution synthesis. We have also observed Soret bands resulting from the reactions of several alkyl adehydes, namely butanal, pentanal, octanal and trimethyl acetaldehyde. The characterization and determination of the yield for these presumable *meso*-alkylsubstituted porphyrin products is ongoing, although the preliminary yields are in the 10–20% range.

Mechanochemical oxidation

Having established the efficacy of a mechanochemical approach for the first step of the porphyrin synthesis (Scheme 1) namely, the condensation between pyrrole and an aldehyde, we set out to find a way of achieving the second step of the synthesis, oxidation, using mechanochemistry. TPP was used as a model compound for these studies. Various oxidizing agents were chosen, some based on the demonstrated success in previous oxidative aromatization reactions,[11] and potassium peroxymonosulfate (Oxone; $2KHSO_5 \cdot KHSO_4 \cdot K_2SO_4$) for its proven success in mechanochemical oxidations.[12] The milling time for the oxidation step was varied from twenty minutes to eight hours. In some cases, a grinding agent was added to the reaction to aid with the grinding process by reducing stickiness, increasing friction, and/or removing water from the reaction. We tested three grinding agents: sodium chloride, silica and magnesium sulfate, with the hopes

Table 1 The results from the mechanochemical synthesis of tetrasubstituted porphyrins, with oxidation in solution using DDQ

RCHO; R =	Milling time (min)	Observations	Soret Band (nm)	Yield[a] (%)
Ph	20	Pink powder	417	27.9
2-ClC_6H_4	20	Pink sticky solid	418	32.7
3-ClC_6H_4	30	Pink powder	419	21.9
4-ClC_6H_4	30	Pink sticky solid	419	20.0
2-FC_6H_4	30	Pink powder	416	28.1
3-FC_6H_4	20	Pink powder	417	25.0
4-FC_6H_4	30	Pink powder	418	9.7
α-naphthyl	30	Brown sticky solid	424	2.7
β-naphthyl	40	Brown sticky solid	425	31.3

[a] The yield is based on the average from three trials of dry weight of the fractions displaying a Soret band in the UV-Vis spectrum, with no other major absorbances, after column chromatography.

they would help drive the reaction forward by removing the water produced from the reaction between benzaldehyde and pyrrole and alleviate the "stickiness" of the reaction mixture (facilitating grinding). After the initial milling of benzaldehyde, pyrrole and toluenesulfonic acid to produce a powder as described above, the grinding jar was opened and an oxidizing agent (1.5 molar equivalents based on benzaldehyde) was added. The mixture was then further milled. The presence or absence of TPP was confirmed by monitoring the absorbance at 417 nm in the UV-Vis spectrum. In order to obtain an approximate yield of porphyrin, a sample of the reaction mixture was weighed and a serial dilution was performed in chloroform to obtain a solution with a concentration of 3.0×10^{-6} g mL^{-1}. The absorbance at 417 nm was compared to the absorbance at 417 nm for a pure TPP solution of the same concentration. The results of selected experiments are presented in Table 2.

Although the yields from these experiments are consistently lower than the yields of TPP from the oxidation in solution using DDQ, milling with some oxidizing agents produces the porphyrin in a shorter time than simple exposure to air. The control experiment with no oxidizing agent added yielded no Soret band after 20 minutes post-condensation milling. Milling with DDQ also produced no result. Molecular iodine produced a low yield of TPP. Nitric acid, while producing the highest yield, is difficult to work with and, as one might expect, caused pitting on the inside of the grinding jars. Sodium perborate, sodium ethoxide and Oxone were moderately successful, producing yields of just under 10% with the right conditions. Lengthening the grinding times produced different results for sodium ethoxide and Oxone. For the former, it destroyed the yield completely while for the latter, no TPP was observed after 20 minutes but the yield increased with time, reaching a maximum of 8.1%. The two oxidizing agents also behaved differently in the presence of various grinding agents: silica and

Table 2 The results from the two-step mechanochemical synthesis of TPP (condensation followed by oxidation)

Oxidizing agent	Grinding agent	Milling time (min) for condensation/oxidation	Yield[a] (%)
None	None	20/20	0[b]
DDQ	None	20/20	0[b]
Iodine	None	20/20	1.6
Nitric acid	None	20/20	11.5
Sodium perborate	None	20/20	8.3
Sodium ethoxide	None	20/20	8.2
Sodium ethoxide	None	20/480	6.7
Sodium ethoxide	Silica	20/20	5.2
Sodium ethoxide	Sodium chloride	20/20	0[b]
Sodium ethoxide	Magnesium sulphate	20/20	9.3
Oxone	None	20/20	0[b]
Oxone	None	20/480	8.1
Oxone	Silica	20/20	5.3
Oxone	Sodium chloride	20/20	4.7
Oxone	Magnesium sulphate	20/20	0[b]

[a] The yields were determined by the absorbance of the reaction mixture at 417 nm, in comparison to pure TPP at the same concentration. [b] No Soret band detected.

Table 3 Comparison of synthetic methodologies for TPP

	Lindsey synthesis	Mechanochemical condensation step	Two-step mechanochemical
E factor	2784	1682	19.7
EcoScale[a]	−46	−33.5	6.64

[a] Tables showing the penalty points for each category of the EcoScale calculation can be found in the Electronic Supplementary Information.†

sodium chloride caused the yield to decrease for sodium ethoxide but increase for Oxone upon 20 minutes of milling, while the addition of magnesium sulphate resulted in the best yield for the sodium ethoxide experiments, but did not help in the case of Oxone.

Changes to longer grinding times for the condensation step did not result in significantly different yields of TPP. Manual grinding was also employed for these experiments and consistently resulted in slightly higher yields (1–2% increase) than those obtained with automated grinding. During automated grinding, the balls may become stuck in the reaction mixture at various times, lowering the efficacy. This problem can be addressed by using larger, heavier grinding balls.

Other methods of oxidation were also tested. Since benzaldehyde itself can be an oxidising agent, the synthesis of TPP was attempted in two ways. Firstly, after obtaining the pink solid from the condensation reaction as described above, an additional three molar equivalents of benzaldehyde were introduced as the oxidizing agent and the mixture was milled for a further 20 minutes. Secondly, instead of beginning with a 1 : 1 molar ratio of benzaldehyde to pyrrole, a 10 : 1 molar ratio was used, and the reaction was milled for 60 minutes, while checking for a Soret band at 10 minute intervals to see if TPP could be isolated in a one-step reaction. Neither of these methods resulted in the production of TPP, as evidenced by the absence of an absorbance at 417 nm in the electronic spectra. The oxidation of the pink solid by exposure to nitric acid vapours over several hours was also attempted, with no positive results.

Purification by sublimation

While column chromatography has been the method of choice for separating the porphyrins from the several by-products that can form during the course of porphyrin synthesis, we have found success by using entrainment sublimation,[13] where a carrier gas such as nitrogen is employed to transport sample vapours away from the heated impure material, so that they may be deposited on a cooler surface. We have isolated pure TPP (as confirmed qualitatively by electronic spectroscopy) from an oxidised reaction mixture using this method, although it was done on a scale too small to allow the determination of the isolated yield. Experiments are underway to scale up this process using a tube furnace.

Green chemistry impacts

As environmental awareness increases both in industry and academia, the incorporation of the 12 Principles of Green Chemistry[14] becomes more and more attractive. Through the elimination of the solvent, the simplification of the technical setup and the reduction of reaction time, the field of mechanochemistry has much to contribute here, however it can sometimes be difficult to determine whether an alternative method is in fact "greener" than the established approach to synthesis. Is it worth employing greener methods for a synthesis, when the yield may be lowered as a result? There are several metrics that have been proposed to evaluate how "green" a reaction is in order to compare the "greenness" of various synthetic methodologies. Here we will use two such metrics, the E factor[15] and EcoScale,[16] to compare three synthetic methodologies to produce TPP: 1) the Lindsey synthesis, 2) mechanochemical condensation followed by oxidation in solution and purification by column chromatography, and 3) two-step mechanochemical synthesis and purification by sublimation. The results are shown in Table 3:

The E factor is a simple measurement of waste produced for the amount of product obtained:

$$E = \frac{\text{total waste}}{\text{total product}}$$

Waste includes the solvent, by-products and consumables such as silica or alumina. We can see from Table 3 that the elimination of the solvent from just one step of the TPP synthesis results in a great reduction in the E factor, while the complete elimination of the solvent results in a huge reduction, even if the yield is lowered.

The ideal reaction has an EcoScale value of 100.[16] Starting there, penalty points are deducted based on the requirements of the synthetic procedure in the categories of the price of reaction components, safety, technical setup, temperature, time, workup and purification. The percent yield is also taken heavily into consideration; penalty points for the yield are (100 − yield)/2. A large, positive EcoScale means a "greener" synthesis. We can see from Table 3 that of the three synthetic methodologies compared, the two-step mechanochemical synthesis has the only positive EcoScale score, in spite of the low yield. This can be attributed to both the elimination of the solvent and the large amount of DDQ needed for the oxidation, and the elimination of column chromatography as a purification method.

Conclusions

Our results show that the synthetically-important *meso*-tetrasubstitued porphyrins can be synthesized without a solvent, through a two-step mechanochemical process: condensation followed by oxidation. Mechanochemical condensation followed by oxidation in solution leads to yields for a series of *meso*-tetrasubstituted porphyrins that are comparable to those obtained from a Lindsey synthesis. Switching to a mechanochemical oxidation lowers the yield, but the elimination of the large quantities of hazardous solvents from the porphyrin synthesis has important implications for the utilization of porphyrin systems in industrial-scale applications. Two green chemistry metrics, the E factor and the EcoScale show that, even with a lower yield, the mechanochemical process is more efficient from an environmental perspective. Additionally, this is one of the few examples to date of carbon–carbon bond formation *via* mechanochemistry, although examples are growing.[8] Mechanistic studies of this reaction will be important in order to understand reaction intermediates and pathways. Studies of the reactions of pyrrole with a mixture of aldehydes are underway, and should shed some light on this.

Acknowledgements

We thank Dr Tomislav Friščić and Dr William Jones for the use of research facilities. This research was supported by an award from the Research Corporation for Science Advancement Cottrell College Science Award No. 19762 and NIH NIGMS R25 GM059244 to Barry University.

References

1 C. M. Drain, A. Varotto and Radivojevic, *Chem. Rev.*, 2009, **109**, 1630–1658; C. J. Medforth, Z. Wang, K. E. Martin, Y. Song, J. L. Jacobsen and J. A. Shelnutt, *Chem. Commun.*, 2009, 7261–7277; I. Goldberg, *CrystEngComm*, 2008, **10**, 637–645; R. L. Halterman, S.-T. Jan, A. H. Abdulwali and D. J. Standlee, *Tetrahedron*, 1997, **53**, 11277–11296.

2 P. Rothemund, *J. Am. Chem. Soc.*, 1935, **57**, 2012–2011; P. Rothemund, *J. Am. Chem. Soc.*, 1939, **61**, 2912–2915; R. G. Little, *J. Heterocycl. Chem.*, 1981, **18**, 833–834; G. M. Badger, R. A. Jones and R. L. Laslett, *Aust. J. Chem.*, 1964, **17**, 1028–1035.

3 A. D. Adler, F. R. Longo, J. D. Finarelli, J. Goldmacher, J. Assour and L. Korsakoff, *J. Org. Chem.*, 1967, **32**, 476; A. D. Adler, L. Sklar, F. R. Longo, J. D. Finarelli and M. G. Finarelli, *J. Heterocycl. Chem.*, 1968, **5**, 669–678; G. H. Barnett, M. F. Hudson and K. M. Smith, *J. Chem. Soc., Perkin Trans. 1*, 1975, 1401–1403.

4 J. S. Lindsey, I. C. Schreiman, H. C. Hsu, P. C. Kearney and A. M. Marguerattaz, *J. Org. Chem.*, 1987, **52**, 827–836; J. S. Lindsey, K. A. MacCrum, J. S. Tyhonas and Y.-Y. Chuang, *J. Org. Chem.*, 1994, **59**, 579–587.

5 C. M. Drain and X. Gong, *Chem. Commun.*, 1997, 2117–2118.

6 A. Petit, A. Loupy, P. Maillard and M. Momenteau, *Syth. Commun.*, 1992, **22**, 1137–1142.

7 S. L. James, C. J. Adams, C. Bolm, D. Braga, P. Collier, T. Friščić, F. Grepioni, K. D. M. Harris, G. Hyett, W. Jones, A. Krebs, J. Mack, L. Maini, A. G. Orpen, I. P. Parkin, W. C. Shearouse, J. W. Steed and D. C. Waddell, *Chem. Soc. Rev.*, 2012, **41**, 413–447.

8 G. Kaupp, M. R. Naimi-Jamal and J. Schmeyers, *Tetrahedron*, 2003, **59**, 3753; E. M. C. Gerard, H. Sahin, A. Encinas and S. Bräse, *Synlett*, 2008, 2702–2704; J. Mack and M. Shumba, *Green Chem.*, 2007, **9**, 328–330; V. P. Balema, J. W. Wiench, M. Pruski and V. K. Pecharsky, *J. Am. Chem. Soc.*, 2002, **124**, 6244–6245; P.-N. Chen, C.-C. Lai and S.-H. Chiu, *Org. Lett.*, 2011, **13**, 4660.

9 J. Bonnamour, T.-X. Métro, J. Martinez and F. Lamatay, *Green Chem.*, 2013, **15**, 1116–1120; D. Tan, V. Štrukil, C. Mottillo and T. Friščić, *Chem. Commun.*, 2014 Advance Article DOI: 10.1039/C3CC47905F.

10 G. Peychal-Heiling and G. S. Wilson, *Anal. Chem.*, 1971, **43**, 550–556; D. Dolphin, *J. Heterocycl. Chem.*, 1970, **7**, 275–283.

11 S. G. Hegde, A. M. Kassim and A. I. Ingrum, *Tetrahedron Lett.*, 1995, **36**, 8395–8398; M. J. Mphalele, *Molecules*, 2009, **14**, 5308–5322.

12 I. Huskić, I. Halasz, T. Friščić and H. Vančik, *Green Chem.*, 2012, **14**, 1597–1600; R. Thorwirth, F. Bernhardt, A. Stolle, B. Ondruschka and J. Asghari, *Chem.–Eur. J.*, 2010, **16**, 13236–13242; G. Cravotto, D. Garella, D. Carnaroglio, E. C. Gaudino and O. Rosai, *Chem. Commun.*, 2012, **48**, 11632–11634.

13 H. J. Wagner, R. O. Loutfy and C.-K. Hsiao, *J. Mater. Sci.*, 1982, **17**, 2781–2791; P. D. Sattsangi, *J. Chem. Educ.*, 1996, **73**, A3.

14 P. T. Anastas and J. C. Warner, in *Green Chemistry: Theory and Practice*, Oxford University Press, New York, 1998, p.30.

15 R. A. Sheldon, *Green Chem.*, 2007, **9**, 1273–1283.

16 K. Van Aken, L. Strekowski and L. Patiny, *Beilstein J. Org. Chem.*, 2006, **2**, 3.

Faraday Discussions

DISCUSSIONS

Mechanochemistry of organic molecules, soft materials and pharmaceuticals: General discussion

DOI: 10.1039/C4FD90005G

Dr Moores opened the discussion of the Introductory Lecture by William Jones: Is there any relationship between the polarity of the solvent added in LAG and the nature of the crystalline forms obtained using it? Thanks again for a great talk.

Professor Jones replied: Various remarks have been made concerning the outcome of the LAG experiment and the polarity of the liquid added although it remains, I believe, far from being understood. Whole the liquid may stabilise one polymorphic form over another because of stabilisation of particular preferred dominant faces associated with a particular polymorphic form it may also be the case that the polarity will influence the solubility of the different components and the nucleation events during a possible crystallization from a solution intermediate phase.

Professor Bolm remarked: I refer to the seeding method reported in your presentation and I wonder if it could be applied for solid state seed-induced resolutions of racemic mixtures [as, for example, reported for deracemizations of naproxen in MeOH by Kaptein and co-workers (*Angew. Chem. Int. Ed.*, 2009, **48**, 4581). If in this particular case the presence of the base (NaOMe) is problematic, one could also think about a deracemization of binaphthyl by solid state seed-induced resolution, as related processes are known]. Do you think that this is possible?

Professor Jones replied: I suspect that there is considerable scope for mechanistic studies based on chiral systems and would mention the work of Viedma (C. Viedma, *Phys. Rev. Lett.*, 2005, **94**, 06544) as one example of the interesting chemistry associated with chiral systems.

Dr Baláž asked: Nowadays, the wet milling for the preparation of drug nano-suspensions is applied in the pharmaceutical industry. What is your opinion on such an approach, which competes with the classical preparation steps in which micronization is applied?

Professor Jones replied: This is certainly a potentially fruitful area and I would mention the developments within the company iCeutica – a company which claims to use a dry milling method to produce submicron drug particles (www.iceutica.com).

Professor James opened the discussion of the paper by Leonard MacGillivray: Nice to see this evolution, to see how cheap and readily available milling equipment can be, I spent £6000 on a ball-mill!

Professor Suslick then addressed Professor MacGillivray: There is a quite general problem with photochemistry in the solid state that your apparatus addresses very nicely. The penetration depth of photons into reactive solids is generally highly problematic. Typically, the starting material and product will have absorbances in similar regions of the UV-vis spectrum. This means that after the photochemical reaction occurs on the outer skin of the solid, the interior is left unreacted. The specific case of a [2+2] cycloaddition is an exception: the cyclobutane product is UV transparent, so the photoreaction can get all the way to the center of the crystals. The use of grinding a solid *in situ* during a solid-state photochemical reaction, however, provides an unthought of way to remove non-transparent product from the solid reactant and provides for continuous generation of a photochemically reactive surface.

Professor MacGillivray replied: I completely agree with your assessment and this is an area in which we need to follow up.

Dr Hamilton said: The reactions you present all involve stoichiometric amounts of template. Since the grinding and photoreaction are taking place simultaneously, might it be possible to conduct these reactions using less template?

Professor MacGillivray answered: Using less template in vortex grinding is our next step.

Professor Bolm asked: I noted that you used an unsymmetrically substituted catechol (with a benzyl group in 4 position) as hydrogen donor. Why did you choose this particular one, and has the possibility of forming regioisomeric 2:1 aggregates (head/head and head/tale) played a role?

Professor Bolm said: You mentioned the possibility of supramolecular catalysis. My feeling is that the turn-over required for the catalysis in the reaction you studied {the [2+2] cycloaddition induced by light} is hampered by product inhibition (as the "atom locations" will be very similar in the starting material and the product). Have you considered the incorporation of molecular switches (such as diazo units, which are light-switchable as well) for "liberating" the catalyst and achieving turn-over?

Professor Jones asked: Have you been able to make amorphous forms of your materials and possibly use the outcome of the reaction as a way of probing local structure compared to the crystalline arrangement?

Professor MacGillivray answered: Yes, in earlier and published work we have reported on the development of Rebek's imide as a template. The system involved a co-crystal solvate. We deliberately removed the included solvent prior to photoreaction in order to gain insight into whether the hydrogen-bonded assemblies were maintained in a loss of crystallinity. The experiments were successful.

Professor Jones remarked: It would be interesting to know how the structural integrity of the product crystals might be related to the mechanical properties of the starting crystals. Is there any literature which discusses this on the likelihood of, say, single-crystal to single-crystal transformations?

Professor MacGillivray answered: We have recently described the use of AFM nanoindendation to study the changes in mechanical properties of a co-crystal before and after photoreaction (*Angew. Chem. Int. Ed.*, 2011, **50**, 8642). Crystal of macro (*i.e.*, millimeter) and nano dimensions were shown to become softer and harder, respectively, after the photodimerization.

Professor Craig asked: Congratulations on not only the very nice chemistry, but the success of your do-it-yourself approach that made it happen! One of the things that is appealing to me about the vortex mixing you describe is that it provides an easy, low-investment entry into solid state or liquid assisted mechanochemistry. Can you compare it to existing techniques, in particular ball-milling? For example, are the co-crystallizations you report here especially easy or difficult? Do you expect that anything done in a conventional ball mill could be accomplished in this same set up?

Professor MacGillivray answered: In terms of comparisons, we are still working to get a sense of where the vortex mixing sits with respect to a ball mill. Our co-crystallizations are relatively straightforward while we still need to apply our approach to additional systems. We have no reason to expect that our approach will not work.

Dr Friščić addressed the meeting: It would be very informative to know if there has been any previous work on combining mechanochemical reactivity with other types of energy input or excitation, such as irradiation (as presented in this and previous papers from MacGillivray group) or heat? To that end I highlight the recent work from the Cuccia laboratory describing facile control of reaction temperature during mechanochemical treatment in a lysis mill (*CrystEngComm*, 2014, **16**, 7180). I would encourage the participants of the meeting to send along any information of such attempts of combining mechanochemical and other reactivity.

Dr Baláž said: What was the exact wavelength of the UV light you used for your experiments?

Professor MacGillivray replied: We generally use a broadband photochemical reactor, which involves light at a number of different wavelengths. We do use selective irradiation from time to time (*e.g.*, 350 nm) when interested in studying single crystal to single crystal transformations.

Dr Blair said: You use borosilicate vessels in your vortex mixer which does not transmit well in the deep UV (below 300 nm). Have you looked at using fused silica and do you think this would improve your yields?

Professor MacGillivray responded: Silica glass is one of our next steps. At this stage, we are still seeing near quantitative yields.

Dr Blair addressed Professor MacGillivray: BBs obtained from a sport store can be copper jacketed (as in your movie). James Mack has shown that the ball material can participate in the reaction. It might be useful to look at 316 stainless steel balls from a supply house like McMaster-Carr. Although not quite as inexpensive as a jar of BBs, they have a known composition and are available in many sizes. Additionally, there are many alloys available in the size range you use.

Professor MacGillivray replied: Thanks greatly for the suggestions.

Professor Boldyreva communicated: I wonder if these reactions can also be observed in the complete dark, if LAG alone is applied. There are examples, when mechanical treatment causes the same transformations as irradiation by light (see, *e.g.*, A. A. Politov, A. P. Chupakhin, V. M. Tapilin, N. N. Bulgakov and A. G. Druganov, To mechanochemical dimerization of anthracene. Crystalline phenanthrene under high pressure and shear conditions, *J. Struct. Chem.*, 2010, **51**(6), 1064–1069; V. M. Tapilin, N. N. Bulgakov, A. P. Chupakhin, A. A. Politov and A. G. Druganov, On mechanochemical dimerization of anthracene. different possible reaction pathways, *J. Struct. Chem.*, 2010, **51**(4), 635–641; V. M. Tapilin, N. N. Bulgakov, A .P. Chupakhin and A. A. Politov, On the mechanism of mechanochemical dimerization of anthracene. Quantum-chemical calculation of the electronic structure of anthracene and its dimer *J. Struct. Chem.*, 2008, **49**(4), 581–586).

Dr Blair addressed Dr Friščić: Does the dielectric constant of the solvent play a role in LAG syntheses? It seems that solvents compatible with the reactants will act as a plasticizer and reduce the effective impact forces realized. http://www.clippercontrols.com/pages/Dielectric-Constant-Values.html

Dr Friščić responded: To the best of my knowledge, a comparison of how the dielectric constant affects LAG processes has not yet been performed. However, some of our studies on ILAG (*Angew. Chem. Int. Ed.*, 2010, **49**, 712) *vs.* use of ionic liquids (not published) to activate metal oxides showed no clear relationship between the ionic nature of the system and reactivity enhancement.

Professor Suslick opened the discussion of the paper by Graeme Day: Is it true that thermodynamic control has been demonstrated in these specific molecules? Just because milling takes you to a specific product, that does not prove that the thermodynamically most stable product has been produced. Shouldn't it matter if it's solvent-free or liquid-assisted? Please elaborate.

Dr Day responded: Our previous study of these molecules[1] addressed equilibration and thermodynamic control of the reactions. Those experiments showed that the final composition after grinding for sufficient time was invariant to the initial conditions; for example, we could start with the statistical mixture of homo- and heterodimers that result from solution equilibration, pure homodimers or pure heterodimer and obtain the same mixture after grinding.

One observation that might be difficult to reconcile is that neat and liquid assisted grinding (LAG) conditions lead to different polymorphs of the 4Cl-2NO_2 heterodimer. How can one polymorph be the thermodynamic product under neat conditions, while a different polymorph is the thermodynamic product under LAG? One possible explanation is that the conditions lead to crystalline products with different particle sizes; we might be in a size regime where the polymorph stability crosses over. Another possibility that we are currently investigating is that there is enough solvent in the LAG experiments to alter the particle surface energies sufficiently to change the relative stability of the polymorphs.

1 A. M. Belenguer, T. Friščić, G. M. Day and J. K. M. Sanders, *Chem. Sci.*, 2011, **2**, 696.

Mr Kochhar said: How can you verify that you are at a global energy minimum in your simulations?

Dr Day responded: It is not generally possible to ensure that the global minimum has been located in these searches; due to the high dimensionality of the search space (unit cell dimensions, molecular position, molecular orientation and molecular conformation), the whole energy surface cannot be mapped out. We sample this surface in as even a way as possible until we believe that the sampling points are close enough together that we do not miss any local energy minima. This is where we use low discrepancy sequences. We examine the convergence of the search for structures by following the number of times we hit each of the structures that we do find. Once all structures are located multiple times, we assume that the sampling density is high enough that we have not missed anything. As it is important to be confident that we have located all relevant minima, we are currently developing tools within our Global Lattice Energy Explorer software for analysing the convergence of the structure search.

Dr Pecharsky asked: Prediction of intermediates and final products in mechanochemical reactions is indeed a very important problem and I am very happy to see an attempt to use density functional theory. Can you comment on how relevant lattice-based calculations are considering that mechanochemical reactions necessarily proceed *via* surfaces that are in contact with one another?

Dr Day responded: Our approach to predicting the equilibrium of these reactions was that the equilibrium is dictated by the energy of the final state of the homo- and heterodimers. The reactions might occur at surfaces or in a mixed amorphous phase, but we observe that the reaction leads to microcrystalline product. So it is the lattice energies of these crystals that dictate the relative stability of homo- and heterodimers and should determine the equilibrium of the reaction. We are initially ignoring the surface energy component of the crystal energies, but please note my earlier response: we are looking at including the surface energies as well. Modelling of reactions at surfaces would also be interesting to gain insight into the kinetics of the reactions.

Professor Craig commented: The point has been made in prior discussions that the mechanical stability of crystals, rather than just the thermodynamic stability, might be a determinant of mechanochemical outcomes. Can you comment on the potential to computationally screen the mechanical stability of thermodynamically favorable crystal forms?

Dr Day replied: This is an interesting idea. It is not something that we have looked at yet in predictions for these grinding experiments, but could do so. For each predicted crystal structure, we are able to predict the elastic properties from the curvature of the lattice energy surface.[1] These calculated elastic constants have helped us understand the observed mechanical properties of known crystal structures, such as understanding the tablettability of pharmaceutical materials from calculations of the weakest shear planes in their crystal structures.[2] We have occasionally considered the mechanical stability of putative structures in crystal structure prediction studies[3,4] and it would be relevant to do so in predictions of the outcome of grinding experiments.

1 G. M. Day, S. L. Price and M. Leslie, *Cryst. Growth Des.*, 2001, **1**, 13.
2 S. Karki, T. Friščić, L. Fabian, P. R. Laity, G. M. Day and W. Jones, *Adv. Mater.*, 2009, **21**, 3905.
3 T. Beyer, G. M. Day and S. L. Price, *J. Am. Chem. Soc.*, 2001, **123**(21), 5086.
4 A. T. Anghel, G. M. Day and S. L. Price, *CrystEngComm*, 2002, **4**(62), 348.

Dr Moores commented: What you describe is purely thermodynamics. Any plan to probe kinetics; how would you do this? Monte-Carlo perhaps?

Dr Day replied: It might be possible to investigate the kinetics of these reactions by computational methods, although we do not have immediate plans to do so. *In situ* monitoring of the grinding experiments shows that they proceed *via* an amorphous intermediate phase. Presumably, this is where the metathesis reaction takes place. So, one aspect that could be investigated is the rate of amorphisation of the initial crystals, for which Monte Carlo or molecular dynamics would probably be necessary. Kinetics of the reaction itself might be studied with QM/MM methods to describe the bond breaking and forming in the amorphous phase. There is then the kinetics of crystallisation of the products, which is very challenging to address using current computational methodologies, due to the timescales involved.

Professor James asked: Initially you were very sceptical that you'd get an insight into the actual process or kinetics of the mechanochemical reaction, is this computationally too hard a problem?

Dr Day responded: It is a hard problem and I was sceptical because there is uncertainty surrounding the mechanism of the reactions. However, it would be possible to look at aspects of the mechanism by computational methods, such as amorphisation of the crystals and modelling of the reaction pathway with some simplified model of the amorphous environment.

Dr Friščić addressed Dr Day: It might be interesting to consider if the calculations could be performed with the intention not only to explain and predict the equilibrium position of reversible mechanochemical reactions, but also to try and explain their kinetic behavior? Presumably, neat grinding (*i.e.*, without added external solvent) reactions proceed through an intermediate amorphous material that provides higher mobility for reactants than their corresponding crystalline states. Perhaps it would be possible to calculate the ease of amorphisation of a crystalline solid from its crystal structure. Such information might then lead to a theoretical description of the kinetics of mechanochemical reactions?

Dr Day replied: These are good suggestions. It should be possible to model amorphisation of the crystalline solid and to model to mobility of molecules in the amorphous material *via* molecular dynamics. Such simulations would rely on having force fields that describe the intermolecular interactions with sufficient accuracy, and we have found that this is a problem for some of the systems that we have studied.

Dr Irikura commented: Are the entropy and the thermal contribution to the enthalpy included in your calculations?

Dr Day replied: These contributions are not included in the calculations that are presented here, in which we only calculate the potential energy of each structure. It is clear from some of the results that we need to improve the description of the potential energy before adding entropic and thermal contributions. Once we have a sufficiently accurate model of the potential energy, then we should be able to include the lattice vibrational contributions to the relative free energies [1], and we plan to look at this both for the crystal structure prediction and reaction energies of these systems. Harmonic lattice vibrational contributions have sometimes been included in crystal structure prediction studies (see, for example [2,3]), but is not currently the norm.

1 G. M. Day, S. L. Price and M. Leslie, *J. Phys. Chem. B*, 2003, **107**, 10919.
2 A.T. Anghel, G. M. Day and S. L. Price, *CrystEngComm*, 2002, **4**(62), 348.
3 B. P. van Eijck, *J. Comput. Chem.*, 2001, **22**(8), 816.

Professor Beyer remarked: How much computing time is required to calculate a crystal structure, *i.e.*, to generate an energy landscape as shown in Fig. 4?

Dr Day answered: The crystal structure prediction studies are almost perfectly parallelisable, as each computer-generated crystal structure is independent and can be sent to a processor to calculate its energy. So, while the total computing time required is quite high, the real time needed comes down to manageable times with access to high performance computing resources. We didn't keep track of exact timings, but these studies took us on the order of weeks to complete all of the molecules using the University of Southampton's Iridis4 high performance computer.

Professor Boldyreva communicated: This is a very interesting and promising study. I have a comment. I think that the authors should not say that their prediction was "unsuccessful" when a co-grinding procedure gives a high-energy polymorph. Co-grinding may give equilibrium products and non-equilibrium products. It is very helpful for understanding the mechanism of a mechanochemical reaction if a computational procedure makes it possible to distinguish between the low- and high-energy forms. In this respect, any result of the calculation is informative and thus "successful".

Professor James opened the discussion of the paper by Tamara Hamilton: The classic solvent-based synthetic methods for these compounds have been around for a long time, and they may not be the most effective or efficient, it seems timely to go back and look at these reactions again and see if they can be done more effectively mechanochemically.

Professor Suslick remarked: From my mis-spent youth doing porphyrin chemistry, the greatest impact of your novel and interesting approach would be in preparing sterically hindered porphyrin. What other hindered aldehydes other than naphthyl- have you used. Of special interest to me is the bis-pocket porphyrin which we made long ago and extremely painfully from 2,4,6-triphenylbenzaldehyde.

Dr Hamilton answered: We are moving in that direction. Since submitting this manuscript we have used mesitaldehyde, which provided 5-6% isolated yield of the corresponding porphyrin in the first trial. The triphenylbenzaldehyde you suggest will be next on our list – thanks for your input.

Mr Julien remarked: One of the main challenges in porphyrin synthesis is the purification of the porphyrin/porphyrinogen after synthesis. This is generally done by directly crystallizing the porphyrin/porphyrinogen out of the reaction mixture. Can you comment on the purification methods applicable to mechanochemically synthesized porphyrins?

Dr Hamilton responded: We have used column chromatography, which works well but of course uses a lot of solvent, and we are trying to get away from that. We have had some success using sublimation –tetraphenylporphyrin sublimes at around 450 °C – and also we are experimenting with entrainment sublimation using a tube furnace and nitrogen as a carrier gas. This has worked well for some porphyrins but not for others.

Professor James asked: It seems that mechanochemistry can provide a way to eliminate or reduce the amount of solvent used in a reaction, but large amounts of solvent are also normally used for purification such as in chromatography or crystallisation. I wonder if innovative ways can be found to separate and purify products without using solvents?

Dr Hamilton replied: We are working on purification by entrainment sublimation using a tube furnace and nitrogen as the carrier gas. We have had success in some cases, the process is still being optimized.

Professor Mack commented: Given the fact that the poryphin synthesis generates water as a by-product and the fact that your starting materials are generally non-polar, do you see a change in polarity over time due to the amount of water released in the reaction?

Dr Hamilton answered: It's hard to see the result of any water formation during the course of this reaction. Our starting materials are both liquids and proceed through a "sticky" phase to give a nice, dry-looking powder as a product. On the scale that we do the reaction formation of water, or any sort of phase separation is not visible. We haven't carried out any sort of polarity measurements.

Dr Moores commented: Have you tried oxidation catalysts for this reaction in any way – iron, copper, iron oxide?

Dr Hamilton answered: We have done the reaction in the presence of various metal salts and oxides, with the idea that perhaps the metal could template the formation of the porphyrinogen macrocycle, and oxidize it as well, ending in the metalloporphyrin. Just as has been found in solution, this doesn't seem to be the case.

Professor MacGillivray asked: You categorise in your paper that your approach is a two-step eco-friendly process. Are you able to quantify the eco-friendliness of each step?

Dr Hamilton answered: The metrics of e-factor and eco-factor allow us to quantify and compare synthetic methodologies 1) without mechanochemistry, 2) with a mechanochemical first step, and 3) when both steps are done mechanochemically. We find that, even though a lower yield is obtained from the two-step mechanochemical process, the elimination of waste makes it more eco-friendly by these metrics.

Mr Stauch said: In Table 2 it is shown that in the presence of sodium ethoxide as the oxidizing agent and with no grinding agent added to the system, the yield of TPP is 8.2% if the milling time for oxidation is 20 min. If the time for oxidation is increased to 480 min, however, the yield decreases to 6.7%. Why does the yield decrease if the time for oxidation increases?

Dr Hamilton replied: This is a good question, and unfortunately we don't really know right now why this could be. TPP is very stable and I wouldn't expect it to decompose under milling conditions. I wouldn't like to speculate too much on exactly what sort of chemical conversions could be going on in this example without further careful study. Possibly some metal from the grinding jars becomes involved after these extended times.

Dr Irikura asked: You have shown improved yield for some difficult, substituted porphyrins. The unsubstituted parent compound, porphine, is also hard to prepare; the yield from Longo's method is disappointing. Have you tried to prepare porphine?

Dr Hamilton replied: Not yet. This is something we will try soon, thanks for your suggestion.

Dr Maini asked: The presence of NaCl or $MgSO_4$ is not so innocent. They gave different results depending on the oxidizing reagents. Do you known if they react with oxidizing reagents? Are you able to detect their presence at the end of the reactions?

Dr Hamilton responded: The results show that absolutely these are not innocent bystanders during the reaction. We don't know what they are doing with the different oxidizing agents. We have not checked for their presence at the end of the reaction, but your suggestion to use XRD to check that is a good one.

Dr Maini said: X-Ray powder diffraction can help you also if you have a sticky material. If they are still crystalline they can be easily detected.

Dr Hamilton answered: Thank you. We have conducted XRPD on some samples, both before and after mechanochemical oxidation, and unfortunately never found any crystallinity.

Professor Jones asked: Would the formation of crystalline hydrates be one way of removing water from the system during reaction? Has this been considered?

Dr Hamilton answered: We have tried the addition of conventional drying agents like $MgSO_4$, but not many others.

Professor Beyer remarked: Is the mechanism of the reaction known? Does the surface of the reaction vessel or the milling balls play a role, possibly as a heterogeneous catalyst?

Dr Hamilton replied: Much work still needs to be done to understand the mechanism of this reaction in the absence of solvent. It's possible that the surface plays a role, however I can mention that we have also performed this reaction using polystyrene vials, as well as manual grinding with an agate mortar and pestle and it works equally well, so the material of the surface doesn't seem to make a difference.

Professor Galembeck commented: Did you attempt the synthesis starting with a small amount of product? Was there any evidence for autocatalytic behavior?

Dr Hamilton answered: We have not attempted this, but it's a good suggestion. We will try it, thanks.

Dr Blair remarked: Your reaction is a condensation reaction and you have had success using both a Retsch mixer mill and a mortar and pestle. You might see improved yields using a mill that allows removal of the water produced. This may drive the reaction forward. We can achieve quite effective water removal using an attrition mill.

Dr Hamilton answered: This would be great to try and would also be important in thinking of scale-up of the reaction. Thanks for your comment.

Dr Baláž commented: Have you tried using different milling media in order to increase the yield of the reactions?

Dr Hamilton answered: We used the Retsch MM200 with stainless steel grinding jars and 5 mm diameter stainless steel balls for the results reported here. We have seen that using 10 mm diameter balls decreases the reaction time to about 10 min, but doesn't seem to affect the yield. We have also used a Wig-L-Bug, with polystyrene vials and grinding balls to do the reaction, and the yield is not affected. We haven't tried any other milling media.

Dr Friščić invited delegates who had presented posters at the meeting to describe their work in a Lightning Session.

Miss Holz opened the Lightning Session by describing her poster: We have synthesized a triazol which was bound to the AFM in order to induce a mechanical cycloreversion. To distinguish the cycloreversion from a rupture elsewhere, this safety loop was integrated. 4% of the experiments douple rupture events were observed. Between these two events the length of the polymer increased 1.3 nm, which equates to theoretical estimates.

Mr Schmidt continued by describing his poster: We investigated the Knoevenagel condensation of vanillin and barbituric acid to analyze the influence and relationship of different parameters related to operation in planetary ball mills. The parameters revolution speed, milling ball diameter and milling ball filling degree were varied and showed high influence on the yield. The experimental results were supported by calculations of the stress frequency and power input. Thus, reaction conditions with maximized yield and minimized energy intensity could be identified.

Mr Jörres described his poster: In our group the use of ball milling techniques in asymmetric organocatalytic transformations has been studied. Lately, we developed a new protocol for the hydrogen bond-mediated organocatalytic addition of nitrocyclohexanone to aryl nitroalkenes under solvent-free conditions in a ball mill providing the corresponding addition products in high yields and with excellent stereoselectivities. A natural amino acid derived organocatalyst was applied with a remarkably low catalyst loading. In comparison to the results achieved in common solvent-phase reactions the ones performed solvent-free in a ball mill were faster and more stereoselective (http://www.beilstein.tv/tvpost/asymmetric-organocatalysis-in-a-ball-mill/).

Dr Baláž described his poster: Within this work, CdS@ZnS QDs with were synthesized by utilizing two consecutive milling procedures. Firstly, cadmium sulphide was prepared from cadmium acetate and sodium sulphide. Secondly, the precursors of zinc sulphide (zinc acetate and sodium sulphide) were co-milled with CdS. The product exhibited the nanocomposite structure, not the generally known core–shell one. The crystallite sizes of both constituents were under

10 nm. The whole series of samples with different molar ratios between CdS and ZnS was prepared, in order to study the influence of ZnS on the leakage of toxic cadmium. The positive effect of zinc sulfide was confirmed by cadmium dissolution tests performed in physiological medium. The whole series was characterized by XRD and N_2 adsorption–desorption isotherms, from which also the pore size distributions were calculated. The small size of nanocrystallites was confirmed by HRTEM measurement. The possibility of using a mechanochemical approach for the capping procedure was also studied in preliminary experiments within this work. The amino acid L-cysteine was used as a capping agent. The synthesized QDs could possibly be used in the future as labelling media in medicine, or they could be utilized as co-agents in anti-cancer drugs.

Professor Cuccia concluded the Lightning Session by describing his poster: Attrition-enhanced deracemization or Viedma ripening is the attrition-induced spontaneous chiral amplification of a conglomerate crystal mixture of achiral or racemizing compounds. In this process, amplification to homochirality relies on chiral recognition between crystals and fragmented chiral crystal clusters. We report efficient enantiomer-specific oriented attachment for suspended crystals of guanidine carbonate to form mesoscale homochiral aggregates under boiling or shaking conditions. These aggregates formed by non-classical crystal growth often display a polyhedral macrostructure and their chirality was monitored using polarized light microscopy and solid-state circular dichroism.

Faraday Discussions

PAPER

Mechanochemical synthesis of an organometallic compound: a high volume manufacturing method

David W. Peters*[a] and Richard G. Blair[b]

Received 22nd December 2013, Accepted 24th February 2014
DOI: 10.1039/c3fd00157a

Chemical vapor deposition (CVD) precursor chemicals are held to some of the highest purity levels in industry. Many metal reagents form stable, unbreakable adducts with the coordinating solvents that are necessary for solvating highly polar reagents. These adducts are undesirable and must be removed prior to usage. Herein we describe a mechanochemical approach to the synthesis of bis(*n*-propyltetramethylcyclopentadienyl)strontium that eliminates the use of strongly coordinating solvents. This method overcomes the solubility problems of the two reagents without the formation of stable, unbreakable adducts. We utilize a unique reactor geometry that facilitates mechanochemical syntheses while simplifying handling and allowing for "one pot" production. The synthesis was scaled to five hundred gram lot sizes in a six liter reactor. This technique is applicable to many syntheses and is linearly scalable – limited only by reactor size.

Introduction

Strontium-based semiconductors may be a key material for the production of several types of advanced memory devices. Thin films of the strained strontium titanate ($SrTiO_3$) perovskite structure can be used in ferroelectric memory devices such as smart cards and may pave the way for hybrid transistors that would allow for "Instant on" computers.[1] Barium Strontium Titanate (BST, $Ba_xSr_{1-x}TiO_3$) is a promising high K dielectric material for advanced DRAM capacitors. In MIM capacitors, the dielectric constant (K) for BST can be tuned from under 100 to over 1000 by varying the ratio of barium to strontium.[2] In addition, it also enjoys high breakdown strength, small dielectric loss, and good thermal stability. Furthermore, strontium ruthenium oxide (SRO, $SrRuO_3$) is the one of the best conductive materials available for use as electrodes in these ferroelectric capacitors.

[a]*ATMI, 706 Houston Clinton Dr., Burnet, TX 78611, USA. E-mail: dpeters@atmi.com; Fax: +1-512-756-2594; Tel: +1-512-715-5321*

[b]*University of Central Florida, NanoScience Technology Center, 12424 Research Parkway Suite 400, Orlando, FL 32826-3250, USA. E-mail: Richard.Blair@ucf.edu; Fax: +1-407-882-2819; Tel: +1-407-823-0639*

Although these materials can be synthesized by sol gel methods[3–9] and deposited *via* solution deposition[10,11] or sputtered from metal targets,[12–14] the aggressive geometry and high aspect ratios of these cutting edge devices will likely require the material to be deposited by chemical vapor (CVD) or atomic layer (ALD) deposition. These methods require a precursor that is volatile yet thermally stable at the process temperatures. Several organometallic precursors for deposition of strontium metal are known in the art but are not ideal.[15–20] The beta-diketonate based precursors are solids and, due to the strong metal–carbon bond, result in carbon contamination of the films.[21] The cyclopentadiene based ligands have a weaker metal–carbon bond, but are solids with low vapor pressures. It is desirable to have a liquid precursor for ease of handling and delivery. By using the bulky ligand *n*-propyltetramethylcyclopentadiene (Cp′) the liquid, bis(*n*-propyltetramethylcyclopentadienyl)strontium ($SrCp'_2$) can be prepared (Fig. 1).

$SrCp'_2$ is an excellent precursor as it is both monomeric and a liquid at room temperature. Typical syntheses of cyclopentadienides utilize alkali metal cyclopentadienides and metal halides to produce the desired product by salt metathesis (eqn (1)). However, strontium iodide is poorly soluble in any solvent compatible with the air- and water-sensitive reagents and final product.

$$SrI_2 + 2\,KCp' \xrightarrow[\text{RT}]{\text{Diethyl ether}} Sr(Cp')_2 : Et_2O + 2\,KI \quad (1)$$

The reaction represented in eqn (1) is thermodynamically and entropically favored. However, the low solubility of the precursor halide severely limits the diffusion rate of the precursors. Vigorous stirring and heating can alleviate this problem. However, the density of SrI_2 (5.46 g cm^{-3}) makes stirring large batches difficult and the maximum reaction temperature is limited to the boiling point of diethyl ether (34.6 °C). Mechanochemical methods reduce diffusion barriers effectively and increase reaction rates through attrition, agitation, and localized high-pressure events.[22–25] High energy milling can produce macroscopic temperatures in excess of 200 °C.[26] Exothermic processes can also increase the macroscopic temperature. $SrCp'_2$ was chosen as a CVD precursor and high macroscopic and microscopic temperatures would lead to undesirable by-products. In order to

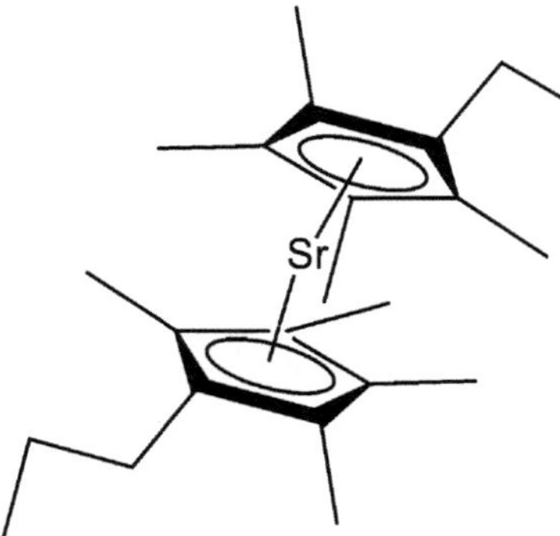

Fig. 1 The bulky *n*-propyltetramethylcyclopentadiene (Cp′) ligand forms a sandwich complex with strontium that is liquid at room temperature.

minimize any detrimental temperature rise, the mechanochemical synthesis of $SrCp'_2$ was undertaken under solvent or liquid assisted grinding (LAG).[22] Using this approach, the solvent serves to remove product from the surface of the insoluble SrI_2 and also serves as a heat sink. Since chemical reactions only occur during a collision and there is a large liquid heat sink, this approach is easily scaled.

Results and discussion

Bis(*n*-propyltetramethylcyclopentadienyl)strontium is readily synthesized in small quantities *via* standard salt metathesis methods (eqn (1)) The two sparingly soluble reagents are magnetically stirred in an ethereal solvent overnight and worked up *via* standard methods. More polar solvents such as dimethyl ether (DME) and tetrahydrofuran (THF) are better at solvating the reagents, but are to be avoided as they form stable, unbreakable adducts with the product. The presence of these adducts during epitaxial growth results in carbon and oxygen incorporation and compromises the integrity of the semiconductor device. Additionally, alkali metal salts of the cyclopentadiene (Cp′) ligand are highly reactive toward water. Anhydrous diethyl ether forms a Lewis acid/base adduct with the product that can be broken *via* a high temperature distillation resulting in pure (solvent-free) product (eqn (2))

$$\mathrm{Sr(Cp')_2 : Et_2O} \xrightarrow{\Delta} \mathrm{Sr(Cp')_2 + Et_2O} \qquad \text{(Eq. 2)}$$

This synthesis does not scale readily since neither the strontium iodide nor the potassium salt of the Cp′ ligand are readily soluble in diethylether. Strontium iodide is a very dense material (5.46 g cm^{-3}) and large quantities cannot be stirred magnetically. Larger scale reactions therefore require overhead stirring with a motor and a PTFE paddle to move the reagents in the reactor. However, none of the synthetic attempts utilizing overhead stirring produced any measureable yield of the desired product. Smaller scale syntheses with magnetic stirring always generated product, although yields did vary. Serendipitously, we discovered that a combination of magnetic and overhead stirring would, eventually, produce the desired product. This method, however, suffered from low yields and was very difficult to set up. The magnetic stir plate and the overhead stirrer had to be synchronized or the reaction failed. In addition, reaction times were inconsistent, varying from three to seven days. We observed that repeatable yields could be obtained if the same glassware, paddle and stir bar were used, but only then. Heating of the reactions did not accelerate the reaction nor did it improve yield.

It was theorized that the magnetic stir bar was not aiding in the stirring of the insoluble reagents but was grinding the two together and that the physical grinding was the force driving the reaction. This hypothesis explained why overhead stirring with a paddle alone did not result in a chemical reaction but coupling with a magnetic stir bar did. As there is necessarily a minimal amount of force between the glass reactor and the PTFE coated magnetic stir bar (by design) we set out to increase the force generated. This was expected to decrease the reaction time and increase the yield.

A commercially available vertical geometry filter reactor was chosen. This allowed for an increase in the stirring friction and for easy reagent filtration. In the filter reactor, KCp′ was prepared from HCp′ and a potassium salt followed by removal of the solvent and by-products *via* the bottom frit. Strontium iodide, 12.7 mm 316 stainless steel balls, and solvent were added to the KCp′ in the reactor. The media–reagent ratio was 0.05 : 1. Once the reaction was complete, the product was removed from the insoluble by-product through the same bottom frit.

Initially, the glass frit at the bottom of the reactor was expected to provide a rough surface for the steel balls to grind against. It became apparent that the stirring speed needed to be high enough that the ball bearings did not remain on the frit but instead were forced against the sides of the reactor. This still provided enough friction that the reagents ground together and the reaction was complete in twenty-four hours. This suggests that there is a minimum impact force required to induce the reaction. This force would be related to the fracture toughness of SrI_2 and KCp′.

With only a PTFE paddle, stirring the reagents in solvent produced no measureable reaction, even with additional heating. Both reagents are effectively insoluble in diethyl ether and remain, unreacted, on the bottom of the flask. No color is observed in the solvent even after days of stirring *via* the PTFE paddle. Upon the addition of milling media reaction products were observed and high yields were obtained after twenty-four hours. A distinct yellow color is observed in the solvent after just a few hours of stirring. The insoluble solid remains white and quickly settles to the bottom of the reactor when stirring is stopped. In this liquid assisted mechanochemical reaction, the grinding action of the steel balls on the insoluble reagents improves the reaction kinetics. Although substantially more solvent is used than in typical liquid assisted grinding[22] the principle is the same.

The solvent is serving at least three functions. It acts as a carrier, removing the soluble product from the reaction area, as a suspending agent for easy intermixture of the two insoluble reagents, and as a heat sink limiting the reaction temperature to the boiling point of the solvent (34.6 °C). Even though this synthesis is slow and the rate is determined by the grinding action, it is still an exothermic reaction. This heat, coupled with the heat of grinding, could exceed the thermal stability of the reagents or product at a local level. The solvent carries away the excess heat and radiates it to the atmosphere through the glass walls of the reactor. A fourth function may be to serve as a lubricant. Media wear is an important issue when a mechanical attrition process is implemented. It can be a major expense when operating large-scale solutions. The use of a solvent can serve as a lubricant and reduce media wear. Additionally, cyclopentadienides readily form from the metals found in the steel milling media (Fe, Cr, Ni, Mo, Mn). Metal impurities would make the product unsuitable for CVD application. Although wear was observed on the steel media, reactor sides (Fig. 2), and stirrer paddle, no appreciable amounts of these metals were found in the final product.

Although other researchers have reported the synthesis of metal–cyclopentadiene complexes *via* mechanochemical salt metathesis,[27] we believe this is the first report of a preparative scale synthesis using these methods. The use of larger quantities of solvent may play an important part in scaling up mechanochemical syntheses. Functioning as a carrier, the solvent removes the product from the reaction area which allows for a localized application of Le Chatelier's principle. In addition, by removing the product from the reaction area, the solvent

Fig. 2 Although wear (arrows) from the media is evident on the sides of the glass reactor, no metals were found in the product.

effectively prevents mechanochemical destruction of the product due to excessive grinding, a problem encountered by Makhaev *et al.*

Conclusion

The mechanochemical synthesis of bis(*n*-propyltetramethylcyclopentadienyl) strontium is six times faster than the standard method and produces a 90 percent yield. The synthesis can be performed using reactors and stirring apparatus developed for traditional solvent-based chemistry. This reduced effectively the capital expenditure needed for implementing this approach. It also allows air and moisture sensitive reactions to be carried out using proven reaction apparatus. Consistent, high yields were obtained on the pre-pilot plant scale (500 g). This yield should scale linearly with the size of the reactor. We have been very conservative in our experiments and limited the overall reaction scale to that of the first step – generating the KCp′ by standard solvent based techniques. As long as the stirring system has sufficient torque to move the solid reagents and the steel balls, a higher concentration of reagents could be placed in the reactor and more product could be generated in the same volume and with the same amount of solvent. An attrition mill modified for inert atmosphere use could grind many times the amount of solid we used with a minimal amount of solvent.

Since this approach scales linearly, the need for extensive reaction engineering as production needs increase is reduced. As we have shown, existing batch reactors can be readily converted to mechanochemical reactors. The use of liquid solvents in a mechanochemical reactor enhances mixing, reduces media wear, and improves thermal regulation. The use of a liquid carrier facilitated the reduction in media mass needed. In fact, the media loading for this reaction is almost two orders of magnitude lower than typical mechanochemical reactions (0.05 : 1) Previous reports of LAG have focused on the minimal amount of solvent needed.[22] Although it is true that less solvent results in a greener process, there are potential benefits to the use of larger quantities of less aggressive and more environmentally friendly solvents especially for larger scale reactions. This synthesis may be further improved by replacing the low-boiling, peroxide forming solvent diethyl ether with toluene. The toluene would solvate the product and aid in suspension but would not form an adduct with the product, eliminating an energy intensive processing step.

Experimental

General Considerations

All manipulations were performed using standard Schlenk and dry box techniques under an atmosphere of nitrogen.[28] Anhydrous solvents and reagents were purchased from catalog vendors and used without further purification. Deuterated solvents were dried over 4 Å sieves prior to use. NMR spectra were referenced to residual protium in the solvent. The filter reactor used is an ACE 6384-137 heavy wall reactor with a medium glass frit. All glass, steel and PTFE parts are dried overnight in a 120 °C oven and purged with nitrogen prior to use.

Reactor assembly

The reactor is assembled hot from the oven following the manufacturer's instructions. Kalrez® O-rings are used in place of the encapsulated ones supplied. The reactor is placed in a ring stand and secured with a chain clamp. It is purged with nitrogen through the gas inlet in the top and out the bottom valve. (Fig. 3.)

Synthesis of KCp′

Approximately 500ml of toluene was added to the reactor to leak check the reactor and dilute the $KN(SiMe_3)_2$. To this 3600g of a 15% solution of $KN(SiMe_3)_2$ in toluene (550g, 2.28 mol of $KN(SiMe_3)_2$) was added. The mixture was stirred *via* an

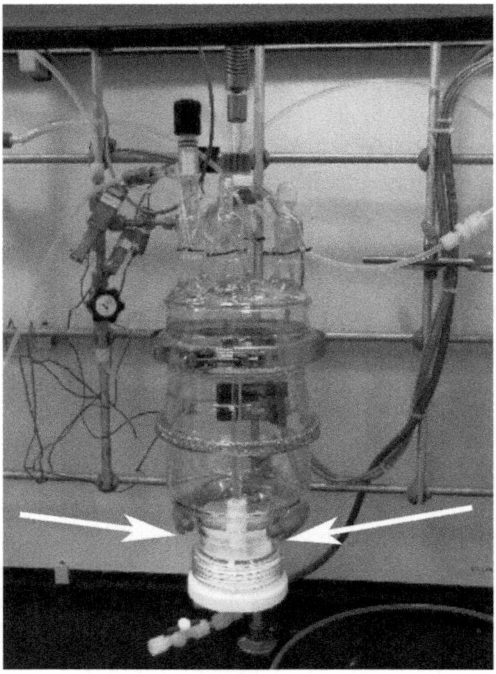

Fig. 3 The reactor assembly is based on commercially available large-scale glassware. The media ride in the region indicated by the arrows.

overhead stirrer @300 rpm and 392g (2.38 mol) of Cp′H was transferred to the reactor using 6.35 mm PTFE tubing under a positive pressure of nitrogen. Addition rate is not critical. The reaction mixture was stirred overnight at room temperature under nitrogen and yielded a yellow solution with a white solid (KCp′). The solid was isolated by removing the solvent through the bottom valve and washed with three 500 ml portions of toluene. Quantitative yield for the KCp′ was assumed for the calculations utilized for the remainder of the synthesis.

Mechanochemical synthesis of $Sr(Cp')_2$:OEt_2

SrI_2 powder (380 g, 1.12 mol) was loaded into a solid addition Schlenk flask inside a glove box and connected to a Schlenk line. Four liters of diethyl ether were added to the KCp′ in the reactor. Six 12.7 mm stainless steel ball bearings (316) were added to the reactor. The stirring speed was adjusted so that the ball bearings remained on the bottom of the reactor. Once stirring had been established, the solid SrI_2 powder was added using a flexible PTFE solid-addition adapter. The reaction was stirred for twenty-four hours at room temperature under a nitrogen overpressure.

The product was collected by filtering through the frit at the bottom of the reactor. The filter cake is washed with three 500 ml portions of diethyl ether and added to the product flask. The filtrate was concentrated under reduced pressure yielding approximately 492 grams of $Sr(Cp')_2$–OEt_2 as a waxy oil (88% yield).

Isolation of $Sr(Cp')_2$

Vacuum distillation of $Sr(Cp')_2$:OEt_2 affords the base free compound. In a typical distillation $Sr(Cp')_2OEt_2$ was distilled using a jacketed short path distillation head. Once the pressure was below 750 microns, the source flask was heated to 145 °C. After 30 minutes the temperature was increased to 160 °C to break the adduct and remove diethyl ether from the material. After 3 hours, all of the ether had been removed and the apparatus could be transferred to the product flask. The external temperature of the source flask was increased to 230 °C in 20° steps. The product begins to distil at a head temperature of ~165 °C. Distillation was carried out until no more material came over. Yield: 380 grams of pure material. ^{1}H NMR (C_6D_6, 15 °C): δ 0.907 triplet, 1.31 multiplet, 1.96 singlet, 2.001 singlet, 2.39 multiplet.

Characterization

Organic purity was determined by NMR. The spectrum was integrated and peaks not assignable to $Sr(Cp')_2$ were integrated and found to be less than 1% of the total spectrum. Metallic purity was determined by ICP-MS. All lots were examined for the presence of elements found in 316SS – iron, chromium, manganese, molybdenum, and nickel. In two of the six lots iron was detectable at 1ppm and 1.09ppm. The other four lots had no detectable iron and none of the other elements were detectable in any of the six lots made by this method. Non-volatile residue was determined by simultaneous thermal analysis and all lots showed less than 2.5% NVR.

References

1 Maitri P. Warusawithana, Cheng Cen, Charles R. Sleasman, Joseph C. Woicik, Yulan Li, Lena Fitting Kourkoutis, Jeffrey A. Klug, Hao Li, Philip Ryan, Li-Peng Wang, Michael Bedzyk, David A. Muller, Long-Qing Chen, Jeremy Levy and D. G. Schlom, *Science*, 2009, **324**, 367–370.

2 R. Balachandran, B. H. Ong, H. Y. Wong, K. B. Tan and M. M. Rasat, *Int. J. Electochem. Sci.*, 2012, **7**, 11895–11903.

3 R. Balachandran, H. K. Yow, B. H. Ong, K. B. Tan, K. Anuar, W. T. Teoh, F. M. N. Ahmad, S. Sreekantan and V. Swaminathan, *J. Mater. Sci.*, 2010, **46**, 1806–1813.

4 S. Fuentes, R. A. Zarate, E. Chavez, P. Munoz, D. Diaz-Droguett and P. Leyton, *J. Mater. Sci.*, 2009, **45**, 1448–1452.

5 W. Li, Z. Xu, R. Chu, P. Fu and J. Hao, *J. Alloys Compd.*, 2010, **499**, 255–258.

6 R. M. Mahani, I. K. Battisha, M. Aly and A. B. Abou-Hamad, *J. Alloys Compd.*, 2010, **508**, 354–358.

7 U. Selvaraj, A. V. Prasadarao, S. Komarneni and R. Roy, *Mater. Lett.*, 1991, **12**, 311–315.

8 A. Tsuzuki, K. Kato, K. Kusumoto and Y. Torii, *J. Mater. Sci.*, 1998, **33**, 3055–3058.

9 Y. Wu, S. Zamani and G. Z. Cao, *MRS Online Proc. Libr.*, 2011, **548**, 587–592.

10 V. Kumar, I. Packiaselvam, K. Sivanandan, M. A. Vahab and A. K. Sinha, *J. Am. Ceram. Soc.*, 2006, **89**, 1136–1139.

11 M. P. Siegal, P. G. Clem, J. T. Dawley, J. Richardson, D. L. Overmyer and T. G. Holesinger, *J. Mater. Res.*, 2005, **20**, 910–921.

12 T. S. Kim, C. H. Kim and M. H. Oh, *J. Appl. Phys.*, 1994, **75**, 7998–8003.

13 B. Laughlin, J. Ihlefeld and J.-P. Maria, *J. Am. Ceram. Soc.*, 2005, **88**, 2652–2654.

14 H.-N. Tsai, Y.-C. Liang and H.-Y. Lee, *J. Cryst. Growth*, 2005, **284**, 65–72.

15 R. Breitkopf, L. J. Meda, T. Haas and R. U. Kirss, *MRS Online Proc. Libr.*, 1998, **495**, 51–55.

16 JP07268634A, 1995.

17 T. Kawahara, M. Yamamuka, M. Tarutani, T. Horikawa and K. Ono, *Proc. - Electrochem. Soc.*, 1998, **98-3**, 190–195.

18 JP2009040707A, 2009.

19 J. A. T. Norman, M. Perez, M. S. Kim, X. Lei, S. Ivanov, A. Derecskei-Kovacs, L. Matz, I. Buchanan and A. L. Rheingold, *Inorg. Chem.*, 2011, **50**, 12396–12398.

20 M. Popovici, K. Tomida, J. Swerts, P. Favia, A. Delabie, H. Bender, C. Adelmann, H. Tielens, B. Brijs, B. Kaczer, M. A. Pawlak, M.-S. Kim, L. Altimime, E. S. Van and J. A. Kittl, *Phys. Status Solidi A*, 2011, **208**, 1920–1924.

21 F. B. P. Timothy and P. Holme, *J. Phys. Chem. A*, 2007, **111**, 8147–8151.

22 S. L. James, C. J. Adams, C. Bolm, D. Braga, P. Collier, T. Friscic, F. Grepioni, K. D. M. Harris, G. Hyett, W. Jones, A. Krebs, J. Mack, L. Maini, A. G. Orpen, I. P. Parkin, W. C. Shearouse, J. W. Steed and D. C. Waddell, *Chem. Soc. Rev.*, 2012, 41.

23 G. Kaupp, *CrystEngComm*, 2009, **11**, 388–403.

24 C. Suryanarayana, *Prog. Mater. Sci.*, 2001, **46**, 1–184.

25 S. M. Hick, C. Griebel, D. T. Restrepo, J. H. Truitt, E. J. Buker, C. Bylda and R. G. Blair, *Green Chem.*, 2010, **12**, 468–474.

26 L. Takacs and J. S. McHenry, *J. Mater. Sci.*, 2006, **41**, 5246–5249.
27 V. D. Makhaev, A. P. Borisov and L. A. Petrova, *J. Organomet. Chem.*, 1999, **590**, 222–226.
28 M. A. Drezdzon and F. Shriver, *The Manipulation of Air-Sensitive Compounds*, 2nd edn, 1986.

Faraday Discussions

PAPER

Mechanochemical preparation of copper iodide clusters of interest for luminescent devices†

Lucia Maini,*[a] Paolo P. Mazzeo,[a] Francesco Farinella,[a] Valeria Fattori[b] and Dario Braga[a]

Received 30th December 2013, Accepted 25th February 2014
DOI: 10.1039/c3fd00164d

The copper iodide complexes are known for their large variety of coordination geometries. Such diversity, while making it difficult to predict the final structure, permits the preparation of a great number of copper iodide complexes based on the same ligand. The target of the research was that of thoroughly exploring the chemistry of CuI and the ligand diphenyl-2-pyridyl phosphine (PN) by varying the stoichiometric ratio and/or the aggregation state. Six different compounds have been identified: $[Cu_4I_4(PN)_2]$, $[Cu_4I_4(PN)_2\cdot(CH_2Cl_2)_{0.5}]$, $[CuI(PN)_{0.5}]_\infty$, $[CuI(PN)_3]$ whose structures have been determined during this study, $CuI(PN)_2$ which was characterized by powder diffraction and $[Cu_2I_2(PN)_3]$ which has been already reported. The preparation routes are also different: synthesis in solution yielded $[Cu_4I_4(PN)_2\cdot(CH_2Cl_2)_{0.5}]$ and $[CuI(PN)_3]$ while $[CuI(PN)_{0.5}]_\infty$ and $CuI(PN)_2$ were obtained only *via* solid state reactions. These two latter examples confirmed that mechanochemistry is a valid route to explore the landscape of the possible structures of CuI derivatives. Crystallization by traditional solution procedures failed to give the desired crystal, so structure determination of the new compounds was tackled in two ways: by attempting crystal growth *via* solvothermal synthesis and by resolving the structure from X-ray powder diffraction data with "direct space" methods. What is more the photophysical properties of the complexes that could be obtained as sufficiently pure powders have also been investigated and are reported herein.

1 Introduction

Complexes based on copper iodide are currently at the forefront of coordination chemistry and crystal engineering research because of the quest for active materials in optoelectronics.[1–8] The interest stems from the increasing demand of

[a]*Dipartimento di Chimica "G. Ciamician", Università di Bologna, Via F. Selmi 2, Bologna, Italy. E-mail: l.maini@unibo.it; Fax: +39 0512099456; Tel: +39 051 2099597*

[b]*Istituto per la Sintesi Organica e la Fotoreattività (ISOF) – CNR, Via Gobetti 101, 40129 Bologna, Italy*

† CCDC reference numbers 979008–979010. For crystallographic data in CIF or other electronic format see DOI: 10.1039/c3fd00164d

more-affordable complexes in preference to luminescent metal complexes based on precious (*i.e.* the platinum group) and rare-earth metals, which are often quite expensive and environmentally problematic.[9] The copper iodide complexes present several advantages: they are characterized by a large variety of coordination geometries which arise from the many possible combinations of coordination numbers (two, three and four) available for copper(I) and geometries that can be adopted by the halide ions (from terminal to μ^2- and up to μ^8-bridging);[10] they present different luminescent levels which can be of a ligand centred, charge transfer or, in the case of polynuclear compounds, even metal-centred nature;[11,12] most of the complexes studied are characterized by a remarkable high quantum yield in solid state;[4,13,14] and finally the reagents are cheap and it is relatively easy to obtain the final products.[4,15]

The luminescent properties of copper(I) halide clusters are strictly related to the geometries adopted by the clusters.[12] In the case of the cubanes $Cu_4I_4L_4$ for example, two distinct emission bands were observed, the high-energy band attributed to halide-to-ligand charge transfer (^{3}XLCT), and the low-energy emission band was attributed to a triplet Cu-I Cluster-Centered (^{3}CC) excited state with the excitation localized to the Cu_4I_4 core, only possible in the presence of an interaction between the metal centres. Recent studies have widened the possibility of the emission properties of copper(I) compounds. Several Cu(I) compounds with P^N ligands present a low singlet–triplet splitting which allows the emission from both the singlet and the triplet excited state depending on the temperature. The singlet excited state, which is slightly higher in energy than the triplet state, can be activated thermally at the expense of the triplet state, leading to the so-called thermally activated delayed fluorescence (TADF). The TADF emission mechanism allows harvesting of both singlet and triplet excited states generated by charge recombination in electroluminescent devices, therefore light generation efficiency is enhanced.[13,16]

Due to the increasing interest in CuI(P^N ligands) complexes for OLED applications and for photochemistry studies,[3,13,15,17] we exploited the reactivity of the CuI towards diphenyl-2-pyridyl phosphine, hereafter PN. Two crucial aspects of the research will be presented: we will first discuss how the synthetic procedure influences the formation of the final product with particular attention to the mechanochemical synthesis; secondly we will address the different strategies to determine the crystal structure when the final product is a crystalline powder.

The crystal engineering of copper halides is particularly difficult since in solution, halocuprate complexes are involved in kinetically fast dissociative and associative equilibria[18,19] and the labile coordination numbers and geometries of both copper(I) and halide ions allows an inner-core variability difficult to predict. These complexes are characterized by a flat energy landscape confirmed by the presence of several different copper halides complexes in the CSD[20] which present dramatically different nuclearity of their Cu_xI_y cores such as discrete dimers, cubane tetramers and stepped cubane tetramers to infinite polymeric chains (including split stairs, zigzag, helical, staircase, rack and columnar structures). Although the final structures of the copper(I) halide complexes in crystals could be hardly designed, the reaction stoichiometry allows some degree of control of the copper(I) halide core: reactions with a stoichiometry ratio of CuI/L $\gg$ 1 higher nuclearity are favoured while CuI/L $\ll$ 1 promote the formation of monomers.[21] However, to explore all the possible crystal forms it is necessary to vary not only the stoichiometry but also other parameters such as solvent, temperature *etc.* To overcome the problem of the low solubility of the CuI which restricts the possibility of changing the solvent, the reactions can be performed in the solid state which can yield crystal forms hardly or even not obtainable in solution.[22] The reactivity of CuI towards nitrogen, sulphur or phosphorus based ligands is very high and preliminary work on mechanochemical synthesis with CuI with N-base satured ligands has been reported;[23,24] more extensive studies have been done for copper thyocyanate complexes,[25–27] in both cases mechanochemical synthesis led to the formation of compounds not obtained by solution synthesis.

The main drawbacks in the mechanochemical reaction methods consist of the formation of microcrystalline products, obviously unsuitable for X-ray single crystal diffraction. Structural characterization is a key point to understand the properties of the compounds and to design new compounds, which is why great effort will be devoted to tackling this problem. Most of the time traditional crystallization from solution, but also triple layer crystallizations or seeding of solutions with microcrystalline powder of the desired compound, fail to yield single crystals of amenable size for single crystal diffraction experiments. However, the structure can be determined from X-ray powder diffraction data, thanks to the development of "direct space" methods.[28] This approach is particularly suited for materials constructed from well-defined modular building units and it is more challenging for copper iodide complexes whose coordination is difficult to control. Spectroscopy techniques such as IR or Solid State NMR can give some insights on the coordination or on the asymmetric units but not guarantee exact structure determination.[26,29]

The quest of suitable methods to obtain crystals for structure determination of copper complexes led us to explore different crystallization processes, including those based on solvothermal synthesis. Recently, several zeolites, commonly obtained by solvothermal synthesis, have been produced also by grinding or ball milling,[30–33] which suggests that the solid state reactions can have several aspects in common with solvothermal conditions.[25] Although, for large scale preparations, nowadays researchers try to move from solvothermal to solvent-free reactions in order to reduce their environmental impact and energy consumption, solvothermal synthesis can still be exploited to obtain single crystals of the desired compounds since the high-temperature and high-pressure environments conditions created would facilitate the crystallization process of complexes with poor solubility.

Herein we report the synthesis of five new crystal forms of copper iodide with the diphenyl-2-pyridyl phosphine (PN): $[Cu_4I_4(PN)_2]$, $[Cu_4I_4(PN)_2\cdot(CH_2Cl_2)_{0.5}]$, $[CuI(PN)_{0.5}]_\infty$, $[CuI(PN)_3]$ and $CuI(PN)_2$; as well as that of the known complex $[Cu_2I_2(PN)_3]$.[14] Mechanochemical synthesis was decisive to obtain the new compounds $[CuI(PN)_{0.5}]_\infty$ and $CuI(PN)_2$ while the monomer $[CuI(PN)_3]$ is formed only by precipitation from solution. Not only the stoichiometry variation but also performing the reaction in solid state or solution allowed us to widen the landscape of the crystal forms of copper iodide complexes with PN. The structures of $[Cu_4I_4(PN)_2]$ and $[CuI(PN)_{0.5}]_\infty$ have been determined by single crystals obtained from solvothermal synthesis, while the bulk product has been obtained by reaction in solid state or solution. The structure of $[CuI(PN)_3]$ was determined by direct methods from X-ray powder diffraction data.

All the compounds obtained are luminescent and the photophysical properties of the complexes that could be obtained as sufficiently pure powders, have been studied and reported.

Experimental

General

All reagents were purchased from Sigma Aldrich and used without further purification.

Synthesis of $[Cu_2I_2(PN)_3]$

Copper(I) iodide (0.380 g; 2 mmol) and the PN ligand (0.789 g; 3 mmol) were suspended under nitrogen atmosphere in dry dichloromethane (10 mL) and stirred for 12 h at room temperature. The yellowish powder obtained was filtered and washed with a saturated solution of KI_{aq} to remove the unreacted CuI and distilled water to remove KI.

The title compound was obtained by grinding CuI (0.190 g; 1 mmol) and PN (0.394 g; 1.5 mmol) with two drops of acetonitrile in a ball mill for 30 minutes at 20 Hz.

Synthesis of $[CuI(PN)_3]$

Copper iodide (0.095 g, 0.5 mmol) was dissolved in saturated aqueous solution of KI (5 mL) then an acetone solution of PN (0.526 g, 2 mmol) was added under stirring. A whitish powder precipitates instantly which was filtered and washed with cold acetone.

Synthesis of $CuI(PN)_2$

Copper iodide (0.190 g; 1mmol) and PN ligand (1.052 g; 4 mmol) were ground with two drops of AcCN in a ball mill for 60 minutes at 20 Hz.

Synthesis of $[Cu_4I_4(PN)_2\cdot(CH_2Cl_2)_{0.5}]$

CuI (0.760 g; 4 mmol) dissolved in 20 mL of CH_3CN at 40 °C was added to a solution of PN (0.263 g; 1mmol) in 20 mL of CH_2Cl_2. The clear solution obtained was left standing at room temperature overnight. The brownish powder obtained was filtered and then washed with an aqueous saturated solution of KI_{aq} to remove the unreacted CuI and with distilled water to remove the KI.

Ball milling of CuI (0.760 g; 4 mmol) and PN (0.263 g; 1 mmol) with two drops of dichloromethane for 60 minutes at 20 Hz led to a mixture of $[Cu_4I_4(PN)_2\cdot(CH_2Cl_2)_{0.5}]$ and $[CuI(PN)_{0.5}]_\infty$.

Synthesis of $[Cu_4I_4(PN)_2]$

$[Cu_4I_4(PN)_2]$ can be easily obtained by heating $[Cu_4I_4(PN)_2\cdot(CH_2Cl_2)_{0.5}]$ at 170 °C.

Crystallization of $[Cu_4I_4(PN)_2]$ and $[CuI(PN)_{0.5}]_\infty$

Crystals of $[Cu_4I_4(PN)_2]$ suitable for SC-XRD analysis were obtained in a 5 mL reactor *via* solvothermal synthesis by mixing 0.2 g of CuI with 0.8 g of PN in 1 mL of ethanol at 170 °C and slowly cooling down to ambient temperature over 7 days.

Crystals of $[CuI(PN)_{0.5}]_\infty$ suitable for SC-XRD were obtained in a 5 mL reactor *via* solvothermal conditions by mixing 0.2 g of CuI with 0.6 g of PN in 1 mL of toluene at 140 °C and slowly cooling down to ambient temperature over 7 days.

TGA measurements

TGA measurements were performed using a Perkin Elmer TGA7 in the temperature range 35–700 °C under N_2 gas flow and heating was carried out at 5 °C min^{-1}.

Photophysics

All determinations made use of powder samples placed between two quartz slides and were done at room temperature. Excitation and emission spectra were obtained with a SPEX Fluorolog fluorometer; excitation spectra were monitored at the maximum emission wavelength and emission spectra were recorded exciting all samples at 350 nm.

X-Ray powder diffraction

X-Ray powder diffractograms were collected on a Panalytical X'Pert PRO automated diffractometer with Cu Kα radiation and an X'Celerator equipped with an Anton Paar TTK 450 low-temperature camera. The program Mercury[34] was used for calculation of X-ray powder patterns. XRPD data of $[CuI(PN)_3]$ were collected over the range 3–70° 2θ with a Bruker D8 Advance diffractometer equipped with a LynxEye detector and focusing mirror.

Crystal structure determination

Crystal data for $[Cu_4I_4(PN)_2]$ and $[CuI(PN)_{0.5}]_\infty$ were collected on an Oxford Xcalibur S with MoKα radiation, $\lambda = 0.71073$, monochromator graphite.† Crystal data and details of measurements are summarized in Table 1. SHELX97[35] was used for the structure solution and refinement based on F^2. Non-hydrogen atoms were refined anisotropically. The uniqueness of the crystal of $[CuI(PN)_{0.5}]_\infty$, synthesized by solvothermal reaction, did not allow the collection of better data. The bad quality data due to the poor crystallinity and high mosaicity of the sample.

The Mercury[34] software package was used for the graphical representation of the resultant structures.

Results and discussion

Synthesis and crystal structures

The copper iodide was reacted with PN with different synthetic procedures which differed mainly in stoichiometric ratio or aggregation state (solid state or solution). The reactions in solution are carried out mainly in acetonitrile or a saturated aqueous solution of KI in air, while the reactions in the solid state were carried out *via* ball milling with a drop of acetonitrile or dichloromethane.

Six different compounds have been isolated so far, $[Cu_2I_2(PN)_3]$ recently described by Zink *et al.*[13] $[Cu_4I_4(PN)_2]$ and its solvate form $[Cu_4I_4(PN)_2\cdot(CH_2Cl_2)_{0.5}]$, $[CuI(PN)_{0.5}]_\infty$, $[CuI(PN)_3]$ (whose structure has been determined during these studies) and $CuI(PN)_2$.

Table 1 Crystal data

	$[Cu_4I_4(PN)_2]$	$[CuI(PN)_{0.5}]_\infty$	$[CuI(PN)_3]^a$
Chemical formula	$C_{34}H_{28}Cu_4I_4N_2P_2$	$C_{17}H_{14}Cu_2I_2N_1P_1$	$C_{51}H_{42}Cu_1I_1N_3P_3$
Formula mass	1288.28	644.15	980.28
Crystal system	Monoclinic	Triclinc	Trigonal
a/Å	16.7222(3)	7.9805(7)	13.6471(2)
b/Å	13.9164(3)	8.5298(8)	13.6471(2)
c/Å	16.4466(3)	14.3667(14)	14.7400(4)
α/°	90	78.092(8)	90
β/°	99.688(2)	85.576(8)	90
γ/°	90	86.494(7)	120
Volume/Å^3	3772.75(13)	953.05(15)	2377.4
Temperature	RT	RT	RT
Space group	P2$_1$/c	P-1	P-3
No. of independent reflections	11078	2594	—
R_{int}	0.0534	0.046	—
Final R_1 values	0.0489	0.1225	
Final $wR(F^2)$ values	0.0933	0.3395	
Final R_1 values (all data)	0.0621	0.1386	

[a] Solved from X-ray powder diffraction.

Table 2 Copper-copper distances

Compound	Distance/Å	
$[Cu_2I_2(PN)_3]$	Cu1–Cu2	2.7694(6)
$[Cu_4I_4(PN)_2]$	Cu6–Cu7	2.7499(9)
	Cu5–Cu7	2.5883(9)
	Cu6–Cu8	2.574(1)
	Cu5–Cu8	2.699(1)
$[CuI(PN)_{0.5}]_\infty$	Cu3–Cu4	2.771(7)
	Cu3–Cu3	2.900(7)
	Cu4–Cu4	3.485(6)

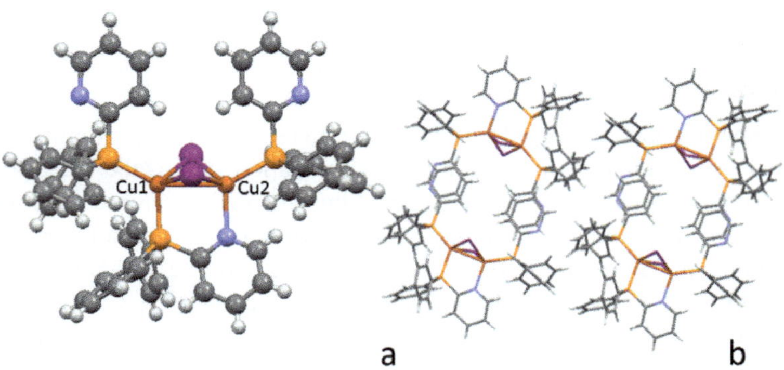

Fig. 1 (a) $[Cu_2I_2(PN)_3]$. (b) π–π interactions present in the crystal packing.

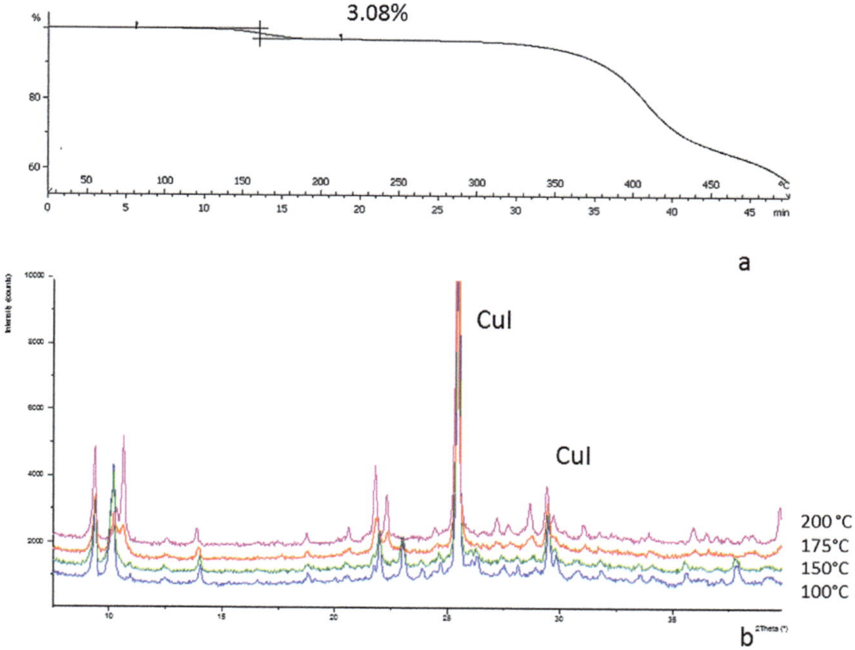

Fig. 2 (a) TGA curve of $[Cu_4I_4(PN)_2 \cdot (CH_2Cl_2)_{0.5}]$, the first step is consistent with the release of half a dichloromethane molecule (experimental 3.08%, calculated 3.20%). (b) Variable temperature X-ray powder diffraction: the patterns reported are at 100 °C (bottom line), 150 °C (medium bottom line), 175 °C (medium top line) and 200° (top line). Bragg reflections of unreacted CuI are highlighted by the labels.

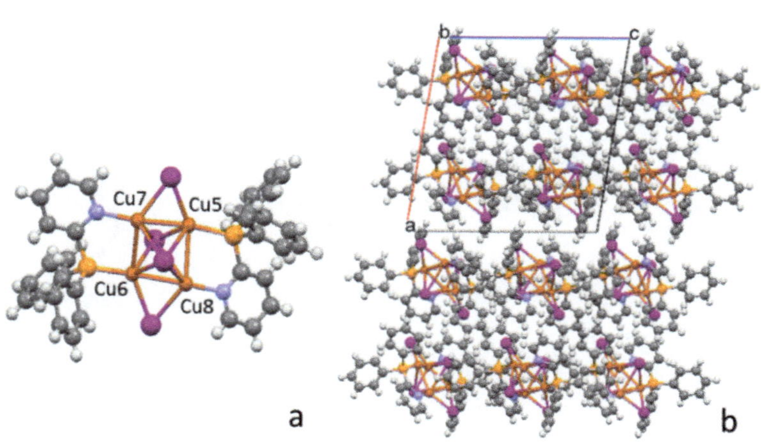

Fig. 3 (a) Compound $Cu_4I_4(PN)_2$; (b) crystal packing view along the *b* axis.

The dimer $[Cu_2I_2(PN)_3]$ was easily reproduced by the synthesis described in the literature[13] and it was also obtained by ball milling of CuI and PN with a stoichiometry ratio 1 : 1.5 with a drop of acetonitrile. This complex seems to be very stable and it was obtained as an undesired product in several other reactions (up to a stoichiometry ratio CuI–PN 3 : 1 or 1 : 3) both in solution and the solid state.

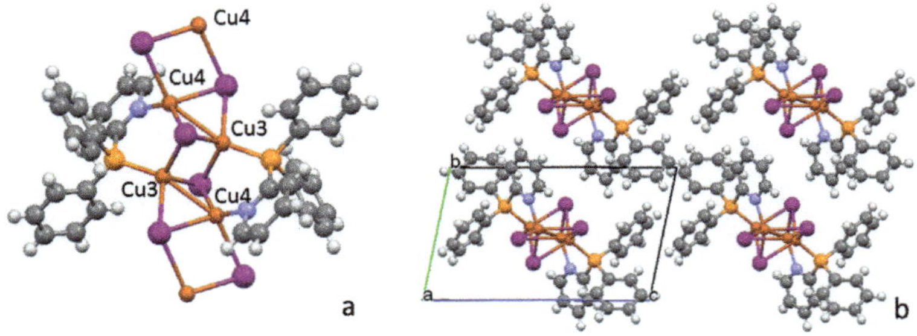

Fig. 4 (a) Polymeric structure of $[CuI(PN)_{0.5}]_\infty$; (b) crystal packing view along the *a* axis. The polymeric chains run parallel to each other.

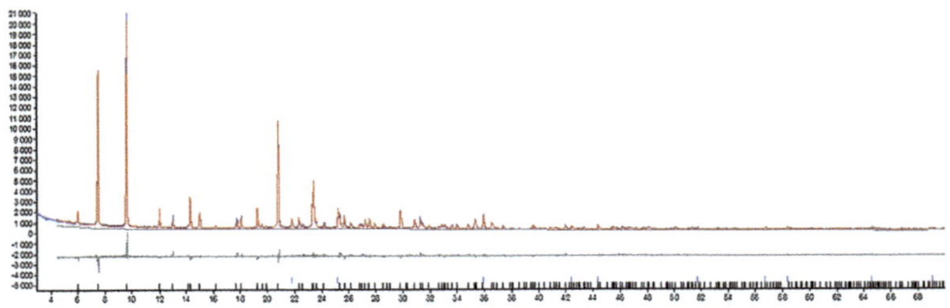

Fig. 5 Rietveld refinement of $[CuI(PN)_3]$, the red line is the calculated diffractogram, the blue line the observed diffractogram and the bottom line the difference plot. Upper ticks correspond to the Bragg peaks of KI, lower ticks correspond to the Bragg peaks of $[CuI(PN)_3]$.

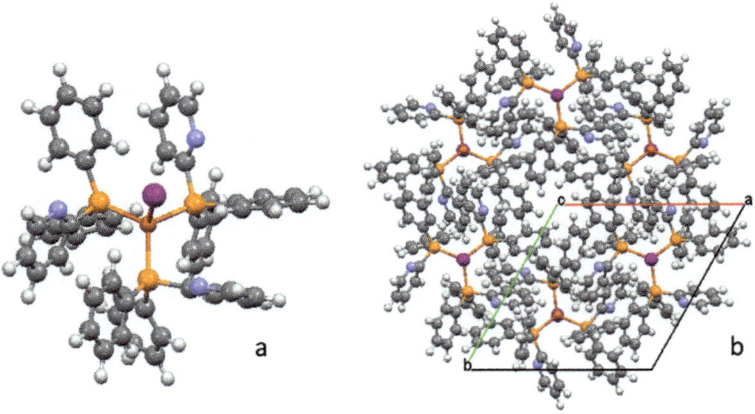

Fig. 6 (a) Single molecule of $[CuIPN_3]$. (b) Packing of $[CuIPN_3]$ viewed along the *c* axis.

In many cases to avoid the presence of $[Cu_2I_2(PN)_3]$ the reactions were run with a great excess of one of the reagents.

The dimeric core of $[Cu_2I_2(PN)_3]$ is characterized by a butterfly shape due to the presence of a bridging ligand coordinating with both N and P atoms, and a short copper–copper distance (2.7694(6) Å) consistent with metallophilic interactions (see Table 2). The two other ligands coordinated only with the P atoms and they

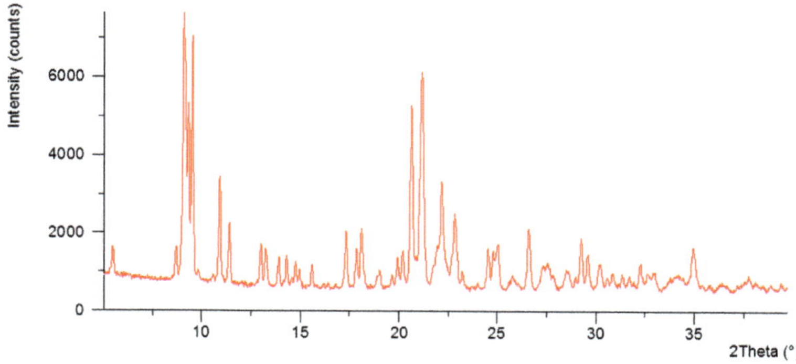

Fig. 7 Powder pattern of CuI(PN)$_2$ obtained by ball milling.

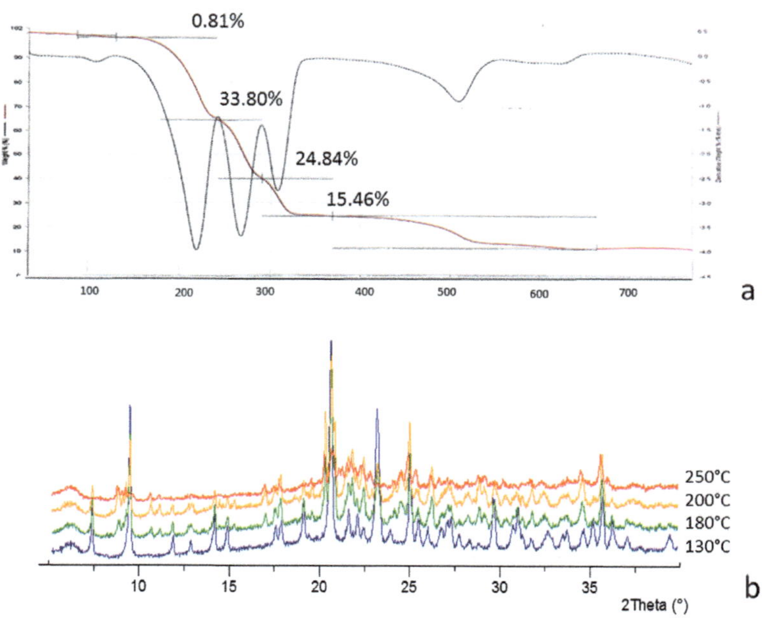

Fig. 8 (a) TGA curve of [CuI(PN)$_3$], the first step (0.81%) is due to some solvent release, the second step can be ascribed to the release of one ligand molecule which is suddenly followed by the release of other ligand molecules; (b) VTXRPD of [CuI(PN)$_3$], at 180 °C the peaks of CuI(PN)$_2$ appear, the complete transformation is characterized and observed at 250 °C. The powder pattern presents a low crystallinity profile.

are described as ancillary ligands. The complexes interact *via* C–H---π and in the case of the ancillary ligands also *via* π–π interactions (see Fig. 1). These weak interactions do not affect the photophysical properties since the compound does not suffer emission quenching.

The reactions carried in excess of CuI (stoichiometry ratio CuI–PN 4 : 1) lead to the formation of $[Cu_4I_4(PN)_2\cdot(CH_2Cl_2)_{0.5}]$, and $[CuI(PN)_{0.5}]_\infty$; the pure phase of $[Cu_4I_4(PN)_2\cdot(CH_2Cl_2)_{0.5}]$, is obtained by reaction in solution, while by ball milling a mixture of $[Cu_4I_4(PN)_2\cdot(CH_2Cl_2)_{0.5}]$ and $[CuI(PN)_{0.5}]_\infty$ is obtained; it was not possible to isolate the pure phase of $[CuI(PN)_{0.5}]_\infty$ Single crystals of $[Cu_4I_4(PN)_2]$

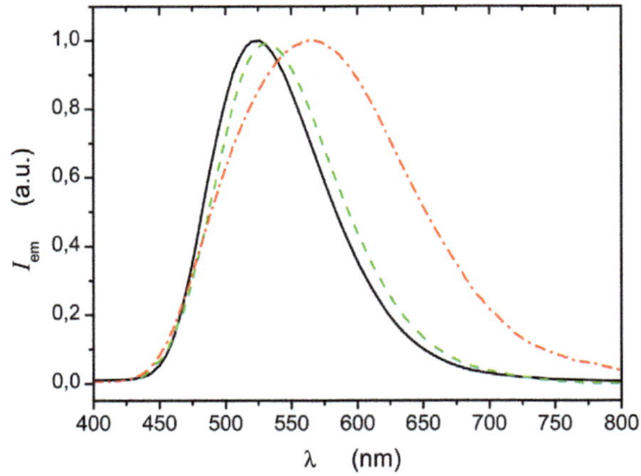

Fig. 9 Emission spectra of $[Cu_2I_2(PN)_3]$ (solid, black), $[CuI(PN)_3]$ (dash, green) and $[Cu_4I_4(PN)_2{\cdot}(CH_2Cl_2)_{0.5}]$ (dash–dot, red) as crystalline powders at room temperature.

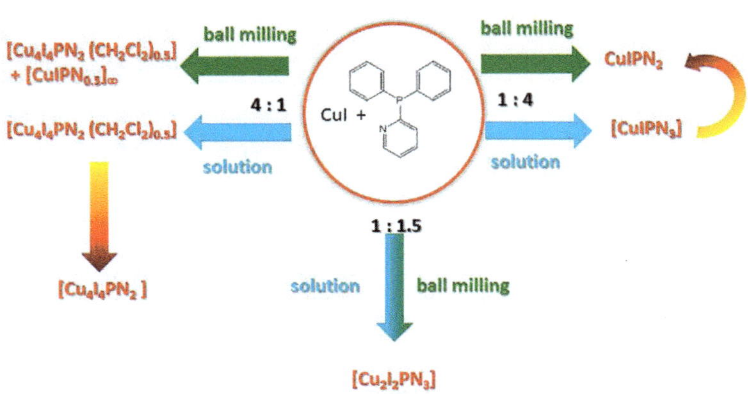

Scheme 1 Scheme of the compounds obtained and their relationship. Plain coloured arrows indicate syntheses in solution and the solid state, gradient coloured arrows indicate thermal treatment.

and $[CuI(PN)_{0.5}]_\infty$ have been obtained by solvothermal reactions while all other attempts to obtain crystals by triple layer crystallization failed. $[Cu_4I_4(PN)_2{\cdot}(CH_2Cl_2)_{0.5}]$ presents a powder diffraction pattern very similar to $[Cu_4I_4(PN)_2]$; while the structure of the unsolvated form was determined by single crystal X-ray diffraction, it was only possible to index the powder pattern of the $[Cu_4I_4(PN)_2{\cdot}(CH_2Cl_2)_{0.5}]$. The Pawley refinement converged to $R_{wp} = 6.65\%$ with a monoclinic cell, space group $P2_1/c$ with parameters $a = 17.4853(5)$ Å, $b = 14.180(2)$ Å, $c = 16.303(2)$ Å, $\beta = 97.07(1)°$ and volume 4011.5(8) Å^3 which is about 239 Å^3 greater than the volume of $Cu_4I_4(PN)_2$. The difference in volume can be ascribed to the presence of dichloromethane molecules in the elementary cell. The presence of the solvent was confirmed by the TGA which shows a 3.08% weight loss at 150 °C corresponding to the release of half a molecule of dichloromethane (calculated 3.20%) (see Fig. 2a). The transformation of $[Cu_4I_4(PN)_2{\cdot}(CH_2Cl_2)_{0.5}]$

into $[Cu_4I_4(PN)_2]$ was observed also with variable temperature X-ray diffraction (see Fig. 2b).

Instead of the common motif of cubane-like clusters, the $[Cu_4I_4(PN)_2]$ cluster adopts an "octahedral" geometry which has been observed recently by Liu *et al.*[4] The copper atoms are arranged in a parallelogram with μ^4-iodides above and below the parallelogram and μ^2-iodides bridging the copper atoms on the short edge. All Cu–Cu distances are lower than the sum of the van der Waals radii (see Table 2). The aromatic ring interacts mainly *via* C–H---π interactions and π–π interactions involving the pyridine coordinating ring and a phenyl ring (see Fig. 3). The crystal packing suggests that the solvent molecules can be placed in the (1 0 0) plane, which is less dense. Further studies will be done to ascertain the crystal packing of $[Cu_4I_4(PN)_2\cdot(CH_2Cl_2)_{0.5}]$.

It is worth noting that $[CuI(PN)_{0.5}]_\infty$ is easily obtained by mechanosynthesis, although not as a pure phase. Its formation was observed also in solid state reactions with a stoichiometry ratio CuI–PN 1 : 1.5. The low solubility of the compound, which characterizes all the coordination polymers, seems to preclude its formation from synthesis in solution, while it seems to be a driving force in the solid state reactions. The solvothermal conditions (high temperature and high pressure) allow the crystallization of phases which are known to be poorly soluble; in addition, solid state reactions are known to reach phases which can be hard to obtain in solution, as observed for the co-crystals.[36] In our case the solvothermal syntheses were explored in order to obtain a single crystal of the desired phase, since structure determination by X-ray powder diffraction (XRPD) would be very challenging due to the lack of information on the molecular geometry of the complexes.

Despite of all the structures of CuI(P^N ligands) described in the literature[1–4,13,15,37] which show discrete complexes, the structure of $[CuI(PN)_{0.5}]_\infty$ is characterized by a one-dimensional copper iodide polymeric structure (see Fig. 4). The infinite double chain of CuI presents an alternation of short and long copper–copper distances (see Table 2). Usually the tetrahedral coordination of the copper(I) ions is fulfilled by multidentate bridging ligands to construct a 2D sheet network;[24,38] in our case the PN ligand bridges two copper atoms but on the same chain and brings the two copper atoms to a separation of 3.0716(3) Å.

The reactions carried in excess of ligand leads to the formation of two new phases: $[CuI(PN)_3]$ and $CuI(PN)_2$. The pure compound $[CuI(PN)_3]$ is obtained only from reaction in solution carried out with a stoichiometry ratio of CuI–PN 1 : 4, but when the same reaction is carried out in the solid state only $CuI(PN)_2$ is obtained. The structure of $[CuI(PN)_3]$ was solved from powder pattern X-ray diffraction data even if the diffractogram acquired shows peaks of unwashed KI salt. The powder pattern is indexed by a trigonal cell with cell axes $a = 13.6471(2)$ Å, $c = 14.7400(4)$ Å volume = 2377 Å^3 and space group *P*-3. The volume of the asymmetric unit, 396 Å^3, is consistent with the volume of one third of $[CuI(PN)_3]$. This solution is possible only if the monomer is located on the 3-fold axis. The structure was solved with a simulated annealing algorithm using as starting model one molecule of the ligand not bonded to the metal atom and the copper atom and the iodide with a site occupancy of 1/3. The Rietveld refinement converged to $R_{wp} = 7.681\%$ and $R_{exp} = 4.351\%$ (see Fig. 5). The position of the nitrogen atoms was assumed from the structure solution since the packing features or the Rietveld refinement gave no hints on the possible position (see Fig. 6). The crystal packing of $[CuI(PN)_3]$ does not present important intermolecular interactions. It is worth noting that along the *c* axis the molecular disposition create cavities which can host solvent molecules as suggested also by the TGA curve (see Fig. 8a).

Several attempts have been made to determine the crystal structure of $CuI(PN)_2$, although as the diffractogram does not present peaks of the reagents or of other known phases we were not able to index it (see Fig. 7). It is not possible to exclude the presence of more than one phase.

The empirical formula is suggested by TGA and VTXRPD (see Fig. 8). X-Ray diffraction at variable temperature shows that $[CuI(PN)_3]$ starts to transform into $CuI(PN)_2$ at 180 °C and the complete transformation accompanied by amorphization is observed at 250 °C. In the TGA curve the first step which corresponds to a weight loss of 0.81% is probably due to some solvent trapped in the crystals, the second step due to the release of one ligand (calculated 26.84% weight loss) which is suddenly followed by decomposition and the release of the second and third ligand.

Photophysical properties

All complexes are luminescent in their solid state at room temperature with emissions spanning from green to orange. The tetrahedral coordination around the copper(I) ion and the stiff molecular structure assured by rigid ligands hamper the Jahn–Teller distortion of the excited state and allow high emission quantum yields.[39,40] Examination of the relevant bond angles reveals only a slightly distorted tetrahedral geometry around the Cu ions, demonstrating the fitting bite angle of the bidentate P^N ligand in the polynuclear complexes.[13] The lack of concentration quenching makes these complexes attractive for applications in solid state devices as OLEDs or LECs. Moreover their emissions stem from excited states populated by fast intersystem crossing, which allow the so-called "triplet harvesting" or "singlet harvesting" that leads to high electroluminescence efficiencies.[13,16,41–43]

The emission lifetimes are relatively short, in the order of microseconds, thus preventing excited state quenching due to triplet–triplet and triplet–charge interaction processes originating from the typical roll-off in the electroluminescence efficiency decay at high currents.[44]

The frontier orbitals involved in electronic transitions have a (X+M)LCT character, only slightly affected by the halogen/metal orbital ratio contributing to the HOMO, and dependent on the Cu_xI_x nuclearity, and by the number of bridging ligands that contribute to the LUMO.

The emission spectra of the complexes that could be obtained as sufficiently pure powders are represented in Fig. 9. The dimeric $[Cu_2I_2(PN)_3]$ complex shows an emission band with maximum at 525 nm which has been assigned to a temperature assisted delayed fluorescence (TADF) from a (X+M)LCT singlet excited state in thermal equilibrium with the triplet state lying at a very close energy level.[13] The monomer $[CuI(PN)_3]$ has a very similar emission band, most probably ascribable to a transition akin to the $[Cu_2I_2PN_3]$ one. None of the two compounds show radiative decay from excited states generated by cuprophilic interactions. Instead, the red-shifted and broader emission band of the $[Cu_4I_4(PN)_2\cdot(CH_2Cl_2)_{0.5}]$ octahedral complex is indicative of a cluster centred (CC) transition at lower energy contributing to the emission.[4]

Conclusions

Copper iodide complexes are known for their large variety of coordination geometries which allows a great deal of different stoichiometries to be attained with compounds based on the same ligand. One of the targets of the research described in this paper was that of exploring as thoroughly as possible the chemistry of CuI and the ligand diphenyl-2-pyridyl phosphine PN. Five new different compounds have been discussed: $[Cu_4I_4(PN)_2]$, $[Cu_4I_4(PN)_2\cdot(CH_2Cl_2)_{0.5}]$, $[CuI(PN)_{0.5}]_\infty$, $[CuI(PN)_3]$ whose structure has been determined during this study, and $CuI(PN)_2$ (see Scheme 1). Two main synthetic parameters were changed, the stoichiometry and the media in which reactions were carried out, whether in solution or the solid state. Regarding the stoichiometry, we have discovered that only by using a great excess of one of the reagents was it possible to move from the dimeric structure $[Cu_2I_2(PN)_3]$ and to obtain new compounds with higher ligand/CuI ratios. In terms of synthetic routes we have shown that solution synthesis yielded $[Cu_4I_4(PN)_2\cdot(CH_2Cl_2)_{0.5}]$ and $[CuI(PN)_3]$ while $[CuI(PN)_{0.5}]_\infty$ and $CuI(PN)_2$ were obtained only *via* solid state reactions, which posed the additional problem of structure determination when only polycrystalline samples were available. One way to circumvent this problem is that of finding an alternative way to grow single crystals of the desired species. This is why in our study we have explored in parallel the solvothermal synthesis with CuI and PN in an attempt to obtain the same species as produced by solid state reactions. In such a way we have obtained single crystals of $[Cu_4I_4(PN)_2]$ and $[CuI(PN)_{0.5}]_\infty$ which allowed full identification of this latter mechanochemical products by comparison of observed and simulated patterns. The observation that $[CuI(PN)_{0.5}]_\infty$ could be obtained by solid state reaction and by solvothermal methods suggests that the two processes can reach similar conditions which are not allowed in the traditional synthesis in solution. $[Cu_4I_4(PN)_2]$ and $[Cu_4I_4(PN)_2\cdot(CH_2Cl_2)_{0.5}]$ present very similar powder patterns which suggests that the solvated compound should maintain the same geometry, and upon thermal treatment $[Cu_4I_4(PN)_2\cdot(CH_2Cl_2)_{0.5}]$ converts into $[Cu_4I_4(PN)_2]$. The crystal structure of $[CuI(PN)_3]$ was solved from powder pattern X-ray diffraction data. Only the empirical formula of $CuI(PN)_2$ was postulated. This compound was obtained by ball milling or upon heating of $[CuI(PN)_3]$ at 180 °C, which corresponds to the TGA curve at the release of one ligand.

All the compounds obtained are luminescent and the photophysical properties of the complexes that could be obtained as sufficiently pure powders have been reported.

References

1 L. Bergmann, J. Friedrichs, M. Mydlak, T. Baumann, M. Nieger and S. Brase, *Chem. Commun.*, 2013, **49**, 6501–6503.

2 D. M. Zink, T. Baumann, J. Friedrichs, M. Nieger and S. Bräse, *Inorg. Chem.*, 2013.

3 D. M. Zink, D. Volz, T. Baumann, M. Mydlak, H. Flügge, J. Friedrichs, M. Nieger and S. Bräse, *Chem. Mater.*, 2013.

4 Z. Liu, P. I. Djurovich, M. T. Whited and M. E. Thompson, *Inorg. Chem.*, 2012, **51**, 230–236.

5 Z. Liu, M. F. Qayyum, C. Wu, M. T. Whited, P. I. Djurovich, K. O. Hodgson, B. Hedman, E. I. Solomon and M. E. Thompson, *J. Am. Chem. Soc.*, 2011, **133**, 3700–3703.
6 V. A. Krylova, P. I. Djurovich, J. W. Aronson, R. Haiges, M. T. Whited and M. E. Thompson, *Organometallics*, 2012, **31**, 7983–7993.
7 I. Roppolo, E. Celasco, A. Fargues, A. Garcia, A. Revaux, G. Dantelle, F. Maroun, T. Gacoin, J.-P. Boilot, M. Sangermano and S. Perruchas, *J. Mater. Chem.*, 2011, **21**, 19106–19113.
8 P. P. Mazzeo, L. Maini, D. Braga, G. Valenti, F. Paolucci, M. Marcaccio, A. Barbieri and B. Ventura, *Eur. J. Inorg. Chem.*, 2013, 4459–4465.
9 A. Barbieri, G. Accorsi and N. Armaroli, *Chem. Commun.*, 2008, 2185–2193.
10 R. Peng, M. Li and D. Li, *Coord. Chem. Rev.*, 2010, **254**, 1–18.
11 M. Vitale and P. C. Ford, *Coord. Chem. Rev.*, 2001, **219–221**, 3–16.
12 P. C. Ford, E. Cariati and J. Bourassa, *Chem. Rev.*, 1999, **99**, 3625–3648.
13 D. M. Zink, M. Bächle, T. Baumann, M. Nieger, M. Kühn, C. Wang, W. Klopper, U. Monkowius, T. Hofbeck, H. Yersin and S. Bräse, *Inorg. Chem.*, 2013, **52**, 2292–2305.
14 R. Czerwieniec, J. Yu and H. Yersin, *Inorg. Chem.*, 2011, **50**, 8293–8301.
15 D. Volz, D. M. Zink, T. Bocksrocker, J. Friedrichs, M. Nieger, T. Baumann, U. Lemmer and S. Bräse, *Chem. Mater.*, 2013, **25**, 3414–3426.
16 M. J. Leitl, F.-R. Küchle, H. A. Mayer, L. Wesemann and H. Yersin, *J. Phys. Chem. A*, 2013.
17 D. Volz, M. Nieger, J. Friedrichs, T. Baumann and S. Bräse, *Langmuir*, 2013.
18 C. H. Arnby, S. Jagner and I. Dance, *CrystEngComm*, 2004, **6**, 257–275.
19 D. J. Fife, W. M. Moore and K. W. Morse, *Inorg. Chem.*, 1984, **23**, 1684–1691.
20 *CSD*, 2013, Cambridge Crystallographic Data Centre, UK.
21 L. Maini, D. Braga, P. P. Mazzeo and B. Ventura, *Dalton Trans.*, 2012, **41**, 531–539.
22 S. L. James, C. J. Adams, C. Bolm, D. Braga, P. Collier, T. Friscic, F. Grepioni, K. D. M. Harris, G. Hyett, W. Jones, A. Krebs, J. Mack, L. Maini, A. G. Orpen, I. P. Parkin, W. C. Shearouse, J. W. Steed and D. C. Waddell, *Chem. Soc. Rev.*, 2012, **41**, 413–447.
23 D. Braga, L. Maini, P. P. Mazzeo and B. Ventura, *Chem.–Eur. J.*, 2010, **16**, 1553–1559.
24 D. Braga, F. Grepioni, L. Maini, P. P. Mazzeo and B. Ventura, *New J. Chem.*, 2011, **35**, 339–344.
25 G. A. Bowmaker, *Chem. Commun.*, 2013, **49**, 334–348.
26 G. A. Bowmaker, J. V. Hanna, R. D. Hart, P. C. Healy, S. P. King, F. Marchetti, C. Pettinari, B. W. Skelton, A. Tabacaru and A. H. White, *Dalton Trans.*, 2012, **41**, 7513–7525.
27 G. A. Bowmaker, J. V. Hanna, B. W. Skelton and A. H. White, *Chem. Commun.*, 2009, 2168–2170.
28 W. I. F. David, K. Shankland, L. M. McCusker and C. Baerlocher, *Structure Determination from Powder Diffraction Data*, Oxford University Press, New York, 2002.
29 G. A. Bowmaker and J. V. Hanna, *Z. Naturforsch., Teil B*, 2009, **64**, 1478–1486.
30 P. J. Beldon, L. Fábián, R. S. Stein, A. Thirumurugan, A. K. Cheetham and T. Friščić&cacute, *Angew. Chem., Int. Ed.*, 2010, **49**, 9640–9643.
31 R. E. Morris and S. L. James, *Angew. Chem., Int. Ed.*, 2013, **52**, 2163–2165.

32 Y. Jin, Q. Sun, G. Qi, C. Yang, J. Xu, F. Chen, X. Meng, F. Deng and F.-S. Xiao, *Angew. Chem.*, 2013, **125**, 9342–9345.

33 J. F. Fernández-Bertrán, M. P. Hernández, E. Reguera, H. Yee-Madeira, J. Rodriguez, A. Paneque and J. C. Llopiz, *J. Phys. Chem. Solids*, 2006, **67**, 1612–1617.

34 C. F. Macrae, I. J. Bruno, J. A. Chisholm, P. R. Edgington, P. McCabe, E. Pidcock, L. Rodriguez-Monge, R. Taylor, J. van de Streek and P. A. Wood, *J. Appl. Crystallogr.*, 2008, **41**, 466–470.

35 G. M. Sheldrick, *SHELX97*, 1997, University of Göttingen, Germany.

36 S. L. Childs, N. Rodriguez-Hornedo, L. S. Reddy, A. Jayasankar, C. Maheshwari, L. McCausland, R. Shipplett and B. C. Stahly, *CrystEngComm*, 2008, **10**, 856–864.

37 D. Volz, M. Nieger, J. Friedrichs, T. Baumann and S. Bräse, *Inorg. Chem. Commun.*, 2013, **37**, 106–109.

38 A. J. Blake, N. R. Brooks, N. R. Champness, P. A. Cooke, M. Crew, A. M. Deveson, L. R. Hanton, P. Hubberstey, D. Fenske and M. Schröder, *Cryst. Eng.*, 1999, **2**, 181–195.

39 M. Hashimoto, S. Igawa, M. Yashima, I. Kawata, M. Hoshino and M. Osawa, *J. Am. Chem. Soc.*, 2011, **133**, 10348–10351.

40 N. Armaroli, G. Accorsi, F. Cardinali and A. Listorti, *Top. Curr. Chem.*, 2007, **280**, 69–115.

41 M. A. Baldo, S. Lamansky, P. E. Burrows, M. E. Thompson and S. R. Forrest, *Appl. Phys. Lett.*, 1999, **75**, 4–6.

42 M. A. Baldo, D. F. O'Brien, Y. You, A. Shoustikov, S. Sibley, M. E. Thompson and S. R. Forrest, *Nature*, 1998, **395**, 151–154.

43 C.-W. Hsu, C.-C. Lin, M.-W. Chung, Y. Chi, G.-H. Lee, P.-T. Chou, C.-H. Chang and P.-Y. Chen, *J. Am. Chem. Soc.*, 2011, **133**, 12085–12099.

44 *Highly Efficient OLEDs with Phosphorescent Materials*, ed. H. Yersin, Wiley-VCH, Weinheim, 2008.

Faraday Discussions

PAPER

Mechanochemical synthesis and characterisation of cocrystals and metal organic compounds

L. Tröbs and F. Emmerling*

Received 30th December 2013, Accepted 26th February 2014
DOI: 10.1039/c3fd00163f

The mechanochemical synthesis of two model compounds, a metal organic framework $(H_2Im)[Bi(1,4\text{-bdc})_2]$ (bdc = benzene dicarboxylate, H_2Im = imidazole cation)) and a cocrystal (carbamazepine : indometacin 1 : 1) were followed ex situ using a combination of two analytical methods. Powder X-ray diffraction (XRD) and Raman spectroscopy data were evaluated for the synthesis of the metal organic framework. The XRD measurements and REM images were analysed for the synthesis of the cocrystal. The measurements revealed that both model compounds were synthesised within minutes. The metal organic framework $(H_2Im)[Bi(1,4\text{-bdc})_2]$ is synthesised *via* an intermediate structure. The cocrystal carbamazepine : indometacin 1 : 1 is formed within a few seconds. The crystallite size decreases during the further milling treatment.

Introduction

Mechanochemistry has gained increasing importance for the synthesis of novel compounds in recent years.[1–3] Mechanochemical reactions provide many advantages over their liquid-phase counterparts in terms of shorter reaction times, higher product yields and the elimination of (harmful) organic solvents. Among other methods, mechanochemistry is now an established method for the synthesis of metal organic frameworks (MOFs) and cocrystals.[4–8]

MOFs are compounds consisting of metal or metal-cluster centres connected *via* organic molecules resulting in a porous coordination system. MOFs offer a wide range of interesting properties and applications.[9,10] In particular, they are characterized by large inner surfaces and pronounced porous systems. Based on these properties, MOFs can be utilized in different sectors, such as drug-carrier processes, catalysis, gas storage and gas purification.[9–14] Because of the enormous number of available organic ligands and metal centers, MOFs are characterized by a great structural diversity. The use of metal cations with higher and flexible

BAM Federal Institute for Materials Research and Testing, Richard-Willstätter-Str. 11, 12489 Berlin, Germany. E-mail: franziska.emmerling@bam.de; Fax: +49-30-8104-1139; Tel: +49-30-8104-1133

coordination numbers such as bismuth(III) can additionally increase the flexibility of the resulting structure.[15,16]

Cocrystals are organic compounds consisting of at least two neutral organic molecules, which are connected *via* intermolecular interactions such as π–π stacking, van der Waals forces and hydrogen bonds. Pharmaceutical cocrystals consist of an active pharmaceutical ingredient (API) and a coformer, typically a small organic molecule.[17–19] Cocrystallisation is used to enhance the physicochemical properties of the APIs, including the stability, solubility, compressibility and bioavailability.[19,20]

For both material classes, mechanochemical reactions offer an effective and environmentally-friendly synthesis pathway, preventing the use of solvents and achieving high material yields. Furthermore, the products are of high purity and the reactions periods are typically completed within minutes. Although, mechanochemistry is applied successfully for the synthesis of these materials, the underlying mechanisms of the milling processes have not been clarified yet. Theories explaining the milling process range from intermediate melts or eutectics to predictions that reactions are occurring *via* plasma or hot-spots which are induced by the collision of the grinding balls and the grinding jar surfaces.[2,21–24]

In an elegant experiment, Friščić, and co-workers succeeded in directly monitoring milling reactions using synchrotron XRD.[25,26] The authors used a modified ball mill equipped with a transparent Perspex jar, which allows them to follow the synthesis of the MOFs and cocrystals *in situ*. Further studies have been conducted in terms of an *ex situ* analysis of grinding reactions using Raman spectroscopy or solid-state fluorescence measurements.[27,28]

In a different approach, we evaluate the possibility of lab scale XRD measurements combined with further methods to elucidate the milling synthesis of two example compounds. The synthesis of the metal organic framework $(H_2Im)[Bi(1,4\text{-bdc})_2]$[29] (bdc = benzene dicarboxylate, H_2Im = imidazole cation) was investigated based on *ex situ* XRD measurements and Raman spectroscopy. The synthesis of the cocrystal carbamazepine : indometacin 1 : 1[30] was followed *via ex situ* XRD and REM measurements at different steps during the grinding process. Based on the obtained data it was possible to derive a mechanistic understanding of the formation processes, occurring intermediates and reaction periods.

Results and discussion

The crystal structures of both model compounds, $(H_2Im)[Bi(1,4\text{-bdc})_2]$ and carbamazepine : indometacin 1 : 1, are shown in Fig. 1. These compounds were chosen to meet the following criteria: i) a typical example for the respective material class, ii) known crystalline structure, and iii) clearly distinguishable scattering areas for educts and products in a limited 2θ range. The mechanochemical syntheses of the compounds were carried out in a conventional ball mill. The reactions were investigated *ex situ* by a combination of different analytical methods. For both syntheses XRD measurements were performed and backed by either Raman measurements ($(H_2Im)[Bi(1,4\text{-bdc})_2]$) or REM investigations (carbamazepine : indometacin 1 : 1). In a typical experiment, the reactions were stopped at different time intervals and a small amount of sample was retrieved for analysis.

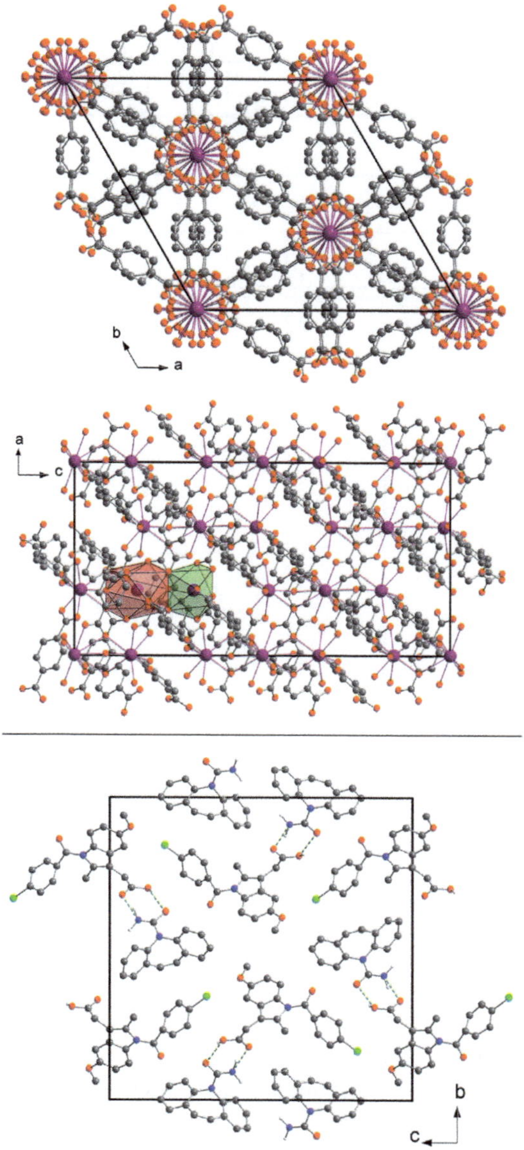

Fig. 1 Top: a view of the unit cell of $(H_2Im)[Bi(1,4\text{-bdc})_2]$ seen along the *c*- and *b*-axes with the coordination polyhedra of both Bi^{3+} cations in the asymmetric unit shown in green (Bi1) and red (Bi2). The hydrogen atoms were omitted for clarity. A detailed description of the crystal structure can be found elsewhere.[29] Bottom: a view of the unit cell of the cocrystal CBZ : indometacin 1 : 1 seen along the *a*-axis. The green dotted lines indicate the hydrogen bonds. The hydrogen atoms not participating in the hydrogen bonds were omitted for clarity. A detailed description of the crystal structure can be found elsewhere.[30]

The XRD patterns obtained during the synthesis of $(H_2Im)[Bi(1,4\text{-bdc})_2]$ are shown in Fig. 2 (reaction 1). Until a reaction period of 3 min was reached, samples were extracted every 10 s. Afterwards, the samples were taken every 30 s. After a reaction period of 20 s, the XRD measurements indicate the presence of an

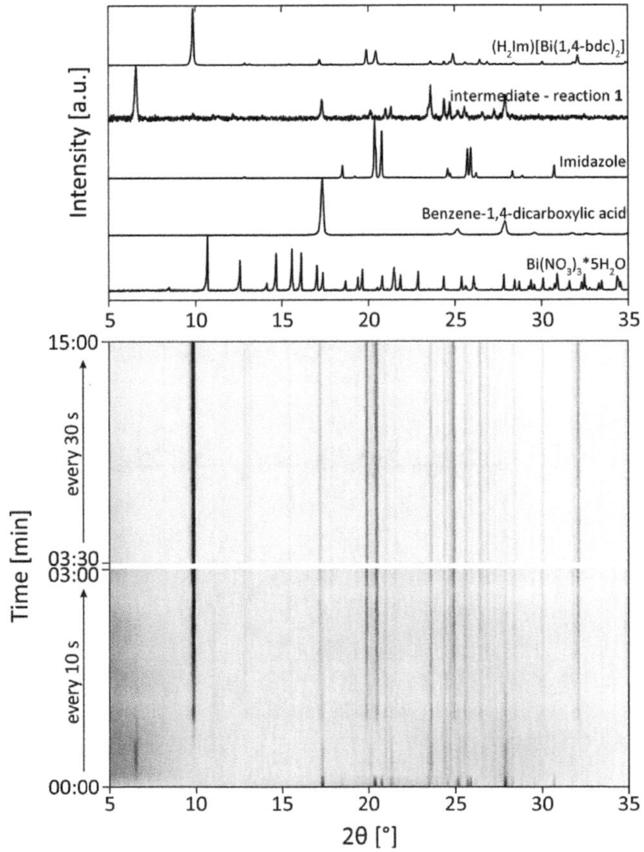

Fig. 2 The diffractograms at milling times between 0–15 min during the mechanochemical synthesis of (H$_2$Im)[Bi(1,4-bdc)$_2$]. For comparison the XRD patterns of (H$_2$Im)[Bi(1,4-bdc)$_2$], the intermediate of reaction 1, bismuth(III) nitrate pentahydrate (Bi(NO$_3$)$_3$*5H$_2$O), benzene-1,4-dicarboxylic acid and imidazole are depicted above.

intermediate which is characterized by an intensive reflection at 6.5° 2θ (intermediate - reaction 1). The corresponding XRD pattern revealed, that the benzene-1,4-dicarboxylic acid is still present in the reaction mixture, whereas the reflections of bismuth(III) nitrate pentahydrate and imidazole vanished (Fig. 2, top). It can be assumed, that an intermediate was formed at the beginning of the reaction. The diffractogram corresponds rather well with the diffraction pattern of the basic bismuth nitrate $Bi_6O_5(OH)_3(NO_3)_5(H_2O)_3$.[31,32] Based on this assignment, it is reasonable that the first step in the formation of the framework consists of a release of H_2O and NO_x from bismuth(III) nitrate pentahydrate along with a protonation of imidazole. The resulting coordination sphere around the bismuth atoms consists solely of water molecules. Afterwards, this intermediate reacts with benzene-1,4-dicarboxylic acid to form the metal organic framework (H$_2$Im)[Bi(1,4-bdc)$_2$]. The product is formed after a reaction period of 2 min. The corresponding crystallite size amounts to 108 nm. The crystallite size of the product at different time periods was calculated based on the Scherrer equation.[33,34] The product remains stable until the end of the grinding process at 15 min, indicating no significant changes in the crystallite size. Obviously, a longer grinding process is not necessary for the completeness of the reaction and the crystallite size is not influenced significantly by a prolonged reaction period.

These findings are supported by the simultaneously collected Raman spectra (Fig. 3). The Raman spectra show that the formation of $(H_2Im)[Bi(1,4\text{-bdc})_2]$ is completed after a milling time of 2 min. The protonation of the imidazole molecule can be verified based on the Raman spectra. Two intensive bands at 1220 cm^{-1} and 1445 cm^{-1} characterize the spectrum of the imidazolium ion (blue bars in Fig. 3).[35] These bands are detectable after 30 s, indicating that the protonation of imidazole proceeds in the first seconds of the milling reaction. The imidazolium molecule might act as a template, triggering the formation of the MOF. The Raman spectra of $(H_2Im)[Bi(1,4\text{-bdc})_2]$ show similar Raman bands compared to the Raman spectra of the educts benzene-1,4-dicarboxylic acid and imidazole. They are clearly shifted in the Raman spectrum of the product due to the different coordination spheres of the participating molecules.

Similar experiments were conducted for the mechanochemical synthesis of the cocrystal carbamazepine : indometacin 1 : 1 (reaction 2a). The diffractograms recorded during the milling process are shown in Fig. 4. At the beginning of the reaction, the XRD measurements were performed every 10 s, after 3 minutes every 30 s, and after 10 minutes every 60 s. The diffractograms indicate that the formation of the cocrystal is completed after 1 min. Obviously, the crystal structure of both educts is destroyed and the new structure is formed within these first seconds of the reaction. Crystalline intermediates could not be detected. A closer

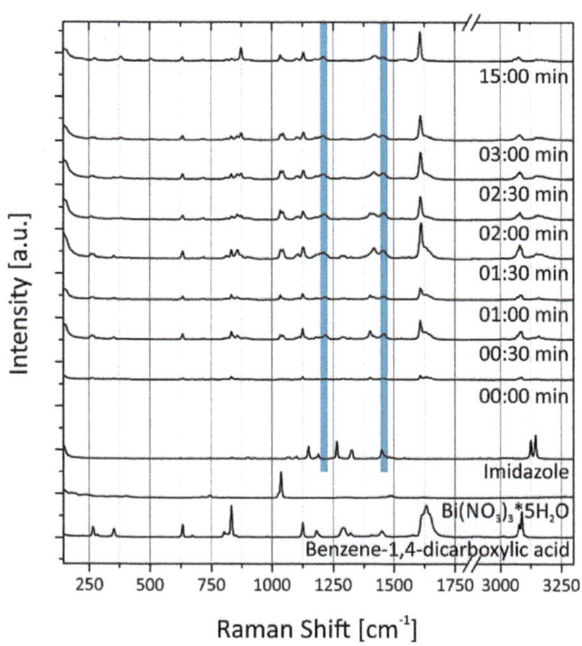

Fig. 3 The Raman spectra recorded during the mechanochemical synthesis of (H_2Im) $[Bi(1,4\text{-bdc})_2]$ in comparison to the Raman spectra of the educts bismuth(III) nitrate pentahydrate ($Bi(NO_3)_3{*}5H_2O$), benzene-1,4-dicarboxylic acid and imidazole. The blue bars indicate the characteristic bands for the imidazolium molecule.

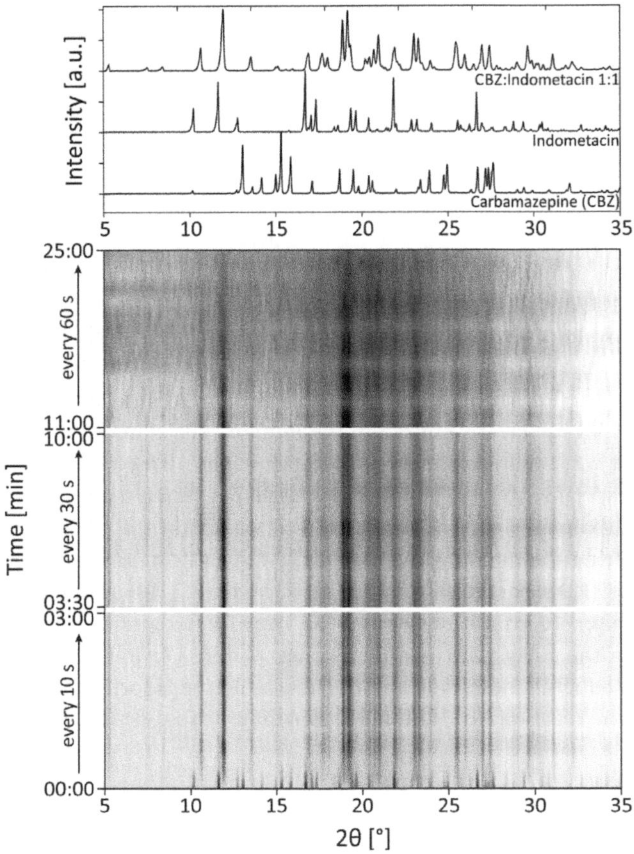

Fig. 4 The diffractograms at milling times between 0–25 min during the mechanochemical synthesis of the cocrystal carbamazepine : indometacin 1 : 1. For comparison the XRD patterns of carbamazepine : indometacin 1 : 1, carbamazepine and indometacin are depicted above.

investigation of the reflections indicates an increasing width of the reflections with a prolonged milling time. After a milling time of 1 min, the crystallite size amounts to 61 nm. After 5 min a crystallite size of 44 nm was determined, and after 15 min a crystallite size of 22 nm was determined. After 25 min, the crystallite size was below 5 nm. These findings indicate that prolonged milling can destroy the intermediately formed crystallites, whereas large crystals are formed within the first seconds of the experiment.

REM images were taken at different steps of the mechanochemical synthesis, to monitor the development of the crystal morphologies (Fig. 5). Before the start of the reaction (0 s), the two cocrystal components can be clearly separated from each other within the reaction mixture. The indometacin crystals are plate-like whereas the carbamazepine crystals appear powder-like with no distinct morphology. After a reaction time of 30 s, the REM images clearly indicate the presence of needle-shaped product crystallites next to educt crystallites, mainly indometacin plates. After 60 s only needle-shaped crystals can be identified in the REM image. This observation is in accordance with the XRD experiment. Here,

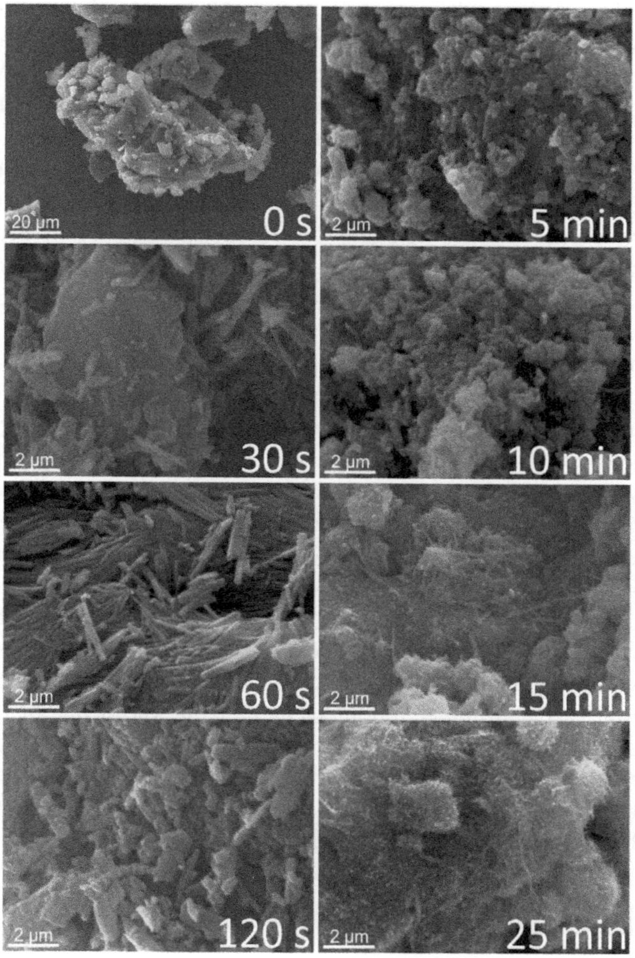

Fig. 5 REM images of different time periods of the mechanochemical synthesis of the cocrystal carbamazepine : indometacin 1 : 1.

the formation of the cocrystal is completed after 60 s. No traces of the educt crystallites could be detected. The length of the needles is in the range of 2000–4000 nm. After 120 s the length of the needles decreases (1500–3000 nm). Prolonged milling leads to a further decrease in the length of the needles (700–1000 nm after 5 min, and 600–700 nm after 10 min). After a milling reaction of 15 min and 25 min, only a few needles are detectable, whereas the majority of the sample consists of small powder-like crystallites. These findings are in accordance with the decreasing crystallite size determined by the XRD measurements. The combination of both analytical methods reveals the formation of needle shaped crystals after a reaction period of 60 s and the destruction of the crystals during the further milling process.

Keeping in mind that a mechanochemical process conducted in a stepwise manner might yield a different result compared to an uninterrupted one,[36] an additional experiment was performed. Here, the reactants carbamazepine and indometacin were ground together for 1 min. Afterwards, the reaction mixture was stored for four days in the closed grinding jar at ambient conditions to allow an uninfluenced crystallisation of the cocrystal (reaction 2b). An REM image of the final product of reaction 2b is shown in Fig. 6 in comparison to an REM image of the cocrystal synthesised in reaction 2a after a milling time of 60 s. The product from reaction 2b consists of long needles, which are interwoven into each other. The thickness of the needles obtained in both reactions is comparable whereas the length of the needles crystallised without interruption (reaction 2b) is significantly larger. The diffractogram of the product derived from reaction 2b indicates the completeness of the reaction. The crystallite size was determined to be 51 nm. These findings suggest that a short initial milling period followed by an uninfluenced crystallisation could lead to the same but more crystalline product.

The analysis of the synthesis of both model compounds revealed that mechanochemistry is a very fast synthesis method. Both products were gained after milling times of 2 min (($(H_2Im)[Bi(1,4\text{-bdc})_2]$) or 1 min (CBZ : indometacin 1 : 1). Especially the synthesis of the cocrystal proceeded within a few seconds. The speed of the cocrystallisation can be attributed to the fact that the intermolecular interactions between the organic molecules are more a coordination of the cocrystal components to each other.

Based on the data obtained from the combination of different analytical methods, it was possible to derive a mechanistic understanding of the formation processes, intermediates and reaction periods. In particular for the synthesis of the cocrystal carbamazepine : indometacin 1 : 1 the knowledge of the optimal reaction periods is essential to receive a product of high crystallinity. Amorphous, liquid, or eutectic intermediates could not be detected for both syntheses. The propensity to form a product could be associated to an increase in the contact areas with decreasing particle size during the grinding process.

Experimental

Materials

Bismuth(III) nitrate pentahydrate $Bi(NO_3)_3 \cdot 5H_2O$ (98%), benzene-1,4-dicarboxylic acid (terephthalic acid) $C_6H_4(COOH)_2$ (98+%) and imidazole $C_3H_4N_2$ (99%) were purchased from Alfa Aesar, Karlsruhe, Germany and used without further purification. Carbamazepine $C_{15}H_{12}N_2O_2$ (≥99.5%, Fluka, Steinheim, Germany),

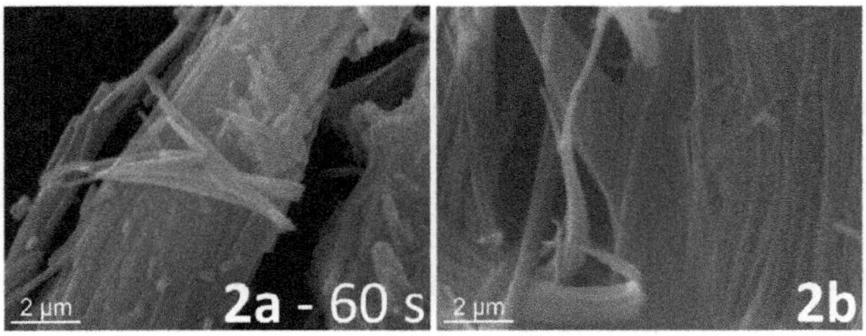

Fig. 6 REM images for comparison of the crystalline products (cocrystal carbamazepine : indometacin 1 : 1) from reaction 2a after 60 s grinding time and from reaction 2b. In the latter reaction, the cocrystal was synthesised by 60 s grinding and four days storage at ambient conditions in the closed grinding jar.

indometacin $C_{19}H_{16}ClNO_4$ (≥99%, Sigma, Steinheim, Germany) and methanol (99.9%, Roth, Karlsruhe, Germany) were also purchased commercially and used without further purification.

Sample preparation

$(H_2Im)[Bi(1,4\text{-}bdc)_2]$ (reaction 1). $Bi(NO_3)_3 \cdot 5H_2O$ (590.0 mg, 1 eq.), benzene-1,4-dicarboxylic acid (404.1 mg, 2 eq.), and imidazole (331.2 mg, 4 eq.) were ground together in a conventional ball mill (Pulverisette 23, FRITSCH GmbH, Idar-Oberstein, Germany) with a frequency of 50 Hz using a 10 mL steel grinding bowl and two steel balls (10 mm diameter). The reaction was interrupted at different points and an appropriate amount of the sample was used for X-ray analysis and Raman spectroscopy. Sampling intervals were chosen as follows: 00:00 min–03:00 min every 10 s; 03:30 min–15:00 min every 30 s. The product was obtained as a white powder.

CBZ : indometacin 1 : 1 (reaction 2). Carbamazepine (0.1989 g, 1 eq.) and indometacin (0.3011 g, 1 eq.) were ground together after adding 250 μL methanol. The reaction was performed in a conventional ball mill (Pulverisette 23, FRITSCH GmbH, Idar-Oberstein, Germany) at 30 Hz using a 10 mL steel grinding jar and two steel balls (10 mm diameter). Two types of experiment were conducted to analyse this reaction:

(a) To follow the reaction by X-ray analysis and REM measurements, the following sampling scheme was applied: 00:00 min–03:00 min every 10 s; 03:30 min–10:00 min every 30 s; 11:00 min–25:00 min every 60 s. At these steps, the reaction was interrupted and a small amount of the sample was taken from the reaction mixture.

(b) The educts were ground for 60 s before the grinding jar was placed under ambient conditions (20 °C) without opening the jar for four days. The product was obtained as a white powder and analysed by XRD and REM measurements.

Methods

Powder diffraction measurements. Powder diffraction measurements were performed on a D8 Discover diffractometer (Bruker AXS, Karlsruhe, Germany) equipped with a Lynxeye detector and operated in transmission geometry (Cu-Kα1 radiation. $\lambda = 0.154056$ nm). The samples were sealed in a glass capillary of 0.5 mm diameter (WJM-Glas, Müller GmbH, Berlin, Germany) and the powder diffraction data were collected at room temperature. Measurements were conducted over a 2θ range of 5–50° with a step size of 0.009° and 0.05 s per step for the *ex situ* measurements and 0.5 to 5.0 s per step for the educts and final products. The crystallite size was calculated using the Scherrer equation.[33,34]

Raman spectroscopy. The Raman spectra were collected using a Raman RXN1™ Analyzer (Kaiser Optical Systems, Inc., Ecully, France) with NIR excitation radiation at 785 nm. The spectrometer is equipped with a non-contact probe head and a CCD camera (1024 × 256 pixels). The working distance is 15 mm and the spot size on the sample has a diameter of 1 mm. The Raman spectra were recorded with an acquisition time of 5 × 5 s and an irradiance of 6.6 W cm^{-2} on the sample.

REM measurements. The REM images were obtained using a scanning electron microscope XL30 ESEM (FEI, Gräfelfing, Germany) equipped with a

W-cathode (20 kV) and adapted with a secondary electron detector (Everhard-Thornlay detector) working in a high vacuum. In addition, the scanning electron microscope is equipped with an energy dispersive X-ray spectrometer (EDAX, Wiesbaden, Germany) using a SiLi detector. Prior to the REM measurements, the samples were sealed on carbon tabs and sputtered with gold (15 nm min^{-1}) using a Scancoat Six (Edwards, Kirchheim, Germany).

Conclusions

The presented *ex situ* approach proved to be completely suitable to determine the formation processes during mechanochemical reactions. The combination of two analytical techniques provides further information for a thorough understanding of the processes. Surprisingly, both model compounds are synthesised very fast within the first minutes of the milling process. The synthesis of the metal organic framework $(H_2Im)[Bi(1,4\text{-bdc})_2]$ proceeds *via* the formation of an intermediate. The analysis of the synthesis of the cocrystal carbamazepine : indometacin 1 : 1 revealed that the initially formed crystallites are destroyed during the prolonged milling times. The *in situ* monitoring analysis has the potential to elucidate the mechanisms of the milling reactions and provide optimized synthesis parameters.

Acknowledgements

We are grateful to Mrs Feldmann for providing the REM measurements.

Notes and references

1 D. Braga, S. L. Giaffreda, F. Grepioni, A. Pettersen, L. Maini, M. Curzi and M. Polito, *Dalton Trans.*, 2006, 1249–1263.

2 S. L. James, C. J. Adams, C. Bolm, D. Braga, P. Collier, T. Friščić, F. Grepioni, K. D. M. Harris, G. Hyett, W. Jones, A. Krebs, J. Mack, L. Maini, A. G. Orpen, I. P. Parkin, W. C. Shearouse, J. W. Steed and D. C. Waddell, *Chem. Soc. Rev.*, 2012, **41**, 413–447.

3 T. Friščić, *Chem. Soc. Rev.*, 2012, **41**, 3493–3510.

4 A. Delori, T. Friščić and W. Jones, *CrystEngComm*, 2012, **14**, 2350–2362.

5 D. R. Weyna, T. Shattock, P. Vishweshwar and M. J. Zaworotko, *Cryst. Growth Des.*, 2009, **9**, 1106–1123.

6 S. Heiden, L. Tröbs, K. J. Wenzel and F. Emmerling, *CrystEngComm*, 2012, **14**, 5128–5129.

7 M. Klimakow, P. Klobes, A. F. Thunemann, K. Rademann and F. Emmerling, *Chem. Mater.*, 2010, **22**, 5216–5221.

8 E. Boldyreva, *Chem. Soc. Rev.*, 2013, **42**, 7719–7738.

9 G. Ferey, *Chem. Soc. Rev.*, 2008, **37**, 191–214.

10 J. L. C. Rowsell and O. M. Yaghi, *Microporous Mesoporous Mater.*, 2004, **73**, 3–14.

11 J. L. C. Rowsell and O. M. Yaghi, *Angew. Chem., Int. Ed.*, 2005, **44**, 4670–4679.

12 A. C. McKinlay, R. E. Morris, P. Horcajada, G. Ferey, R. Gref, P. Couvreur and C. Serre, *Angew. Chem., Int. Ed.*, 2010, **49**, 6260–6266.

13 P. Horcajada, C. Serre, M. Vallet-Regi, M. Sebban, F. Taulelle and G. Ferey, *Angew. Chem., Int. Ed.*, 2006, **45**, 5974–5978.

14 M. Eddaoudi, J. Kim, N. Rosi, D. Vodak, J. Wachter, M. O'Keeffe and O. M. Yaghi, *Science*, 2002, **295**, 469–472.
15 A. Thirumurugan and A. K. Cheetham, *Eur. J. Inorg. Chem.*, 2010, 3823–3828.
16 A. C. Wibowo, M. D. Smith and H. C. zur Loye, *Cryst. Growth Des.*, 2011, **11**, 4449–4457.
17 N. Qiao, M. Li, W. Schlindwein, N. Malek, A. Davies and G. Trappitt, *Int. J. Pharm.*, 2011, **419**, 1–11.
18 N. Schultheiss and A. Newman, *Cryst. Growth Des.*, 2009, **9**, 2950–2967.
19 T. Friščić and W. Jones, *J. Pharm. Pharmacol.*, 2010, **62**, 1547–1559.
20 S. Karki, T. Friščić, L. Fabian, P. R. Laity, G. M. Day and W. Jones, *Adv. Mater.*, 2009, **21**, 3905–3909.
21 P. Balaz, *Mechanochemistry in Nanoscience and Minerals Engineering, Springer-Verlag*, Berlin, Heidelberg, 2008.
22 J. F. Fernandez-Bertran, *Pure Appl. Chem.*, 1999, **71**, 581–586.
23 F. K. Urakaev and V. V. Boldyrev, *Powder Technol.*, 2000, **107**, 93–107.
24 D. Braga, L. Brammer and N. R. Champness, *CrystEngComm*, 2005, **7**, 1–19.
25 T. Friščić, I. Halasz, P. J. Beldon, A. M. Belenguer, F. Adams, S. A. J. Kimber, V. Honkimaki and R. E. Dinnebier, *Nat. Chem.*, 2013, **5**, 66–73.
26 I. Halasz, A. Puskaric, S. A. J. Kimber, P. J. Beldon, A. M. Belenguer, F. Adams, V. Honkimaki, R. E. Dinnebier, B. Patel, W. Jones, V. Strukil and T. Friščić, *Angew. Chem., Int. Ed.*, 2013, **52**, 11538–11541.
27 M. Frenette, G. Cosa and T. Friščić, *CrystEngComm*, 2013, **15**, 5100–5106.
28 S. Rehder, M. Klukkert, K. A. M. Löbmann, C. J. Strachan, A. Sakmann, K. Gordon, T. Rades and C. S. Leopold, *Pharmaceutics*, 2011, **3**, 706–722.
29 L. Tröbs, M. Wilke, W. Szczerba, U. Reinholz and F. Emmerling, *CrystEngComm*, 2014, **16**, 5560.
30 L. Tröbs, N. Zientek, D. Rump, M. Maiwald and F. Emmerling, 2014, in progress.
31 F. Lazarini, *Acta Crystallogr., Sect. B: Struct. Crystallogr. Cryst. Chem.*, 1978, **34**, 3169–3173.
32 B. Liu, W. W. Zhou, Z. Q. Zhou and X. Y. Zhang, *Inorg. Chem. Commun.*, 2007, **10**, 1145–1148.
33 P. Scherrer, in *Nachrichten von der Gesellschaft der Wissenschaften*, Akademie der Wissenschaften, Göttingen, 1918, vol. 2, pp. 98–100.
34 H. P. Klug and L. E. Alexander, *X-Ray Diffraction Procedures: For Polycrystalline and Amorphous Materials*, 2nd edn, Wiley-Interscience, New York, 1974
35 L. M. Markham, L. C. Mayne, B. S. Hudson and M. Z. Zgierski, *J. Phys. Chem.*, 1993, **97**, 10319–10325.
36 V. Strukil, L. Fabian, D. G. Reid, M. J. Duer, G. J. Jackson, M. Eckert-Maksic and T. Friščić, *Chem. Commun.*, 2010, **46**, 9191–9193.

Faraday Discussions

PAPER

The mechanically induced structural disorder in barium hexaferrite, $BaFe_{12}O_{19}$, and its impact on magnetism

V. Šepelák,†*[ab] M. Myndyk,[c] R. Witte,[a] J. Röder,[d] D. Menzel,[e] R. H. Schuster,[bf] H. Hahn,[a] P. Heitjans[bg] and K.-D. Becker[bh]

Received 13th December 2013, Accepted 15th January 2014
DOI: 10.1039/c3fd00137g

The response of the structure of the M-type barium hexaferrite ($BaFe_{12}O_{19}$) to mechanical action through high-energy milling and its impact on the magnetic behaviour of the ferrite are investigated. Due to the ability of the ^{57}Fe Mössbauer spectroscopic technique to probe the environment of the Fe nuclei, a valuable insight on a local atomic scale into the mechanically induced changes in the hexagonal structure of the material is obtained. It is revealed that the milling of $BaFe_{12}O_{19}$ results in the deformation of its constituent polyhedra (FeO_6 octahedra, FeO_4 tetrahedra and FeO_5 triangular bi-pyramids) as well as in the mechanically triggered transition of the Fe^{3+} cations from the regular 12k octahedral sites into the interstitial positions provided by the magnetoplumbite structure. The response of the hexaferrite to the mechanical treatment is found to be accompanied by the formation of a non-uniform nanostructure consisting of an ordered crystallite surrounded/separated by a structurally disordered surface shell/interface region. The distorted polyhedra and the non-equilibrium cation distribution are found to be confined to the amorphous near-surface layers of the ferrite nanoparticles with the thickness extending up to about 2

[a]*Institute of Nanotechnology, Karlsruhe Institute of Technology (KIT), Hermann-von-Helmholtz-Platz 1, D-76344 Eggenstein-Leopoldshafen, Germany. E-mail: vladimir.sepelak@kit.edu; Fax: +49-721-60826368; Tel: +49-721-60828929*

[b]*Center for Solid State Chemistry and New Materials, Leibniz University Hannover, Callinstr. 3-3a, D-30167 Hannover, Germany*

[c]*Department of Inorganic Chemistry, Dresden University of Technology, Mommsenstr. 6, D-01062 Dresden, Germany*

[d]*Physics Department, European Council for Nuclear Research (CERN), CH-1211 Geneva 23, Switzerland*

[e]*Institute of Condensed Matter Physics, Braunschweig University of Technology, Mendelssohnstr. 3, D-38106 Braunschweig, Germany*

[f]*German Institute of Rubber Technology (DIK), Eupener Str. 33, D-30519 Hannover, Germany*

[g]*Institute of Physical Chemistry and Electrochemistry, Leibniz University Hannover, Callinstr. 3-3a, D-30167 Hannover, Germany*

[h]*Institute of Physical and Theoretical Chemistry, Braunschweig University of Technology, Hans-Sommer-Str. 10, D-38106 Braunschweig, Germany*

† On leave from the Slovak Academy of Sciences, Košice, Slovakia.

nm. The information on the mechanically induced short-range structural disorder in $BaFe_{12}O_{19}$ is complemented by an investigation of its magnetic behaviour on a macroscopic scale. It is demonstrated that the milled ferrite nanoparticles exhibit a pure superparamagnetism at room temperature. As a consequence of the far-from-equilibrium structural disorder in the surface shell of the nanoparticles, the mechanically treated $BaFe_{12}O_{19}$ exhibits a reduced magnetization and an enhanced coercivity.

Introduction

Mechanically induced chemistry (the so-called mechanochemistry) as a branch of chemistry, being concerned with the chemical and physical transformations of solids induced by mechanical action, continues to be of high importance.[1] Among the inorganic materials susceptible to mechanical action, oxides exhibit a wide range of responses.[2] In particular, complex oxides possessing more than one cation sublattice, such as spinels, olivines and mullites (with two cation sublattices), have been considered as convenient model systems for the investigation of mechanically induced processes in ionic systems, because of their structural flexibility providing a wide range of physical and chemical behaviours.[3] Despite their deceptively simple structure, many oxides exhibit complex disordering phenomena under the influence of mechanical action. These involve, for example, (i) the mechanically induced redistribution of cations over non-equivalent cation sublattices provided by a complex oxide structure, (ii) the formation of canted spin arrangements in the case of magnetic compounds, (iii) the changes in the geometry of constitutive polyhedra, and (iv) the mechanically triggered formation of defective cation centers with an unsaturated oxygen coordination.[3,4]

Although there is a surge of investigations in the field of mechanochemistry of oxides, studies on the mechanically induced response of complex oxides possessing more than two cation sublattices, such as M-type barium hexaferrite, $BaFe_{12}O_{19}$ (with five cation sublattices), are very scarce in the literature.[5,6] As a result of its specific magnetic properties, $BaFe_{12}O_{19}$ is widely used in permanent magnets, magnetic recording media and microwave applications. Its derivatives are currently magnetic materials with great scientific and technological interests due to their relatively high Curie temperature, high coercive force and high magnetic anisotropy field as well as their excellent chemical stability and corrosion resistivity. Recently, multiferroic properties have been reported for $BaFe_{12}O_{19}$ ceramics.[7] In this article, for the first time, detailed information is obtained on the response of the local (short-range) structure of $BaFe_{12}O_{19}$ to mechanical action through high-energy milling. In addition to ^{57}Fe Mössbauer spectroscopy, the evolution of the mechanically induced structural disorder as well as the morphology and macroscopic magnetic behaviour of the ferrite were monitored with comprehensive techniques including X-ray diffraction (XRD), high-resolution transmission electron microscopy (TEM), and superconducting quantum interference device (SQUID) magnetometry.

Experimental

M-type $BaFe_{12}O_{19}$, which was used as the starting material for subsequent mechanochemical treatment, was synthesized using a chemically reliable co-precipitation route.[8] The precursor materials of barium and iron chlorides (99.9% purity; Merck, Darmstadt, Germany) were dissolved in deionized water, and subsequently solutions of Na_2CO_3 and NaOH were added to achieve pH = 6. The precipitated precursor was sintered at 1173 K to obtain bulk $BaFe_{12}O_{19}$ (which is further referred to as the non-treated material). Fig. 1 shows the morphology of the as-prepared bulk $BaFe_{12}O_{19}$ powder, which served as a reference sample in the present study. The particles of the co-precipitated ferrite were found to be uniform in shape; the majority of them are hexagonal platelet crystals. Whereas the thickness of the ferrite platelets is in the 20 – 100 nm range, their length ranges from 300 to 400 nm. The corresponding selected area electron diffraction (SAED) pattern of the ferrite (Fig. 1b) is dominated by the discrete diffraction spots originating from the well crystalline hexagonal platelets.

10 grams of the bulk material were mechanically treated for various times t_m (up to 8 h) in a high-energy planetary mill Pulverisette 6 (Fritsch, Idar-Oberstein, Germany) at room temperature. A grinding chamber (250 cm^3 in volume) and balls (10 mm in diameter) made of tungsten carbide were used. The ball-to-powder weight ratio was 20 : 1. The milling experiments were performed in air at 600 rpm.

The XRD patterns were measured using a PW 1820 X-ray diffractometer (Philips, Netherlands), operating in Bragg configuration and using Cu Kα radiation (λ = 1.54056 Å). The XRD scans were collected from 10° to 80° (2Θ), using a step of 0.02° and a data collection time of 5 s. The JCPDS PDF database[9] was utilized for phase identification using the STOE software. The hexagonal structure of $BaFe_{12}O_{19}$ was visualized using the Diamond program[10] and Java Structure Viewer software.[11]

The ^{57}Fe Mössbauer spectra were taken at 293 K in transmission geometry using a ^{57}Co/Rh γ-ray source. Recoil spectral analysis software[12] was used for the quantitative evaluation of the Mössbauer spectra. The velocity scale of the spectra was calibrated relative to ^{57}Fe in Rh.

The morphology of the powders was studied using a combined field-emission (scanning) transmission electron microscope (S)TEM (JEOL JEM-2100F) with a high-resolution pole piece that provides a point resolution better than 0.19 nm at 200 kV. Prior to the TEM investigations, the powders were crushed in a mortar, dispersed in ethanol, and fixed on a copper-supported carbon grid.

The magnetic measurements were performed using a SQUID magnetometer (Quantum Design MPMS-5S, USA). The samples were filled in a small container

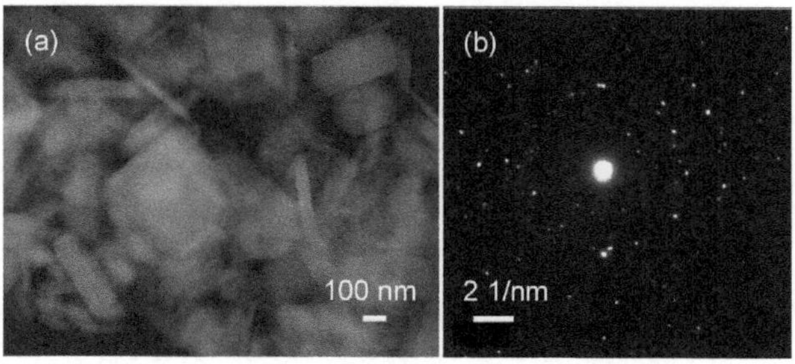

Fig. 1 (a) The electron microscopy image and (b) the corresponding SAED pattern illustrating the presence of hexagonal platelet crystals in the co-precipitated $BaFe_{12}O_{19}$.

made of polyvinyl chloride, whose diamagnetic moment was subtracted from the measured magnetization values. Magnetic hysteresis loops were recorded at 5 K in external magnetic fields from 0 to ±5 T.

Results and discussion

Prior to any characterization of the structural disorder and functional properties of the mechanically treated $BaFe_{12}O_{19}$, the atomic configuration of the non-treated bulk material has to be known. An effective way to do this is by means of nuclear spectroscopic techniques such as ^{57}Fe Mössbauer spectroscopy, which makes possible observations on a local atomic scale. This spectroscopic method has been proven to be well suited for the investigation of the charge state, the local coordination, and the magnetic state of iron ions in various ferrites.[13] Fig. 2 illustrates the room temperature ^{57}Fe Mössbauer spectrum of the $BaFe_{12}O_{19}$ standard sample. As can be seen, the spectrum of the material is well-fitted by a superposition of five magnetically split spectral components (sextets) indicating the presence of five different atomic environments around the iron nuclei. Three sextets with isomer shift IS > 0.19 mm s^{-1} (see Table 1) correspond to octahedrally coordinated ferric (Fe^{3+}) ions in the $4f_2$, 12k and 2a crystal sites of the hexagonal structure of the ferrite.[14] The spectral component with the lowest value of IS (~ 0.14 mm s^{-1}) is typical for Fe^{3+} cations in tetrahedral ($4f_1$) coordination of oxygen anions.[15] The sextet with a relatively large quadrupole splitting (QS ~ 1.1 mm/s), indicating the presence of a large electric field gradient acting on the iron nuclei, corresponds to Fe^{3+} ions in the trigonal bi-pyramidal (2b) sites of $BaFe_{12}O_{19}$. The Mössbauer parameters resulting from the least-squares fitting of the spectrum of bulk $BaFe_{12}O_{19}$ (Table 1) are in reasonable agreement with those determined in previous work.[14] From the relative intensities of the sextets, the number of Fe^{3+} cations located on the $4f_2$, 12k, 2a, $4f_1$ and 2b sublattices was calculated to be 2, 6, 1, 2 and 1 per formula unit (f.u.) of $BaFe_{12}O_{19}$, respectively. Based on the present Mössbauer results, the structural formula of $BaFe_{12}O_{19}$, emphasizing the site occupancy at the atomic level, may be written as Ba

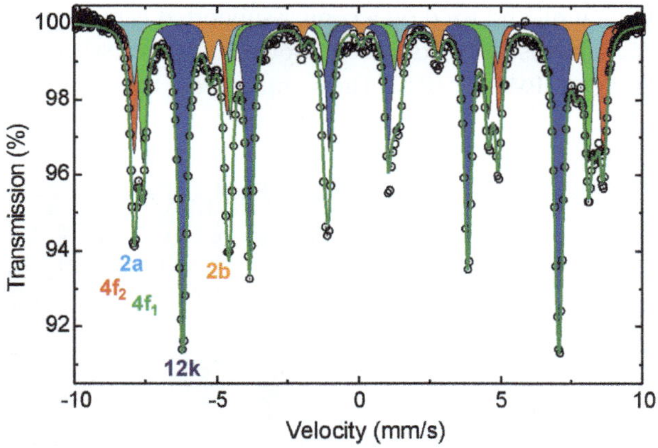

Fig. 2 The room temperature ^{57}Fe Mössbauer spectrum of the standard $BaFe_{12}O_{19}$ sample. The red, blue, cyan, green and orange colours denote the subspectra corresponding to the Fe^{3+} cations in the 4f2, 12k, 2a, 4f1, and 2b sublattices, respectively.

Table 1 The parameters obtained by fitting the room temperature ^{57}Fe Mössbauer spectrum of polycrystalline $BaFe_{12}O_{19}$[a]

Site	Coordination of Fe^{3+} ions	IS (mm s^{-1})	QS (mm s^{-1})	H (T)	Number of Fe^{3+} ions per f.u.
$4f_2$	Octahedral	0.252(7)	0.102(8)	51.21(8)	2
12k	Octahedral	0.218(1)	0.210(1)	41.10(1)	6
2a	Octahedral	0.196(2)	0.022(2)	50.40(2)	1
$4f_1$	Tetrahedral	0.139(4)	0.111(3)	48.66(3)	2
2b	Bi-pyramidal	0.142(1)	1.083(1)	39.99(9)	1

[a] IS: isomer shift; QS: quadrupole splitting; H: magnetic hyperfine field. A Lorentzian line width of 0.26 mm s^{-1} resulted from the fit of the spectrum of $BaFe_{12}O_{19}$.

$[Fe_2]_{4f2}[Fe_6]_{12k}[Fe]_{2a}(Fe_2)_{4f1}\{Fe\}_{2b}O_{19}$, where the square brackets, parentheses and curly brackets enclose cations in sites of octahedral, tetrahedral and trigonal bi-pyramidal coordination, respectively. The hexagonal structure of $BaFe_{12}O_{19}$ with the five different Fe nearest-neighbour configurations is shown in Fig. 3.

The mechanically induced evolution of $BaFe_{12}O_{19}$ was followed by XRD. Fig. 4 shows the XRD patterns of the ferrite milled for various times. The XRD pattern of the starting powder is characterized by sharp diffraction peaks corresponding to $BaFe_{12}O_{19}$ with the magnetoplumbite structure and space group $P6_3/mmc$ (JCPDS PDF 27-1029).[9] With increasing milling time, XRD reveals a gradual decrease in the intensity and an associated broadening of the Bragg peaks of the oxide. This reflects a continuous fragmentation of the material accompanied by the refinement of its crystallite size (D) to the nanometer range; with the prolongation of t_m, a monotonous reduction of the average crystallite size of the ferrite to $D = 14$ nm

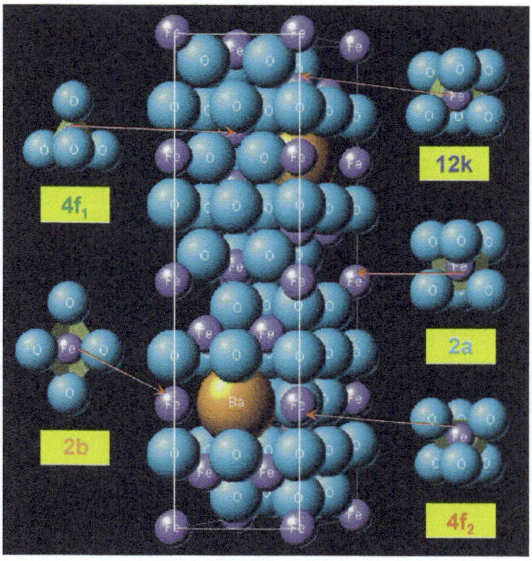

Fig. 3 The crystal structure of the M-type hexaferrite. Fe^{3+} cations are distributed over the sites of octahedral ($4f_2$, 12k, 2a), tetrahedral ($4f_1$) and trigonal bi-pyramidal (2b) coordination.

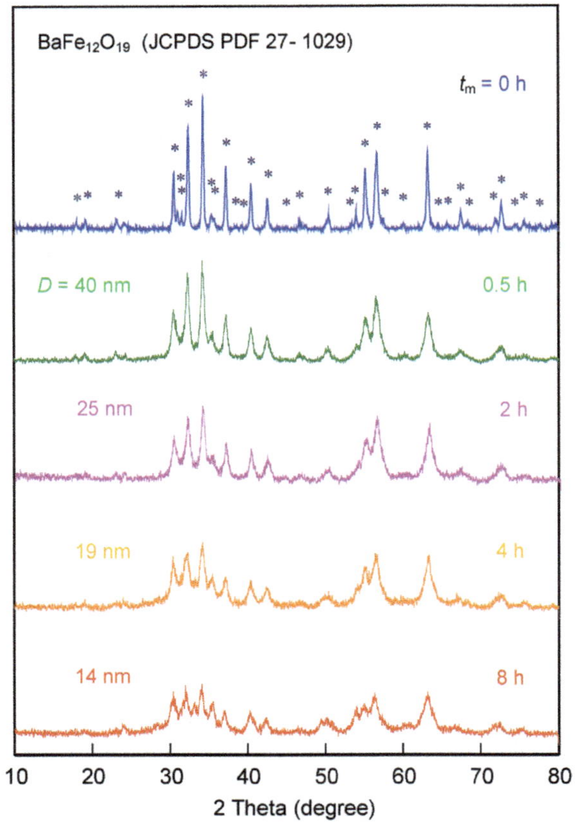

Fig. 4 The XRD patterns of $BaFe_{12}O_{19}$ milled for various milling times. All diffraction peaks of bulk $BaFe_{12}O_{19}$ (denoted by stars) can be indexed to the hexagonal magnetoplumbite structure with the space group $P6_3/mmc$. The milling times (t_m) and the corresponding average crystallite sizes (D) are shown in the figure.

(for $t_m = 8$ h) is observed, see Fig. 4. Simultaneously, the high-energy milling process leads to the formation of a broad diffraction maximum in the range of about 30–40° (2Θ) indicating a partly amorphization of the structure. The superimposition of the relatively broad diffraction reflections of the $BaFe_{12}O_{19}$ phase on a broad diffraction maximum in the range of about 30–40° (2Θ) reflects a typical morphology of the mechanochemically prepared nanostructured oxides[3] consisting of small crystalline regions (often called nanograins or nanocrystallites) surrounded/separated by structurally disordered internal interfaces (grain boundaries) and/or external surfaces (near-surface layers); the detailed TEM analysis of the mechanochemically prepared $BaFe_{12}O_{19}$ nanoparticles is given below. Note that the atomic arrangement in internal interfaces/external surfaces of the mechanically treated materials may lack any long- or short-range order.[3,4] Because of the sensitivity to medium- and long-range structural order, the applied XRD technique loses much of its resolving power in such nanoscale and disordered systems. Therefore, the nature of the mechanically induced structural disorder in $BaFe_{12}O_{19}$ will be analyzed concurrently with the discussion of Mössbauer data (see next paragraph).

To determine the nature of the mechanically induced structural disorder in $BaFe_{12}O_{19}$, the evolution of its structure on the local atomic scale was followed by ^{57}Fe Mössbauer spectroscopy. The room temperature ^{57}Fe Mössbauer spectra of the material milled for various times are presented in Fig. 5. As can be seen, with increasing t_m, the sextets corresponding to the Fe^{3+} ions located on the five sublattices of the hexaferrite become asymmetric toward the inside of each line, slowly collapse, and are gradually replaced by a broad central doublet with the isomer shift of about 0.21 mm s^{-1} characteristic of Fe^{3+} ions. It should be mentioned in this context that the central doublet is clearly visible after only 30 min of milling (see Fig. 5). Further milling leads to a gradual increase of its relative intensity. After 8 h of mechanical treatment, the sextets disappear completely and the Mössbauer spectrum of the milled $BaFe_{12}O_{19}$ is dominated by the doublet.

The relatively broad shape of the Mössbauer spectral lines for milled $BaFe_{12}O_{19}$, in contrast to the non-treated ferrite, provides clear evidence of a wide distribution of hyperfine interactions experienced by the Fe^{3+} nuclei in the

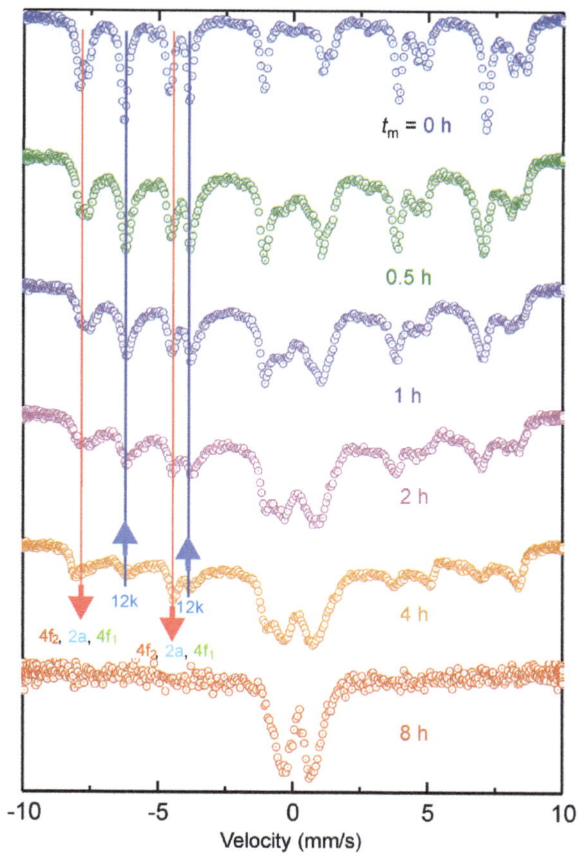

Fig. 5 The room temperature ^{57}Fe Mössbauer spectra of $BaFe_{12}O_{19}$ milled for various milling times (shown in the figure). The arrows emphasize the redistribution of the spectral intensities corresponding to the Fe^{3+} cations in the $4f_2$, 2a, $4f_1$ (red arrow) and 12k (blue arrow) crystal sites of the ferrite.

material. This feature may be explained by the presence of a broad distribution of local environments around the Fe nuclei due to the mechanically induced deformation of the constituent polyhedra (FeO_6 octahedra, FeO_4 tetrahedra and FeO_5 triangular bi-pyramids) of the hexaferrite. To elucidate the origin of the broad central doublet in the spectrum of the milled ferrite, we should recall the effect of superparamagnetism.[16] The latter arises if the particle sizes of a material are so small that thermally induced energy fluctuations can overcome the anisotropy energy and change the direction of the magnetization of a particle from one easy axis to another. In line with this, the central doublet in the Mössbauer spectrum of the milled $BaFe_{12}O_{19}$ is related to its particles of such small size that they behave superparamagnetic on the time scale of Mössbauer spectroscopy (about 10^{-9} to 10^{-10} s).[16] Thus, the refinement of the crystallite size of the ferrite results in an increase of the fraction of the superparamagnetic phase at the expense of the ferrimagnetic one; the $BaFe_{12}O_{19}$ sample with the average crystallite size $D = 14$ nm ($t_m = 8$ h) exhibits a pure superparamagnetic behaviour at room temperature.

The important result derived from the present Mössbauer data is that mechanical action on $BaFe_{12}O_{19}$ is accompanied by the changes in the relative intensities of the sublattice subspectra (Fig. 5). It is clearly visible that the relative intensity of the 12k subspectrum (I_{12k}) decreases with increasing milling time; for the bulk $BaFe_{12}O_{19}$, the intensity ratio $I_{12k}/(I_{4f2} + I_{2a} + I_{4f1})$ is 6/5, whereas it takes a value of about 4/5 for the material milled for 4 h. This variation can be explained by a decrease of the population of Fe^{3+} ions on the 12k crystal sites. In this context, it should be noted that similar redistributions of Mössbauer and/or NMR spectral intensities have also been reported for other mechanochemically treated complex oxides, such as spinels,[17] and are consistently explained in terms of the mechanically induced cation redistribution over the available sites of tetrahedral and octahedral coordination provided by the spinel structure. Taking into account that in the magnetoplumbite structure of $BaFe_{12}O_{19}$, in contrast to the spinel structure, all the regular sites are occupied by the cations, the decrease of the intensity ratio $I_{12k}/(I_{4f2} + I_{2a} + I_{4f1})$ from 6/5 to 4/5 can be associated with the mechanically triggered transition of 2 of a total of 6 Fe^{3+} cations (per formula unit of $BaFe_{12}O_{19}$) from the regular 12k sites into the interstitial positions (further denoted by 12k′) of the hexagonal structure. This structural variation has important implications for the macroscopic magnetic properties of the hexaferrite; the detailed discussion of magnetic behaviour of $BaFe_{12}O_{19}$ is given below.

The application of the JSV program[11] enables us to visualize possible 12k′ interstitial sites (cavities) with various radii in the magnetoplumbite structure, where the Fe^{3+} cations could be located. Taking into account the ionic radii of O^{2-}, Fe^{3+}, and Ba^{2+} to be $r_O = 1.24$ Å, $r_{Fe} = 0.785$ Å, and $r_{Ba} = 1.56$ Å, respectively,[18] Fig. 6 illustrates the interstitial sites of various sizes for $BaFe_{12}O_{19}$. The largest cavities in the crystal lattice of the hexaferrite were found to have a radius of $r =$ 0.881 Å (Fig. 6a) which is larger than the ionic radius of the Fe^{3+} cations (0.785 Å). This provides evidence that, from the geometrical point of view, the above-mentioned 12k→12k′ structural transition of the Fe^{3+} cations in the hexagonal crystal lattice is generally possible. In this respect, it should be mentioned that the appearance of the Fe^{3+} cations in the interstitial positions of the hexagonal crystal lattice can result in a local distortion of the structure. The latter is supported by

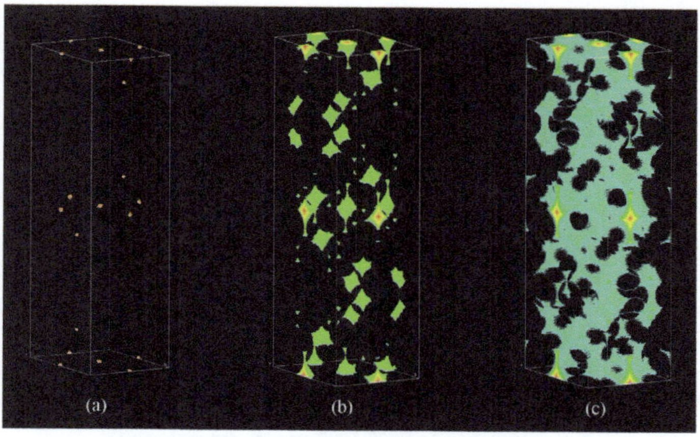

Fig. 6 The visualization of the atomic interstitial sites (cavities) with radii of (a) 0.8, (b) 0.5, and (c) 0.4 Å in the magnetoplumbite structure.

the present Mössbauer study providing clear evidence of a wide distribution of hyperfine interactions experienced by the Fe^{3+} nuclei in the milled hexaferrite due to the distortion of the geometry of its constituent polyhedra (see above). Similar disordering effects, *i.e.*, the nonequilibrium cation distribution and the deformed polyhedron geometry, have already been observed in mechanochemically prepared oxides, such as spinels,[19] olivines[20] and perovskites,[21] as well as orthorhombic complex oxides.[22] In the case of mechanosynthesized perovskites (*e.g.*, $BiFeO_3$) and trigonal nanostructured oxides (*e.g.*, $LiNbO_3$), their interface/surface regions have been found to be even amorphous.[21,23] Note that the presence of the amorphous phase in the milled $BaFe_{12}O_{19}$ samples, registered in the present case by XRD and TEM (see below), could be a consequence of a broadly distorted geometry of the constituent polyhedra. Eventually, mechanical action on $BaFe_{12}O_{19}$ can also cause the displacement of the relatively large Ba^{2+} cations from their regular positions leading to additional modifications of the local structure of the ferrite and, finally, to its amorphization.

Representative TEM micrographs of the milled $BaFe_{12}O_{19}$ ($t_m = 8$ h) at low and high magnifications are shown in Fig. 7. It is revealed that the milled material consists of nanoparticles with a size distribution ranging from about 8 to 20 nm, which consistent with the average crystallite size determined by XRD. The nanoparticles are found to be roughly spherical. An interesting observation is that the amorphous phase, evidenced by XRD, is confined to the near-surface layers/interfaces of the $BaFe_{12}O_{19}$ nanoparticles. As clearly seen, the milled hexaferrite consists of crystalline regions, represented by lattice fringes in Fig. 7b, and amorphous rim regions with the thickness of about 2 nm. Based on the present high-resolution TEM observations as well as on the analogy with previous work on mechanochemically prepared complex nanooxides,[3] we can state that a far-from-equilibrium distribution of the Fe^{3+} cations in the milled $BaFe_{12}O_{19}$, its deformed polyhedra, and the amorphous phase, evidenced by Mössbauer spectroscopy and XRD, respectively, are confined to the particle's near-surface layers with the thickness extending up to about 2 nm. The shell thickness in the milled $BaFe_{12}O_{19}$ nanoparticles is comparable to that observed for other nanooxides

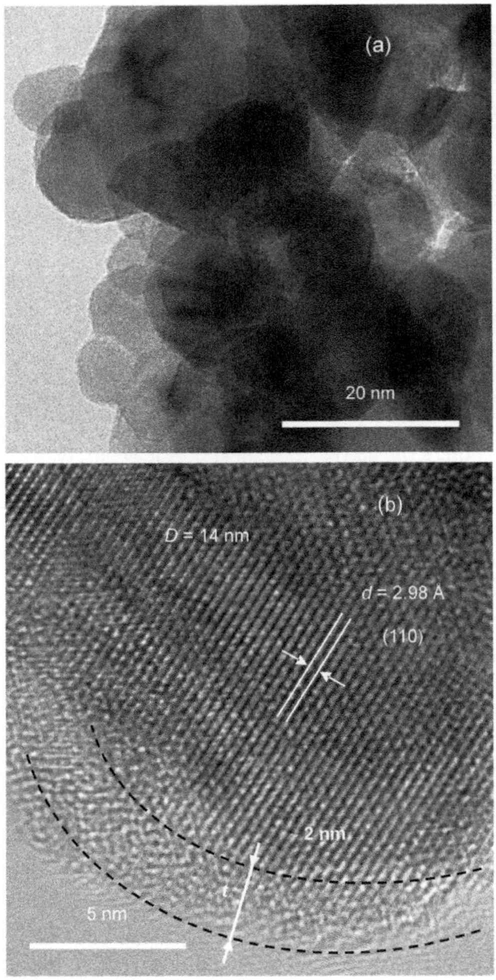

Fig. 7 (a) Bright-field and (b) high-resolution TEM images of the nanocrystalline $BaFe_{12}O_{19}$ prepared by high-energy milling of the corresponding bulk material. The high-resolution TEM micrograph demonstrates the inhomogeneous structure of the produced nanoparticles with the average size $D = 14$ nm which exhibit a clearly defined rim region and a core region. The so-called core–shell configuration of the nanoparticles with the thickness t of the surface shell of about 2 nm is evident. The lattice fringes correspond to the crystallographic plane (110) ($d = 2.98$ Å) of the $BaFe_{12}O_{19}$ phase (JCPDS PDF 27-1029).

prepared by mechanochemical routes. We note that, in general, up to 2 nm is a typical thickness of the grain boundary/surface shell regions in nanostructured mechanochemically prepared oxides.[3,4]

Assuming a spherical shape of $BaFe_{12}O_{19}$ nanoparticles and taking both their average diameter ($D = 14$ nm) and the thickness of their surface shell ($t = 2$ nm) as determined experimentally by XRD and TEM, respectively, one can easily deduce quantitative information on the volume fraction of disordered surface shell

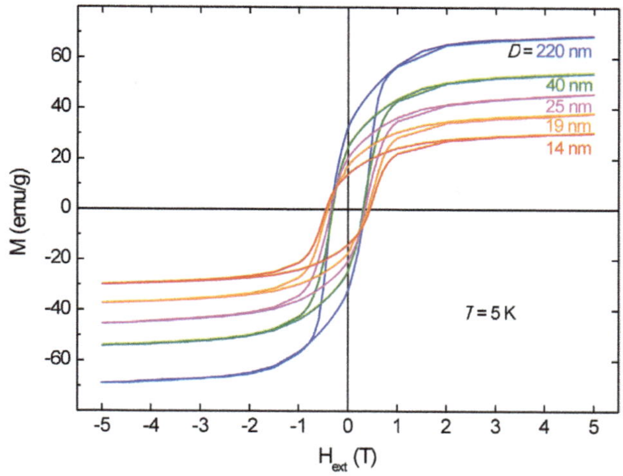

Fig. 8 Magnetization hysteresis loops for bulk and nanoscale mechanically treated $BaFe_{12}O_{19}$. Hysteresis loops were measured at 5 K after field cooling with $H_{ext} = 5$ T. The average crystallite sizes of the $BaFe_{12}O_{19}$ samples are shown in the figure.

regions, w, to the volume of whole particles ($w = V_{shell}/(V_{core} + V_{shell})$) in the nanomaterial.‡ The estimated value of $w \approx 0.636$ indicates that about 64% of atoms in the milled ferrite are in a structurally disordered state located in the surface shell of nanoparticles. In this context, it should be noted that nanomaterials, in general, and especially those prepared by mechanochemical processing exhibit a high volume fraction of structurally disordered regions.[3,4]

The far-from-equilibrium structural state, confined to the particle's near-surface layers of the mechanically treated ferrite, has significant implications for its magnetism. Fig. 8 compares the hysteresis loops measured at 5 K for bulk and milled $BaFe_{12}O_{19}$. An interesting observation is that the saturation magnetization (M_{sat}) of the ferrite decreases with increasing t_m (*i.e.*, with decreasing D). By extrapolating the high-field region ($H_{ext} > 3.5$ T) of the $M(H_{ext})$ curves to infinite field, we estimated the M_{sat} values of hexaferrites with various particle sizes, see Table 2. They span over a wide interval from $M_{sat} = 68.601(1)$ emu g^{-1} (for bulk $BaFe_{12}O_{19}$) to $M_{sat} = 30.093(6)$ emu g^{-1} (for nano-$BaFe_{12}O_{19}$ with $D = 14$ nm). Another feature observed is that the nanoscale $BaFe_{12}O_{19}$ exhibits an enhanced magnetic hardness, *i.e.*, the coercive field of the nanomaterial ($H_c = 0.449(5)$ T) is about 34% larger than that of the bulk material ($H_c = 0.334(3)$ T), see Table 2. Obviously, these large variations in magnetization and coercivity offer an ample opportunity to manipulate and tailor the functional properties of this magnetically hard material.

The information on the nonequilibrium distribution of the Fe^{3+} cations and the distorted polyhedron geometry in the milled $BaFe_{12}O_{19}$, obtained from the analysis of Mössbauer data, is very helpful in the interpretation of its reduced magnetization. Because Ba^{2+} ions possess no magnetic moment, the total magnetic moment μ of $BaFe_{12}O_{19}$ is entirely due to the uncompensated magnetic

‡ $w = [1 - (1 - 2t/D)^3]$.

Table 2 Average crystallite size (D), saturation magnetization (M_{sat}), coercive field (H_c), and remanent magnetization (M_r) derived from the hysteresis loops measured at 5 K for $BaFe_{12}O_{19}$ milled for various times (t_m)[a]

t_m (h)	D (nm)	M_{sat} (emu g^{-1})	H_c (T)	M_r (emu g^{-1})
0	220	68.601(1)	0.334(3)	32.433(8)
0.5	40	54.021(1)	0.312(2)	24.513(2)
2	25	45.610(2)	0.344(4)	20.246(9)
4	19	37.902(2)	0.411(8)	16.972(9)
8	14	30.093(6)	0.449(5)	13.706(5)

[a] Values of the saturation magnetization are obtained by linear extrapolation of the high-field region (H_{ext} > 3.5 T) of the $M(1/H_{ext})$ curves to infinite field.

moments of the ferric ions. Thus, based on the structural formula of bulk $BaFe_{12}O_{19}$ emphasizing its spin arrangement, $Ba[Fe_2\downarrow]_{4f2}[Fe_6\uparrow]_{12k}[Fe\uparrow]_{2a}(Fe_2\downarrow)_{4f1}\{Fe\uparrow\}_{2b}O_{19}$, we can calculate the effective magnetic moment (per formula unit) of the hexaferrite as $(-2 + 6 + 1 - 2 + 1) \times \mu_{Fe} = 20\ \mu_B$, where μ_{Fe} is the magnetic moment of the Fe^{3+} ion; $\mu_{Fe} = 5\ \mu_B$ (μ_B is Bohr magneton (see also Table 3). On the other hand, taking into account the mechanically triggered transition of 2 of a total of 6 Fe^{3+} cations from the regular 12k octahedral sites into the interstitial 12k′ positions, the effective magnetic moment of the milled hexaferrite can be expressed as $\mu = (-2 + 4 + 1 - 2 + 1) \times \mu_{Fe} = 10\ \mu_B$/f.u. (the calculation was made under the assumption that the spins of the Fe^{3+} cations in the interstitial sites are randomly aligned, resulting in zero net magnetic moment). Thus, the cation disorder in $BaFe_{12}O_{19}$ generally tends to reduce its magnetic moment. However, it should be noted that the presence of nonequilibrium cation distribution and/or distorted polyhedron geometry in ferrites effects their spin alignments on all available sublattices leading to the formation of canted spin arrangements as it was demonstrated for mechanochemnically prepared magnetic spinels and perovskites.[3] Thus, a more realistic picture of the milled $BaFe_{12}O_{19}$ is that the magnetic moment of each sublattice is affected by spin canting. Moreover, the confinement of the structural disorder to the

Table 3 The calculated effective magnetic moment (μ) per formula unit of bulk and nanocrystalline $BaFe_{12}O_{19}$ materials

Material	Site	Number of Fe^{3+} ions/f.u.	Spin orientation	μ per f.u.
Bulk $BaFe_{12}O_{19}$ ($t_m = 0$ h, $D = 220$ nm)	4f$_2$	2	↓	$(-2 + 6 + 1 - 2 + 1) \times 5\ \mu_B = = 20\ \mu_B$
	12k	6	↑	
	2a	1	↑	
	4f$_1$	2	↓	
	2b	1	↑	
Nanocrystalline $BaFe_{12}O_{19}$ ($t_m = 4$ h, $D = 19$ nm)	4f$_2$	2	↓	$(-2 + 4 + 1 - 2 + 1) \times 5\ \mu_B = = 10\ \mu_B$
	12k	4	↑	
	12k′	2	random	
	2a	1	↑	
	4f$_1$	2	↓	
	2b	1	↑	

particle's interfaces/near-surface layers can even results in the formation of magnetically inactive (the so-called "dead") regions with zero net magnetic moment.[24]

In the following, we will estimate the shell thickness using the experimentally determined D and M_{sat} values of $BaFe_{12}O_{19}$ milled for various times (see Table 2). Assuming that t is independent of D and that the shell is magnetically "dead" ($M_{shell} = 0$), the variation of M_{sat} with D will then be described by

$$M_{sat}^{1/3} = M_{core}^{1/3}(1 - 2t/D),$$

where M_{core} is the saturation magnetization of the core of nanoparticles.[24] As can be seen in Fig. 9, the present experimental data $M_{sat}^{1/3}$ and $1/D$ indeed show a good linear relationship. Note that the intercept at $1/D = 0$ and the slope of the straight line correspond to $M_{core}^{1/3}$ and $2tM_{core}^{1/3}$, respectively. From a linear fit to the data points, the saturation magnetization of the particle core and the thickness of the shell were estimated to be $M_{core} \approx 71.7$ emu g^{-1} and $t \approx 1.8$ nm, respectively. The value of M_{core} thus obtained is close to the saturation magnetization measured for bulk $BaFe_{12}O_{19}$, 68.6 emu g^{-1}. The value of the shell thickness obtained is in reasonable agreement with that estimated directly from high-resolution TEM experiments (2 nm). Based on this agreement, we can state that the magnetism of the milled $BaFe_{12}O_{19}$ is controlled by the mechanically induced far-from-equilibrium structural disorder, which is confined to the interface/surface regions of the nanoferrite.

Conclusions

The structural and magnetic evolution of $BaFe_{12}O_{19}$ caused by high-energy milling has been studied. During the milling process, the material is subjected to a continuous fragmentation accompanied by the reduction of its crystallite size to the 8–20 nm size range. In addition to the XRD method being sensitive to

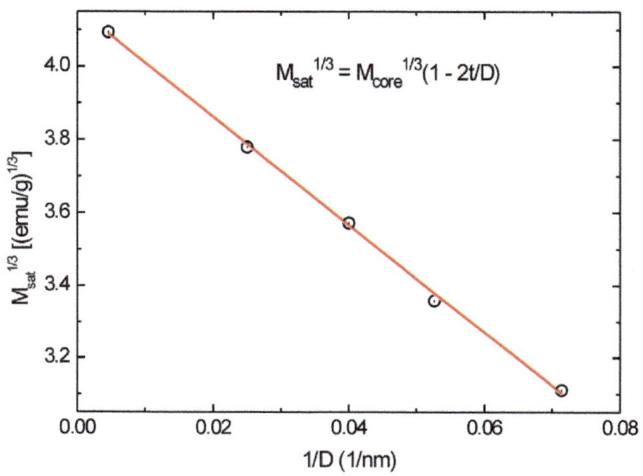

Fig. 9 $M_{sat}^{1/3}$ vs $1/D$ plot, where M_{sat} is the saturation magnetization and D is the diameter of the $BaFe_{12}O_{19}$ nanoparticles.

medium- and long-range structural order, the mechanically induced evolution of $BaFe_{12}O_{19}$ was followed by ^{57}Fe Mössbauer spectroscopy, which made possible observations on a local atomic scale. The latter method revealed the presence of a broad distribution of local environments around the Fe^{3+} cations in the nanoferrite due to the mechanically induced deformation of its constituent polyhedra (FeO_6 octahedra, FeO_4 tetrahedra and FeO_5 triangular bi-pyramids). The response of $BaFe_{12}O_{19}$ to mechanical action is found to be accompanied by the mechanically triggered transition of Fe^{3+} cations from the regular 12k octahedral sites into the interstitial positions of the hexagonal structure. The refinement of the crystallite size of the ferrite results in an increase of the fraction of the superparamagnetic phase at the expense of the ferrimagnetic one; the $BaFe_{12}O_{19}$ sample with the average crystallite size of about 14 nm exhibits a pure superparamagnetic behaviour at room temperature. The as-prepared nanoparticles are found to be roughly spherical, consisting of a crystalline inner core surrounded/separated by an amorphous surface shell/interface region with the thickness and the volume fraction extending up to about 2 nm and 64%, respectively. The far-from-equilibrium structural state, confined to the particle's near-surface layers of the mechanically treated ferrite, has significant implications for its magnetism. It is found that the saturation magnetization of $BaFe_{12}O_{19}$ decreases with decreasing particle size from about 69 emu g^{-1} (for bulk $BaFe_{12}O_{19}$) to about 30 emu g^{-1} (for nanoferrite with $D = 14$ nm). On the other hand, the coercive field of the nanomaterial is about 34% larger than that of the bulk sample. The macroscopic magnetic behaviour of the milled $BaFe_{12}O_{19}$ is attributed to the effects of the mechanically induced far-from-equilibrium structural disorder, which is located in the interface/surface regions of the nanoferrite.

Acknowledgements

The present work is supported by the DFG within the framework of the Priority Program "*Crystalline Nonequilibrium Phases*" (SPP 1415). Partial support by the APVV (project 0528-11) and the VEGA (2/0097/14) is gratefully acknowledged.

Notes and references

1 *Mechanochemistry*, ed. S. L. James and T. Friščić, *Chem. Soc. Rev.*, 2013, **42**, 7487–7740.
2 *Mechanoresponsive Materials*, ed. C. Weder, *J. Mater. Chem.*, 2011, **21**, 8217–8476.
3 V. Šepelák, A. Düvel, M. Wilkening, K.-D. Becker and P. Heitjans, *Chem. Soc. Rev.*, 2013, **42**, 7507.
4 V. Šepelák, S. Bégin-Colin and G. Le Caër, *Dalton Trans.*, 2012, **41**, 11927.
5 S. J. Campbell, E. Wu, W. A. Kaczmarek and K. D. Jayasuriya, *Hyperfine Interact.*, 1994, **92**, 933.
6 K. Haneda and H. Kojima, *J. Am. Ceram. Soc.*, 1973, **57**, 68.
7 G. Tan and X. Chen, *J. Magn. Magn. Mater.*, 2013, **327**, 87.
8 D. Belušova, T. Alshuth, R. H. Schuster, M. Myndyk, V. Šepelák and I. Hudec, *Kaut. Gummi Kunstst.*, 2008, **3**, 118.

9 Joint Committee on Powder Diffraction, *Standards (JCPDS) Powder Diffraction File (PDF)*, International Centre for Diffraction Data, Newtown Square, PA, 2004.

10 K. Brandenburg and H. Putz, *Diamond – Crystal and Molecular Structure Visualization Software, Version 3.0a*, Crystal Impact GbR, Bonn, Germany, 2004.

11 S. Weber, *J. Appl. Crystallogr.*, 1999, **32**, 1027; S. Weber, *Java Structure Viewer (JSV), 1.08lite*, 1999.

12 K. Lagarec and D. G. Rancourt, *Recoil – Mössbauer Spectral Analysis Software for Windows, Version 1.02*, Department of Physics, University of Ottawa, Ottawa, ON, 1998.

13 V. Šepelák and K. D. Becker, *J. Mater. Synth. Process.*, 2000, **8**, 155.

14 F.-M. Gao, D.-C. Li and S.-Y. Zhang, *J. Phys.: Condens. Matter*, 2003, **15**, 5079.

15 F. Menil, *J. Phys. Chem. Solids*, 1985, **46**, 763.

16 G. J. Long, *Mössbauer Spectroscopy Applied to Inorganic Chemistry*, Plenum Press, New York, 1987, vol. 2.

17 V. Šepelák, I. Bergmann, S. Indris, A. Feldhoff, H. Hahn, K. D. Becker, C. P. Grey and P. Heitjans, *J. Mater. Chem.*, 2011, **21**, 8332; V. Šepelák, D. Baabe, D. Mienert, D. Schultze, F. Krumeich, F. J. Litterst and K. D. Becker, *J. Magn. Magn. Mater.*, 2003, **257**, 377; V. Šepelák, D. Baabe, F. J. Litterst and K. D. Becker, *J. Appl. Phys.*, 2000, **88**, 5884; V. Šepelák, D. Baabe and K. D. Becker, *J. Mater. Synth. Process.*, 2000, **8**, 333; P. Druska, U. Steinike and V. Šepelák, *J. Solid State Chem.*, 1999, **146**, 13; V. Šepelák, K. Tkáčová, V. V. Boldyrev, S. Wißmann and K. D. Becker, *Phys. B*, 1997, **234–236**, 617.

18 R. D. Shannon, *Acta Crystallogr., Sect. A: Cryst. Phys., Diffr., Theor. Gen. Crystallogr.*, 1976, **32**, 751.

19 V. Šepelák, A. Feldhoff, P. Heitjans, F. Krumeich, D. Menzel, F. J. Litterst, I. Bergmann and K. D. Becker, *Chem. Mater.*, 2006, **18**, 3057; V. Šepelák, S. M. Becker, I. Bergmann, S. Indris, M. Scheuermann, A. Feldhoff, C. Kübel, M. Bruns, N. Stürzl, A. S. Ulrich, M. Ghafari, H. Hahn, C. P. Grey, K. D. Becker and P. Heitjans, *J. Mater. Chem.*, 2012, **22**, 3117.

20 V. Šepelák, M. Myndyk, M. Fabián, K. L. Da Silva, A. Feldhoff, D. Menzel, M. Ghafari, H. Hahn, P. Heitjans and K. D. Becker, *Chem. Commun.*, 2012, **48**, 11121.

21 K. L. Da Silva, D. Menzel, A. Feldhoff, C. Kübel, M. Bruns, A. Paesano, Jr, A. Düvel, M. Wilkening, M. Ghafari, H. Hahn, F. J. Litterst, P. Heitjans, K. D. Becker and V. Šepelák, *J. Phys. Chem. C*, 2011, **115**, 7209.

22 L. J. Berchmans, M. Myndyk, K. L. Da Silva, A. Feldhoff, J. Šubrt, P. Heitjans, K. D. Becker and V. Šepelák, *J. Alloys Compd.*, 2010, **500**, 68; V. Šepelák, K. D. Becker, I. Bergmann, S. Suzuki, S. Indris, A. Feldhoff, P. Heitjans and C. P. Grey, *Chem. Mater.*, 2009, **21**, 2518.

23 P. Heitjans and M. Wilkening, *MRS Bull.*, 2009, **34**, 915.

24 V. Šepelák, I. Bergmann, A. Feldhoff, P. Heitjans, F. Krumeich, D. Menzel, F. J. Litterst, S. J. Campbell and K. D. Becker, *J. Phys. Chem. C*, 2007, **111**, 5026.

Faraday Discussions

PAPER

Dry mechanochemical synthesis of alane from LiH and $AlCl_3$†

Ihor Z. Hlova,[ac] Shalabh Gupta,*[a] Jennifer F. Goldston,[ab] Takeshi Kobayashi,[ab] Marek Pruski*[ab] and Vitalij K. Pecharsky*[ac]

Received 29th December 2013, Accepted 17th February 2014
DOI: 10.1039/c3fd00161j

A mechanochemical process for the synthesis of alane (AlH_3) starting from lithium hydride (LiH) and aluminium chloride ($AlCl_3$) at room temperature and the underlying reaction pathway have been studied. In contrast to a conventional process using the same two reactants dissolved in diethyl ether, our approach enables a solvent-free synthesis, thereby directly leading to adduct-free alane. The method described here is quick and efficient, resulting in the quantitative conversion of all aluminium in the starting mixture to alane. Both the intermediate compounds formed during the reaction and the final products have been characterized by powder X-ray diffraction, solid-state ^{27}Al NMR spectroscopy, and temperature programmed desorption analysis of the as-milled mixtures. We show that excess LiH in the starting mixture (with an optimal ratio of 9LiH : $1AlCl_3$) is essential for the formation and stability of Al–H bonds, initially in the form of alanates and, eventually, as alane. Further processing of this mixture, gradually adding $AlCl_3$ to reach the ideal 3LiH : $1AlCl_3$ stoichiometry, appears to restrict the local accumulation of $AlCl_3$ during the ball-milling process, thereby preventing the formation of unstable intermediates that decompose to metallic Al and molecular hydrogen. We also demonstrate that under the milling conditions used, a moderate hydrogen pressure of *ca.* 300 bar is required to suppress competing reactions that lead to the formation of metallic Al at room temperature. The identification of the reaction intermediates at each stage of the synthesis provides significant insight into the mechanism of this solid-state reaction, which may potentially afford a more rational approach toward the production of AlH_3 in a simple solvent-free process.

[a]*Ames Laboratory, Iowa State University, Ames, IA 500011-3020, USA. E-mail: shalabh@ameslab.gov; Fax: +515 294 9579; Tel: +515 294 9158*

[b]*Department of Chemistry, Iowa State University, Ames, IA 500011-3111, USA. E-mail: mpruski@iastate.edu; Fax: +515 294 4709; Tel: +515 294 2017*

[c]*Department of Materials Science and Engineering, Iowa State University, Ames, IA 50011-2300, USA. E-mail: vitkp@ameslab.gov; Fax: +515 294 9579; Tel: +515 294 8220*

† Electronic Supplementary Information (ESI) available: ^{27}Al DPMAS reference spectra for $LiAlCl_4$, $LiAlH_4$, $AlCl_3$, and Al_2O_3; the integrated peak intensities obtained from the ^{27}Al DPMAS spectra collected for Stages 1–3; the ^{27}Al 5QMAS spectra obtained for $LiAlCl_4$ and $LiAlH_4$; the XRD patterns of the products obtained after TPD; the XRD patterns comparing the progress of the 3 : 1 and 9 : 1 reaction mixtures during the first 20 min of milling; the ^{27}Al DPMAS and ^{27}Al{^{1}H} CPMAS spectra for a 3 : 1 LiH : $AlCl_3$ sample milled for 20 h at 150 rpm under 350 bar H_2. See DOI: 10.1039/c3fd00161j

1. Introduction

The advantages of using hydrogen as an alternative to fossil fuels are manifold and well established, owing to its benign oxidation product (water) and gravimetric energy density that is up to three times that of gasoline.[1,2] However, in order to enable the transition to hydrogen fuel for consumer use, the safe handling and transportation of hydrogen is of paramount concern. Solid-state hydrogen storage, in this regard, offers significant advantages over cryogenic and high-pressure approaches, since the safety risks associated with solid storage media are inherently lower than those posed by the latter options. Furthermore, solid materials also boast a significantly enhanced volumetric storage capacity relative to those typically encountered in cryogenic or high-pressure approaches. Despite these obvious advantages, however, more than five decades of research dedicated specifically to the development of materials-based hydrogen storage has led to only a few candidate materials that exhibit performance parameters commensurate with requirements estimated for on-board vehicle applications. Of the materials identified, aluminium trihydride (AlH_3, also commonly known as alane) has attracted much interest due primarily to its excellent gravimetric and volumetric hydrogen capacities (10.1 wt% and 148 g L^{-1}, respectively),[3] both of which are theoretically compliant with the ultimate full fleet targets set by the U.S. Department of Energy (7.5 wt% and 70 g L^{-1}, respectively) to be achieved by 2017.[4] In addition, AlH_3 completely dehydrogenates with an appreciable rate at temperatures close to 100 °C, which is an essential trait for compatibility with PEM fuel cells.[5] Finally, AlH_3 releases pure hydrogen with only metallic Al remaining, the latter of which can be recycled relatively easily.

Notwithstanding its promising attributes, however, alane has not yet been widely utilized as a hydrogen storage material due mainly to the lack of a convenient process for its bulk synthesis. AlH_3 was first synthesized in 1942, as an amine complex.[6] Later, Finholt *et al.* prepared an ethereal solution of AlH_3 by carrying out an exchange reaction between LiH and $AlCl_3$.[7] It was not until 1955 that the non-solvated form of alane was reported, initially by Chizinsky *et al.*,[8] and subsequently by Brower *et al.*[9] *via* an organometallic route at the Dow Chemical Company. It was also discovered that AlH_3 could be synthesized from the hydrides or tetrahydroaluminates of alkali and alkaline earth metals *via* wet chemical processes,[10–12] which usually involve the formation of alane–ether complexes such as those shown in eqn (1) and (2):[7]

$$3LiH + AlCl_3 + nEt_2O \rightarrow AlH_3{\cdot}nEt_2O + 3LiCl \quad (1)$$

$$3LiAlH_4 + AlCl_3 + nEt_2O \rightarrow 4AlH_3{\cdot}nEt_2O + 3LiCl \quad (2)$$

Although at least seven polymorphic forms of AlH_3 have been discovered and characterized so far, only the α, α', β and γ forms have been broadly studied, since these can be synthesized reproducibly and with quantitative yields.[9] While all seven polymorphs decompose cleanly into the constituent elements, only the α-form does so in a single step (*i.e.*, without undergoing any phase transformation prior to the decomposition). Such a one-step desorption of hydrogen from a solid is an important material property from an engineering standpoint.

Based on thermodynamic considerations, the direct hydrogenation of metallic Al to form AlH_3 is highly unfavourable, and would hence require extremely high hydrogen pressures to proceed (in the range of a few GPa according to the phase diagram reported by Konovalov and Bulychev[13]). Although this has been achieved in a laboratory setting, such a process holds no merit for the large-scale production of AlH_3. In addition to the high-pressure solid-gas reaction, solvent-based processes have been developed, in which a range of organic amines are utilized as donor ligands to achieve direct hydrogenation of metallic Al under low hydrogen pressures.[14] Aside from requiring large amounts of organic solvents, these approaches often only produce alane in the form of an adduct, which must then be subjected to purification processes often requiring ligand-exchange and heating, leading to significant decomposition of alane. Thus, novel procedures aimed at recycling metallic Al, albeit suitable only for off-board regeneration, are constantly sought to overcome the limitations of existing methods. Among those currently being explored, the electrochemical regeneration of aluminium hydride and the hydrogenation of metallic Al in supercritical fluids are notable.[15,16]

Apart from the direct hydrogenation of metallic Al, the synthesis of AlH_3 from $LiAlH_4$ and $AlCl_3$ has also been studied *via* alternative routes. Recent examples of adduct-free alane syntheses include the direct thermochemical conversion reported by Dinh *et al.*[17] and several convenient, scalable and energy-efficient mechanochemical methods developed by various groups.[18–20] The solvent-free mechanochemical reaction between $LiAlH_4$ and $AlCl_3$ under ambient temperature conditions indeed leads to the production of AlH_3 (and LiCl), but not without suffering a net loss in hydrogen capacity from the formation of metallic Al, perhaps caused by the *in situ* thermal decomposition of nascent AlH_3 generated during the processing.[18,19] One way to avoid this pitfall is to reduce the reaction temperature, preferably into the cryogenic range, but this would likely prove to be a complicated and expensive procedure during the scale-up.

Considering that $LiAlH_4$ is prepared either from LiH and $AlCl_3$ in diethyl ether,[7] or through an ion-exchange reaction between $NaAlH_4$ and LiCl,[21] it may be advantageous to be able to carry out a one-pot mechanochemical reaction starting from LiH. As far as we are aware, there is only one earlier report that describes the solid-state mechanochemical reaction between LiH and $AlCl_3$. According to the report, the mechanical agitation of a mixture with $LiH : AlCl_3 = 3.6$ for 4.5 h yielded only $LiAlH_4$ and LiCl with no evidence for the formation of alane.[22] Herein, we report an efficient, room-temperature mechanochemical procedure for the synthesis of alane *via* a solid-state metathesis reaction between LiH and $AlCl_3$. The key element of this new strategy is adding one of the reagents, $AlCl_3$, in three discrete steps (rather than all at once) so as to direct the reaction pathway toward the desired product (AlH_3) and away from unstable intermediates. We also demonstrate that the gas pressure applied during the mechanochemical processing may significantly affect the course of the reaction, which (to the best of our knowledge) has not hitherto been established.

2. Experimental

2.1. Materials and mechanochemical processing

LiH (Aldrich, 98%) and $AlCl_3$ (Aldrich, 99.99%) were used as starting materials without further purification. Because of the air and/or moisture sensitivity of the starting and the final compounds, all manipulations were performed in a continuously purified, argon-filled glove box with the oxygen and moisture levels controlled at less than 5 ppm (v/v). In a typical mechanochemical reaction, a total of *ca.* 1 g of starting material was weighed out in the desired molar ratio, and then transferred to a high-pressure stainless steel container that was filled with twenty chrome steel balls (11.9 mm dia.), each of which weighed *ca.* 7 g. The container was sealed under argon, and moved out of the glove box followed by evacuation and refilling with Zero grade H_2 (Linweld, 99.999%) until the desired pressure ranging between 1 to 350 bar was achieved. Ball-milling was carried out for various time intervals at 150 and 300 rpm with a ball-to-powder mass ratio (b : p) of ~140 : 1 in a two-station horizontal planetary mill (Fritsch, Pulverisette7). The milling sequence alternated between forward and reverse directions (2 min each) with an intermittent pause of 1 min to keep the average temperature of the vial as close to room temperature as possible during the entire ball-milling period (τ_{BM}).

2.2. X-ray powder diffraction analysis

The reaction products were characterized by X-ray powder diffraction (XRD) at room temperature on a PANalytical X'PERT powder diffractometer using Cu $K_{\alpha 1}$ radiation with a 0.02° 2θ step, in the range of Bragg angles 2θ from 10° to 80°. In order to protect the samples from air and moisture the sample was covered by a thin polyimide (Kapton) film during the measurements. The presence of the film resulted in an amorphous-like background in the XRD patterns in the 2θ range of 13–20°.

2.3. Solid-state NMR spectroscopy

Common solid-state (SS)NMR parameters, which are referred to in the description below, are defined as follows: ν_{RF} is the magnitude of the applied radiofrequency (RF) field, t_P is the duration of the applied RF pulse, τ_{CP} is the contact time for cross polarization (CP), τ_{RD} is the relaxation delay between transients, NS is the number of transients collected for signal averaging and ν_R is the magic angle spinning (MAS) rate.

All of the SSNMR data were collected on a 400-MHz (9.4 T) Agilent DD2 spectrometer equipped with a 3.2-mm Otsuka double resonance MAS probe using frequencies of 104.2 and 400.0 MHz for ^{27}Al and ^{1}H, respectively. ^{27}Al direct polarization (DP)MAS experiments were performed using single-pulse excitation under the following conditions: $\nu_R = 16$ kHz; ν_{RF} (^{27}Al) = 125 kHz with $t_P = 0.2$ μs; ν_{RF} (^{1}H) = 64 kHz during two-pulse phase-modulation (TPPM) decoupling;[23] τ_{RD} = 1 s; and NS = 512. Note that a short excitation pulse was employed in the DPMAS experiments to ensure that the observed signal intensities remained quantitative. The resulting numerical data obtained *via* the integration of the relative signal intensities are given in the discussion below (represented as S_{DP}) as well as in Table S1†. The ^{27}Al{^{1}H} CPMAS experiments were collected under the following conditions: $\nu_R = 16$ kHz; ν_{RF} (^{1}H) = 100 kHz during excitation (with $t_P = 2.5$ μs) and 64 kHz during CP and TPPM decoupling; ν_{RF} (^{27}Al) = ~48 kHz during CP; τ_{CP} = 0.2 ms; τ_{RD} = 60 s; and NS = 512. All of the ^{27}Al spectra were referenced with respect to a 1.0 M aqueous solution of $Al(NO_3)_3$, which exhibits a single sharp resonance at 0 ppm.

Due to the sensitive nature of the materials involved in this study, all samples were packed within an argon-filled glove box using zirconia MAS rotors equipped with "gas-tight" accessories (*i.e.*, a cap and spacer components that featured silicon rubber O-rings). These rotor accessories, when combined with careful packing protocols, were found to provide sufficient protection to the packed samples from air and/or moisture contamination, such that they could be removed from the glove box and analysed under ambient conditions, without spinning with dry nitrogen or using other precautionary measures (Fig S1†).

2.4. Temperature programmed desorption

Thermal desorption experiments were performed using a Setaram PCTPro-2000 automatic volumetric Sievert's type gas sorption instrument that was coupled to an RGA100 residual gas analyser. The as-prepared powder samples (*ca.* 0.15–0.2 g) were loaded in a custom built autoclave followed by volume calibration of the free sample space, which consisted of three helium absorption–desorption cycles over a period of 30–40 min. The thermal decomposition behaviour of the samples was studied by continuously monitoring the pressure change during heating at a rate of 4 °C min^{-1} from room temperature to 200 °C. This was followed by soaking at this temperature until saturation was achieved. A qualitative analysis of the gases released during the experiment was performed after each measurement.

3. Results and discussion

In the following we first describe our efforts to optimize the synthetic conditions by analysing the single-step reactions for different LiH : $AlCl_3$ ratios (3 : 1, 4 : 1, 6 : 1, 9 : 1 and 12 : 1) with hydrogen pressures between 1 and 350 bar. The lowest LiH : $AlCl_3$ ratio that did not result in the accumulation of metallic Al (9 : 1) was then used to produce a series of samples for detailed spectroscopic examination and subsequent mechanochemical processing with additional $AlCl_3$ to yield AlH_3 and LiCl. We conclude by describing a series of follow-up experiments performed to refine the understanding of the reaction mechanisms.

3.1 Determination of the initial reaction conditions

Based on the reported ethereal reaction between LiH and $AlCl_3$ (eqn (1)), a 3 : 1 molar mixture of LiH and $AlCl_3$ was prepared and processed mechanochemically under dry conditions (*i.e.* in the absence of diethyl ether). After 48 min of processing at 300 rpm under 350 bar of H_2 pressure (the maximum hydrogen pressure rating of our vials) only metallic Al and LiCl were obtained as major products. The XRD pattern and the corresponding reaction are shown in Fig. 1a. Thereafter, an approach was attempted in which a slight excess of LiH (4LiH : 1$AlCl_3$) was used to test whether the following reaction could provide a thermodynamically favorable pathway against the decomposition to metallic Al:

$$4LiH + AlCl_3 \rightarrow LiAlH_4 + 3LiCl \quad (3)$$

$$3LiAlH_4 + AlCl_3 \rightarrow 4AlH_3 + 3LiCl \quad (4)$$

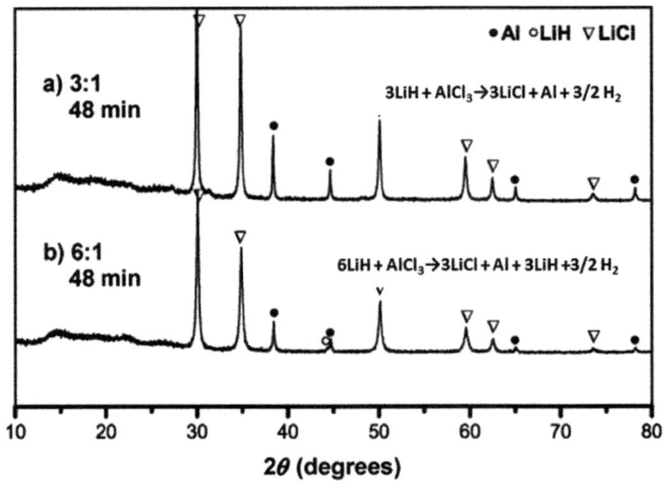

Fig. 1 The X-ray diffraction patterns corresponding to the as-milled samples with LiH : $AlCl_3$ starting ratios of a) 3 : 1, and b) 6 : 1, after ball-milling for 48 min under 350 bar H_2 with a milling speed of 300 rpm and a b : p of ~140.

However, this attempt was also ineffective, yielding for $\tau_{BM} = 48$ min only metallic Al, LiCl and unreacted LiH (XRD data not shown). Considering the very different reaction conditions presented by the dry mechanochemical approach *versus* those encountered in solvent-based reactions, the failed attempts described above suggest that 1) the formation of alane–ether adducts in the solution-based approach is essential for the stabilization of the hydride phase in the (nearly) stoichiometric reactions, and 2) in each case, the reactions proceed *via* thermodynamically controlled pathways, which are distinct under such dissimilar conditions.

We next hypothesized that the presence of a larger excess of hydride source in the starting mixture may prevent the formation of unstable intermediates that decompose to metallic Al. Indeed, when an arbitrarily chosen reaction mixture containing LiH and $AlCl_3$ in a molar ratio of 12 : 1 is processed under the same high hydrogen pressure, the reaction proceeds *without the formation of metallic* Al. Instead, lithium aluminium hexahydride (Li_3AlH_6), LiCl and unreacted LiH were obtained as the major products ($\tau_{BM} = 48$–60 min). Encouraged by these results, we set out to systematically explore the optimal composition of the starting mixture, and determined it to be 9LiH : $1AlCl_3$ under the milling parameters used. All reactions carried out using a LiH content less than this optimum resulted in metallic Al along with unreacted LiH, as shown in Fig. 1b for the starting mixture of 6 : 1. The processing of a 9 : 1 mixture was also tested at lower pressures, and it was found that the minimum pressure required to avoid the formation of metallic Al is about 300 bar under the milling parameters used. Again, all reactions carried out at less than 300 bar lead to metallic Al, as well as LiCl and LiH (Fig. S2†).

3.2 The 9 : 1 reaction of LiH and $AlCl_3$ (stage 1)

To investigate further the results observed at the starting ratio of 9 : 1, the reaction mixture was sampled as a function of ball-milling time and the resulting products and intermediates were analysed by XRD in Fig. 2 (top panel) and SSNMR (bottom panel), both of which provided clear and consistent mechanistic insight into the progression of the 9 : 1 reaction, as discussed below.

According to the XRD data obtained for $\tau_{BM} = 16$ min (pattern a), there exist three sets of Bragg peaks, including those from the reaction products, LiCl (the most intense peaks are at $2\theta \approx 30$, 35 and 50°)[24] and the monoclinic $LiAlCl_4$ (two sets of clearly identifiable Bragg peaks are at $2\theta \approx 15$–20° and 27–32°),[25,26] as well as the unreacted LiH ($2\theta \approx 38$ and 44°).[27]

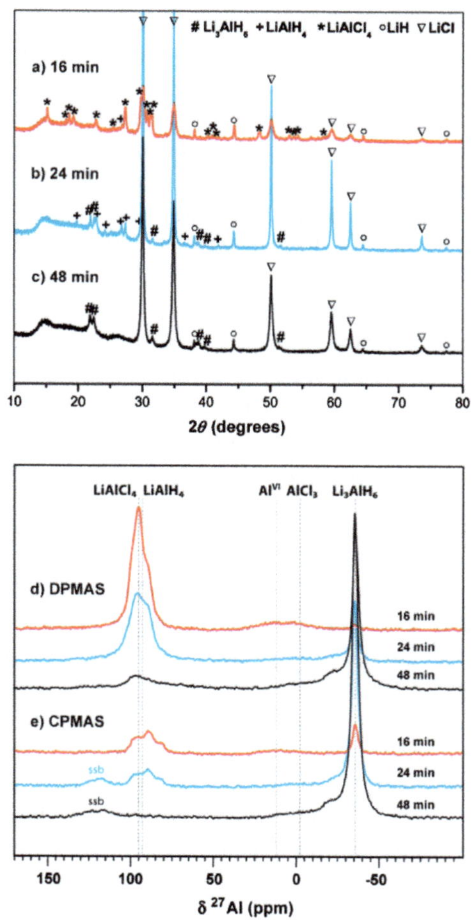

Fig. 2 The X-ray diffraction patterns (a–c) and SSNMR (^{27}Al DPMAS and $^{27}Al\{^1H\}$ CPMAS) spectra (d–e) obtained for samples of the 9 : 1 LiH : $AlCl_3$ reaction mass after ball-milling for the indicated time under 350 bar H_2 with a milling speed of 300 rpm and a b : p of ~140. Note that this reaction corresponds to the first stage of the stepwise mechanochemical synthesis of alane, as described in section 3.2. In the bottom two spectra of (e), 'ssb' is used to label a spinning sideband, which arises from the Li_3AlH_6 species.

The SSNMR measurements revealed the presence of five Al-containing species at $\tau_{BM} = 16$ min. These include a small amount ($S_{DP} \cong 9\%$) of unreacted $AlCl_3$, represented in the ^{27}Al DPMAS spectrum by a broad peak centred around −2 ppm. (Note that $AlCl_3$ cannot be detected by $^{27}Al\{^1H\}$ CPMAS.) The dominant spectral band between 80 and 105 ppm ($S_{DP} \cong 84\%$) consists of two superimposed central transition powder patterns ascribed to $LiAlH_4$ and $LiAlCl_4$. The above assignments are based on the reference spectra of $AlCl_3$, $LiAlH_4$ and $LiAlCl_4$ in neat form shown in Fig. S3†. The relative contributions of $LiAlH_4$ and $LiAlCl_4$ are difficult to quantify due to severe spectral overlap. Indeed, the ^{27}Al quintuple-quantum (5Q) MAS measurements of the neat $LiAlH_4$ and $LiAlCl_4$ materials have shown that the chemical shift and quadrupolar line shape parameters are very similar for these two species (Fig. S4†). Nonetheless, as noted several times in the ensuing discussion, useful information (*i.e.*, semi-quantitative approximations) can still be deduced from the changes in the apparent lineshape of the superposition signal since the constituent powder patterns are known from the abovementioned reference spectra. Another hydride observed at this stage is Li_3AlH_6, represented by a peak near −35 ppm, which comprises only a minor fraction ($S_{DP} \cong 2\%$) of the quantitative DPMAS intensity but is featured quite prominently in the CPMAS spectrum. This highlights the selectivity associated with the CPMAS experiment, whereby the ^{27}Al signals of species in close proximity to protons have the potential to be significantly amplified. By the same token, the absence of signals emanating from species such as $AlCl_3$ and $LiAlCl_4$ can be understood in terms of the lack of sufficiently nearby protons (*i.e.*, located within less than ~0.5 nm), without which $^{27}Al\{^1H\}$CP cannot operate. Consequently, the only species observed *via* CPMAS in the present study are those containing protons (*i.e.*, various aluminium hydrides).

Finally, the broad, low-lying resonance occurring in the neighbourhood of 13 ppm in the DPMAS and CPMAS spectra presents a strong case for the presence of a small fraction ($S_{DP} \cong 5\%$) of six-coordinated Al^{VI} in AlH_3 and/or chloride-substituted $AlCl_{3-x}H_x$-type species that form at even earlier times.[28,29] Indeed, it is quite possible that AlH_3 forms *via* the following reaction at early ball-milling times:

$$3/4\ LiH + AlCl_3 \rightarrow 3/4\ LiAlCl_4 + 1/4\ AlH_3 \tag{5a}$$

Eqn (5a) likewise supports the formation of $LiAlCl_4$, which, according to both the SSNMR and XRD analyses, also appears to be prominent at short milling times. Moreover, because the Al^{VI} hydrides are not observed near 13 ppm in the later samples obtained for stage 1, they appear to exist only as transient intermediates, which quickly vanish upon further processing. (Note that such short-lived hydride intermediates are also observed in the SSNMR spectra corresponding to samples obtained early in stage 2.) One plausible explanation for the disappearance of these Al^{VI} intermediate hydrides is based on their transformation into $LiAlH_4$, as shown for alane in the following reaction:[30]

$$AlH_3 + LiH \rightarrow LiAlH_4. \tag{5b}$$

Thus, eqn (5a) and (5b) not only account for the initial formation of AlH_3 and its subsequent disappearance but are also consistent with the early preponderance of $LiAlCl_4$ and $LiAlH_4$, directly in accordance with the SSNMR and XRD observations.

At $\tau_{BM} = 24$ min, the XRD pattern clearly shows the formation of $LiAlH_4$[31] and Li_3AlH_6[32] with a concomitant increase in LiCl and the disappearance of the $LiAlCl_4$ phase (Fig. 2, pattern (b)). The corresponding DPMAS and CPMAS spectra (Fig. 2d and e, middle traces) confirm the presence of $LiAlH_4/LiAlCl_4$ ($S_{DP} \cong 73\%$). The peak centred at 95 ppm has now a lineshape that appears more characteristic of $LiAlH_4$, suggesting that this species is largely responsible for the observed intensity at $\tau_{BM} = 24$ min. Accordingly, $LiAlCl_4$ is suspected to exist as only a minor component at 24 min. Also confirmed in the SSNMR spectra is the formation of Li_3AlH_6 ($S_{DP} \cong 24\%$) and the presence of unreacted $AlCl_3$ ($S_{DP} \cong 3\%$). Note that there is no indication of the aforementioned transient hydride species at 13 ppm. At this point, the relative abundance of Li_3AlH_6 can be explained by the following well-known transformation:[33]

$$LiAlH_4 + 2LiH \rightarrow Li_3AlH_6. \quad (6)$$

At $\tau_{BM} = 48$ min, the species detected by XRD were Li_3AlH_6, LiCl and unreacted LiH (Fig. 2, pattern (c)). The corresponding SSNMR experiments revealed, as expected, an intense resonance from Li_3AlH_6 ($S_{DP} \cong 90\%$), and the apparent complete loss of $LiAlH_4$. The minor resonance around 95 ppm ($S_{DP} \cong 10\%$) is assigned exclusively to $LiAlCl_4$, as suggested by the absence of a CP signal. The disappearance of $LiAlH_4$ can likewise be understood on the basis of eqn (6), with the complete consumption of this species signifying the end of stage 1. To ensure that this was the case, the milling time was extended up to 3 h and no significant change in the product profile was observed, thus providing further confirmation that the reaction is, indeed, complete at $\tau_{BM} = 48$ min under the milling conditions prescribed above.

With the exception of a few minor differences resulting mainly from the sensitivity of the XRD only to the crystalline phases, and the detection of both crystalline and amorphous phases by SSNMR, both techniques provided consistent and complementary results describing in detail the overall mechanochemical transformation sequence for the reaction of 9LiH : $1AlCl_3$ carried out at 350 bar H_2 pressure.

3.3 Further addition of $AlCl_3$ (4 : 1 cumulative LiH : $AlCl_3$ ratio, stage 2)

Thereafter, 2 molar equivalents of $AlCl_3$ were added to the product mixture of stage 1 (Li_3AlH_6, LiCl and LiH at $\tau_{BM} = 48$ min) to reach the desired LiH : $AlCl_3$ ratio of 3 : 1. When the resulting mixture was processed under 350 bar of H_2 pressure, a dark grey powder was obtained after 15–20 min of ball-milling, indicating the formation of metallic Al, which was later confirmed by XRD analysis (not shown). However, when the amount of $AlCl_3$ added during the this first attempt was slightly reduced to 1.25 molar equivalents instead of 2 (resulting in an overall LiH : $AlCl_3$ ratio of 4 : 1 rather than 3 : 1), the reaction proceeded without the formation of metallic Al. Again, the progress of the reaction was monitored using XRD and SSNMR, the results of which are shown in Fig. 3 and described below.

For $\tau_{BM} = 8$ min, the XRD pattern shows only $LiAlCl_4$ and LiCl as the crystalline phases present in the sample. However, the ^{27}Al DPMAS spectrum of this sample shows signals at ~95, 13, −2 and −35 ppm representing $LiAlH_4/LiAlCl_4$ ($S_{DP} \cong 66\%$), six-coordinated Al^{VI} ($S_{DP} \cong 8\%$), $AlCl_3$ ($S_{DP} \cong 8\%$), and Li_3AlH_6

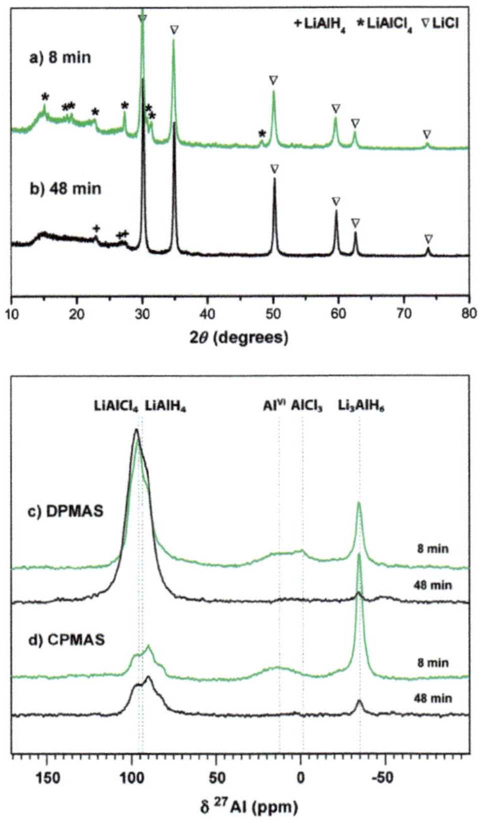

Fig. 3 The X-ray diffraction patterns (a–b) and SSNMR (^{27}Al DPMAS and $^{27}Al\{^1H\}$ CPMAS) spectra (c–d) obtained for samples of the 4 : 1 LiH : $AlCl_3$ reaction mass after ball-milling for the indicated time under 350 bar H_2 with a milling speed of 300 rpm and a b : p of ~140. Note that this starting composition (obtained by adding 1.25 moles of $AlCl_3$ to the as-milled 9 : 1 products shown in Fig. 2) corresponds to the second stage of the stepwise mechanochemical synthesis of alane, as described in section 3.3.

(S_{DP} ≅ 17%), respectively (Fig. 3, c–d for τ_{BM} = 8). Although the relative contributions of $LiAlH_4$ and $LiAlCl_4$ are difficult to quantify, the lineshape of the superposition signal at 95 ppm is strikingly similar to that observed at τ_{BM} = 8 min during stage 1, suggesting that the reactions occurring in each case are initiated by a similar mechanism, despite their overall stoichiometric differences. In addition, in both stages the spectra contain a similarly small, yet conspicuous, signal arising from the Al^{VI} species around 13 ppm, which is absent at later processing times. Further still, these spectra also contain similar amounts of unreacted $AlCl_3$ at −2 ppm. In fact, apart from all of the similarities mentioned above, these spectra essentially differ only in the concentration of Li_3AlH_6.

After τ_{BM} = 48 min, the resonance observed at ~95 ppm dominates both the DPMAS (S_{DP} ≈ 97%) and CPMAS spectra. The lineshape of this resonance in the DPMAS spectrum indicates a larger contribution from $LiAlH_4$, which is supported by a similar increase observed in the corresponding CPMAS signal. The only other species observed after 48 min of milling is Li_3AlH_6, although it constitutes only ~3% of the overall DPMAS intensity. Thus, notwithstanding this and some relatively small fraction of $LiAlCl_4$, the major product observed upon completion of stage 2 is $LiAlH_4$, which is in accordance with eqn (3) given the 4 : 1 ratio of LiH : $AlCl_3$ in the mixture.

3.4 The final addition of $AlCl_3$ (3 : 1 cumulative LiH : $AlCl_3$ ratio, stage 3)

Lastly, in order to obtain AlH_3 as the final product, another 0.75 molar equivalent of $AlCl_3$ was added to the mixture obtained upon completion of stage 2. The resulting mixture, having a net composition of 3LiH : 1$AlCl_3$, was then processed as in the previous two stages. As was observed in stage 2 at $\tau_{BM} = 8$ min, the XRD pattern of the products at $\tau_{BM} = 10$ min shows only Bragg peaks corresponding to

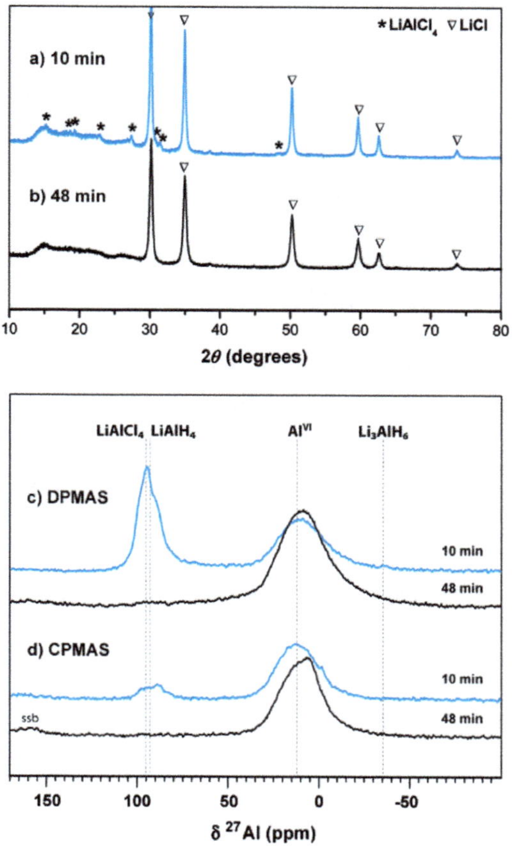

Fig. 4 The X-ray diffraction patterns (a–b) and SSNMR (^{27}Al DPMAS and ^{27}Al{^{1}H} CPMAS) spectra (c–d) obtained for samples of the 3 : 1 LiH : $AlCl_3$ reaction mass after ball-milling for the indicated time under 350 bar H_2 with a milling speed of 300 rpm and a b : p of ~140. Note that this starting composition (obtained by adding 0.75 moles of $AlCl_3$ to the as-milled 4 : 1 products shown in Fig. 3) corresponds to the third and final stage of the stepwise mechanochemical synthesis of alane, as described in section 3.4. In the last spectrum of (d), 'ssb' is used to label a spinning sideband, which arises from the Al^{VI} species.

LiCl and $LiAlCl_4$ (Fig. 4, pattern (a)). It may be noted that the intensities for the $LiAlCl_4$ phase appear weak because of the relatively high concentration of LiCl in the sample. Although the expected Bragg peaks for AlH_3 are not observed in the diffraction patterns, together with the SSNMR results, the absence of metallic Al and the starting components indicate that the reaction proceeded as intended. Considering that the corresponding DPMAS spectrum exhibits a pronounced increase in the intensity of the signal at ~13 ppm assigned to the Al^{VI} species ($S_{DP} \approx 55\%$), the absence of Bragg peaks attributable to AlH_3 likely reflects the amorphous or nanocrystalline nature of the resulting aluminium trihydride. In addition to the resonance assigned to Al^{VI}, there is yet a substantial contribution from the signal occurring at ~95 ppm after 10 min, which encloses the remaining 45% of the total DPMAS intensity. Interestingly, the lineshape of this signal again closely resembles that observed at early points in the previous two stages. As such, it appears that comparable amounts of $LiAlCl_4$ and $LiAlH_4$ exist with the latter likely making a somewhat larger contribution. Regardless, it is quantitatively clear from the DPMAS spectrum that after 10 min of milling most of the aluminium species in the reaction mixture are now present as Al^{VI} hydrides.

At $\tau_{BM} = 48$ min, only the Bragg peaks corresponding to LiCl are observed by XRD. The corresponding SSNMR spectra now show only one resonance band centered around 13 ppm suggesting that all available Al has transformed to the hydrogenated Al^{VI} species. Note, however, that the line shape of the Al^{VI} signal in this sample differs slightly between the DPMAS and CPMAS experiments, suggesting that the final product may contain several polymorphic forms of AlH_3 or other structurally related species, most likely Cl-containing derivatives of AlH_3, such as $AlClH_2$,[28,29] though only in relatively small proportions.

3.5 A summary of the overall reaction scheme

Based on the XRD and SSNMR results, the reaction progress during stages 1–3 may be summarized as follows:

$$1^{st}\ \textbf{stage}:\ 9LiH{:}1AlCl_3 \rightarrow 9LiH + AlCl_3 \rightarrow \underbrace{3LiCl + Li_3AlH_6 + 3LiH}_{(\mathbf{a})} \quad (7)$$

$$2^{nd}\ \textbf{stage}:\ 4LiH{:}1AlCl_3 \rightarrow (\mathbf{a}) + 5/4AlCl_3 \rightarrow \underbrace{27/4LiCl + 9/4LiAlH_4}_{(\mathbf{b})} \quad (8)$$

$$3^{rd}\ \textbf{stage}:\ 3LiH : 1AlCl_3 \rightarrow (\mathbf{b}) + 3/4\ AlCl_3 \rightarrow 36/4\ LiCl + 12/4\ AlH_3 \quad (9)$$

3.6 Temperature-programmed desorption (TPD) analysis

Furthermore, the products that were obtained after the completion of each stage, and their compositions were verified from the TPD curves. Fig. 5 compares the decomposition profiles of the products obtained from eqn (7)–(9) after $\tau_{BM} = 48$ min. The single-step desorption observed for the 9 : 1 sample (Fig. 5, curve (a)) has a desorption onset temperature (T_d) of *ca.* 180 °C, which is characteristic for the decomposition of pristine Li_3AlH_6.[34,35] The net hydrogen evolution of

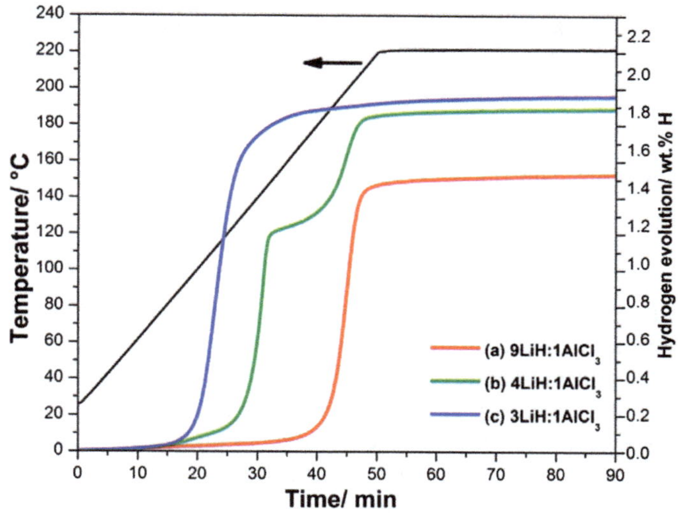

Fig. 5 The TPD curves obtained for samples of the reaction mixtures resulting from LiH : $AlCl_3$ starting ratios of a) 9 : 1, b) 4 : 1 and c) 3 : 1 after 48 min of ball-milling under 350 bar H_2 with a milling speed of 300 rpm and a b : p of ~140. Note that these samples correspond (respectively) to the final products obtained upon completion of the first, second and final stages of the stepwise mechanochemical synthesis of alane, as described in the text and shown in eqn (7), (8) and (9).

ca. 1.4 wt% is in good agreement with the ratio of the products obtained in eqn (7), considering that the decomposition temperature of LiH is much higher than 200 °C.[34] It may also be noted that at 200 °C Li_3AlH_6 releases only 1.5 mole of H_2 gas (50% of the total hydrogen content by weight) according to eqn (11) below.

Curve (b) in Fig. 5 represents the decomposition profile of the products obtained after the second stage (eqn (8)). The two-step decomposition with T_d values close to 130 °C and 180 °C is in excellent agreement with the decomposition pathway of $LiAlH_4$ as shown below.[34]

$$LiAlH_4 \rightarrow 1/3Li_3AlH_6 + 2/3Al + H_2 \text{ (theory 5.3 wt\% H)} \tag{10}$$

$$1/3Li_3AlH_6 \rightarrow LiH + 1/3Al + 1/2H_2 \text{ (theory 2.6 wt\% H)} \tag{11}$$

The measured hydrogen capacity of *ca.* 1.8 wt% also agrees well with the theoretical value of 1.87 wt%, assuming that eqn (8) proceeds to completion. Finally, the desorption profile of the products from eqn (9) shows a one-step decomposition with a T_d of *ca.* 90 °C, as typically expected for AlH_3,[36] and a net hydrogen evolution corresponding to 1.9 wt%, which confirms the completion of the reaction. In all three cases, the mass-spectroscopic analyses showed that H_2 constituted 99.7% or more of the gases released, the rest being residual air. The XRD analyses of the desorbed samples from the 9 : 1 and 4 : 1 mixtures showed Bragg peaks corresponding to metallic Al, LiH and LiCl. On the other hand only metallic Al and LiCl were observed in the desorbed 3 : 1 sample, which suggests that LiH was completely consumed during the mechanochemical processing and that all of the alanates formed as intermediates during the first two stages were transformed into alane (Fig. S5†). To summarize, the desorption profiles of the products obtained after stages 1–3 are in excellent agreement with the XRD and SSNMR analyses, and thereby further validate the suggested pathway for this mechanochemical transformation, as given by eqn. (7)–(9).

3.7 Reaction mechanisms

To explain why the reaction between LiH and $AlCl_3$ in a 3 : 1 molar ratio leads to metallic Al even under H_2 pressure, products from this reaction were analysed for $\tau_{BM} = 6$, 12 and 18 min. It was observed that the 3 : 1 reaction does not yield metallic Al right away, but rather proceeds to form a small amount of $LiAlCl_4$ at $\tau_{BM} = 18$ min, in a pathway similar to that of the 9 : 1 reaction. (Caution! This product, which likely contains $AlCl_2H$ and $AlClH_2$, is highly unstable and reacts vigorously under argon even upon a slight impact to yield metallic Al and H_2 gas). However, in contrast to the 9 : 1 reaction after 18 min, a significant amount of unreacted $AlCl_3$ was also noticed in the 3 : 1 mixture, which suggests that the 3 : 1 reaction is slower under similar milling conditions. A comparison of the progress of the 3 : 1 and 9 : 1 reactions within the first 20 min of milling is shown in Fig S6†. Whereas the formation of Li_3AlH_6 is already evident in the 9 : 1 mixture for $\tau_{BM} = 24$ min, no alanate peaks are observed in the 3 : 1 reaction. Thus, one may infer that in the presence of excess LiH in the mixture, the formed intermediates (likely a mixture of $AlCl_2H$ and $AlClH_2$) are readily transformed to alanate before decomposing into metallic Al. In parallel to the observations made by Mikheeva *et al.* during their studies of the reaction between LiH and $AlCl_3$ (in ether), the following argument may explain to some extent the failure of the stoichiometric 3 : 1 and 4 : 1 reactions.[37] Because of the relatively low volume fraction in these mixtures compared to 9 : 1, LiH may not be sufficiently dispersed, leading to high local $AlCl_3$/LiH ratios. This in turn may lead to the decomposition of unstable chlorine-containing derivatives such as $AlCl_2H$ and $AlClH_2$ (or AlH_3AlCl_3) into metallic Al with the evolution of hydrogen, rather than the formation of alanate. It was also noticed by Mikheeva *et al.* that the slow addition of an ethereal solution of $AlCl_3$ improved the yield of $LiAlH_4$, which is akin to the required sequential addition of $AlCl_3$ in the present case.

To obtain further insight, another reaction was performed in which the 3 : 1 LiH : $AlCl_3$ reaction mixture was processed at half the milling speed (150 rpm) and, as before, at 350 bar H_2 pressure. It was observed that the reaction progresses without the formation of metallic Al, albeit at a very slow pace, so much so that it was incomplete even after 20 h of milling, as shown by the SSNMR results (Fig S7†). More interestingly, this latter reaction could also be performed at a relatively low H_2 pressure of less than 10 bar, although the reaction time remained unaffected. However, when this reaction was performed under argon at a pressure slightly over 1 bar (*i.e.*, glove box pressure), metallic Al was obtained within 60 min of milling. Based on these results one may naively conclude that at low milling speeds the local temperature spikes are insignificant when compared to those at high milling speeds, which prevents the decomposition of AlH_3. However there may be two caveats to this view. Firstly, it is well known that AlH_3 is kinetically metastable at room temperature with an equilibrium pressure of >1 GPa. It is quite unlikely in this scenario that a moderate gas pressure of 350 bar could prevent the decomposition of nascent AlH_3. Secondly, this view does not support the formation of metallic Al in the 4 : 1 mixture since it would only lead to $LiAlH_4$ as a final product (eqn (3)), which is known to be stable under milling unless carried out in the presence of a catalyst such as titanium.[38] It is quite intriguing to note that the 9 : 1 composition, which is far from the equilibrium composition of 3 : 1, still requires a pressure of about 300 bar to prevent the formation of metallic Al at 300 rpm. Although the minimum pressures required during the second and third stages were not determined independently in this work, it is known that a certain critical pressure (*ca.* 200 bar) is needed to prevent the formation of metallic Al during the reaction given by eqn (9).[39] Although the role of the pressure during the processing is not clearly understood, it is postulated that the gas under pressure acts as a fluid medium that may assist in the long-range solid-state diffusion of ions that is essential for preventing the destabilization of Al–H bonds. Alternatively, it is also likely that the high-pressure gas may serve as an effective heat transport medium and hence alter the mechanochemical reaction pathway.

It is also clear that the manipulation of the gas pressure alone may not be sufficient to avoid the Al–H decomposition pathway. Indeed, the minimum required pressure for the 9 : 1 reaction is 300 bar, whereas the minimum for the 3 : 1 reaction (if any) is higher (likely, much higher) than 350 bar. Thus, the presence of excess LiH in the mixture appears to be essential for the preservation of the newly formed Al–H bonds *via* the formation of intermediate, thermodynamically stable alanates.

4. Conclusion

A solvent-free mechanochemical route has been developed for the preparation of alane starting from LiH and $AlCl_3$ at room temperature. Although this reaction has been known to occur in solvents such as diethyl ether, it is important to discover methods that are simple to carry out on a large scale. In contrast to the wet-chemical route however, the direct solvent-free mechanochemical reaction of LiH and $AlCl_3$ at a ratio of 3 : 1 or 4 : 1 fails to yield AlH_3 or $LiAlH_4$, respectively. Instead, the use of excess LiH in the starting mixture leads to conditions that favour the formation of Al–H bonds resulting in the intermediate alanates. The results described in this work establish this mechanochemical route as a versatile and very useful alternative to solution-based methods, which reduces or eliminates the use of potentially hazardous solvents. We note, however, that in addition to non-solvated alane, the resulting material contains about 80% of dead weight in the form of a LiCl by-product, which reduces the net gravimetric hydrogen density by the same amount. The separation of alane from LiCl is thus critical for realizing its full potential as a hydrogen storage material. Although different solubilities of alane and LiCl in various organic solvents have been employed with varying success to achieve the separation, future research must focus on finding a suitable solvent, or a combination of solvents, that are both inexpensive and benign. Other separation techniques (including solvent-free) may become practical if similar reactions are performed using different starting hydrides and/or aluminium salts leading to more easily separable by-products. It is also demonstrated that the mechanochemical reaction pathway can be altered by adjusting the applied gas pressure. In the present case, this was demonstrated by manipulating the H_2 pressure such that the reaction pathway leading to metallic Al was completely avoided. Although the exact role of the gas pressure could not be clearly elucidated, it is postulated that the gas under high pressure acts as a fluid medium, which may have significant effects on the underlying solid-state diffusion processes. Alternatively, the high-pressure gas may serve as an effective heat transfer medium, which may alter the local temperature profile and thereby affect the reaction pathway. The ability to successfully synthesize alane at room temperature also eliminates the need for cryogenic cooling to suppress the formation of metallic Al. A better understanding of the mechanistic pathway for the synthesis of alane is expected to provide the much needed basic scientific insight necessary for the development of more refined approaches not only toward the production of AlH_3, but also related reactions, including the direct mechanochemical hydrogenation of metallic Al, which remains elusive to this day.

Acknowledgements

This work was supported by the U.S. Department of Energy (DOE), Office of Science, Basic Energy Sciences, Materials Science and Engineering Division. The research was performed at the Ames Laboratory, which is operated for the U.S. DOE by Iowa State University under contract No. DE-AC02-07CH11358.

References

1 L. Schlapbach and A. Züttel, *Nature*, 2001, **414**, 353.
2 D. Abbott, *Proc. IEEE*, 2010, **98**, 42.
3 J. Graetz, J. Reilly, V. Yartys, J. Maehlen, B. Bulychev, V. Antonov, B. Tarasov and I. Gabis, *J. Alloys Compd.*, 2011, **509**, S517.
4 http://www1.eere.energy.gov/hydrogenandfuelcells/storage/pdfs/targets_onboard_hydro_storage_explanation.pdf.
5 J. M. Pasini, C. Corgnale, B. A. van Hassel, T. Motyka, S. Kumar and K. L. Simmons, *Int. J. Hydrogen Energy*, 2013, **38**, 9755.
6 O. Stecher and E. Wiberg, *Ber. Dtsch. Chem. Ges. B*, 1942, **75**, 2003.
7 A. E. Finholt, A. C. Bond and H. I. Schlesinger, *J. Am. Chem. Soc.*, 1947, **69**, 1199.
8 G. Chizinsky, *J. Am. Chem. Soc.*, 1955, **77**, 3164.
9 F. Brower, N. Matzek, P. Reigler, H. Rinn, C. Roberts, D. Schmidt, J. Snover and K. Terada, *J. Am. Chem. Soc.*, 1976, **98**, 2450.
10 E. C. Ashby, J. R. Sanders, P. Claudy and R. Schwartz, *J. Am. Chem. Soc.*, 1973, **95**, 6485.
11 A. E. Finholt, G. D. Barbaras, G. K. Barbaras, G. Urry, T. Wartik and H. I. Schlesinger, *J. Inorg. Nucl. Chem.*, 1955, **1**, 317.
12 B. M. Bulychev, A. V. Golubeva and P. A. Storozhenko, *Russ. J. Inorg. Chem.*, 1998, **43**, 1765.
13 S. K. Konovalov and B. M. Bulychev, *Inorg. Chem.*, 1995, **34**, 172.
14 J. Graetz, *Chem. Soc. Rev.*, 2009, **38**, 73.
15 R. Zidan, B. L. Garcia-Diaz, C. S. Fewox, A. C. Stowe, J. R. Gray and A. G. Harter, *Chem. Commun.*, 2009, **25**, 3717.
16 G. S. McGrady, US Patent US 2008/0241056 A1, 2 October, 2008.
17 L. V. Dinh, D. A. Knight, M. Paskevicius, C. E. Buckley and R. Zidan, *Appl. Phys. A: Mater. Sci. Process.*, 2012, **107**, 173.

18 H. W. Brinks, A. Istad-Lem and B. C. Hauback, *J. Phys. Chem. B*, 2006, **110**, 25833.

19 M. Paskevisius, D. A. Sheppard and C. E. Buckley, *J. Alloys Compd.*, 2009, **487**, 370.

20 Y. C. Luo, S. K. Mao, R. X. Yan, L. B. Kong and L. Kang, *Acta Physico-Chimica Sinica*, 2009, **25**, 237.

21 E. C. Ashby, G. J. Brendel and H. E. Redman, *Inorg. Chem.*, 1963, **2**, 499.

22 N. N. Mal'tseva, K. G. Myakishev, A. I. Golovanova, V. V. Volkov and N. Kuznetsov, *Zh. Neorg. Khim.*, 1989, **34**, 40.

23 A. E. Bennett, C. M. Rienstra, M. Auger, K. V. Lakshmi and R. G. Griffin, *J. Chem. Phys.*, 1995, **103**, 6951.

24 PDF 74-1972 (ICDD, 2000).

25 PDF 70-435 (ICDD, 2000).

26 G. Mairesse, P. Barbier, J. P. Wignacourt and P. Baert, *Cryst. Struct. Commun.*, 1977, **6**, 15.

27 PDF 78-839 (ICDD, 2000).

28 E. Wiberg, M. Schmidt, H. Graf and R. U. Lacal, *Z. Naturforsch*, 1952, **7b**, 578.

29 E. Wiberg and M. Schmidt, *Z. Naturforsch*, 1951, **6b**, 333.

30 G. Sandrock, J. Reilly, J. Graetz, W.-M. Zhou, J. Johnson and J. Wegrzyn, *J. Alloys Compd.*, 2006, **421**, 185.

31 PDF 73-461 (ICDD, 2000).

32 PDF 27-282 (ICDD, 2000).

33 I. Zaluski, A. Zaluska and J. O. Storm-Olsen, *J. Alloys Compd.*, 1999, **290**, 71.

34 J. A. Dilts and E. C. Ashby, *Inorg. Chem.*, 1972, **11**, 1230.

35 V. P. Balema, V. K. Pecharsky and K. W. Dennis, *J. Alloys Compd.*, 2000, **313**, 69.

36 G. Sandrock, J. Reilly, J. Graetz, W.-M. Zhou, J. Johnson and J. Wegrzyn, *Appl. Phys. A: Mater. Sci. Process.*, 2005, **80**, 687.

37 V. I. Mikheeva, E. M. Fedneva and Z. L. Shnitkova, *Zh. Neorg. Khim.*, 1956, **1**, 2440.

38 V. P. Balema, J. W. Wiench, K. W. Dennis, M. Pruski and V. K. Pecharsky, *J. Alloys Compd.*, 2001, **329**, 108.

39 S. Gupta, T. Kobayashi, I. Z. Hlova, M. Pruski and V. K. Pecharsky, unpublished.

Faraday Discussions

PAPER

Mechanochemical synthesis of Au, Pd, Ru and Re nanoparticles with lignin as a bio-based reducing agent and stabilizing matrix†

Monika J. Rak, Tomislav Friščić and Audrey Moores

Received 26th March 2014, Accepted 31st March 2014
DOI: 10.1039/c4fd00053f

A versatile, low-energy and solvent-free method to access nanoparticles (NPs) of four different transition metals, based on a bottom-up mechanochemical procedure involving milling of inorganic precursors, is presented. Lignin, a biomass waste, was used effectively as a reducing agent, for the first time in a mechanochemical context, to access MNPs where M = Au, Pd, Ru, Re. A series of metal precursors was used for this reaction and their nature was shown to be integral in determining whether NPs became incorporated within the organic lignin matrix, M@lignin, or not. Specifically, organometallic precursors resulted in extensive encapsulation of the NPs, as well as improved control over their size and shape, while ionic precursors afforded matrix-free NPs. The resulting NP-containing composites were characterized through Fourier-transform infrared attenuated total reflectance (FTIR-ATR) spectroscopy, transmission electron microscopy (TEM), X-ray photoelectron spectroscopy (XPS) and powder X-ray diffraction (PXRD). This mechanochemical grinding method for accessing M@lignin (M = Au, Pd, Ru and Re) is significantly more sustainable than the traditional solvent batch syntheses of metal NPs because it relies on the use of a biomass-based polymer, it is highly atom economical, it eliminates the need for solvents and it reduces drastically the energy input.

Introduction

Mechanical grinding or milling is one of the simplest and industrially relevant methods to obtain materials in the form of finely divided particles of micro- and nano-scale. This top-down approach to access nanoparticles (NPs) has been highly popular since 1960s and proceeds *via* comminution, *i.e.* mechanically crushing

Centre for Green Chemistry and Catalysis, Department of Chemistry McGill University, 801 Sherbrooke St. W., H3A 0B8 Montreal, Canada. E-mail: audrey.moores@mcgill.ca; tomislav.friscic@mcgill.ca; Fax: +1(514)398 3797; Tel: +1(514)398 3959

† Electronic supplementary information (ESI) available: Experimental details; TEM measurements; FTIR-ATR, XPS and PXRD data. See DOI: 10.1039/c4fd00053f

bulk material until a sufficiently small size is obtained (Scheme 1a).[1,2] Despite its simplicity, this approach relies on a significant input of mechanical energy to achieve the breakdown of hard inorganic materials and affords polydisperse NPs in a wide range of sizes and shapes and invariably leads to the appearance of surface defects and amorphization.[3,4] This surface deteriorating effect is the basis of mechanical activation of solids and has been recently studied extensively by the Šepelak group, who established the formation of 2–5 nm thick amorphous layers upon mechanical milling of oxide NPs.[5] Recently, however, an alternative and innovative use of mechanical milling has been inaugurated for the synthesis of NPs, through a bottom-up process based on chemical reactions (Scheme 1b).[6,7]

This novel approach first centered on the synthesis of binary inorganic NPs and has largely been pioneered by the Baláž group, who conducted the simple metathesis reactions between soft metal salt solids (*e.g.* cadmium, zinc or lead chlorides and acetates) and soft solids acting as ionic or molecular sulfide sources (*e.g.* sodium sulfide, cystine *etc*) under solvent-free mechanochemical milling conditions.[8–10] This method afforded monodisperse NPs of group II–VI metal chalcogenides. Such a bottom-up mechanochemical approach to NP synthesis is based on self-assembly processes, rather than particle comminution. Specifically the nucleation of the new phase and the growth of nanocrystals occurs in the presence of an effective capping agent that hinders NP agglomeration and prevents their growth beyond a certain size. The same principles were used by the group of Luque to synthesize ball-milled composites of magnetic iron oxide-based nanocatalysts into mesoporous structures and exploit them in catalysis.[11,12]

The self-assembly approach to NP synthesis is highly appealing as it proceeds under very low energy input, compared to both top-down mechanochemical and solvent-based self-assembly approaches.[7] Moreover, the bottom-up mechanochemical approach ensures the formation of highly ordered structures with a high degree of monodispersity, in a similar fashion to solvent-based approaches which remain popular solutions to high-end applications. Mechanochemistry, however, has a much higher potential for scale-up than solvent-based techniques and can

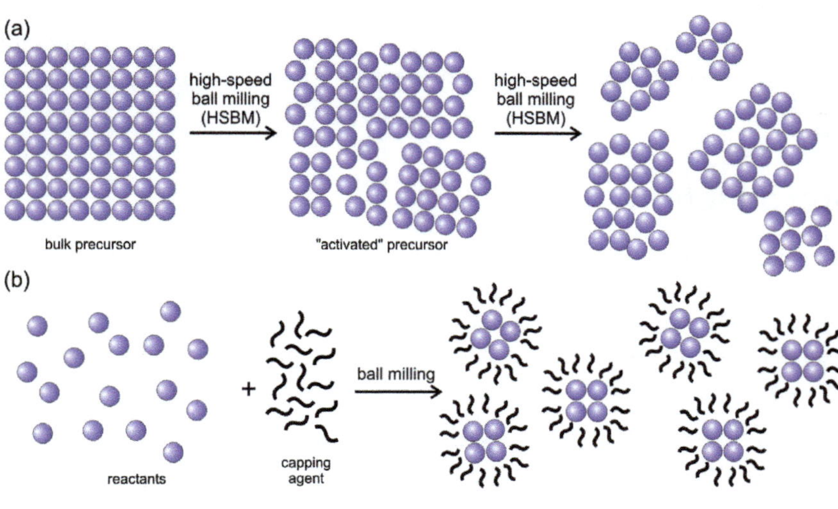

Scheme 1 Illustrations of the (a) top-down comminuting and (b) bottom-up self-assembly approaches to the mechanosynthesis of NPs.

also drastically reduce waste generation.[13,14] In short, mechanochemical bottom-up synthesis of NPs has a strong potential to provide industry-compatible solutions, and their scope urgently needs to be widened to more materials.

Bottom-up mechanosynthesis of metal NPs

Besides for the synthesis of binary inorganic NPs, it was recently demonstrated that mechanochemical bottom-up assembly synthesis was also applicable to metal NPs. In this scheme, a metal salt is used as precursor and reduced during the mechanosynthesis. One of the first examples was the formation of AuNPs by reduction of potassium tetrachloroaurate(III) ($KAuCl_4$) with sodium borohydride ($NaBH_4$).[15] The solvent-free reaction was conducted by mechanochemical milling with poly(vinylpyrrolidone) as a protecting polymer, yielding AuNPs with sizes from 6–28 nm diameter, depending on the ratio of components in the reaction mixture and milling time. This pioneering investigation of metal NPs synthesis through mechanochemical reduction was followed by a report of mechanochemical synthesis of gold-silver nanoclusters *via* a two-step process in which a chitosan biopolymer was used both as a protecting agent and a reducing agent in the presence of sodium hydroxide.[16] Manual grinding of silver nitrate with chitosan using a mortar and pestle leads to the formation of AgNPs which galvanically reduce tetrachloroaurate(III) acid ($HAuCl_4$) with further grinding and form NPs of a gold–silver alloy. Consequently, this work demonstrated the possibility of using galvanic reduction for the synthesis of metal NPs, as well as the use of a bio-based polymer, chitosan, both as a reducing and NP protecting agent.

We have recently demonstrated the mechanochemical synthesis of monodisperse AuNPs by mechanochemical galvanic reduction of $HAuCl_4$ in the presence of terminal monoalkylamines as capping agents (Fig. 1). The reduction of $HAuCl_4$ was achieved without the need for external reductants by the action of the stainless steel milling assembly. The size of the resulting ultra-small AuNPs (diameter from 1–2 nm) was found to be controlled readily by the length of the alkyl chain on the capping agent, as well as by milling time.

Lignin as a mechanochemical reduction agent and matrix for NP dispersion

The current search for non-petroleum chemical feedstocks has focused much attention to biomass-derived polymers: lignin, cellulose, or hemicellulose.[17,18] The particular potential of lignin lies in a structure rich in aromatic residues with the potential to provide a source for a wide variety of aromatic compounds (Scheme 2).[19] The exploitation of lignin as an aromatic chemicals feedstock is hampered by the robustness of its extensive three-dimensional network structure, which renders this material intractably insoluble. However, it has been recently shown by Bolm's group that this polymeric structure can be broken down by mechanical grinding, milling or rubbing in the presence of a solid base, enabling subsequent dissolution.[20] The mechanical degradation of biomass polymers is thought to proceed *via* radical intermediates, as speculated in the early 1990s by Sumimoto *et al.* for the transformation of lignin into chromophores.[21,22] Mechanical breakdown of the lignin structure is thought to cause the formation of phenoxy radicals of considerable stability and, although the exact reaction is not fully understood, it is believed that the phenol groups adjacent to the methoxy groups in lignin structures can be oxidized and therefore used as reducing agents.

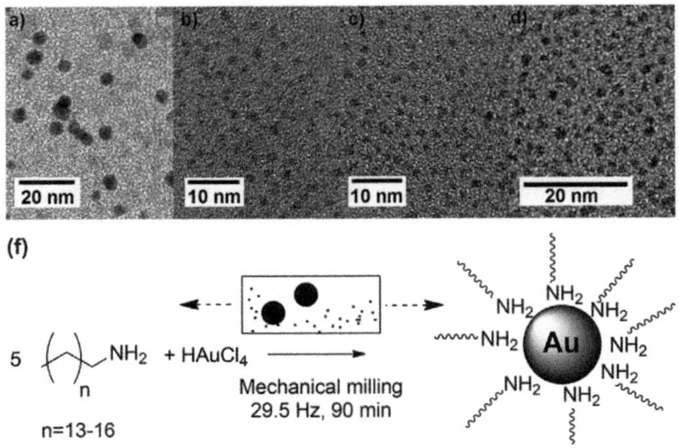

Fig. 1 (a)–(e) Transmission electron microscopy (TEM) images of amine-protected AuNPs prepared by mechanochemical galvanic reduction of $HAuCl_4$ by stainless steel milling assembly and (f) schematic of the underlying reaction.[7]

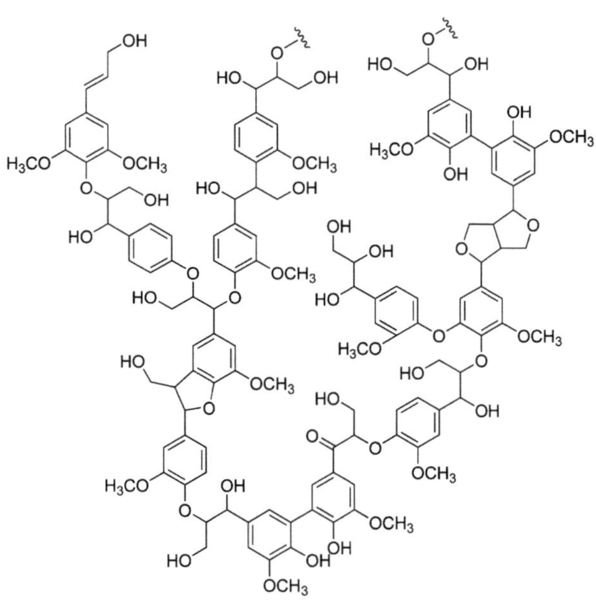

Scheme 2 Proposed structure of Kraft pine lignin.[24]

The increased interest in exploitation of biomass to form value-added products[23,24] has led to the synthesis of composite materials containing metal NPs and lignin (MNP@lignin). AuNPs@lignin was applied to Surface Enhanced Resonance Spectroscopy (SERS) detection of trinitrotoluene (TNT), as lignin could concentrate TNT in the vicinity of the AuNPs. In 2010, a proof of concept article by Lucas *et al.* showed that silver and gold NPs between 20 and 100 nm may be incorporated in wood samples.[25] Pd and PtNPs@lignin were also produced and applied to catalytic oxidation and reduction reactions.[26] In these examples, the

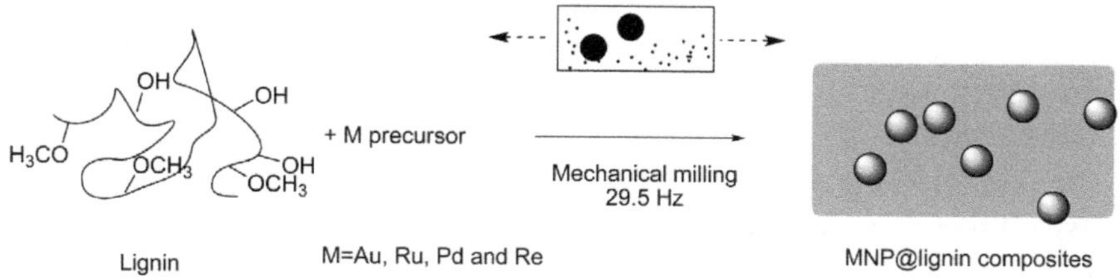

Scheme 3 Mechanochemical synthesis of MNP@lignin composites with M = Au, Pd, Ru and Re.

three-dimensional structure of lignin does not only provide reducing abilities for the generation of the composite, but also makes lignin a suitable solid support for NPs, allowing for easy separation of the catalyst through filtration.

So far, most applications of lignin and lignin-based materials in NP technology have focused on its role as an abundant and inexpensive support material, and synthetic methods are based on conventional techniques. We now describe the first exploration of lignin as both a reduction agent and a NP dispersion matrix for the direct mechanochemical synthesis of NP–lignin composites involving gold, palladium, ruthenium and rhenium (Scheme 3). In doing so, we also drastically extend the scope of existing mechanochemical synthesis of metal NPs, affording access to catalytically relevant materials. The reductive properties of lignin, that can be exploited in a solvent-free mechanochemical environment, have been utilized for *in situ* milling reduction of different metal precursors into NPs. Interestingly, the incorporation of the resulting NPs within lignin appears to be dependent on the nature of the metal precursor and its ability to mix readily with lignin. The produced NP–lignin composites were characterized through transmission electron microscopy (TEM), X-ray photoelectron spectroscopy (XPS), powder X-ray diffraction (PXRD), and Fourier-transform infrared attenuated total reflectance spectroscopy (FTIR-ATR).

Experimental

Materials and reagents

All chemicals and solvents were purchased from commercial sources and used without purification. Hydrogen tetrachloroaurate(III) hydrate, hexakis[μ-(acetato-O:O′)]-triaqua-μ_3-oxotriruthenium(III) (referred to as 'basic Ru(III) acetate' thereafter), palladium(II) acetate, palladium(II) acetylacetonate, and rhenium pentacarbonyl bromide were purchased from Strem Chemicals, Inc., ruthenium(III) chloride hydrate and palladium(II) chloride were purchased from Pressure Chemical. A Westvaco Chemical Division, Indulin AT Kraft pine lignin was purchased and used without further purification. This is a 99% lignin content free-flowing brown powder lignin. 400-mesh carbon supported TEM grids were obtained from Electron Microscopy Science.

Equipment

High-resolution TEM and EDAX were performed using a Philips CM200 200 kV TEM. XPS was performed on a VG ESCALAB 3 MKII spectrometer (VG, Thermo Electron Corporation, UK) equipped with an Mg Kα source. PXRD was performed using the Bruker D2 Phaser diffractometer using a Cu Kα source. A Retsch Mixer Mill MM 400 was used to perform milling experiments. A Perkin Elmer Spectrum Two FTIR with a single bounce diamond ATR was used for all FTIR measurements.

Typical procedure (synthesis of AuNP@lignin)

In a typical reaction a 10 mL stainless steel milling jar was filled with a total of 200 mg of solid reagent material, consisting of 0.1 mmol of the metal precursor with the remainder being Kraft lignin powder. To this reaction mixture were added two stainless steel balls of 7 mm diameter (1.34 grams weight, total ball-to-sample weight ~13). The jar was then closed and loaded onto the Retsch MM400 mixer mill. The reactions were conducted over 90 minutes, milling at a frequency of 29.5 Hz. At the end of the reaction, the solid product was scraped out of the jar with a spatula and weighed. The product was placed onto a Kimwipe® filter plug in a Pasteur pipette and the solid was washed using three 1 mL aliquots of water, followed by three 1 mL aliquots of acetone. The solid was allowed to dry under vacuum overnight and analyses were run on the dried samples. TEM grids were prepared by suspending the powders in acetone and depositing the resulting suspensions onto a 400-mess carbon supported copper TEM grid. XPS samples were run by supporting dry samples onto carbon tape. The FTIR-ATR and PXRD measurements samples were conducted on solid dry samples.

Results

Mechanochemical milling of precursors of Au, Pd, Ru and Re with lignin was performed to establish the possibility to form the corresponding MNPs@lignin whereby NPs are embedded inside the lignin matrix to allow future applications. A summary of the results obtained is provided in Table 1.

Preparation of AuNP@lignin composites

Mechanochemical milling of $HAuCl_4$ with Kraft lignin produced a dark purple loose powder. Characterization of the milled sample with TEM revealed the

Table 1 Summary of the results obtained by TEM and XPS after the mechanochemical reduction of metal precursors by lignin

Metal precursor	Size (standard deviation)/nm[a]	Morphology of composite	Oxidation state (XPS)
$HAuCl_4$	14.8 (6.8)	AuNPs in lignin	Au(0)
$PdCl_2$	3.2 (1.6)	NPs outside of lignin[b]	Pd(0) and PdO
Pd(acetate)$_2$	2.8 (0.7)	PdNP in lignin	Pd(0) and PdO
Pd(acetylacetonate)$_2$	3.6 (0.9)	PdNP in lignin	Pd(0) and PdO
Basic Ru(III) acetate	3.1(1.5)	RuNP in lignin	RuO
$RuCl_3$	6.7 (4.2)	NPs outside of lignin[c]	RuO
$Re(CO)_5Br$	6.2 (1.5)	Snail eye morphology	ReO_2

[a] Based on TEM measurment. [b] PdNPs were observed on the TEM grids of washed, solid phase product, outside of lignin. [c] No RuNPs were observed with washed, solid phase product. RuNPs were visible on TEM grids produced with the washings of reaction product.

formation of a composite material in which high contrast metal-containing NPs were dispersed throughout a low contrast matrix presumed to be lignin. Importantly, TEM analysis clearly demonstrated that the NPs are uniformly distributed throughout the volume of lignin (Fig. 2a,b).

The chemical composition of the NPs was established by X-ray photoelectron spectroscopy (XPS) which revealed the 4f signals characteristic for Au(0),[27] therefore confirming that the NPs are metallic gold (Fig. 2c). The formation of AuNPs was also confirmed by powder X-ray diffraction (PXRD) analysis which revealed strong X-ray reflections consistent with (111), (200), (220), and (311) reflections of metallic gold (Fig. 3a).[28]

The AuNPs embedded in the lignin matrix have an average diameter of 14.8 ± 6.8 nm and are larger than the ones we produced using amines.[26] This size is consistent with the Scherrer particle size analysis based on the PXRD (Fig. 3a) pattern, which gives a size of 14.3 (±2.1) nm (assuming a spherical particle shape). In all other samples explored herein the metal NPs were of considerably smaller sizes and it was, consequently, difficult to identify by PXRD which were generally dominated by a strong background generated by the amorphous lignin support (see the ESI†). In this reaction, we predicted that lignin would play the role of a reducing agent to afford Au(0) species. In order to verify this hypothesis, FTIR-ATR was used to detect oxidation of the lignin matrix. Whereas the FTIR-ATR spectrum of the Kraft lignin is dominated by extensive phenolic O–H stretching bands distributed around 3200 cm^{-1}, the spectra of AuNP@lignin were characterized by a strong C=O absorption band around 1700 cm^{-1} resulting from lignin oxidation (Fig. 3b).

Preparation of PdNP@lignin composites

Next, we explored the possibility of mechanochemically synthesising lignin composites containing Pd NPs. Palladium(II) chloride ($PdCl_2$), palladium(II) acetate and palladium(II) acetylacetonate were selected as metal precursors for this experiment. Analysis by TEM revealed a significant difference in the morphology of the materials obtained with $PdCl_2$ *vs* the other two precursors (Fig. 4).

Whereas all samples demonstrated the presence of PdNPs, the samples generated from $PdCl_2$ largely consisted of NPs clearly located outside of the lignin matrix. In contrast, both the acetate and the acetylacetonate precursors yielded composites in which the PdNPs were distributed throughout the organic matrix. The choice of the precursor only affected marginally the size of the resulting NPs, with average sizes measured to be 3.2, 2.8 and 3.6 nm with $PdCl_2$, $Pd(acetate)_2$ and $Pd(acac)_2$ respectively. The NPs size dispersity and shape were also dependent on the precursor. The NPs formed from $PdCl_2$ outside of the organic matrix exhibited a diversity of shapes and a standard deviation of 1.6 nm. An improved monodispersity was observed for particles obtained from $Pd(acetate)_2$ and $Pd(acac)_2$ with values of 0.7 and 0.9 respectively. The obtained particles were also more spherical in shape. In all cases, XPS analysis revealed that the NPs consist of Pd^0 and $Pd^{II}O$ species[29] and FTIR-ATR confirmed the oxidation of the lignin structure and thus its reducing role in the process (see the ESI†).[30] These results reveal a clear correlation between the size and shape control of the resulting PdNPs and the ability for the system to embed them properly inside the lignin matrix. It is

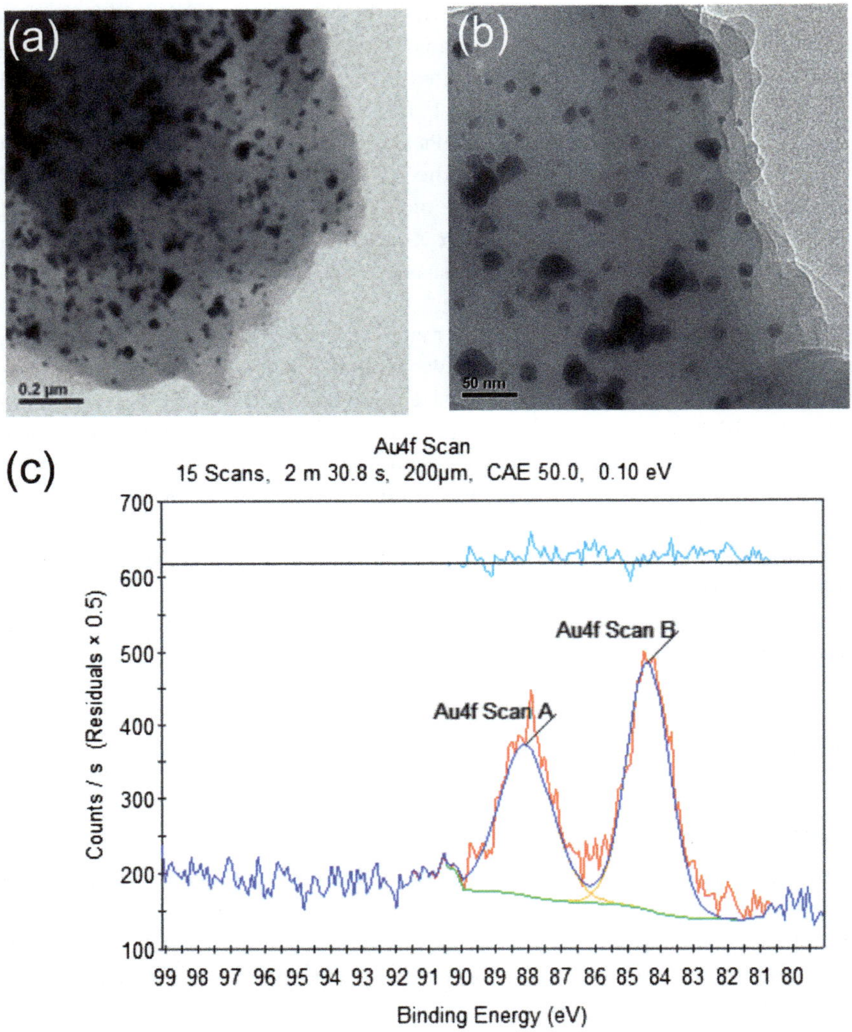

Fig. 2 (a)–(b) Transmission electron microscopy (TEM) images of the mechanochemically prepared AuNP@lignin composite at different magnifications. (c) X-Ray photoelectron spectrum of the AuNP@lignin composite.

also likely that lignin here acts not only as a reducer, but also as a stabilizer, since its presence in high concentration at the PdNP surface affords better control over the growth process. We hypothesize that the difference between the chloride-based and the metal–organic precursors resides in their compatibility with the organic matrix. The metal–organic precursors demonstrate a clear superiority to incorporate within the non-polar lignin structure.

Preparation of RuNP@lignin composites

Attempts to synthesize lignin composites containing embedded RuNPs revealed similar behaviour to Pd-based systems. The synthesis of the composites was explored using two commercially and readily accessible metal precursors, the

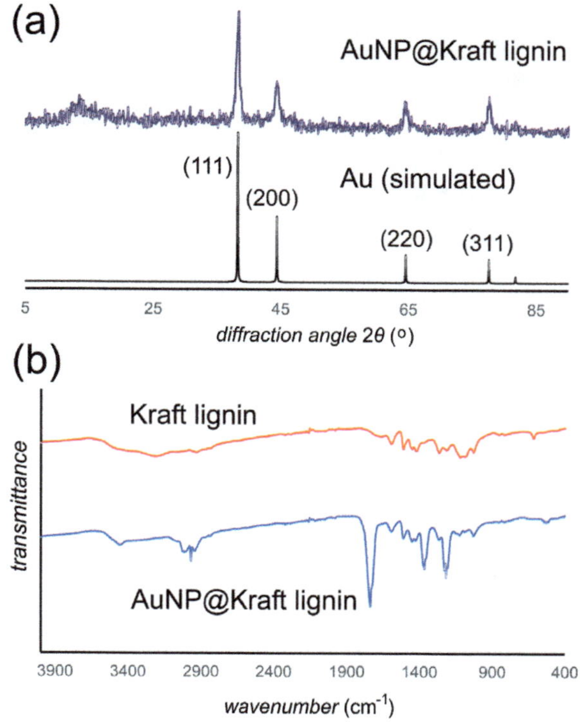

Fig. 3 (a) Overlay of powder X-ray diffraction pattern of the prepared AuNP@lignin composite (top) and the one simulated for gold metal (bottom). (b) Overlay of FTIR-ATR spectra of precursor Kraft lignin after milling for 90 min (top) and a mixture of $HAuCl_4$ and Kraft lignin after milling for 90 minutes, corresponding to the AuNP@Kraft lignin composite.

inorganic, ionic ruthenium(III) chloride hydrate ($RuCl_3 \cdot nH_2O$) and the trinuclear μ-oxo metal-organic complex, basic Ru(III) acetate. Both precursors produced NPs upon milling with Kraft lignin. However, samples prepared using $RuCl_3$ did not exhibit any NPs under TEM analysis of the washed sample, despite the oxidation of lignin evidenced by FTIR-ATR (ESI). Next, we analysed the washings of this reaction product by TEM and indeed detected RuNPs of 6.7 + 4.2 nm diameter besides the organic matter composed presumably of nano- and micron-sized lignin fragments (see the ESI†). On the other hand, the metal–organic complex precursor yielded lignin samples with RuNPs distributed throughout the volume of the organic matrix (Fig. 5a,b). As previously, oxidation of lignin was confirmed by FTIR-ATR (see the ESI†).

In both instances the Ru3d XPS spectra of mechanochemical products were identical, exhibiting 3d emission lines consistent with the presence of $Ru^{II}O$ (Fig. 5c,d).[31] The appearance of RuO is not surprising, as the formation of an oxide layer on ruthenium NPs was previously observed and established to become increasingly prominent with the reduction in particle size.[31] Similar to the results obtained with the Pd series, RuNPs produced by mechanochemical synthesis in the presence of lignin have a very different morphology depending on the nature of the Ru precursor. Basic Ru(III) acetate, an organometallic precursor, could mix

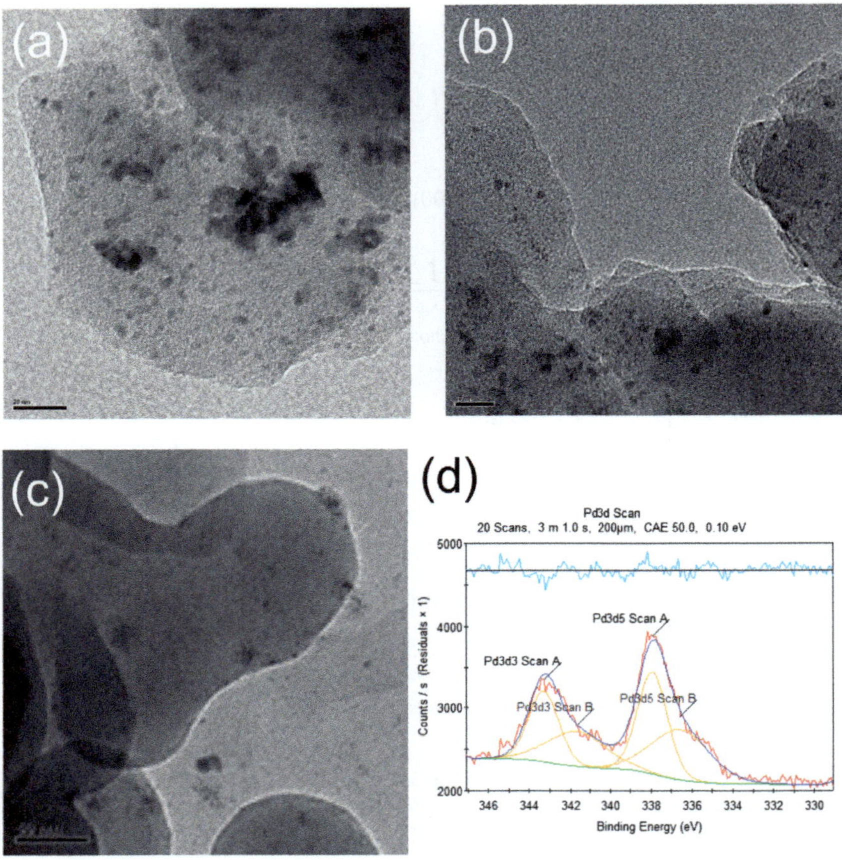

Fig. 4 TEM images of samples prepared by mechanochemical reaction of Kraft lignin with: (a) palladium(II) acetate; (b) palladium(II) acetylacetonate and (c) palladium(II) chloride. (d) Sample XPS spectrum for the mechanochemically prepared NP-lignin composite from the $PdCl_2$ precursor.

easily within the lignin upon milling and afforded well-embedded RuNPs@lignin composites, while the chloro-metallic species affords RuNPs besides the matrix. Interestingly, the fact that lignin appears oxidized in all cases shows that lignin does reduce Ru precursors even when their uptake within the lignin matrix is negligible. The versatility of this synthetic method is exemplified by the possibility to access either free NPs or matrix supported NPs by a careful choice of precursor, while using the same bio-based reducer: lignin.

Mechanochemical reduction of rhenium(I) precursors

We have also performed preliminary experiments in generating Re-based nanoparticles through mechanochemical lignin-induced reduction. Milling of $Re(CO)_5Br$ with Kraft lignin for 90 minutes at 29.5 Hz led to the formation of ReNPs, as demonstrated by TEM analysis (Fig. 6). The use of $Re(CO)_5Br$ as a metal precursor is particularly interesting due to the presence of a non-ionic halide covalently bound to the metal centre as well as the presence of organometallic CO

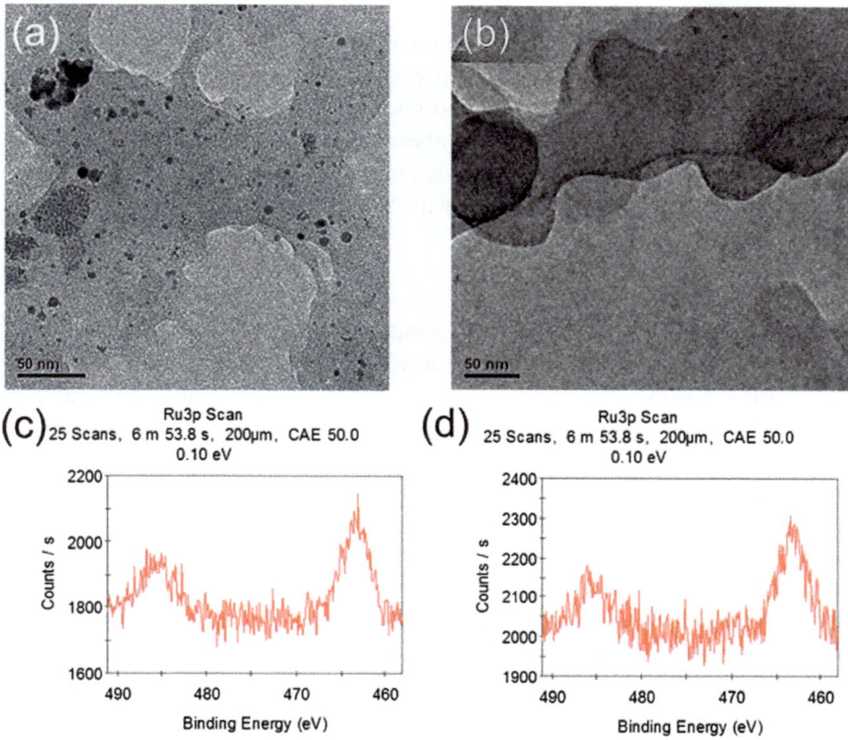

Fig. 5 TEM images of RuNPs obtained by milling of Kraft lignin with: (a) basic ruthenium(III) acetate and (b) hydrated ruthenium(III) chloride. XPS spectra of the RuNPs generated by grinding Kraft lignin with: (c) basic ruthenium(III) acetate and (d) hydrated ruthenium(III) chloride.

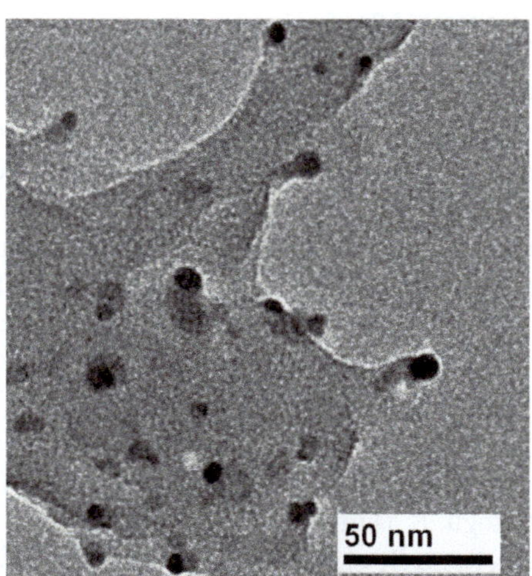

Fig. 6 TEM image of the ReNP@lignin composite obtained by mechanical milling of $Re(CO)_5Br$ and lignin.

ligands. The generated ReNPs were found to be highly monodisperse in size with a diameter of 6.2 ± 1.5 nm and distributed neither fully inside the matrix, nor detached from it. The TEM images clearly reveal the ReNPs lying at the surface of the organic matrix, which adopts a morphology similar to snail eyes. FTIR-ATR confirmed, as in all samples described above, the oxidation of the lignin structure (see the ESI†). However the XPS analysis did not confirm the reduction of the Re species and showed peaks consistent with ReO_2.

Conclusion

We have shown, for the first time, the viability of mechanochemical milling with lignin as a means to producing high quality, and highly monodisperse metal NPs. The mechanochemical methodology presented herein was readily applicable to four distinct noble metals: Au, Pd, Ru and Re. In most explored examples lignin was demonstrated to participate as a reducer of metal salts to afford metal NPs which, in some cases, were partially oxidized. An interesting variability of the product structure was observed in the dependence on the nature of the metal precursor used. For Pd and Ru, metal chloride precursors afforded metal NPs detached from the lignin matrix, while the use of metal–organic precursors yielded a metal NP@lignin composite. This observation opens up the possibility to use a mechanochemical lignin-based reduction procedure to prepare either free or lignin-supported NPs using essentially the same methodology, simply by selecting the right precursor. We perceive the presented study, based on four different metals, as a robust proof-of-concept for a potentially general methodology to synthesize metal nanoparticles. In addition to the generality, which is currently being explored in our laboratory, this methodology also offers several attractive advantages over conventional solution-based NP syntheses, such as reducing the amount of waste, eliminating the use of solvent during synthesis, and enabling the use of an inexpensive reducing agent generated from biomass.

Acknowledgements

We thank the Natural Science and Engineering Research Council of Canada (NSERC) Discovery Grant program, the Canada Foundation for Innovation (CFI), the Canada Research Chairs (CRC), the Fonds de Recherche du Québec – Nature et Technologies (FRQNT) Nouveau Chercheur and Equipe programs, the Centre for Green Chemistry and Catalysis (CGCC), NSERC-Collaborative Research and Training Experience (CREATE) in Green Chemistry and McGill University for their financial support.

References

1 P. Balaz, M. Achimovicova, M. Balaz, P. Billik, Z. Cherkezova-Zheleva, J. M. Criado, F. Delogu, E. Dutkova, E. Gaffet, F. J. Gotor, R. Kumar, I. Mitov, T. Rojac, M. Senna, A. Streletskii and K. Wieczorek-Ciurowa, *Chem. Soc. Rev.*, 2013, **42**, 7571–7637.

2 C. Suryanarayana, *Prog. Mater. Sci.*, 2001, **46**, 1–184.

3 V. Sepelak, A. Duvel, M. Wilkening, K.-D. Becker and P. Heitjans, *Chem. Soc. Rev.*, 2013, **42**, 7507–7520.

4 V. Sepelak, S. Begin-Colin and G. Le Caer, *Dalton Trans.*, 2012, **41**, 11927–11948.

5 V. Sepelak, I. Bergmann, S. Indris, A. Feldhoff, H. Hahn, K. D. Becker, C. P. Grey and P. Heitjans, *J. Mater. Chem.*, 2011, **21**, 8332–8337.

6 L. Russo, F. Colangelo, R. Cioffi, I. Rea and L. D. Stefano, *Materials*, 2011, **4**, 1023–1033.

7 M. J. Rak, N. K. Saade, T. Friscic and A. Moores, *Green Chem.*, 2014, **16**, 86–89.

8 P. Baláž, M. Baláž, M. Čaplovičová, A. Zorkovska, L. Čaplovič and M. Psotka, *Faraday Discuss.*, 2014, DOI: 10.1039/C3FD00117B.

9 P. Baláž, E. Boldižárová, E. Godočíková and J. Briančin, *Mater. Lett.*, 2003, **57**, 1585–1589.

10 P. Baláž and E. Dutková, *J. Therm. Anal. Calorim.*, 2007, **90**, 85–92.

11 M. Ojeda, A. M. Balu, V. Barron, A. Pineda, A. G. Coleto, A. A. Romero and R. Luque, *J. Mater. Chem. A*, 2014, **2**, 387–393.

12 A. Pineda, A. M. Balu, J. M. Campelo, R. Luque, A. A. Romero and J. C. Serrano-Ruiz, *Catal. Today*, 2012, **187**, 65–69.

13 J. A. Dahl, B. L. S. Maddux and J. E. Hutchison, *Chem. Rev.*, 2007, **107**, 2228–2269.

14 C. J. Murphy, *J. Mater. Chem.*, 2008, **18**, 2173–2176.

15 D. Debnath, S. H. Kim and K. E. Geckeler, *J. Mater. Chem.*, 2009, **19**, 8810–8816.

16 A. Murugadoss, N. Kai and H. Sakurai, *Nanoscale*, 2012, **4**, 1280–1282.

17 A. K. Mohanty, M. Misra and L. T. Drzal, *J. Polym. Environ.*, 2002, **10**, 19–26.

18 P. Anastas and N. Eghbali, *Chem. Soc. Rev.*, 2010, **39**, 301–312.

19 D. Stewart, *Ind. Crops Prod.*, 2008, **27**, 202–207.

20 T. Kleine, J. Buendia and C. Bolm, *Green Chem.*, 2013, **15**, 160–166.

21 Z.-H. Wu, D.-Y. Lee, M. Matsuoka and M. Sumimoto, *Mokuzai Gakkaishi*, 1991, **37**, 164–171.

22 Z.-H. Wu and M. Sumimoto, *Mokuzai Gakkaishi*, 1992, **38**, 277–284.

23 J. Virkutyte and R. S. Varma, *Chemical Science*, 2011, **2**, 837–846.

24 W. Thielemans, E. Can, S. S. Morye and R. P. Wool, *J. Appl. Polym. Sci.*, 2002, **83**, 323–331.

25 M. Lucas, B. A. Macdonald, G. L. Wagner, S. A. Joyce and K. D. Rector, *ACS Appl. Mater. Interfaces*, 2010, **2**, 2198–2205.

26 F. Coccia, L. Tonucci, D. Bosco, M. Bressan and N. d'Alessandro, *Green Chem.*, 2012, **14**, 1073–1078.

27 N. H. Turner and A. M. Single, *Surf. Interface Anal.*, 1990, **15**, 215–222.

28 C.-H. Wang, C.-C. Chien, Y.-L. Yu, C.-J. Liu, C.-F. Lee, C.-H. Chen, Y. Hwu, C.-S. Yang, J.-H. Je and G. Margaritondo, *J. Synchrotron. Radiat.*, 2007, **14**, 477–482.

29 M. Brun, A. Berthet and J. C. Bertolini, *J. Electron Spectrosc. Relat. Phenom.*, 1999, **104**, 55–60.

30 T. Schalow, B. Brandt, D. E. Starr, M. Laurin, S. K. Shaikhutdinov, S. Schauermann, J. Libuda and H.-J. Freund, *Angew. Chem., Int. Ed.*, 2006, **45**, 3693–3697.

31 M. E. Grass, Y. Zhang, D. R. Butcher, J. Y. Park, Y. Li, H. Bluhm, K. M. Bratlie, T. Zhang and G. A. Somorjai, *Angew. Chem., Int. Ed.*, 2008, **47**, 8893–8896.

Faraday Discussions

PAPER

The dual role of sulfur-containing amino acids in the synthesis of IV–VI semiconductor nanocrystals: a mechanochemical approach

Peter Baláž,*[a] Matej Baláž,[a] Mária Čaplovičová,[b] Anna Zorkovská,[a] Ľubomír Čaplovič[c] and Miroslav Psotka[d]

Received 2nd December 2013, Accepted 22nd January 2014

DOI: 10.1039/c3fd00117b

PbS@cystine nanocrystals were synthesized mechanochemically, with lead acetate and L-cystine being used as the lead and sulfur precursors, respectively. The resulting nanocrystals are 22–34 nm in size, well-faceted and octahedral in shape. Characterization by XRD, FT-IR, NMR, FE-SEM, EDS, TEM (HRTEM) and surface area measurement methods showed that the particles are single, defect-free crystals with a high crystallinity. Furthermore, the crystals were prepared using a solvent-free procedure that was performed under ambient temperature and atmospheric pressure.

1 Introduction

Initiated by original works from the early 1980s,[1–5] the research on semiconductor nanocrystals has made considerable progress.[6] Lead sulfide PbS is a IV–VI semiconductor that has a narrow (0.41 eV) band gap and a relatively large (18 nm) Bohr radius at room temperature. It has been reported that the band gap of PbS can be widened to the visible range by forming nanoclusters. Hence, PbS nanoparticles may have potential applications in electroluminescent devices, such as light-emitting diodes (LED).[7]

To date, many synthetic methods have been reported for the preparation of PbS nanoparticles of various morphologies, such as hollow spheres,[8] wires,[9] rods,[10] tubes,[11] cubes[12] and star-like structures.[13] Syntheses starting from solid

[a]*Institute of Geotechnics, Slovak Academy of Sciences, Watsonova 45, 04353 Košice, Slovakia. E-mail: balaz@saske.sk; balazm@saske.sk; zorkovska@saske.sk; Fax: +421557922604; Tel: +421557922603*

[b]*Comenius University, Faculty of Natural Sciences, Mlynská dolina, 84215 Bratislava, Slovakia. E-mail: caplovicova@fns.uniba.sk; Fax: +4210265429064; Tel: +4210260296111*

[c]*Slovak University of Technology, Faculty of Materials Science and Technology, Paulínska 16, 91724 Trnava, Slovakia. E-mail: lubomir.caplovic@stuba.sk; Fax: +421906068299; Tel: +421906068000*

[d]*University of P. J. Šafárik, Faculty of Science, Moyzesova 11, 04001 Košice, Slovakia. E-mail: miroslav.psotka@gmail.com; Fax: +421556222124; Tel: +421552341198*

state high-energy milling are governed by the mechanochemical methodology.[14–19] Quite recently, the mechanochemical synthesis starting from eggshell membrane and lead acetate was used for the preparation of PbS nanocrystals.[20] The reactions occur at the interfaces of the nanometer-sized grains, which are continuously regenerated during milling. Ambient temperatures, atmospheric pressure and solvent-free techniques are usually sufficient for performing both inorganic and organic mechanochemical reactions with total conversion.[21–28]

L-Cystine is characterized by its disulfide linkage (-S-S-). These disulfide bridges occur in many natural proteins secreted by cells, and the role of these bridges is the stabilization of the generally fragile protein structures. Disulfide bonds are broken at temperatures above 150 °C, especially at low moisture levels (below 20%).[29] Upon their cleavage, it is possible to utilize the free sulfur for the synthesis of interesting compounds.

This contribution belongs to the group of mechanochemical bottom-up syntheses of inorganic nanoparticles, which are of big interest.[30–33] We report the direct synthesis of PbS using L-cystine and lead acetate as precursors. The hallmarks of the synthesis are the presence of crystalline water in the acetate, which mediates the mechanochemical synthesis, and L-cystine, which plays a dual role as the sulfur precursor and surface-functionalizing agent.

2 Experimental

2.1 The synthesis of the PbS@cystine nanocomposites

Materials. Lead acetate $Pb(CH_3COO)_2\cdot 3H_2O$ (ITES, Slovakia) and L-cystine (Alfa Aesar, Germany) were used as precursors for the mechanochemical synthesis of PbS nanocrystals.

Mechanochemical synthesis. The mechanochemical synthesis of the PbS nanocrystals was performed in a Pulverisette 6 planetary mill (Fritsch, Germany). The milling conditions were used as follows. The loading of the mill: 50 balls of 10 mm diameter; the ball charge in the mill: 360 g; the material of the milling chamber and balls: tungsten carbide; the rotation speed of the planet carrier: 500 rpm; an argon atmosphere as a protective medium in the mill; laboratory temperature; milling time: 15–240 minutes. The precursors were added into the mill in a lead acetate : L-cystine molar ratio of 2 : 1, because of the presence of two sulphur atoms in the structure of L-cystine and only one lead atom in the structure of lead acetate (stoichiometric ratio). The product was washed with 0.003M HCl.

2.2 Characterization methods

XRD. The X-ray powder diffraction measurements were carried out in Bragg–Brentano geometry using the D8 Advance diffractometer (Bruker, Germany), working with Cu Kα radiation and a scintillation detector. For the data treatment the commercial Bruker processing tools have been used. Concretely, for the phase identification the Diffrac ^{plus} Eva and for the Rietveld analysis and microstructure characterization (crystalline size) the Diffrac ^{plus} Topas software has been utilized. For phase analysis the ICDD-PDF2 database has been used. The crystalline size was estimated by the "double-Voight" method and using the integral breadth, since this characteristic is the most comparable to values observed by TEM. The conversion degree α has been estimated according to the formula:

$$\alpha = (1 - \frac{I_{cystine}}{I_{PbS} + I_{cystine}}) \times 100\ [\%],$$

where $I_{cystine}$ and I_{PbS} are the values of the intensities of the strongest peaks in the diffraction patterns of L-cystine and PbS, respectively.

FT-IR. The infrared spectra in the frequency range 4000–649 cm^{-1} were obtained by using an IR spectrophotometer Tensor 29 (Bruker, Germany) using the ATR method.

NMR. The ^1H NMR spectra of the compounds dissolved in D_2O (with the addition of a small amount of NaOH in order to make cystine form a soluble salt) were recorded at ambient temperature on a Varian PremiumCOMPACT 600 (599.87 MHz, Varian, USA) spectrometer using tetramethylsilane (TMS) as the internal reference. The chemical shifts are given in δ relative to TMS. The coupling constants (J) were obtained by first-order analysis and measured in Hz.

TEM and HRTEM. Sample preparation: the powder sample was added to ethanol and sonicated for 15 minutes. Then, the suspension was dropped on the Cu-grid with a hole carbon film. After drying in air, the sample was examined in the transmission electron microscope. Low magnification TEM images were recorded by a JEOL JEM 2000FX microscope at 160 kV. The nanocrystal sizes were measured by Image J software. The HRTEM images were taken by a Philips CM300 microscope operated at 300 kV (point-to point resolution 0.205 nm, line resolution 0.144 nm). GATAN Digital Micrograph Software was used for the image acquisition and processing and JCrystal software was employed for a schematic drawing of the nanocrystal shapes.

FE-SEM. The study was performed on the JEOL Field Emission Scanning Electron Microscope JSM-7600F at accelerating voltage 5kV.

Specific surface area, S_A. The specific surface area was determined by the low temperature nitrogen adsorption method in Gemini 2360 sorption apparatus (Micromeritics, USA).

EDS. The powder sample was stuck to a carbon conductive adhesive tape and then coated with carbon because of conductivity. The Energy Dispersive X-ray (EDS) microanalysis was performed at 20 kV on an Oxford instruments X-Max detector attached to a JEOL FE-SEM 7600F microscope.

3 Results and discussion

The progress of the mechanochemical formation of PbS was pursued by XRD. The patterns obtained for all of the milled samples are shown in Fig. 1. As can be seen from the figure, after 15 min of milling, only the cystine peaks can be clearly identified. However, after 60 min of milling, the peaks of lead sulfide begin to show. Lead acetate is consumed very quickly, because no peaks belonging to it were identified. The mechanochemical process is finished after 105 min, because no change in the XRD patterns is evidenced for the samples milled for a longer time.

Fig. 2 comprises the XRD patterns (A–C) and FT-IR (D–F) spectra of both precursors and the product. Fig. 2C shows a typical XRD pattern of lead sulfide PbS (ICDD 65-9496) formed by milling. The diffraction peaks showed that the lead sulfide is well-crystallized; however, PbS was not the only component identified in the product. Peaks from crystalline L-cystine were present (ICDD 23-1663), despite

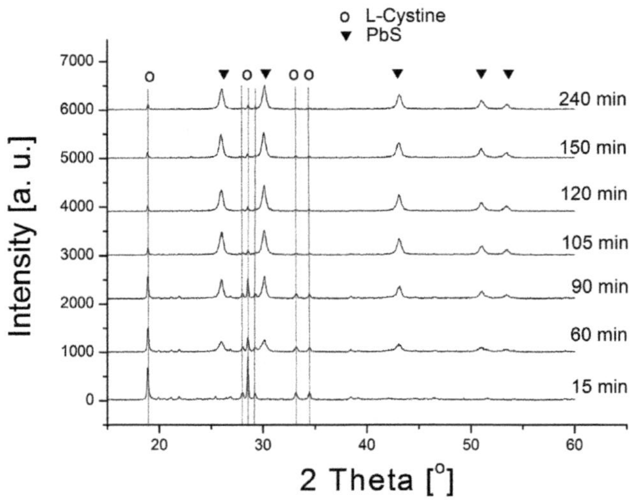

Fig. 1 The XRD patterns of the mechanochemically synthesized PbS nanocrystals (milling time is given in the figure).

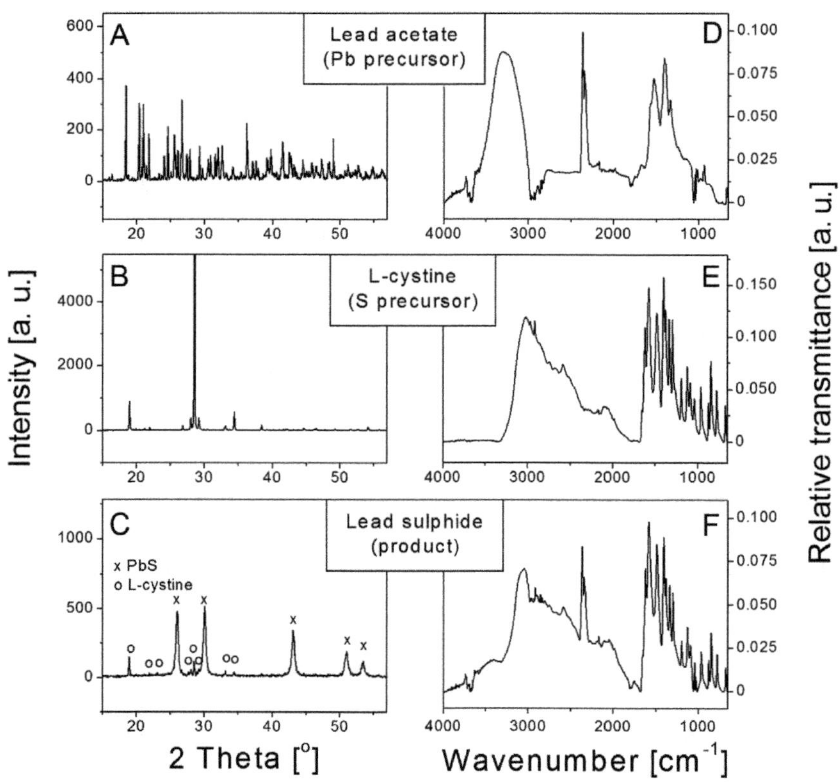

Fig. 2 The XRD patterns (A, B, C) and FT-IR spectra (D, E, F) of the precursors and the product obtained after 105 min of milling.

the acidic post-treatment of the product by which any residual L-cystine should be removed. For comparison, the XRD spectra for pure lead acetate (Fig. 2A) and L-cystine (Fig. 2B) are also given. The FT-IR spectra (Fig. 2D–F) verified the presence of L-cystine in the product because the spectra of L-cystine (E) and the product (F) are quite similar. However, there are several differences. The peaks below 1700 cm^{-1} are characteristic for L-cystine and are all present in the product, although a slight change in intensity was observed. Two sharp peaks at 2341 and 2361 cm^{-1} appear both in the spectra of the product and of lead acetate, but they correspond to the atmospheric CO_2. No peaks corresponding to lead acetate were observed in the spectrum of the product (Fig. 2F) and therefore we claim that it is not present. The results of the FT-IR measurements suggest that cystine is bound to the PbS nanoparticles, however the details about type of bond and exact location within the molecules are not clear yet. Although the majority of both precursors were consumed in the formation of PbS, these results provide evidence that the conversion was incomplete (see next part). The residual cystine forms a PbS@cystine composite and the residual lead acetate is washed out together with the organic debris from the consumed cystine.

The dependence of the degree of conversion α and the surface area S_A on the milling time t_M for the synthesized PbS nanocrystals is given in Fig. 3. A sharp increase was observed in both α and S_A until $t_M = 90$–105 min, and a steady state was achieved after $t_M = 105$ min, where the maximum values for α (84%) and S_A (18.1 m^{2}g^{-1}) were obtained.

The dependence of the crystallite size, D on α is shown in Fig. 4. After 105 min of milling, the size of the PbS nanocrystals was in the range between 32–34 nm. Because α, S_A and D reach their equilibrium values after $t_M = 105$ min, the given experimental conditions are sufficient to complete the mechanochemical synthesis.

FE-SEM (Fig. 5A) and TEM (Fig. 5B) were used to characterize the morphology of the PbS nanocrystals. As can be seen from the SEM image (Fig. 5A), the sample consists of aggregates of uniformly grown, well-faceted, ~30 nm nanocrystals and cystine irregular objects. The low-magnification TEM image (Fig. 5B) shows that the predominant nanocrystal shapes are square- and rhombus-like. The analysis of the rings in the SAED pattern (inset in Fig. 5B) also confirmed that the

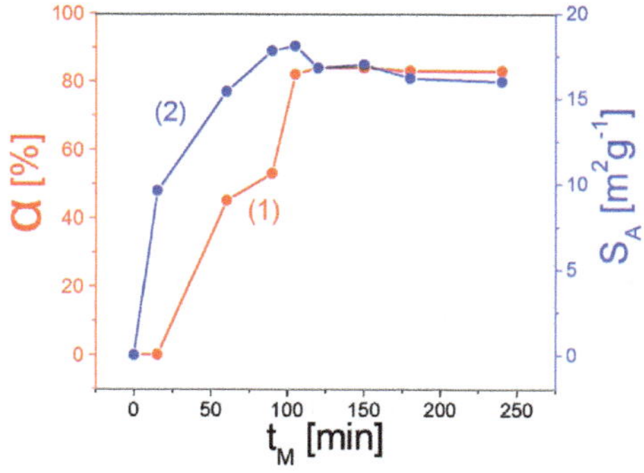

Fig. 3 The conversion degree, α (1) and specific surface area, S_A (2) *vs.* milling time, t_M = 15–240 min.

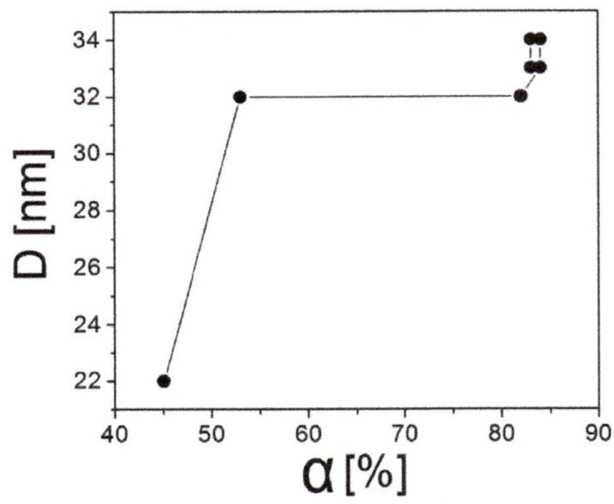

Fig. 4 The crystallite size, D vs. the conversion degree, α.

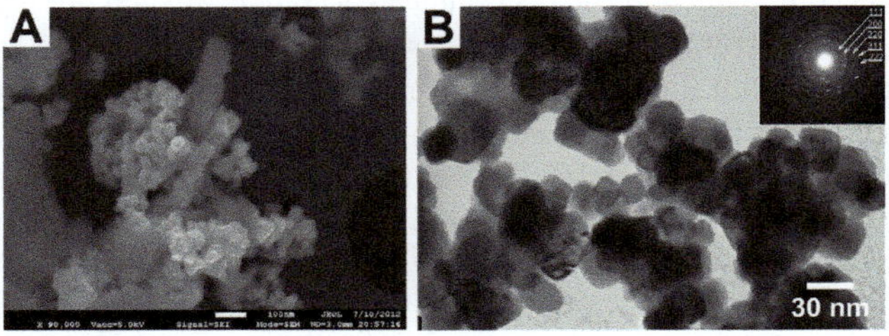

Fig. 5 (A) An FE-SEM image; (B) a bright field TEM image (with SAED inset) of the sample milled for 105 min.

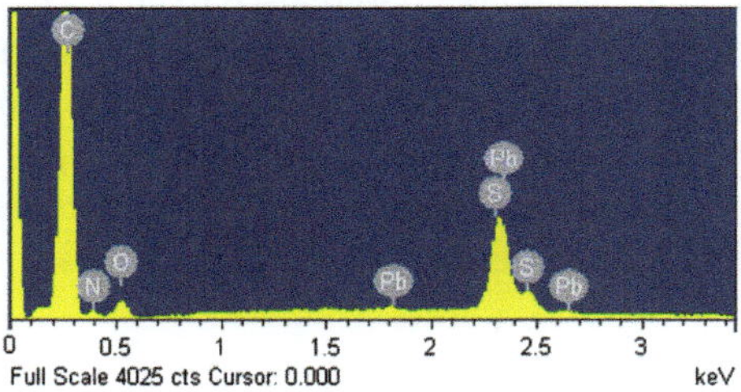

Fig. 6 The EDS spectrum of the PbS nanocrystals milled for 105 min.

nanocrystals are PbS with a cubic crystal structure and $a = 5.9362$ nm. According to ICDD 65-9496, the 111, 200 and 220 diffractions can be assigned to the most intense rings, which exhibit interlayer distances of 0.343 nm, 0.297 nm and 0.21 nm, respectively.

An EDS spectrum of the PbS nanocrystals milled for 105 min is shown in Fig. 6. The EDS spectrum unambiguously confirms that there is cystine ($C_6H_{12}N_2O_4S_2$) in the sample, because the elements C, N, O and S are present. The occurrence of Pb and S is due to its presence in the main product.

Another proof of the presence of cystine in the samples was obtained by application of ^{1}H NMR spectroscopy. The spectra of pure L-cystine and the PbS@cystine product are presented in Fig. 7.

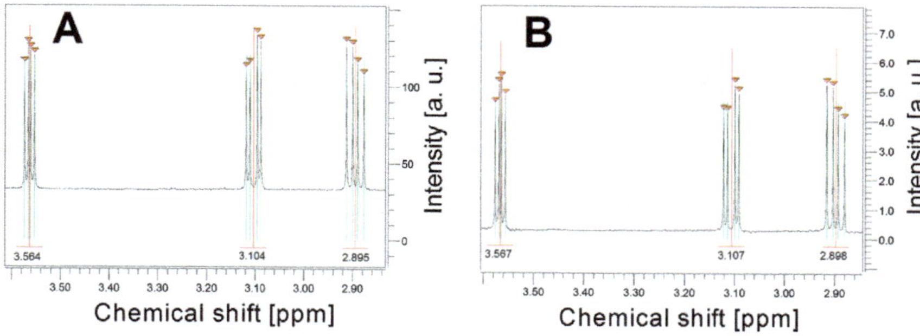

Fig. 7 The NMR spectra: (A) pure L-cystine; (B) PbS@cystine nanocrystals milled for 105 min. (A): ^{1}H NMR (600 MHz, D_2O/NaOH), δ: 3.56 (dd, 1H, $J = 4.7$ Hz, $J = 7.6$ Hz), 3.10 (dd, 1H, $J = 4.7$ Hz, $J = 13.6$ Hz), 2.90 (dd, 1H, $J = 7.6$ Hz, $J = 13.6$ Hz). (B): ^{1}H NMR (600 MHz, D_2O/NaOH), δ: 3.57 (dd, 1H, $J = 4.7$ Hz, $J = 7.6$ Hz), 3.11 (dd, 1H, $J = 4.7$ Hz, $J = 13.6$ Hz), 2.90 (dd, 1H, $J = 7.6$ Hz, $J = 13.6$ Hz).

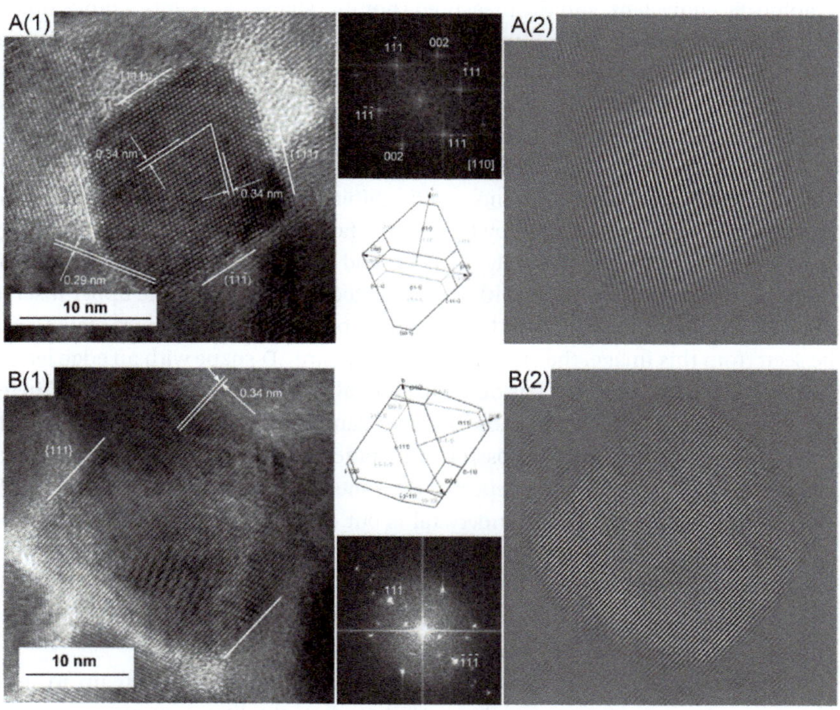

Fig. 8 HRTEM image of the PbS nanocrystals milled for 105 min: A(1) – a truncated octahedron exhibiting rhombic projection, evidencing {100} and {111} type planes edging the nanocrystal. The FFT pattern and relevant schematic representation in the upper and lower insets, respectively, demonstrate the <110> zone axis. The solid lines mark the {111} type surface planes parallel to the primary electron beam. A(2) – the IFFT pattern of the nanocrystal showing the ($\bar{1}$11) and (1$\bar{1}\bar{1}$) planes extended in the crystal, which is in accordance with the FFT image and the depicted nanocrystal. B(1) – the square-like projection and the {111} plane faceting truncated octahedral PbS nanocrystal. The upper inset is a schematic representation of the PbS nanocrystal oriented by a similar way as that depicted by TEM. The lower inset is the relevant FFT image. B(2) – the IFFT pattern of the nanocrystal highlighting the {111} type planes. The lattice fringes extend continuously through the crystal, which indicates that the PbS is a single crystal with no structural defects.

The peak signal at the highest ppm corresponds to the H-3 proton from the CH– group and the other two signals correspond to the protons H-2 from the $-CH_2$ group. All the signals are multiplets (dd) and the coupling constants match with each other. A detailed analysis of the multiplets is given under the figure. The only difference between the two spectra is in the intensity. The intensities of the peaks of the product (Fig. 7B) are much lower than in the case of pure L-cystine (Fig. 7A), in accordance with the amount of cystine present in the sample.

Because the features recorded in the TEM images are only 2D-projections of 3D-nanocrystals, they were examined by HRTEM (Fig. 8). The determined surface planes were compared with the schematic depiction of the proposed 3D-morphology of the crystals. Fig. 8A shows that the PbS crystals with a rhombic shape exhibit faceting along the {111} planes (1), because the distance of 0.34 nm between the neighboring lattice fringes is in agreement with the interplanar distance of these facets. Moreover, the measured angles 109.4° and 70.6° are consistent with the theoretical values between the {111} planes in a cubic crystal structure. The {111} planes along the edges of the nanocrystals are crystallographically equivalent, and the measured rhombic shape represents a projection of an octahedron along the <110> type zone axis as depicted in the fast Fourier transform (FFT) image (upper inset in Fig. 8A). The lattice fringes spaced by 0.29 nm were assigned to the 002 reflections. Hence, the surfaces parallel to them that cause the truncation of the octahedron are the {100} planes. A schematic representation of a truncated octahedron in the same orientation as that depicted in Fig. 8A(1) is shown in the lower inset. The nanocrystal exhibited a high crystallinity without any crystal defects, as corroborated by the inverse fast Fourier transform (IFFT) image (Fig. 8A(2)), which highlights only the $(\bar{1}11)$ and $(1\bar{1}\bar{1})$ planes running through the crystal (which is in accordance with the evaluated FFT pattern in the upper inset in Fig. 8B). Fig. 8B(1) presents another truncated octahedral PbS nanocrystal. As can be seen from this image, the crystal exhibits a square 2D-shape with an edge length of approximately 20 nm and has only two sides aligned parallel to the lattice fringes separated by 0.34 nm. Hence, these surfaces can be determined to be of the {111} type. The scheme displays the possible orientation of truncated octahedra to make clear the {111} faceting. It is evident that the zone axis cannot be determined exactly from this image because the nanocrystal is out of alignment with the <110> zone axis. The corresponding IFFT pattern in Fig. 8B(2) shows that the lattice fringes extend continuously through the crystal, which indicates that the PbS is a single crystal with no structural defects. The HRTEM study suggests that the PbS nanoparticles are wrapped by L-cystine (light regions in Fig. 8).

The mechanism of the PbS nanocrystals formation is not clear up to now. However, some suggestions can be made based on the literature and our experimental observations. We assume that the mechanochemical synthesis of PbS comprises steps (1)–(4):

$$Pb(CH_3COO)_2{\cdot}3H_2O \rightarrow Pb(CH_3COO)_2 + 3H_2O \quad (1)$$

$$R\text{–}S\text{–}S\text{–}R \text{ (cystine)} + H_2O \rightarrow R\text{-}SH \text{ (cysteine)} + H_2S + S \quad (2)$$

$$2R\text{–}SH + Pb(CH_3COO)_2 \rightarrow (R\text{–}S)_2Pb + 2CH_3COOH \quad (3a)$$

$$(R\text{–}S)_2Pb \rightarrow PbS + R\text{–}S\text{–}R \text{ (only proposed reaction)} \quad (3b)$$

$$H_2S + Pb(CH_3COO)_2 \rightarrow PbS + 2CH_3COOH \quad (4)$$

The first step is the dehydration of lead acetate. According to Rajaram,[34] crystalline water is released from lead acetate at the temperature of 75 °C. As stated in,[35] such temperature can be achieved during high-energy milling. This happened also in our case, and therefore crystalline water was released (eqn (1)). Therefore the mechanochemical synthesis under consideration constitutes an example of liquid-assisted grinding (LAG)[25] where in situ-produced water plays an important role. The released water is responsible for the decomposition of cystine (eqn (2)), because according to Routh,[36] it is decomposed into cysteine, hydrogen sulphide and elemental sulfur if it is heated. This equation is not stoichiometrically correct, it just illustrates the products of the decomposition. Also undefined side-products are formed during this process.[36] In the work by Stahl,[37] who used milling for the decomposition of cystine for the first time, it is stated that the necessary amount of water for the decomposition of cystine should be in the range of 8.4–11.5%. In our case, the amount of water released from the hydrated lead acetate was 10.82%, which is in perfect accordance with the mentioned work. Continuing with the mechanism, both cysteine and hydrogen sulfide can react with water-free lead acetate yielding lead sulphide (eqn (3) and (4)). In both cases, acetic acid is formed, the smell of which was detected already after 60 min of milling, when the first traces of PbS were detected (Fig. 1). In the case of shorter milling times, only the less aggressive smell of lead acetate was detected. According to Andrews,[38] the formation of PbS from cysteine is a two-step process (eqn (3a) and (3b)). The organometallic intermediate formed in the eqn (3a) together with acetic acid is decomposed into lead acetate and lanthionine (containing only one sulphur atom) (eqn (3b)). The reaction (3b) is only proposed, not yet experimentally confirmed. Although this mechanism is possible, there are two obscure points: (i) according to ref. 38, the reactions 3a and 3b take place in alkaline conditions and (ii) the role of the elemental sulphur formed in eqn (2) is not clear. We do not have the proof of the presence of mentioned reaction intermediates. Although this mechanism is based on experimental results, it has to be elucidated in further studies.

4 Conclusions

In summary, we have developed a new controllable method to prepare inorganic semiconductor nanoparticles from organic salt and amino acid as precursors using a simple solvent-free, high-energy milling technique. The mechanochemically generated PbS nanocrystals range in crystallite sizes from 22 to 34 nm, depending on milling time. The amino acid L-cystine has a dual role: (1) acting as a source of sulfur for the synthesis, and (2) binding to the synthesized PbS, which enables the nanoparticles to be biofunctionalized for medical purposes. The mechanochemically mediated synthesis of bioconjugated semiconductor nanocrystals has also a great potential in bio-engineering (electroluminescent devices).

Acknowledgements

The work was supported through the projects of the Slovak Research and Development Agency (APVV-0189-10), the Slovak Grant Agency VEGA (2/0027/13) and the Slovak Academy of Sciences (SAS Centre of Excellence: CFNTMVEP). The support of the projects of the European Regional Development Fund (ERDF)-nanoCEXmat I (ITMS 26220120019), nanoCEXmat II (ITMS 26220120035) and Center for development and application of advanced diagnostic methods in processing of metallic and nonmetallic materials (ITMS 26220120014) of Research & Development Operational Programme is also gratefully acknowledged. The authors thank Dr. Martin Fabián for preliminary XRD measurements.

References

1 A. L. Efros, *Sov. Phys. Semicond.-USSR*, 1982, **16**, 772–775.
2 A. Henglein, *Ber. Bunsen-Ges. Phys. Chem.*, 1982, **86**, 301–305.
3 R. Rossetti and L. Brus, *J. Phys. Chem.*, 1982, **86**, 4470–4472.
4 R. Rossetti, L. E. Nakahara and L. E. Brus, *J. Chem. Phys.*, 1983, **79**, 1086–1088.
5 C. B. Murray, D. J. Norris and M. G. Bawendi, *J. Am. Chem. Soc.*, 1993, **115**, 8706–8715.
6 A. L. Rogach, ed., *Semiconductor Nanocrystal Quantum Dots*, Springer, Wien New York, 2008.
7 Z. L. Xiu, S. W. Liu, J. X. Yu, F. X. Xu, W. N. Yu and G. J. Feng, *J. Alloys Compd.*, 2008, **457**, L9–L11.
8 S. F. Wang, F. Gu and M. K. Lu, *Langmuir*, 2006, **22**, 398–401.
9 D. Yu, Z. Wang, J. Meng, Y. Lu and Y. Qian, *J. Mater. Chem.*, 2002, **12**, 403–405.
10 S. M. Zhou, Y. S. Feng and L. D. Zhang, *J. Mater. Res.*, 2011, **18**, 1188–1191.
11 S. F. Wang, F. Gu, M. K. Lu, G. J. Zhou and A. Y. Zhang, *J. Cryst. Growth*, 2006, **289**, 621–625.
12 Q. Liu, Y. Ni, G. Yiu, J. Hong and Z. Xu, *Mater. Chem. Phys.*, 2005, **89**, 379–382.
13 S. M. Lee, Y. W. Jun, S. N. Cho and J. Cheon, *J. Am. Chem. Soc.*, 2002, **124**, 11244–11245.
14 T. Tsuzuki, J. Ding and P. G. McCormick, *Phys. B*, 1997, **239**, 378–387.
15 T. Tsuzuki and P. G. McCormick, *Nanostruct. Mater.*, 1999, **12**, 75–78.
16 P. Baláž, L. Takacs, T. Ohtani, D. E. Mack, E. Boldižárova, V. Soika and M. Achimovičová, *J. Alloys Compd.*, 2002, **337**, 76–82.
17 P. Baláž, E. Boldižárová, E. Godočíková and J. Briančin, *Mater. Lett.*, 2003, **57**, 1585–1589.
18 P. Baláž, P. Pourghahramani, E. Dutková, M. Fabián, J. Kováč and A. Šatka, *Cent. Eur. J. Chem.*, 2009, **7**, 215–221.
19 P. Baláž, P. Pourghahramani, M. Achimovičová, E. Dutková, J. Kováč, A. Šatka and J. Z. Jiang, *J. Cryst. Growth*, 2011, **332**, 1–6.
20 M. Baláž, P. Baláž, M. J. Sayagués and A. Zorkovská, *Mater. Sci. Semicond. Process.*, 2013, **16**, 1899–1903.
21 P. Baláž, *Mechanochemistry in Nanoscience and Minerals Engineering*, Springer, Berlin Heidelberg, 2008.
22 F. Delogu and G. Mulas, eds., *Experimental and theoretical studies in modern mechanochemistry*, Transworld Research Network, Kerala, 2010.
23 T. Friščić, *J. Mater. Chem.*, 2010, **20**, 7599–7605.
24 A. Stolle, T. Szuppa, S. E. S. Leonhardt and B. Ondruschka, *Chem. Soc. Rev.*, 2011, **40**, 2317–2329.
25 S. L. James, C. J. Adams, C. Bolm, D. Braga, P. Collier, T. Friščić, F. Grepioni, K. D. M. Harris, G. Hyett, W. Jones, A. Krebs, J. Mack, L. Maini, A. G. Orpen,

I. P. Parkin, W. C. Shearouse, J. W. Steed and D. C. Waddell, *Chem. Soc. Rev.*, 2012, **41**, 413–447.

26 P. Baláž, M. Achimovičová, M. Baláž, P. Billik, Z. Cherkezova-Zheleva, J. M. Criado, F. Delogu, E. Dutková, E. Gaffet, F. J. Gotor, R. Kumar, I. Mitov, T. Rojac, M. Senna, A. Streletskii and K. Wieczorek-Ciurowa, *Chem. Soc. Rev.*, 2013, **42**, 7571–7637.

27 E. Boldyreva, *Chem. Soc. Rev.*, 2013, **42**, 7719–7738.

28 V. Šepelák, A. Düvel, M. Wilkening, K. D. Becker and P. Heitjans, *Chem. Soc. Rev.*, 2013, **42**, 7507–7520.

29 M. Aslaksen, O. H. Romarheim, T. Storebakken and A. Skrede, *Anim. Feed Sci. Tech.*, 2006, **128**, 320–330.

30 M. J. Rak, N. K. Saadé, T. Friščić and A. Moores, *Green Chem.*, 2014, **16**, 86–89.

31 M. Jalaly, M. S. Bafghi, M. Tamzifar and F. J. Gotor, *J. Alloys Compd.*, 2014, **588**, 36–41.

32 P. Baláž, M. J. Sayagués, M. Baláž, A. Zorkovská, P. Hronec, J. Kováč Jr., E. Dutková, G. Mojžišová and J. Mojžiš, *Mater. Res. Bull.*, 2014, **49**, 302–309.

33 S. Shahbazi and S. Afshar, *Mater. Lett.*, 2014, **115**, 190–193.

34 P. K. Rajaram and J. K. Mohana Rao, *Z. Kristallogr.*, 1982, **160**, 225–233.

35 L. Takacs and J. S. McHenry, *J. Mater. Sci.*, 2006, **41**, 5246–5249.

36 J. I. Routh, *J. Biol. Chem.*, 1938, **126**, 147–154.

37 W. H. Stahl, B. McQue and R. G. H. Siu, *J. Biol. Chem.*, 1949, **177**, 69–73.

38 J. C. Andrews, *J. Biol. Chem.*, 1928, **80**, 191–210.

Faraday Discussions

DISCUSSIONS

Mechanochemistry of inorganic compounds and coordination-based materials: General discussion

DOI: 10.1039/c4fd90006e

Professor Suslick opened the discussion of the paper by David Peters: There may be an initial intermediate formed on the way to your final product that forms a passivation layer. Have you characterized intermediates or the surface during the reaction?

Dr Peters responded: No characterization of intermediates was attempted nor was any surface analysis performed.

Professor Suslick remarked: There are other ways of removing passivation layers: ultrasound, most notably. Have you tried using an ultrasonic bath or horn?

Dr Peters replied: We attempted to drive the reaction with an ultrasonic bath but were unsuccessful. I did not try an ultrasonic horn.

Professor Baláž asked: Did you register the presence of quartz in your sample?

Dr Peters answered: The final step is a distillation. We expect any quartz would be left behind. We do not test for quartz as silicates are not important to our customer.

Professor MacGillivray remarked: The chemistry is being conducted on a very large scale. How big is the stir bar?

Dr Peters replied: We used a Chemglass CG-2005-14 stir bar. It measures 40×20 mm

Dr Stolle asked: (a) Have other methods which are able to generate high shear forces (like ultrasound, *etc.*) been tested for the purpose?

(b) Regarding overhead stirrers, how far does the stirrer type (tangential, axial mixing) influence the whole process?

(c) Please comment on the actual stirring speed/rpm. In the paper it is written only that the stirrer speeds were adjusted, without providing any numbers.

Dr Peters answered: (a) Alternate solvents, higher temperatures and an ultrasonic bath were all tried in unsuccessful efforts to drive the reaction. As mentinoed above, an ultrasonic horn was not tried.

(b) The overhead stirrer was a semicircular paddle on a glass shaft. It was lifted above the bottom of the flask to allow clearance for the magnetic stir bar.

(c) The stirring speed was not directly measured. The speed was adjusted so that the very dense strontium iodide was kept in suspension.

Dr Friščić commented: It is very interesting in the reaction that you have presented that the reactants themselves are not soluble, but a solvate does form. This might provide insight into how the liquid interacts with the solid reactants, suggesting, among other mechanisms, those of dissolution and recrystallisation to a small extent. To prove or disprove this possibility (an alternative would, presumably, be the molecule-by-molecule interaction of the solvent with the reagents) it might be interesting to attempt using worse solvents.

Professor MacGillivray remarked: It would be interesting to study systematically solvents of higher boiling points so as to provide an opportunity to increase temperature in the preparation.

Professor Beyer asked: Why does the magnetic stirrer need to be synchronized with the paddle? What happens if they are not synchronized?

Dr Peters replied: If they are not synchronized the magnetic stir bar will jump around the bottom of the flask and make inconsistent contact with it. It is the grinding action of the magnetic stir bar that drives the reaction.

Professor MacGillivray said: You have discovered an application of mechanochemistry to a problem of industrial importance. To you foresee applications in related areas in industry?

Dr Peters responded: I certainly hope that there are applications for this method in industry. It would be liberating for the chemist to be able to disconnect himself from solvent selection in designing a synthesis. It is my hope that we will be able to avoid toxic and aggressive solvents using this method as well as avoid undesirable byproducts or adducts formed by some of the solvents.

Professor Galembeck said: Can you use the same stirring bar over-and-over or is there some wear?

Dr Peters replied: The stir bars are PTFE coated and no wear was detectable. Nor was any wear detected on the bottom of the round bottom flask. When we switched to the vertical reactor with steel ball bearings we observed wear on both the glass reactor and the steel bearings.

Dr Maini said: When you will further scale-up your synthesis will you expect to scale up also the strength of your grinding?

Dr Peters answered: I have not seen anything in the synthesis that indicates additional grinding strength would be needed for scale-up.

Professor Jones asked: Scaling a mechanochemistry process from bench-top milling/grinding equipment to a larger production unit is always challenging. How large a machine have you ended up with and how does the energy consumption per unit mass of product scale?

Dr Peters answered: We took this synthesis to a 500 g scale in a six liter reactor. The mechanical action is provided by a "Lightning" brand overhead stirrer which is standard for bench-top chemical reactions. The energy consumed is minimal and in line with normal solution based syntheses.

Dr Stolle also added: We are planning to scale up our reactions from 0.25-L-scale (planetary ball mill) to a 20-L-scale (horizontal stirred media mill originally designed for high kinetic processing) at the end of our project. The maximum operating power for this machine will be 11 kW. However, in our experience with a similar model (2 L volume, maximal power = 2.7 kW) we do not need the maximal power input, since under these conditions the energy input will be too high and destroy our product. However, the experiments have started a few months ago and today I am not able to present any specific data.

Dr Blair then noted: If attrition mill technology is used then energy consumption is high until 500 L mills are used. After that energy consumption/MT of product is constant. I have included a Figure showing energy usage for 1 h of processing at 50% conversion. We have run unoptimized on a 4 L machine. We will be running on a 160 L machine in the coming month.

In the United States the average price per kWh is 0.0699 (as of March 2014; http://www.eia.gov/electricity/monthly/epm_table_grapher.cfm?t=epmt_5_6_a). Assuming 50% conversion, this gives a production cost of 107 to 48 per MT of glucose produced from pure cellulose or 134-60 per ton of saccharides from biomass. Other technologies such as the mills offered by ZOZ and ISAMill offer

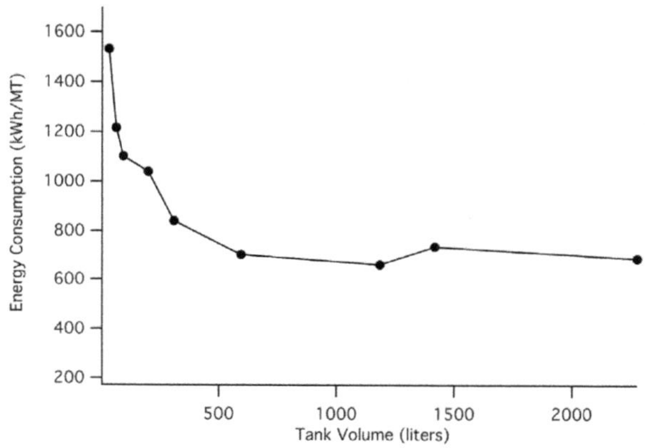

Fig. 1

the promise of lower energy consumption, but more testing would be needed (http://www.isamill.com/EN/Pages/default.aspx http://www.zoz-gmbh.de/_ENGLISCH/content/view/36/49/lang,en/).

Professor Bolm commented: We discuss this transformation as if it were a special case. I wonder, however, if that is true or if there is some generality. Perhaps we can find other transformations in, for example, organic chemistry where we achieve a reaction of two insoluble starting materials upon "grinding" them in an inert solvent that does *not* dissolve both of the educts. We shall try!

Dr Friščić asked: Have you measured the solubilities of the starting materials and the product?

Dr Peters replied: I have determined, to the best of my ability, that neither starting material is soluble in diethylether. The product is very soluble in diethylether. Recall that there is no reaction from just stirring the starting materials in diethylether which would support the conclusion that they are insoluble.

Professor James asked: How soluble is the potassium iodide side-product? I thought some iodides were soluble in ether.

Dr Peters replied: No insoluble material was detected after filtration. No halides were detected after distillation.

Professor MacGillivray remarked: How did the original discovery transpire in terms of the timeframe and realization of the problem?

Dr Peters responded: It took about six months to figure out that the reaction was being driven by the grinding of the solids by the magnetic stir bar. It took another two months to settle on the reported preparation method with the vertical reactor and the steel ball bearings.

Professor MacGillivray said: How long did it take your team in industry to determine that the nature of the stir bar was such a critical factor?

Dr Peters replied: Sadly, it took us about 8 months to understand that the stir bar was grinding the material and to design a better process as a result. Keep in mind that this is all going on at the same time we are under pressure to supply a finished product to customers.

Professor Baláž asked: Did you use also combination of various ball sizes in your experiment?

Dr Peters answered: No. The ball size used was standard. It might be interesting to try an assortment of ball sizes and see if these increase the rate of the reaction.

Dr Friščić addressed Professor Baláž: It would be interesting to know if there exists a simple, empirical method that would allow the reactivity to be established and optimized (in terms of yield and energy efficiency) in a shaker mill to be quickly transferred to a mill with a different geometry, *i.e.* a planetary mill or an attritor. Presumably, such an empirical relationship would greatly facilitate the development of large-scale industrial processes. Are any of the meeting participants familiar with such an empirical methodology?

Professor Baláž responded: In my opinion, this question has to be addressed to Dr Peters and Dr Blair. We do not do such calculations in our department. However, colleagues at IPAT (Braunschweig, Germany) are doing such calculations based on the specific energy of milling.

Dr Blair commented: Increasing the amount of media in this reaction may improve yields. But, it is a balance of forces and mixing. Larger mills (ton to multiton) often use quite small media, such as in a sand mill (http://www.raymerengineering.com/sand-mill.html). Although mixing is improved, the impact energies are lowered. This may be alleviated to some extant by using denser media such as tungsten carbide. It also may not be as critical for reactions with low activation energies. Many mechanochemical organic syntheses use quite small media.

Professor Jones said: We have heard about a diversity of mechanochemistry processes from non-covalent bond interactions through covalent, ionic and catalytic processes. Can we identify the key mechanisms that are responsible for these remarkable reactions. Are we thinking that: (1) highly energetic impact forces are needed; (2) there is some degree of breakdown of the reactants to a small enough size; (3) intimate mixing must be achieved; (4) reactions can take place if a threshold temperature is achieved; or (5) some other factor?

Professor Galembeck responded: Yes, evidence in favor of the factors that you listed was given in one or more presentations during this Discussion but they are not necessarily relevant in every situation. For instance, in the tribocharging apparatus described in my paper shearing forces are used instead of impact.

Professor Takacs responded: Mechanochemistry tends to defy generalization. If we consider every process that uses some form of mechanical action for the induction or modification of a chemical process "mechanochemical", very few, if any, general requirements remain valid:

(1) Very high impact energy is essential when a mixture of hard oxides is milled to obtain a single mixed oxide phase, but only very mild grinding is needed to facilitate the formation of organic cocrystals. The mechanical loading need not even be in the form of impact, like in the classic high pressure-torsion experiments of Bridgman. In a few cases the minimum impact energy has been measured. For example, the hcp cobalt transforms to fcc only if at least 1.7×10^{-2} J impact energy is provided [F. Delogu, *Acta Mater.*, 2008, **56**, 905].

(2) When we consider reactions in a mixture of powders, size reduction is probably a general requirement. But in tribochemistry or the chemical effects of cleaving a crystal, size reduction plays no role.

(3) As long as we restrict our interest to reactions between two or more components, mixing is probably a general feature. But even here, diffusion acts

together with the mechanical mixing. It depends on the system. The mechanical alloying of metals is probably very different from the reaction of large organic molecules in this respect.

(4) The role of temperature is an interesting question. On the one hand, we know very well that most of the mechanical energy input from milling is converted to heat and the temperature increase causes chemical changes. But on the other hand, we want to believe that in "true" mechanochemical reactions the primary bond scission is affected by the mechanical force directly. (By this requirement, the number of proven mechanochmical reactions is quite limited.) While it is difficult to identify principles that are equally applicable to all mechanochemical reactions, it is useful to find general principles for individual classes of mechanochemical processes. Size reduction, mixing and sufficient energy to cause plastic deformation are often identified as important requirements.

Dr Lamaty opened the discussion of the paper by Lucia Maini by asking: Did you observe oxidation of the copper salts during the ball-milling process?

Dr Maini answered: The CuI is quite stable and it can be handled in air without problems, however the oxidation of the copper salt depends on the nature of the ligand. In the case of PN ligand the oxidation of copper iodide is not promoted, nevertheless we have observed that in some cases the obtained powder is slightly green or blue, which is due to the presence of copper (II). The reactions carried out in an inert atmosphere result in products with an higher quantum yield than the reactions carried out in air while no difference is observed from a crystallographic point of view.

Dr Lamaty then enquired: Did you analyse the oxidation state of the metals?

Dr Maini replied: In the products we have copper (I) but we do not exclude the presences of traces of copper (II) and copper (0). The copper (II) can be easily detected by its colour. We have not done further analysis.

Professor Suslick said: Why do you get different compounds under different conditions? In solution, one's concentrations are fixed, but during ball milling the concentrations will be different at the point of impact than the overall stoichiometry of the ingredients put into the mill. Is there any sense of the free energies of formation of all these different copper iodide clusters?

Dr Maini replied: The grinding process involves local higher pressure and temperature and this is why different products are obtained by milling with respect to solution. However, as you point out, during the grinding process the stoichiometry ratio at the point of impact is different from the overall stoichiometry ratio; maybe this can explain the presence of mixture of [CuIPN0.5]8 and [CuI4PN2] during the milling process? However the stoichiometry ratio influences the final product since we obtain different compounds by varying the Cu : I ratio from 1 : 4 to 4 : 1. I do not exclude, for example, that compounds with low nuclearity can also be produced in the presence of an excess of CuI, and these compounds can further react with CuI to yield the final product. *In situ* experiments can probably give interesting insights into the processes.

Dr Friščić added: The presentation of the results on making different stoichiometries of the coordination polymers using different methodologies encourages me to mention that we have observed excellent control of product composition in mechanochemical reactions simply by varying reaction stoichiometry. We have noted this effect in the context of cocrystals (Karki *et al.*, *CrystEngComm*, 2009, **11**, 470-481), coordination polymers (see Strukil *et al.*, *Chem. Commun.*, 2010, **46**, 9191–9193), as well as small organic molecules (see Strukil *et al.*, *Chem. Commun.*, 2012, **48**, 9705–9707). Similar results were also reported by the Stolle group for the halogenation of organic compounds (see Schmidt *et al. Green Chem.*, 2012, **14**, 1673–1679).

Consequently, I would like to suggest that improved (or excellent) control of product composition and, therefore, structure could be an emergent and unexpected benefit of mechanochemical reactions in addition to their solvent-free, "green" nature. I would very much appreciate any comments or further examples that the delegates might have on this.

I would like to note that the topology of the coordination polymers formed when using different metal–ligand ratios might have an influence on the reaction course. Specifically, the 1 : 2 milling reactions of cadmium chloride and cyanoguanidine are expected to give a linear 1-D coordination polymer. However, the reactions were sometimes found to halt at an intermediate 3-dimensional coordination polymer stage with a 1 : 1 composition. However, under harsher grinding (simply using heavier balls) the 3-dimensional structure transformed into a 1-dimensional coordination polymer expected from the reaction stoichiometry (see Strukil *et al.*, *Chem. Commun.*, 2010, **46**, 9191–9193). Tentatively, we would like to explain this observation by the greater kinetic stability of a 3-dimensionally connected structure to that of a 1-dimensional one. A similar explanation was also used to explain the mechanochemical behavior of cadmium iodide towards cyanoganidine. Any comments, critiques or related examples of this would be greatly appreciated.

Professor Baláž asked: Is the synthesis from iodine and copper elements also viable?

Dr Maini answered: We have not tried this.

Dr Blair said: What are the parameters of your milling: type of mill; size of media? How do the mechanical properties of your reagents affect the yield? Nanoindenting may be a way of measuring important processing parameters such as softness/hardness which would be important for the transfer of kinetic energy. The mechanics of plastic deformation may be different in all three instances. Introducing silica could help normalize the mechanical properties.

Dr Maini responded: We use a Retsch MM200 with 1.5 mL stainless jar with two balls and vibration at 20 Hz. It is difficult to answer to your last question. The two reagents have a really different hardness and I would not exclude that the reactions go through an amorphization of the organic powder induced also by the impact with the inorganic powder. We observed this phenomenon during the grinding of barbituric acid with KBr. Your suggestion about using silica is really interesting and we will try it in our future studies.

Mr Zhang asked: In your study, the synthesis of copper iodide complexes was performed using both solvothermal and milling methods. In the milling process, you've found that the final products would vary because of the different ratio between the reactants. Actually I am wondering if the milling energy (*i.e.* revolution speed) also shows significant influence on the final products. For example, with the molar ratio CuI : PN = 1 : 4, $CuI(PN)_2$ was produced by milling under your experimental conditions. But if you change the milling speed, will the final product be different somehow?

Dr Maini answered: During our experiment we have not modified the milling speed and we have increased the duration of milling only when both unreacted reagents were detected. I will not be surprised if energy of further milling will give different products. Different conditions of milling will be studied in the future.

Dr Irikura asked: I'm curious about the role of the solvent in liquid-assisted grinding. In these inorganic materials, could the solvent serve to increase the hardness of the sample, and therefore the energy of grinding, by dissolving soft crystallites selectively?

Professor Galembeck answered: I believe that solvent may play different roles, including your hypothesis. Our results show that tribocharge is easily removed from a polymer surface with solvent. This should prevent further action (*e.g.* catalytic) of these surface charges.

Dr Maini answered: In our case the use of acetonitrile, a coordinating solvent, improves the solubility of the CuI, thus we can talk about a "softness" of the inorganic sample.

Professor Jones asked: The issue of seeds has been discussed here and by others in terms of controlling polymorph outcome. Seeds will only work though if they do not dissolve. I wonder if you could comment on whether you have observed any seeding effects with regard to solubility factors?

Dr Maini replied: Our CuI(PN) compounds are characterized by low solubility which make these seeds stable in different solvent. We have tried to drive the reactions by seeding the solution of CuI before adding the ligand without success. I suspect that the almost instantaneous reaction between CuI and the ligand nullifies the role of the seeds. We observed an interesting case of unintentional seeding with the complex $CuI(PPh_3)$ described in the article "Polymorph and isomer conversion of complexes based on CuI and PPh3 easily observed *via* luminescence", L. Maini, D. Braga, P. P. Mazzeo and B. Ventura, *Dalton Trans.*, 2012, **41**, 531–539.

Professor Jones enquired: I wonder if Dr Day could comment on whether the prediction methods he has successfully applied to organic materials might be applied to the types of Cu complexes described by Dr Maini?

Dr Day answered: In principle, the methods that we have developed and applied to organic molecular crystals could be applied to any molecular crystal. One requirement is to have a model that describes the intramolecular energy as a function of molecular geometry, as well as the intermolecular interactions, to sufficient accuracy to provide a ranking of the possible crystal structures. Models for the interactions between some of the elements in these complexes would have to be carefully evaluated or developed specifically for these systems before structure prediction could be attempted. Alternatively, instead of the approach that we use for energies, where electronic structure methods are applied for the intramolecular energies combined with interatomic potentials for intermolecular interactions, one could apply solid state electronic structure methods (*e.g.* periodic density functional theory) to the problem of assessing relative energies. The other aspect of the problem that differs from what we must consider for organic molecular crystals is the requirement to predict which clusters could form. A complete structure prediction method would need to assess the range of crystal packing possibilities for all feasible clusters. Consideration of all possible clusters would be challenging in this case because of the different possible stoichiometries. We have investigated the problem of stoichiometry prediction for crystal structure prediction of multicomponent organic crystals[1,2] and similar considerations in comparing the energies of structures with different compositions would have to be applied here. Adding to the complexity is the inclusion of solvent (acetonitrile) in one structure. The paper suggests that acetonitrile in $[Cu_4I_4(PN)_2 \cdot CH3CN)]$ is incorporated in the host $[Cu_4I_4(PN)_2]$ framework. This type of behaviour can be anticipated from an analysis of all structures on the predicted crystal energy landscape to identify structures that could accommodate inclusion of guest molecules.[3] Overall, I believe that the structural landscape of crystal structures made up of discrete molecular complexes could be investigated using a very similar methodology to what has been developed for organic molecular crystals. However, polymeric structures, such as $[CuI(PN)_{0.5}]_8$, would require significant development, since our methods are based on the assumption that we can build up all potential crystal structures from discrete molecular units.

1 A. J. Cruz-Cabeza, G. M. Day and W. Jones, *Chem. – Eur. J.*, 2008, **14**, 8830–8836.
2 A. J. Cruz-Cabeza, S. Karki, L. Fabian, T. Friščić, G. M. Day and W. Jones, Chem. Commun., 2010, **46**, 2224–2226.
3 A. J. Cruz-Cabeza, G. M. Day and W. Jones, *Chem. – Eur. J.*, 2009, **15**, 13033–13040.

Professor MacGillivray asked: We have had a lot of discussion on stoichiometry. The ligand here is unusual in that there are two donor groups. Do you have any thoughts on whether the degree of coordination is affected by the mechanochemical preparation?

Dr Maini answered: Thank you for the interesting question. We have not explored this aspect. The two donor groups are different and both are involved in coordination except for $CuIPN_3$. In this case only the phosphorus atom is coordinating. Unfortunately we have no information on the structure of $CuIPN_2$ (obtained by synthesis with the same stoichiometry ratio as $CuIPN_3$ but a different process) to see if the product obtained by grinding presents a different coordination and/or nuclearity.

Dr Maini opened the discussion of the paper by Lisa Tröbs: Your *ex situ* studies are really interesting. Regarding the synthesis of the MOF: is it possible to isolate your intermediate product? Is it stable?

Ms Tröbs responded: In this case, we gained our intermediate together with unreacted terephthalic acid. It was not possible to isolate the pure intermediate product. The intermediate mixed with terephthalic acid is stable for at least five hours.

Dr Maini said: Have you tried to obtain the intermediate by reacting only the imidazole with the bismuth nitrate without the benzene dicarboxylic acid?

Ms Tröbs answered: We have not tried to obtain the intermediate solely but we will keep this in mind for future experiments.

Professor Suslick commented: You are using imidazole as a Brønsted base, but imidazole can also coordinate to bismuth ions. There are a variety of sterically hindered imidazoles commercially available (*e.g.*, 2-methylimidazole, 1,2-dimethylimidazole (which can't form the imidazolate), 2-ethyl and 2-isopropal imidazoles) that can probe the relative roles of Brønsted *vs.* Lewis basicity in these systems.

Ms Tröbs replied: We will keep this aspect in mind. Nevertheless, we also have to keep in mind that this bismuth-containing MOF shows very small pores. Therefore, the mechanochemical synthesis of this MOFs using sterically hindered imidazoles might not yield the formation of the MOF $(H_2Im)[Bi(1,4\text{-bdc})_2]$.

Professor MacGillivray said: The observation of the protonated species in the metal–organic framework is very interesting. I wonder what the 'pK_a' of the acid would be under the grinding conditions. Here, it is likely that we do not have ionization in an aqueous medium.

Dr Baláž remarked: In the paper by Halasz, *et al.* the possibility of monitoring *in situ* the course of mechanochemical reactions is outlined. In another recent paper by this group Raman spectroscopy was also used for that purpose. From my point of view, such approach would be also very beneficial in the case of your work.

Ms Tröbs responded: The benefit of the *in situ* approach described in the mentioned publication is without any doubt. Here, we demonstrated the analysis of the synthesis using exclusively standard laboratory techniques. We proved that these reactions can be analysed without the need of synchrotron-radiation or specially prepared grinding jars.

Professor Jones asked: In Fig. 5 of your paper and in one of your slides you drew attention to a nice spherical particle (*e.g.* at 25 min, in the top left hand corner of the photograph). Can you comment on the likely crystallite size of the individual particles making up the cluster?

Ms Tröbs replied: Based on the XRD pattern we determined the crystallite size to be below 5 nm at the end of the reaction. Assuming that one particle has a diameter of 2 μm we would expect about 125 million crystallites within one particle.

Professor Jones commented: There seemed to be a suggestion that you have the possibility of stable crystals of about 50 Å in diameter? Is this correct? These would present a significant surface area, subsequently impacting upon any properties that might be observed. Do you agree?

Ms Tröbs responded: We agree with this comment.

Professor Suslick asked: To probe the effective surface pH of the solid, one can add a series of pH indicators (separately) and measure the colorimetric response, preferably of the solid, by diffuse reflectance optical spectroscopy.

Ms Tröbs answered: The addition of a series of pH indicators might help to prove the presumed mechanism. We will try this approach, although the reaction might be affected by the presence of additional substances.

Professor Baláž asked: In your REM images (Fig. 5), the tendency towards the formation of agglomerates can be clearly seen. It could be useful to measure also particle size distribution, in order to get a more complex picture of the mechanochemical synthesis.

Ms Tröbs answered: We are grateful for this comment and will include the analysis of the particle size distribution in future publications.

Professor Jones asked: It would be interesting to know how preformed cocrystals of carbamazepine and theophylline would behave under similar milling conditions. Have you studied this effect?

Ms Tröbs replied: We agree with this, unfortunately our attempts to synthesise a cocrystal composed of carbamazepine and theophylline mechanochemically were not successful.

Professor Jones remarked: I return to the issue of particle size, since 5 nm particles would likely contain a small number of molecules – perhaps in the thousands? Do you think such particles would therefore have a significant amorphous character?

Ms Tröbs replied: Based on the evaluation of the crystallite size of the final product (using the Scherrer equation) it seems reasonable that the "particles" visible in the REM measurements consist of crystalline and amorphous material (for further details please see below).

Dr Maini asked: Regarding the formation of the co-crystal, could you exclude that your co-crystal is not formed by the eutectic melting?

Ms Tröbs answered: Assuming that a eutectic melt would not show any Bragg reflections in the XRD pattern, we can exclude the formation *via* a eutectic melt.

Professor MacGillivray commented: You mention destruction of crystals. In what way do you envisage the crystals being destroyed?

Ms Tröbs responded: The mechanical treatment of the educt mixture leads to a constant decrease in the crystallite sizes. It is plausible that the crystals are destroyed upon the impact of the steel balls.

Dr Friščić addressed the meeting: My aim here is to open a discussion on what are the mechanisms underlying mechanochemical reactions, specifically the role of an intermediate eutectic phase. namely, it is known that some mechanochemical reactions actually take place in an eutectic melts (see Rothenberg *et al.*, *J. Am. Chem. Soc.*, 2001, **123**, 8701–8708, or Chadwick *et al.*, *CrystEngComm* 2007, **9**, 732–734) and the deliberate formation of low-melting eutectics by addition of suitable auxiliaries has, in certain cases, been used to facilitate mechanochemical reactions. It has, however, also been argued that such reactions are not "real" mechanochemical reactions, as the reactivity takes place in an intermediate liquid phase.

It can be imagined (and such claims have been made in published work) that sometimes an eutectic is so short-lived that it cannot be observed by the naked eye, *i.e.* so short-lived that we cannot prove the existence of a bulk liquid phase. In what way can we establish the existence of an intermediate liquid phase in such cases, and do we then consider such reactions as different from those in which the eutectic is more long lived (but, presumably, still metastable) and a bulk liquid is observable? My feeling is that we should not be artificially distinguishing whether the reaction is mechanochemical or not based on how long-lived a potential eutectic intermediate is (to an extent, this depends on the rate of nucleation of the product phase which can be largely stochastic). Instead, it might be better to consider eutectic formation as one of the fundamental mechanisms behind mechanochemical reactions as, for example, is also amorphisation. I would very much appreciate any comments or critiques from the delegates on this view.

Professor Suslick commented: The meanings of states of matter is an important can of worms that must be addressed in our discussions. Amorphous, glassy, liquid, solid, solid-state solution, alloy, these all have an implied length scale. The domain size of interest is not necessarily well-defined. Materials can be amorphous on the micron scale but still crystalline on the 10 nm scale. We must be careful when using these terms to specify the length scale of the technique that we are using to probe their structures.

Dr Maini noted: Following this comment of Professor Suslick I would like to focus attention on the X-ray powder diffraction technique. This technique is really powerful in the presence of crystalline material and gives information on the presence of different crystalline phases and their amounts, on preferential orientation and on the microstructure of the different phases. Unfortunately the microstructure parameters such as crystallite size, lattice strain and defects as well the instrument geometry all affect the peak width making the "straightforward extraction of quantitative information on size and strain from the shape ("width") data impossible" (citing Eric J. Mittemeijer and U. Welzel, in *Z. Kristallogr. Int. J.*, 2008, **223**, 552).

Professor Jones added: I would agree with the comment made by Dr Suslick and would suggest that it might be useful to consider within the mixture that we do not simply have separate amorphous or crystalline particles but that in general each particle will consist of domains of fairly regular crystallinity but the domains separated by less-crystalline (amorphous-like regions) with the overall degree of crystalline and amorphous being determined by the volume element of each within a particle. During milling therefore the "boundary" between crystalline and amorphous would simple move changing the relative proportions. In such a way one might also envisage a mechanism by which amorphous particles gradually form crystalline particles by the passage of the boundary through the particle.

Ms Tröbs responded: We agree with this comment; the exact contribution of amorphous and crystalline regions within the bigger (2 μm) particles is not possible. It seems reasonable that each particle consists of crystalline domains and domains with a lower degree of order between the molecules forming the cocrystal.

Professor Galembeck addressed Ms Tröbs: You are looking at crystals that have relatively low melting points. Thus, the mobility of molecules at the surfaces should be significant and surfaces could be important reactive sites. Can you comment on this?

Ms Tröbs answered: It is an interesting idea to differentiate between the surface and the bulk mobility of the molecules but so far no experiments were performed to elucidate the difference.

Professor MacGillivray said: We have talked a lot about shearing of crystals. Another possibility is fragmentation, which is known to occur in sonochemistry. Any thoughts on whether fragmentation may be occurring?

Professor Suslick answered: For molecular crystals, the shockwaves created by cavitation do indeed create fragmentation of single crystals without the necessity of inter-particle collisions, as we demonstrated in our paper: B. W. Zeiger and K. S. Suslick, "Sonofragmentation of molecular crystals", *J. Am. Chem. Soc.*, 2011, **133**, 14530–14533. If the crystals grow with any significant aspect ratio (*i.e.*, needle or plate), then fragmentation is very likely to occur.

Professor MacGillivray asked: Can you offer additional thoughts on whether the fragmentation of crystals may also be occurring?

Ms Tröbs answered: Besides the fragmentation (destruction of the crystallinity of the reactants in the initial phase of the experiment and decrease in the crystallinity of the cocrystal upon prolonged milling time) of the crystals discussed, no further fragmentation was observed.

Professor Suslick remarked: One thing you can do here to gain some mechanistic insight is to grind the reactants separately, characterize their fragmentation, and examine the speed of the reactivity of the pre-ground reagents when ground again together.

Ms Tröbs replied: This is an interesting approach that we will include in our further experiments.

Dr Maini commented: The broadening of the peaks in X-ray powder diffraction is due to several parameters: such as crystallite strain; defects; instrument. The milling process of the cocrystal clearly influences the peak width and probably would be sufficient to describe the trend of the peak width with the FWHM. Further steps such as using the Scherrer equation to determine the crystallite size should be done with care to avoid misleading interpretations.

Ms Tröbs replied: We agree that the Scherrer equation should be used carefully since the broadening of the powder diffraction peaks is influenced by different parameters. Within these measurements none of those parameters were changed. Therefore, we concluded that at least the general trend of the change of the peak width gives reliable results.

Mr Huskic asked: Have you observed the formation of different polymorphs of indomethacin during milling, and also, have you tried starting from different polymorphs and if so, do you observe the same product?

Ms Tröbs answered: During the milling experiment, we observed a direct reaction of the educts indomethacin and carbamazepine to build up the cocrystal. No crystalline polymorphs of indomethacin or carbamazepine could be detected. So far, we did not perform any experiments starting from different polymorphs, but we will keep this in mind for future investigations.

Professor Boldyreva communicated: This is a very interesting paper. It would be interesting to test, if the same mixture reacts if stored (under dry and in humid atmosphere), as well as to test if the same mixture reacts on heating. This would help to understand the mechanism which could account for such a rapid reaction on co-grinding.

Dr Pecharsky opened the discussion of the paper by Vladimir Šepelák: Can you comment on the effect of processing you described on the magnetism of the title compounds?

Professor Šepelák answered: As it is described in the manuscript, the far-from-equilibrium structural state, confined to the particle's near-surface layers of the mechanically treated ferrite, has significant implications for its magnetism. The saturation magnetization of the ferrite decreases with increasing milling time (*i.e.*, with decreasing crystallite size) from about 69 emu g^{-1} (for bulk $BaFe_{12}O_{19}$) to about 30 emu g^{-1} (for nano-$BaFe_{12}O_{19}$ with the crystallite size of 14 nm). On the other hand, the coercive field of the nanomaterial is about 34% larger than that of the bulk material; see Table 2 in the manuscript. These large variations in magnetization and coercivity offer an ample opportunity to manipulate and tailor the functional properties of this magnetically hard material.

Professor James commented: The key thing I think is to be sure of what the sampling depth is for the Raman technique. It can be regarded as a bulk technique (*i.e.* providing information which is representative of the bulk) if the sampling depth is larger than the particle size and if the sample is well mixed, as it should be in milling experiments as long as the material remains free flowing. In our system, for example, the microscope is not confocal, it has a very large 110 micron diameter excitation spot and the collection aperture is set by an optical fibre. Basically that means our collection depth is hundreds of microns (the exact value depends on how you define it) even with a transparent sample. These powders aren't transparent so the laser bounces about inside the sample, this means we also pick up scattering from deep within the sample and from regions that lie outside what would be the illuminated volume in a simple transparent sample, this also averages out local variations in composition. Basically that means we sample from volume which is at least 110 microns across (actually much larger than that) and from a depth that is hundreds of microns deep. Given that this is at least a factor of 100 larger than the likely particle size it shows that Raman can be used effectively as a bulk technique.

Professor Šepelák answered: I fully agree with this comment that Raman spectroscopy can be used effectively as a "bulk" technique. We have the same experiences – see our results of Raman investigations of nonequilibrium structure of Zn_2SnO_4 spinel nanoparticles prepared by mechanosynthesis, published in *J. Mater. Chem.*, 2102, **22**, 3117.

Professor James said: You looked at your system with Raman, is the sampling depth is greater than your particle size?

Professor Šepelák replied: In the present work, we did not apply Raman spectroscopy. Instead of this method, we employed ^{57}Fe Mössbauer spectroscopy to give an insight into the mechanically induced changes in the hexaferrite. I would refer to the comment by S. James above, emphasizing that Raman spectroscopy can effectively be used as a "bulk" technique.

Professor Jones said: I wonder if you or others have considered conversion electron Mössbauer spectroscopy to perhaps determine any surface specific ordering that might be present in your samples?

Professor Šepelák replied: Yes, in addition to ^{57}Fe Mössbauer spectroscopy measurements in transmission mode, we have performed conversion electron Mössbauer spectroscopy (CEMS) measurements to determine the surface specific ordering of Fe cations in mechanosynthesized iron-containing complex oxides (ferrites). It should be noted that internally converted electrons are emitted from distances of *ca.* 300 nm below the exterior surface,[1] leading to a value of the information depth of the CEMS technique that is about 300 times larger than the thickness of the structurally disordered surface shell (of about 1 nm) of mechanosynthesized nanooxides.[2] Thus, the obtained results of CEMS measurements have reflected not only the surface structural disorder of Fe cations but also the interior structure of mechanosynthesized nanoparticles, *i.e.*, they have provided the same information on ordering of Fe cations as that derived from Mössbauer spectroscopy measurements in transmission mode.

1 W. Jones, J. M. Thomas and R. K. Thorpe, *Appl. Surf. Sci.*, 1978, 388.
2 V. Šepelák, A. Düvel, M. Wilkening, K.-D. Becker and P. Heitjans, *Chem. Soc. Rev.*, 2013, **42**, 7507.

Professor Suslick opened the discussion of the paper by Vitalij Pecharsky and Shalabh Gupta: You are proposing using alane as the fuel for a fuel-cell, so I presume that makes the alane a sacrificial reagent, and a pretty expensive one. Is there a mechanochemical way to take aluminum and hydrogen back to alane?

Dr Gupta replied: That is correct. At present, alane can only be utilized as a sacrificial source of hydrogen for fuel-cells yielding metallic aluminum as by-product. To date, there is no simple and direct route to rehydrogenate metallic aluminum except under extremely high hydrogen pressures that are on the order of 1 GPa or greater. Our ultimate goal is to develop practical mechanochemical approaches to convert the byproduct, *i.e.* metallic aluminum, to alane under reasonable hydrogen pressure.

Dr Blair commented: This is essentially a mechanically driven solid state metathesis reaction. LiCl is acting as a heat-sink through its significant enthalpy of fusion. Reaction engineering could solve these problems, such as reactor design for improved heat transfer. Monitoring the ball-mill temperature while it's running would give a lot of insight.

For solid-state metathesis work look at Ric Kaner's work out of UCLA: R. W. Cumberland, R. G. Blair, C. H. Wallace, T. K. Reynolds, and R. B. Kaner, "Thermal control of metathesis reactions producing GaN and InN", *J. Phys. Chem. B*, 2001, **105**(47), 11922–11927. Monitoring reaction temperature of a moving vial can be difficult. Maxim Integrated makes a product called an iButton. This is a small temperature probe and datalogger combined (http://www.maximintegrated.com/en/products/comms/ibutton/temperature-humidity-data-logging.html). James Mack successfully used one on a SPEX mill: K. S. McKissic, J. T. Caruso, R. G. Blair, and J. Mack, "Comparison of shaking *versus* baking: further understanding the energetics of a mechanochemical reaction", *Green Chem.*, 2014, **16**(3), 1628–1632.

Dr Blair then asked: What is the role of hydrogen in this implementation? Is it suppressing the decomposition of alane?

Dr Gupta replied: We are currently investigating the role of gas pressure in controlling the reaction pathway, and agree that reaction enthalpy may be a factor, but only one of them. However, we also realize that other physical parameters such as viscosity of pressurized gas and its thermal properties, such as heat capacity and thermal conductivity play a role. Further, we know that there is no gas–solid exchange of hydrogen in the mechanochemical transformations that we reported. In a different study related to solid synthesis of alane from different precursors we show that inert gases such as argon and helium are also effective in controlling the reaction pathway, even more so than hydrogen since critical pressures of helium and, especially argon, needed to suppress formation of metallic aluminum, are lower. Hence, it is not specifically pressurized hydrogen that is suppressing the decomposition of alane – it is a pressurized gas acting as a fluid process control medium.

Dr Moores asked: Could you please comment on the possibility to scale up your process. The strictly inert conditions seem to make it difficult.

Dr Gupta One of the generally accepted advantages of a mechanochemical over a thermal process is the scalability of the former. While we have not assessed the economics of scaling up the process that we report here, we do believe that maintaining inert conditions on a large scale is a challenge. Overcoming this challenge, however, is within the technological capabilities of an established chemical manufacturing company.

Dr Stolle asked: (a) Please comment on the filling degree of the milling vessels. (b) What are the main barriers regarding scale-up of the process?

Dr Gupta replied: (a) In a typical reaction performed during this study, milling containers (internal volume of ~92 ml) were nearly half-filled with twenty chrome steel balls weighing about 8.3 g each in addition to about one gram of the reaction mixture. This is a typical laboratory scale. (b) The metathesis reaction involved in the process seems to be sensitive to milling intensity, and therefore the supply of mechanical energy may have to be precisely controlled. As pointed out by Dr Moores, maintaining inert conditions and operating large vessels at pressures of 200–300 bar could pose engineering challenges. Finally, the economics of batch mechanochemical processing need to be looked at.

Dr Baláž opened the discussion of the paper by Audrey Moores and Monika Rak: Do you have any idea what the exact interaction between the molecules of amine and your gold nanoparticles might be? Do you suggest some kind of electrostatic interaction, or do you think some chemical bond is created? We encounter the same issue in our research, in which we are up till now unable to figure out the exact type of interaction between the nanoparticle and capping agent.

Ms Rak replied: I'm afraid to say that we find ourselves in the same boat. We never did do a complete study of the amine-AuNP bond and therefore I feel I can't help you elucidate your problem. It's a very good question though.

Dr Stolle asked: Does the lignin influence the optical properties of the Au-nanoparticles (in comparison to amine-based NPs)?

Dr Moores answered: The amine particles are too small to be plasmon-active so we could not do anything. The lignin based ones are of the appropriate size and they feature a nice color, characteristic of their plasmon band. We've not been able to record a meaningful spectrum however because it is challenging to find an appropriate solvent to disperse the NPs. We are working on it, however, as it could be of great interest.

Dr Stolle asked: What are the possibilities to synthesize bimetallic/structured bimetallic particles with this method?

Dr Moores replied: Yes it's perfectly feasible. In fact bimetallic particles by milling have already been reported. See this article:

A. Murugadoss, N. Kaia and H. Sakurai, *Nanoscale*, 2012, **4**, 1280–1282.

Professor Bolm commented: You described an access to metal nanoparticles in (and on) lignin. What is the "holding power" of such particles in the biomass framework? How do they interact with lignin?

Dr Moores responded: We believe that the interaction between the nanoparticles and the organic matter happens through the oxygen atoms (hydroxy and carbonyl functionalities). Our results show as well that this interaction is not sufficient in some cases, and we present precursors that lead to particles outside the matrix and not inside. In further work, we envisage using a second precursor to help building a solid interaction.

Professor Bolm asked: You reported the use of Kraft lignin as reductant. Does the source of lignin and the purity of the biomass have any effect on such transformations?

Dr Moores responded: We used two sources of lignin. One is industrial Kraft lignin (indulin AT, pine wood) that we purchased from Westvaco Chemical division. We have also tested home-made lignin (this sample should be less refined) and we did not observe any modification in the syntheses. We did not report these results.

Professor Suslick enquired: There is a common theme among the questions here: "What is the reductant?" Why not add a solid reductant? A relatively weak one should do, such as a borohydride salt.

Ms Rak responded: We believe that the lignin is the reductant. The reason for this is that in the FTIR comparison of milled lignin *vs.* milled lignin + metal we can see a C=O peak for the latter that is missing in the former indicating the oxidation of lignin. We were seeing reduction of the metal by XPS and therefore we did not think to add an external reducing agent to the reaction. It is an interesting point that we will consider for future reactions however. I hope this answers your question sufficiently.

Professor Baláž then asked: Do you have the information about time scale of your milling experiments?

Ms Rak answered: The reactions were 90 min long for the M@lignin work presented in our paper. The amine-Au reactions were run at 15, 30, 60, 90, and 180 min – the data for which can be found in the supplementary information in *Green Chem.* 2014, **16**, 86.

Professor Beyer said: Ball milling of lignin in the absence of metal precursors will also induce chemical chance. Therefore, the change in the FTIR-ATR spectra shown in your Fig. 3b may very well arise from the effect of ball milling on lignin, and not necessarily from the interaction with the AuNP, if the precursor Kraft lignin was not ball milled before taking the spectra. The figure caption is ambiguous to me, so please clarify: Does the red trace in Fig. 3b show the precursor lignin without ball milling, or after ball milling without metal precursor?

Dr Moores responded: The FTIR recorded on lignin was made on milled lignin (milling time and conditions identical to the reaction). It is true that it would be clearer to add it in the caption directly.

Ms Brekalo remarked: Since there is obvious ambiguity in what is being oxidized, only lignin (as shown by IR studies), or lignin and steel from the grinding jars/balls, have you done studies to check if there is iron in the final product, and have you tried using non-steel media (zirconia, tungsten carbide…)?

Dr Moores replied: Thank you for this important question. By XPS we could see that there is no iron in the final product. IR as well was a strong hint, suggesting that lignin acts as the reducer. Finally, we could not observe macroscopically the wear on the steel apparatus that was observed with the gold–amine system. Because of time constraints we did not get a chance to test other non-steel media, although we recognize it is an important thing to test. This will be performed shortly.

Mr Do opened the discussion of the paper by Peter Baláž: Traditional synthesis of semiconductor nanocrystals is done in the presence of an excess ligand to ensure adequate surface passivation. However, you perform your synthesis of CdS nanocrystals in a stoichiometric ratio with cystine acting both as your source of sulfur and source of ligand. Have you done property measurements on your nanocrystals, such as fluorescence or luminescence, where surface passivation has a significant impact on performance? Are they sufficiently passivated and how well do they compare to more traditionally synthesized nanocrystals?

Professor Baláž answered: Your question probably deals with PbS, not CdS. The main aim of our work was to perform solvent-free synthesis of capped PbS nanocrystals and their structural characterization. Fluorescence and luminescence measurements should be also performed, but were not realized prior to the *Faraday Discussions* meeting. Your comment about using excess ligand is highly appreciated.

Dr Moores asked: Have you tried elemental sulfur in place of cystine?

Professor Baláž answered: No, because we pursued the dual role of the amino acid.

Dr Friščić asked: I was wondering if the chemistry presented in your work is only lead-specific or if it could be extended to other metals?

Professor Baláž answered: The chemistry could be extended also to other metals, like Zn and Cd. However, the exact chemistry was studied only for he presented system.

Dr Baláž added: The mechanism presented in this paper is based on a combination of: (i) the experimental evidence of the final products; and (ii) proposed reaction pathways with possible intermediates, the formation of which is possible according to the literature survey. I encourage all the participants of this meeting to express their view on this particular problem, in order to improve the mechanism.

Dr Lamaty said: In the mechanism you propose, the second step is known in the literature to occur at 150 °C. Do you think this could happen in your process?

Professor Baláž answered: Yes, because on the contact sites between milling media and reactants the temperature can be much higher than 150 °C (*e.g.* the "hot-spot" theory in mechanochemistry).

Dr Lamaty asked: To further support your mechanism, did you use as starting material cysteine (with one sulfur) instead of cystine (with two sulfurs)?

Professor Baláž answered: In the present study we did not use cysteine, however it was used in our contemporary study on CdS@ZnS system. It is much more suitable for potential applications in medicine than cystine, because it is water-soluble and therefore can be used for bioconjugation with further biological objects.

Dr Maini commented: While you follow the reaction of the formation of the PbS nanocrystals it is clear that the profile of the lead acetate disappears quite quickly (after 15 min), while the peaks due to lead sulphide begin to show only after 60 min. Is the grinding process promoting the dehydration of the lead acetate? If you dehydrate the lead acetate trihydrated by heating do you obtain an amorphous or crystalline material?

Professor Baláž responded: We suppose that the milling process definitely promotes the dehydration of the lead acetate trihydrate, as this process is the first step of the proposed mechanism [see our eqn (1)]. We did not perform the thermal dehydration of this compound. However, according to literature (see M. A. Mohamed, *et al.*, *Thermochim. Acta*, 1994, **236**, 249–262), lead acetate trihydrate was found to undergo melting concurrently with the deyhydration process, which takes place in two steps between 30 °C and 110 °C.

Dr Maini asked: Have you tried to start with anhydrous lead acetate to ascertain the role of the water molecules?

Professor Baláž answered: No, lead acetate trihydrate was applied in the mechanochemical synthesis. However, the idea of using the anhydrous form is highly appreciated.

Dr Friščić commented: Is the presented chemistry specifically dependent on using an acetate starting material? Could similar products be obtained, for example, from lead (II) chloride?

Professor Baláž responded: The other precursors like lead (II) chloride can be also used for the synthesis of PbS, however the mechanism would surely be different. Moreover, the chloride medium is not very environmentally friendly.

Dr Khripin asked: This question is somewhat tangential to the main thrust of your work, but does touch on the mechanism and the role of water in the reaction. In your paper you write, "disulfide bonds are broken at temperatures above 150 °C, especially at low moisture levels (below 20%)". This implies water stabilises the –S–S– bond. This seems to contradict the proposed reaction mechanism, where water acts to hydrolyse the –S–S– bond. Could you please comment on this apparent contradiction?

Professor Baláž replied: In our experiments, 10.82% H_2O from lead acetate was released. Since this amount was sufficient for our synthesis and was in accordance with our ref. 37, we did not pay further attention to what could possibly happen in case of higher moisture levels.

Professor Boldyreva communicated: My congratulations with this beautiful work. My comment is that it is hardly appropriate to term such a reaction "solvent-free". You have clearly shown that the hydrate water which is present in a large molar ratio in acetate is involved, and obtaining perfectly-shaped single-crystals also suggests a mechanism involving crystallization from a solution. We have observed similar effects in the system L-serine + oxalic acid (E. A. Losev and E. V. Boldyreva, The role of a liquid in "dry" co-grinding: A case study of the effect of water on mechanochemical synthesis in a "L-serine–oxalic acid" system, *CrystEngComm*, 2014 **16**(19), 3857–3866).

Faraday Discussions

PAPER

Quantitative *in situ* and real-time monitoring of mechanochemical reactions

Ivan Halasz,*[a] Tomislav Friščić,*[b] Simon A. J. Kimber,[c] Krunoslav Užarević,[a] Andreas Puškarić,[a] Cristina Mottillo,[b] Patrick Julien,[b] Vjekoslav Štrukil,[ab] Veijo Honkimäki[c] and Robert E. Dinnebier[d]

Received 16th February 2014, Accepted 19th February 2014
DOI: 10.1039/c4fd00013g

An experimental technique for *in situ* and real-time monitoring of mechanochemical reactions in a shaker ball mill was recently described, which utilises highly penetrating X-ray radiation available at the ID15B beamline of the European Synchrotron Radiation Facility. Herein, we describe the first attempts to perform such reaction monitoring in a quantitative fashion, by introducing an internal X-ray diffraction standard. The use of silicon as an internal standard resolved the issue with variations of the amount of the sample in the X-ray beam due to the non-uniform distribution of the sample in the reaction jar and allowed, *via* Rietveld analysis, the first quantitative estimate of the amorphous phase content in a mechanochemical reaction as it is being milled. We also highlight problems associated with the non-ideal mixing of the reaction mixture.

Introduction

Mechanochemistry,[1,2] *i.e.* chemical transformations achieved through the application of mechanical actions, such as grinding, milling, rubbing, shearing, *etc.*, has recently attracted the attention of a wide audience of researchers in organic synthesis and catalysis, the synthesis of and screening for inorganic, metal-organic and pharmaceutical materials and nanoparticle preparation.[3–6] Despite such broad and growing interest, and in contrast to a long history[7] of mechanochemistry and milling in a number of technological areas (mechanical alloying, mineral processing, lubrication),[8] the mechanisms of the mechanochemical reactions have remained largely unknown. Such a lack of mechanistic knowledge is explained by the difficulties associated with monitoring milling reactions, as they take place under the violent impact of hard grinding media, typically

[a]Ruđer Bošković Institute, Bijenička 54, 10000 Zagreb, Croatia. E-mail: ihalasz@irb.hr; Tel: +38514561184
[b]Department of Chemistry and FRQNT Centre for Green Chemistry and Catalysis, McGill University, 801 Sherbrooke St. W., H3A 0B8 Montreal, Canada. E-mail: tomislav.friscic@mcgill.ca; Tel: +5143983959
[c]European Synchrotron Radiation Facility (ESRF), Grenoble cedex, France
[d]Max-Planck-Institute for Solid State Research, Heisenbergstr.1, 70569 Stuttgart, Germany

grinding balls of steel, zirconia or tungsten carbide, in a rapidly spinning or oscillating vessel (a grinding jar). The direct monitoring of mechanochemical reactions is further hindered for those mechanochemical reactions that develop rapidly, lasting 30 minutes or less, such as the recently introduced liquid-assisted grinding (LAG) and ion- and liquid-assisted grinding (ILAG) reactions,[2] in which the synthesis of organic or metal-organic products is accelerated and directed by a catalytic amount of a liquid phase, or a combination of a liquid phase with a catalytic ionic additive, respectively. Recently, we have presented the first technique for the real-time and *in situ* observation of mechanochemical transformations by powder X-ray diffraction (PXRD).[9] This methodology is enabled by the use of highly penetrating synchrotron X-rays ($\lambda = 0.14$ Å) and has provided the first direct insight into the mechanochemical formation of metal–organic frameworks known as zeolitic imidazolate frameworks (ZIFs),[9,10] and of pharmaceutical cocrystals.[11] The mechanistic information, however, was mostly qualitative due to difficulties associated with the non-uniform time-dependent distribution of the sample throughout the milling jar, the inability to evaluate the amount of amorphous, non-diffracting material and the inability to quantitatively estimate the X-ray scattering behavior of porous frameworks containing highly mobile and chemically diverse guests.

Herein, using as model reactions the LAG and ILAG milling transformation of zinc oxide and imidazole (**HIm**)† into a non-porous ZIF structure (Fig. 1), we present the results of the first attempt to resolve the above mentioned limitations and achieve quantitative, real-time and *in situ* X-ray diffraction monitoring of mechanochemical reactions, including the assessment of the amorphous content, by adding crystalline silicon to the reaction mixture as an inert internal scattering standard.

Recent *ex situ* mechanistic studies of mechanochemical milling reactions

Mechanistic studies of mechanochemical reactions have recently been reviewed.[12,13] Most mechanistic studies of milling reactions rely on step-by-step analysis,[14] *i.e.* on the *ex situ* characterization of samples extracted from the mechanochemical reaction mixtures at different milling times. Over the past decade, such studies revealed the unexpectedly dynamic nature of the mechanochemical reactions of organic or metal–organic solids, involving either metastable amorphous phases or quickly interconverting intermediate crystalline phases. The fact that mechanochemical milling can lead to amorphization is well known in the context of inorganic solids, where the amorphous phase may be relatively long-lived.[15] In contrast, the detection and quantification of the metastable amorphous phases in the mechanochemical transformations of "softer" organic and metal-organic substances is highly challenging.[16] Indeed, the importance of the amorphous content in the mechanochemical cocrystallization of organic solids began to be evaluated only recently, for example by a combined use of PXRD and terahertz spectroscopy or by *ex situ* investigation of samples prepared by milling at liquid nitrogen temperatures (cryomilling).[17,18] However, none of the performed studies provided any knowledge of the amorphous phase

† The reaction of ZnO and imidazole was first reported by manual grinding, see: J. Fernández-Bertrán, L. Castellanos-Serra, H. Yee-Madeira and E. Reguera, *J. Solid State Chem.* 1999, **147**, 561.

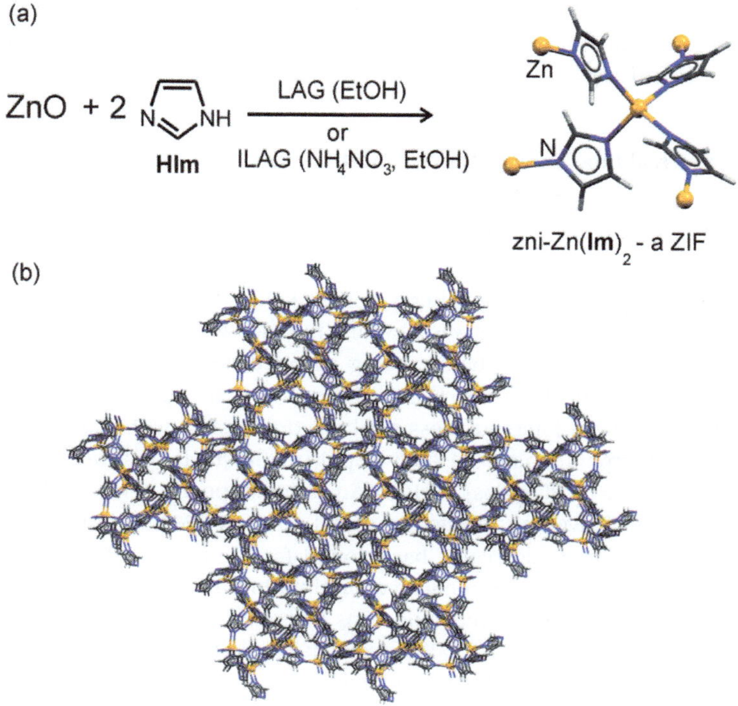

Fig. 1 (a) A schematic representation of the herein studied mechanochemical reaction using either liquid-assisted grinding (LAG) with ethanol (EtOH) or ion- and liquid-assisted grinding (ILAG) with EtOH and the ionic additive NH_4NO_3. (b) A view of the reaction product, a non-porous ZIF with a *zni*-topology, viewed down the crystallographic *c*-axis.

content and its structural evolution while the reaction mixture is being milled. Also, due to the metastable nature of the amorphous organic materials and their general sensitivity to moisture and temperature, reliable *ex situ* analyses of amorphous intermediates are difficult to achieve.

Several recent systematic studies have suggested tentative explanations for the appearance of intermediate crystalline phases in mechanochemical reactions, specifically the competition of supramolecular synthons,[19,20] or mass action effects in the early stages of liquid-assisted grinding reactions.[21,22] However, while revealing a dynamic nature for mechanochemical reactions, such studies also made it clear that *ex situ* step-by-step reaction monitoring is too limited for a complete understanding of the mechanochemical processes, requiring the detection and, ideally, the quantification of the short-lived crystalline and amorphous intermediate phases. Such intermediates are likely to have a highly fleeting nature, lasting only seconds or minutes, and their detection and quantification are not compatible with the delays often required for the sample preparation in step-by-step *ex situ* diffraction or spectroscopic analyses. Furthermore, *ex situ* analysis is also inadequate for systems in which sample extraction causes an evaporative loss of a liquid phase, or reactions sensitive to air moisture or CO_2.[23,24] Finally, it was also demonstrated that dividing a mechanochemical reaction into segments, including the opening of the reaction jar, may lead to a different product as compared to continuous milling.[14b,25] Thus, the development of analytical methods capable of monitoring milling reactions without disturbing them is necessary for a proper and reliable understanding of mechanochemistry.

In situ and real-time X-ray diffraction studies of mechanochemical reactions

With the above considerations in mind, we have recently developed[9] the first *in situ* and real-time X-ray diffraction technique for monitoring the course of mechanochemical reactions. This novel and unprecedented methodology[10] is based on the diffraction of high-energy ($E \approx 90$ keV) synchrotron X-rays which are able to penetrate the reaction vessel walls, interact with the sample and again penetrate out of the vessel before being detected on a two-dimensional area detector. Measurements using this technique are also greatly facilitated by replacing the conventional stainless steel reaction vessels with those made of a light-atom, amorphous material such as poly(methyl)methacrylate (PMMA, also known as Perspex or Plexiglas).[9–11] PMMA is a much poorer X-ray scatterer than most metals and gives no diffraction signal. Thus, PMMA milling equipment provides a much more agreeable background for the *in situ* diffraction experiments than the more conventional steel or aluminium alternatives.

This reaction monitoring technique was first used to follow the course of the mechanochemical formation of ZIFs.[9,26] Specifically, the ILAG reaction of ZnO and 2-ethylimidazole (**HEtIm**), which ultimately yields a non-porous close-packed β-quartz (*qtz*) topology zinc(2-ethylimidazolate) framework, was observed to first yield an open framework with zeolite ρ (RHO) topology. Upon further milling this open structure is subsequently replaced by a more dense analcime (ANA) framework (Fig. 2a), which later collapses into the final *qtz*-topology product. As the framework density, and therefore thermodynamic stability,[27] increases in the observed reaction sequence RHO → ANA → *qtz*, this stepwise transformation clearly follows the Ostwald's rule of stages,[28] in which a thermodynamically less stable system (*e.g.* the reactant mixture) converts to the stable one (the ultimate close-packed product) *via* a series of intermediate steps (open frameworks).

The *in situ* and real-time diffraction monitoring technique was also applied to the mechanochemical formation of model pharmaceutical cocrystals.[11] In particular, the group of Rodríguez-Hornedo was the first to propose, on the basis of *ex situ* analysis of samples prepared by cryomilling, that mechanochemical cocrystallisation of carbamazepine and saccharin proceeds *via* an amorphous phase.[18] The application of *in situ* and real-time diffraction monitoring enabled this amorphization to be confirmed[11] by directly observing the gradual loss of the scattering signal of the reaction mixture upon milling (Fig. 2b).

Although the above described proof-of-principle studies provided new and previously inaccessible insights into the mechanisms of the mechanochemical formation of MOFs and cocrystals, this initial monitoring methodology suffered from two significant shortcomings. These are the significant artefactual variation in the intensity of the diffraction signal, which results from the non-uniform distribution of the sample throughout the milling jar, and the inability to quantitatively measure the reaction mixture composition, which prevented the assessment of the amorphous content during milling.

We now demonstrate how our initial *in situ* and real-time monitoring technique can be modified by the addition of a known amount of crystalline silicon as

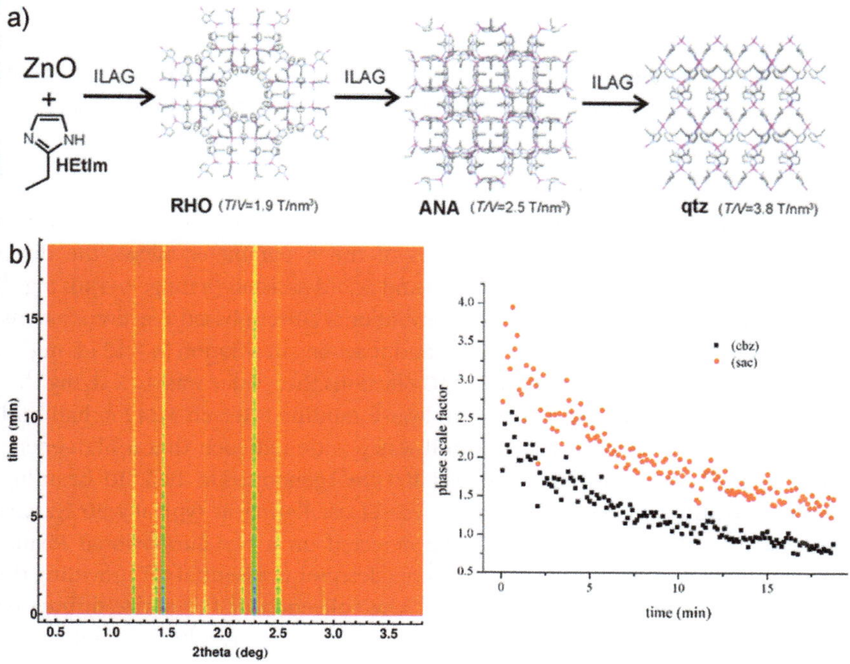

Fig. 2 a) The ILAG synthesis of a zinc(2-ethylimidazolate) ZIF from 2-ethylimidazole and ZnO proceeds through a series of intermediates with decreasing porosity (expressed as *T*/*V*, ratio of tetrahedral centers per unit volume). This stepwise mechanism, observed by *in situ* and real-time X-ray diffraction monitoring, is consistent with the stepwise formation of phases with increasing thermodynamic stability. The formation of the final, close-packed product with α-quartz topology involves the expulsion of the liquid phase from the pores of the zeolite ρ (RHO) and analcime (ANA) intermediates. b) The mechanochemical amorphisation of a mixture of saccharin (sac) and carbamazepine (cbz) during neat milling. Amorphisation is evidenced by the gradual decrease of the diffraction signal from both components (left) and illustrated by the decrease in the phase scale factors extracted from Rietveld refinements (right). The slight random deviations in the scale factor evolution are due to differing amounts of the sample in the X-ray beam in consecutive powder diffraction patterns.

a non-reactive internal X-ray diffraction standard into the reaction mixture. This internal standard technique enabled us to conduct quantitative reaction monitoring, evaluate the reproducibility of mechanochemical reaction kinetics, address the problem of non-uniform distribution of the sample in the reaction jar and, most importantly, establish the extent of the amorphisation of the reaction mixture during mechanochemical milling.

Materials and methods

The details of the sample preparation, measurement and data processing for the *in situ* and real time diffraction monitoring technique have been reported previously[10] and will be only briefly addressed here. The reaction mixtures were prepared for milling by placing the solid reactants in one half of the reaction vessel, while the liquids and ionic additives were placed in the other along with one stainless steel ball of 7 mm in diameter (ball mass ≈1.3 g). By careful operation it is possible to join the two halves of the reaction vessel without bringing their contents to contact before milling was initiated. The LAG and ILAG reactions are conveniently described through the η parameter, defined as;

$$\eta = \frac{V}{m} \tag{1}$$

where V is the volume of the liquid additive (in microliters) and m is the mass of solid reactants (in milligrams). The use of the η parameter allows the direct comparison of neat (dry, $\eta = 0$ μL mg^{-1}) and liquid-assisted (η range 0–1 μL mg^{-1}) mechanochemical reactions to more conventional solvent-based reaction systems, such as slurries (η range 2–12 μL mg^{-1}) and reactions in solution ($\eta > 12$ μL mg^{-1}). For the herein investigated LAG and ILAG reactions the η value was 0.35 μL mg^{-1}.

Milling was performed using an in-house modified Retsch MM200 ball mill, which was operated at 30 Hz. The mill was positioned such that a high-energy X-ray beam ($\lambda = 0.14202$ Å, beam size of 300 μm^2) generated at the ESRF beamline station ID15B could pass through the vessel oscillating at typically 30 Hz and interact with the sample before being detected on a two-dimensional Pixium charge-coupled detector. The use of an area detector is essential for providing the time resolution in seconds, as it allows the collection of the entire diffraction pattern at once. This, however, comes at an expense of the resolution of integrated diffraction patterns since there are no X-ray optics in the path of the diffracted beam. Each diffractogram was typically obtained by summing 10 frames, each collected with an exposure time of 0.4 seconds. Consecutive diffractograms were typically separated by 9 seconds.

The integration of the raw two-dimensional diffraction patterns was performed using the program Fit2d.[29] Rietveld analysis was performed using the program Topas.[30] The amorphous vessels give rise to two broad peaks which were modelled using the "Peaks phase" of Topas. Further modelling of the background was achieved by a shifted Chebyshev polynomial. The structure models for the crystalline phases were taken from the Cambridge Structure Database (CSD)[31] or the Inorganic Crystal Structure Database (ICSD)[32] and, except for their unit cell parameters, were not refined. The peak shape for each phase was described independently using the Voigt function, taking into account the particle size broadening. The instrumental contribution to peak broadening was estimated by measuring the pattern of corundum measured from a blank milling experiment involving only two corundum balls in an otherwise empty jar.

Since each time-resolved diffraction experiment typically provides several hundreds of individual diffraction patterns, automatic Rietveld refinement greatly facilitates data treatment (Fig. 3). This was achieved by using the command-line version of the program Topas,[30] using the same initial input file for each diffraction pattern. The starting parameter values (unit cell, particle size, scale factor) for specific phases were determined from patterns where this particular phase had a clear diffraction signal. The weight fractions, scale factors or other variable characteristics for a specific diffractogram were conveniently output to a textual file suitable for further analysis and data plotting. Further details of the data treatment have been described previously.[9,10]

The weight fractions for all of the crystalline components with fully known crystal structures can be assessed from a Rietveld refinement. Should there be an

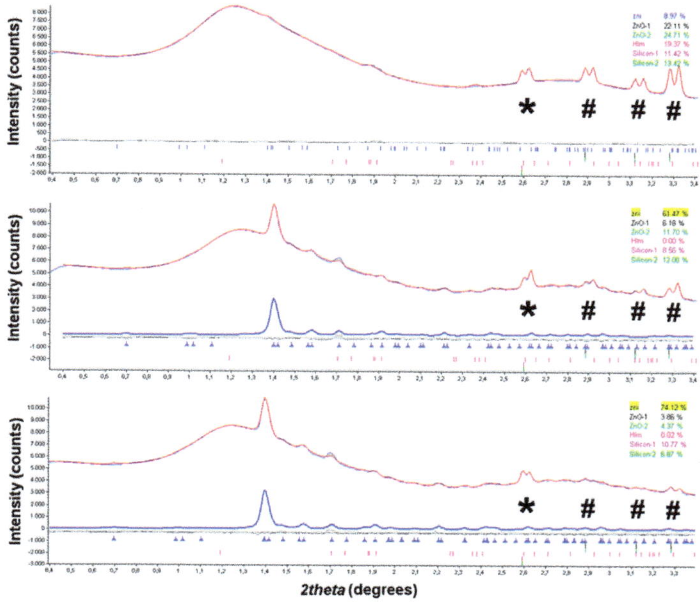

Fig. 3 Example Rietveld refinements of the *in situ* collected diffraction patterns: (top) 10 seconds into the reaction; (middle) 1 minute into the reaction and (bottom) 3 minutes into the reaction. The silicon signal is designated by (*), while (#) designates the ZnO signal. Both are split due to diffraction from the mixture adhering to the two reaction jar walls and were modelled as two phases with the same unit cell, but one having an added refineable shift proportional to tan(θ). Since the material is being continuously redistributed throughout the vessel, the relative intensities of the split peaks vary. The contribution of the zni-topology ZIF is shown with a thick blue line below the experimental patterns in the middle and bottom panels. The weight fraction of silicon in the mixture is 14.3%.

amorphous phase present in the mixture, the Rietveld-refined weight fractions of the crystalline components will be overestimated (Fig. 4) by the same factor each. This factor can be estimated by adding a known mass of a highly crystalline and chemically inert standard, which is assumed to retain its crystallinity during milling (in this case silicon). The factor by which the silicon weight fraction is overestimated is the same for all other crystalline components and they all need to be scaled down. After an appropriate scaling and summation of the scaled weight fractions, the difference to a 100% is attributed to the weight fraction of the amorphous phase.

This procedure is applicable if the full crystal structures of all crystalline components are known. This is not the case for a porous material which can accommodate guest species in its pores. Even though the structure of the framework may be well known, the details of the guest species occupying the pores are still an unresolved structural issue. Porous materials thus cannot be included for the calculation of weight fractions using the formula:

$$w_j = \frac{S_j\,(M\;V)_j}{\sum_i S_i\,(M\;V)_i} \tag{2}$$

where M is the molecular mass of the unit cell, V is the unit cell volume, S is the Rietveld-refined scale factor for the particular phase and the sum in the

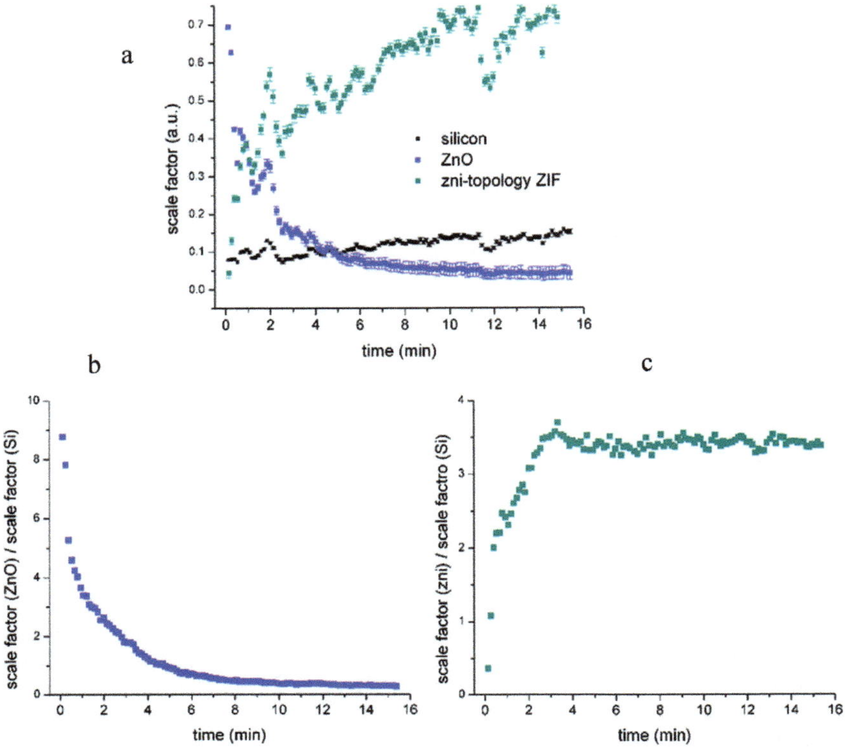

Fig. 4 (a) Scale factors of the internal standard silicon, reactant ZnO and the product zni-Zn(**Im**)$_2$, extracted from time-resolved diffraction patterns (see Fig. 5, top) collected during a LAG (using ethanol as the liquid additive) reaction. The scale factors are affected by the variation of the amount of sample exposed to the X-ray beam during the collection of one diffractogram. The scale factors of (b) ZnO and (c) zni-Zn(**Im**)$_2$ divided by the scale factor of the silicon standard. These plots show that the strength of the diffraction signal may be scaled by that of the silicon signal to reduce the effects of the non-uniform sample distribution and indicate the homogeneity of the mixtures.

denominator runs over all crystalline phases.[33] Porous materials cannot be included since M_i is not known due to the undetermined composition of the pores.

A quantitative analysis requires a reaction mixture to be homogeneous, *i.e.* the sample exposed to the beam for collecting a specific X-ray diffraction pattern must be representative of the entire reaction mixture. Plotting the phase scale factors against time reveals random variations which are a consequence of varying the amount of the sample in the beam during the 4 seconds exposition in the X-ray beam of 300 μm^2 in size (Fig 4a). However, if the individual phase scale factors are first divided by the scale factor of silicon, the plots become smooth and monotonous and decrease (for ZnO, Fig 4b) and increase (for the zni-topology ZIF, Fig 4c) with time. Thus, the representative diffraction signal intensity for each component may be obtained by scaling with the intensity of the diffraction signal of silicon. The success of this scaling procedure also indicates that the reaction mixture is homogenous.

The refined scale factors allow the representative weight fractions of the mixture components to be calculated and, as the time-dependent evolution of the thus calculated weight fractions shows no large oscillations, we conclude that the problem of a non-uniform sample distribution can be overcome by using an internal scattering standard (*i.e.* silicon). Nevertheless, mixture inhomogeneity is still likely to occur at the very beginning of milling. Our findings indicate that the homogeneity of the reaction mixture is most likely achieved within the first 10–20 seconds of milling at 30 Hz (Fig. 4). However, to minimize the effect of the mixture inhomogeneity, the data collected in the first 30 seconds of milling have been excluded from the kinetic analyses.

Results and discussion

Quantification of the reaction mixture and reaction kinetics

In order to facilitate the use of Rietveld refinement for the quantitative *in situ* and real-time analysis of a mechanochemical reaction we have focused on the mehanochemical synthesis of the low-porosity (close-packed) zinc imidazolate framework with the zni-topology (Fig. 1). Namely, it was previously reported that LAG and ILAG reactions of ZnO and unsubstituted imidazole lead to the formation of the zni-Zn(**Im**)$_2$ framework (CSD refcode IMIDZB01) without any observable crystalline intermediates,[23] as long as the grinding liquid is ethanol. This provides a relatively simple system consisting of at most three crystalline phases: the reactants ZnO and **HIm** and the product framework zni-Zn(**Im**)$_2$. As all of these phases can be considered non-porous, their weight fractions should be readily accessible *via* Rietveld refinement of the *in situ* collected diffractograms. In order to evaluate the reproducibility of the mechanochemical reaction kinetics, we have performed each LAG and ILAG reaction twice.

Analysis of the ILAG reaction

The time-resolved diffractograms for the two identical ILAG reactions are shown in Fig. 5, along with the associated reaction kinetics curves obtained by Rietveld analysis. The inspection of the time-resolved diffractograms immediately reveals the rapid formation of the zni-Zn(**Im**)$_2$ product, which appears within less than 30 seconds of milling. The time-resolved diffractograms also display considerable variations in the intensity of the diffraction signal of the reaction mixture. These variations, as noted in our first report,[9] are a result of the variation of the amount of sample in the path of the X-ray beam and would normally prevent the quantitative monitoring of the reaction course through Rietveld analysis. In order to circumvent this obstacle, we normalised the X-ray diffraction signals of ZnO, **HIm** and Zn(**Im**)$_2$ using the diffraction signal of silicon (Fig. 4). The effectiveness of such a normalization procedure for eliminating the fluctuations of the diffraction signal is illustrated by plotting against time the weight fractions of ZnO, **HIm** and zni-Zn(**Im**)$_2$ obtained by quantitative Rietveld analysis (Fig. 5). The resulting kinetic plots reveal a largely smooth increase in the content of the zni-Zn(**Im**)$_2$ product, and similarly smooth decreases in the content of reactants ZnO and **HIm**.

Even though the time-dependent composition of the reaction mixture is fairly similar (within *ca.* 5 weight%) for the two ILAG reactions, the shapes of the curves

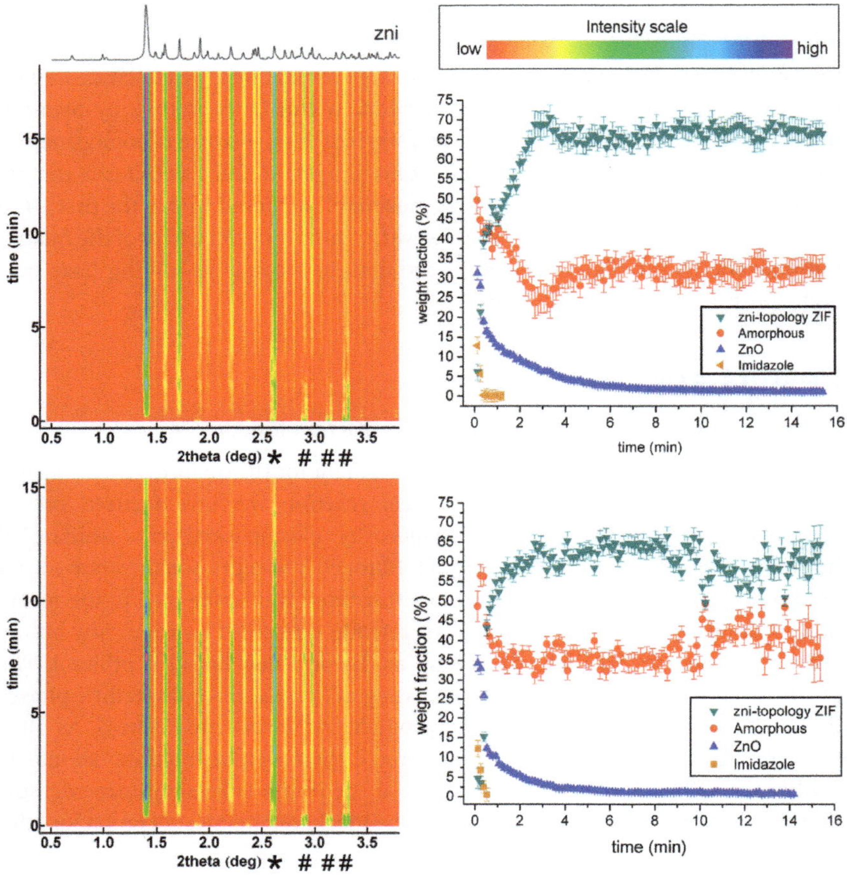

Fig. 5 The time-resolved diffractograms (on the left) and Rietveld-extracted weight fractions (on the right) of the two times repeated ILAG reaction of ZnO (160 mg) and HIm (272 mg) using ethanol (150 μL) as the grinding liquid and ammonium nitrate (18 mg) as the ionic additive. On the left are the time-resolved difractograms, on the right are the Rietveld-extracted weight fractions. The calculated PXRD pattern of the zni-topology ZIF and the colour code for the time resolved diffractogram are given at the top of the figure. The asterisks (*) designate the silicon signal while the hash sign (#) designates the ZnO signal. The weight fractions have been recalculated as if there was no silicon in the reaction mixture.

representing ZIF formation in two experiments are different. In one experiment the ZIF forms at an apparently constant rate, following an initial rapid rise in its weight fraction. In the other, ZIF formation follows a more common sigmoidal curve.[9] We note that this may be due to non-ideal mixing and the distribution of the reaction mixture throughout the jar at the very start of milling. Such effects are particularly likely in LAG and ILAG reactions where the reaction mixture might be sticky and adhere to the reaction vessel walls. In such cases the measured weight fractions and reaction kinetics are influenced not only by the chemical reaction taking place in the free-flowing powder but also by the removal of material adhering to the reaction vessel walls.

The time-resolved diffractograms reveal the rapid and complete disappearance of the diffraction signal of crystalline **HIm** within 30 seconds, after which this phase (CSD code IMAZOL14) was excluded from the Rietveld analysis. The rapid loss of crystalline **HIm** is assigned to several parallel processes, including the dissolution in the grinding liquid and amorphization, as well as the chemical reaction with the salt additive and with ZnO. The formation of zni-Zn(**Im**)$_2$, in agreement with our earlier qualitative study, is rapid and reaches full extent within *ca.* 4 minutes for the two identically prepared ILAG reactions.

The Rietveld analysis allowed us to quantitatively evaluate, for the first time, the amount of the amorphous phase present in a mechanochemical reaction while it is being milled. The sum of the weight fractions of the crystalline reactants and products in the reaction mixture dropped very quickly (within a minute) from the initial 100% to a relatively constant value of 60–65%, indicating that, during milling, as much as *ca.* 35–40% of the reaction mixture weight is present in an amorphous form. Previous *ex situ* investigations of the mechanochemical reactivity established that neat grinding reactions can involve *ca.* 25–35% of the reaction mixture in an amorphous form.[17] Analogous liquid-assisted reactions always gave a fully crystalline (100%) product and were too fast to allow the evaluation of the amorphous content by *ex situ* studies. Thus, the herein described quantitative Rietveld analysis is the first clear demonstration that the amorphous phase can be relevant in a liquid-assisted mechanochemical reaction. This result is consistent with the work of Yuan *et al.*, who proposed that the mechanochemical transformations of different MOF topologies might be assisted by amorphization during LAG.[35] The kinetic plots show that, after *ca.* 4 min milling, the weight fraction of the crystalline zni-Zn(**Im**)$_2$ remains relatively constant whereas ZnO is still being depleted (Fig. 5). This observation suggests the direct transformation of crystalline ZnO into an amorphous ZIF.

As noted in the introduction, the disappearance of the diffraction signal of crystalline **HIm** from the time-resolved diffractograms is a result of several physical and chemical processes and cannot be used for quantitative monitoring of the reaction kinetics. Similarly, the formation of crystalline zni-Zn(**Im**)$_2$ involves at least three processes: the reaction of ZnO with **HIm**, the nucleation of the ZIF and its growth. Consequently, neither **HIm** nor zni-Zn(**Im**)$_2$ is a suitable target for the quantitative diffraction monitoring of the chemical reaction between ZnO and **HIm**. In contrast, the disappearance of the diffraction signal of the harder inorganic substance ZnO is expected to be less affected by the amorphization under the employed milling conditions, offering an opportunity to evaluate the reaction kinetics. Recently, Ma and co-workers have used Raman spectroscopy to monitor *ex situ* the fraction of unreacted **HIm** at different stages of the mechanochemical reaction of ZnO and **HIm** in the presence of DMF.[34] These measurements revealed "pseudo-fluid" behaviour in which the kinetics of disappearance of **HIm** resembled a 2nd-order solution reaction. Here, the non-linear fitting of the time-dependent weight fraction of ZnO (Fig. 6a,b) revealed an excellent fit to the 1st order reaction model. Attempts of non-linear fitting to the 2nd order reaction kinetics gave poorer results (Fig. 6c,d). In both cases the fitting was performed without taking into account the data acquired during the first 30 seconds of milling, due to non-ideal mixing at the very onset of milling. Another reason to ignore those data is because the consumption of ZnO at the onset of milling is likely to be strongly affected by the surface and particulate properties of the commercial ZnO reagent. Although the two identically performed ILAG experiments exhibited an excellent fit to 1st order reaction kinetics, the two

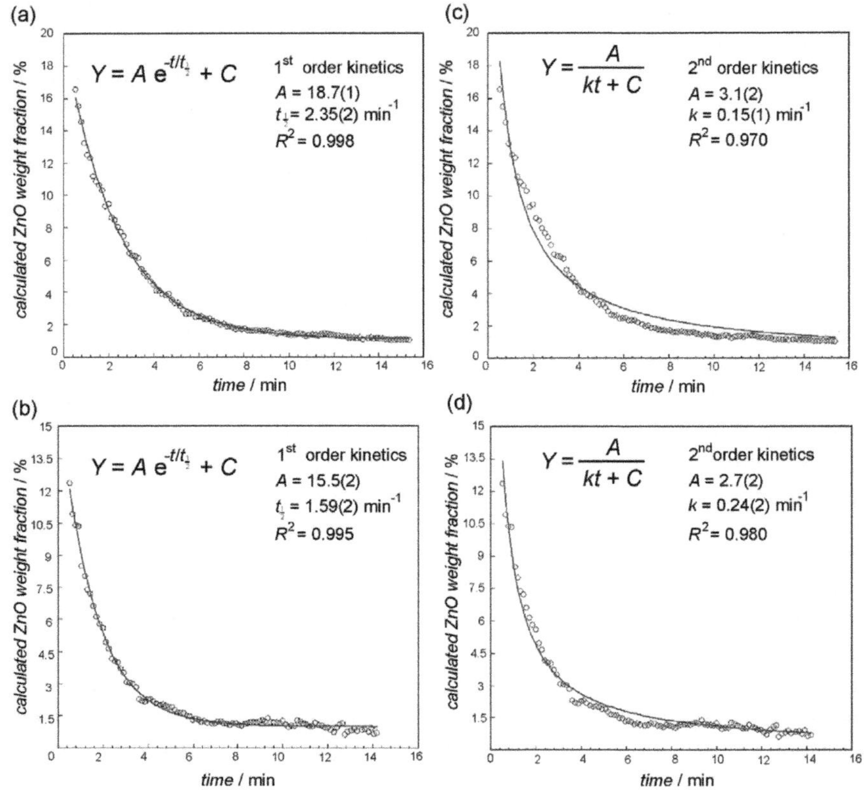

Fig. 6 The non-linear fitting of the time-dependent weight fraction of ZnO, obtained by automated Rietveld refinement of time-resolved diffraction data, to solution-like reaction kinetics laws. The data for the two identically prepared ILAG reactions in (a) and (b) display a superior fit to first-order reaction kinetics. The same two sets of data give a visually and numerically poorer fit to second-order kinetics in (c) and (d). For all fitting procedures the data from the first 30 seconds of milling were excluded, in order to minimise the effects of mixture inhomogeneity and of the initial reactant sample texture.

experiments yielded considerably different values of kinetic parameters (Fig. 6) for ZnO consumption, thus indicating the complexity of the mechanochemical processes being studied. Whereas the possibility that the mechanochemical reaction follows 1st order reaction kinetics conflicts the *ex situ* spectroscopic work,[34] we note there are also significant differences between the currently and previously studied systems. The previously studied reaction involved a significantly smaller content of liquid ($\eta = 0.05\ \mu L\ mg^{-1}$), suggesting a lower level of molecular mobility compared to the herein explored systems ($\eta = 0.35\ \mu L\ mg^{-1}$). Presumably, the higher liquid content in the current study could facilitate contact between ZnO and **HIm**, which was identified as the rate-determining process in previous work. Thus, with a higher liquid content the mechanochemical reaction might be controlled by a different process, such as the rate of activation of ZnO by amorphisation or dissolution, therefore leading to reaction kinetics resembling a 1st order rate law. However, such interpretations remain highly speculative due to a significant oversimplification of the mechanochemical reactivity of ZnO which not only consists of several events, including amorphisation, dissolution and particle comminution, but also requires efficient product removal‡ to ensure the exposure of fresh reactant surface. Finally, it is also important to keep in mind that Raman spectroscopy and PXRD are complementary methods, as the former provides largely information about a thin layer of the sample, whereas the latter provides insight into the sample bulk.

Analysis of the LAG reaction

Compared to the analogous ILAG reactions, the two identically prepared LAG reactions (Fig. 7) proceeded more slowly, which is consistent with the previous observation[2b,9] that salt additives improve the mechanochemical reactions of metal oxides.

The comparison of the Rietveld analyses for the ILAG and LAG reactions reveals that NH_4NO_3 accelerated not only the ZIF formation, but also the disappearance of the reactant **HIm**. This is consistent with a previously proposed reaction model in which the catalytic activity of the ammonium salts involves the activation of the imidazole ligand by protonation.[36] In particular, it took over 1 minute for the diffraction signals of **HIm** to disappear from the PXRD patterns of the LAG reaction mixture. The appearance of crystalline zni-Zn(**Im**)$_2$ was sufficiently slow to exhibit sigmoidal growth behaviour, which is characteristic for processes involving nucleation and crystal growth from an amorphous phase.[9,37] Rietveld analysis reveals that a significant fraction (*ca.* 35% by weight) of the LAG reaction is amorphous during milling. Whereas the content of ZnO in the ILAG reactions decreases steadily by milling and almost completely disappears in less than 10 minutes, in the LAG reaction the weight fraction of ZnO was found to drop to $\sim$15% (*ca.* one-third of its initial value) within *ca.* 4 minutes and after that remained fairly constant. Consequently, it is reasonable to assume that the LAG reaction reaches its maximum extent within 5 minutes.‡ Kinetic analyses were, therefore, conducted only on the data collected within that timeframe. In this case, attempts to analyse the reaction course through non-linear fitting of the data to solution-type kinetic rate laws clearly indicated that the reaction progress is best described by 1st order reaction kinetics (Fig. 8).

Prospects of amorphous phase characterisation *via* pair distribution function and its time evolution

The *in situ* X-ray diffraction technique offers another possibility which we wish to discuss. Namely, the amorphous phase, which is abundantly present in the herein studied mechanochemical transformation, gives no diffraction signal but does have a scattering contribution which was thus far neglected. It seems reasonable to assume that the short-range order in the amorphous phase will have an influence on the nucleation and crystal growth of the bulk crystalline species. Therefore, it seems imperative to develop a means to obtain structural

‡ A recent transmission electron microscopy study of the dry grinding reaction of ZnO and 2-methylimidazole to form the popular framework ZIF-8 revealed the entrapment of the zinc oxide particles within a shell of the ZIF product, hindering the advance of the reaction. The apparently poor advance of the LAG reaction might be tentatively explained by a similar entrapment of ZnO reactant particles within a shell of the product zni-Zn(**Im**)$_2$. As the framework zni-Zn(**Im**)$_2$ is non-porous, it could be efficient in preventing a further mechanochemical reaction of ZnO. See: S. Tanaka, K. Kida, T. Nagaoka, T. Ota and Y. Miyake, *Chem. Commun.* 2013, **49**, 7884.

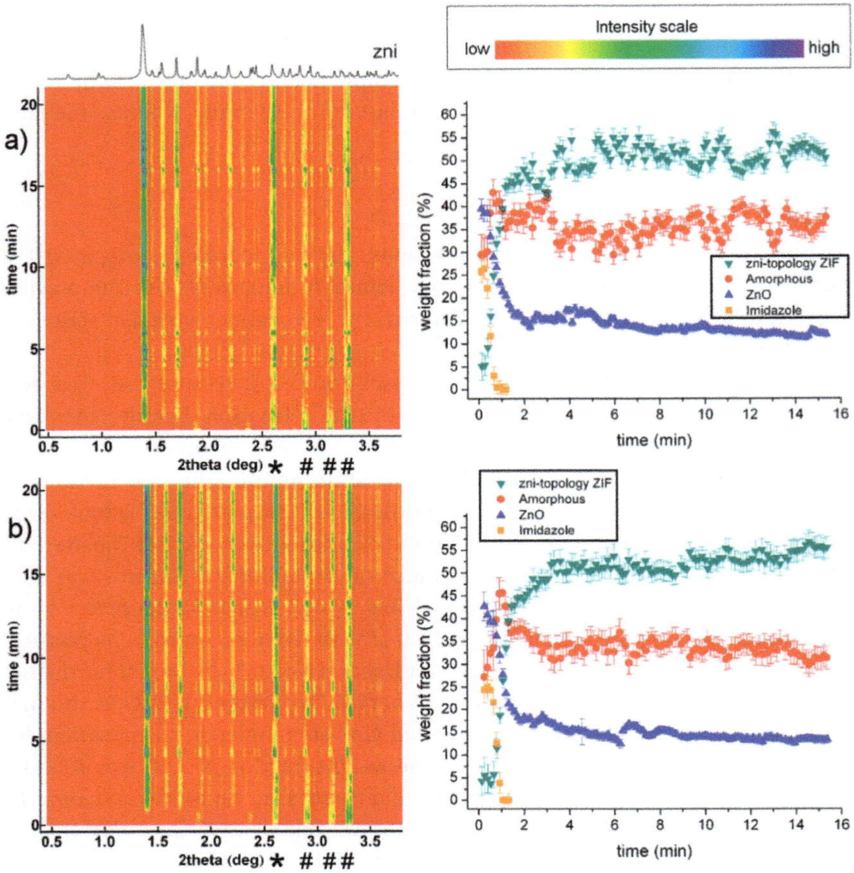

Fig. 7 (a) and (b) Time-resolved diffractograms (left) and weight fractions of the crystalline and amorphous reaction components extracted by Rietveld refinement (right) for two identically prepared LAG reactions of ZnO (160 mg, 2 mmol) and **HIm** (272 mg, 4 mmol) using ethanol (150 μL, 2.6 mmol, $\eta = 0.35$ μL mg^{-1}) as the grinding liquid. The calculated PXRD pattern of the reaction product zni-topology Zn(**Im**)$_2$ framework and the colour scheme for the time-resolved diffractogram are given at the top of the figure. The time-dependent variations in the intensity of the diffraction signal of the reaction mixture are related to differing amounts of the sample in the beam and illustrate well the need for an internal diffraction standard. The asterisks (*) designate the silicon signal while the hash sign (#) designates the ZnO signal. The weight fractions have been recalculated as if there was no silicon in the reaction mixture. The weight fraction of ZnO rises slightly between 6 and 9 minutes which is correlated with the stronger diffraction signal of the whole reaction mixture. This indicates that the standard deviations of the weight fractions as determined from the Rietveld refinement are underestimated.

information of the amorphous intermediate and determine if it is changing during milling. The X-ray scattering technique suitable for such analysis is the pair distribution function (PDF). High-energy X-rays available at ID15B are well suited to collect scattering data to high Q values as needed for reliable PDF determination. In such an experiment the background contribution of the moving amorphous reaction jar would need to be carefully subtracted in order to leave only the sample's scattering contribution in the pattern. Should this be

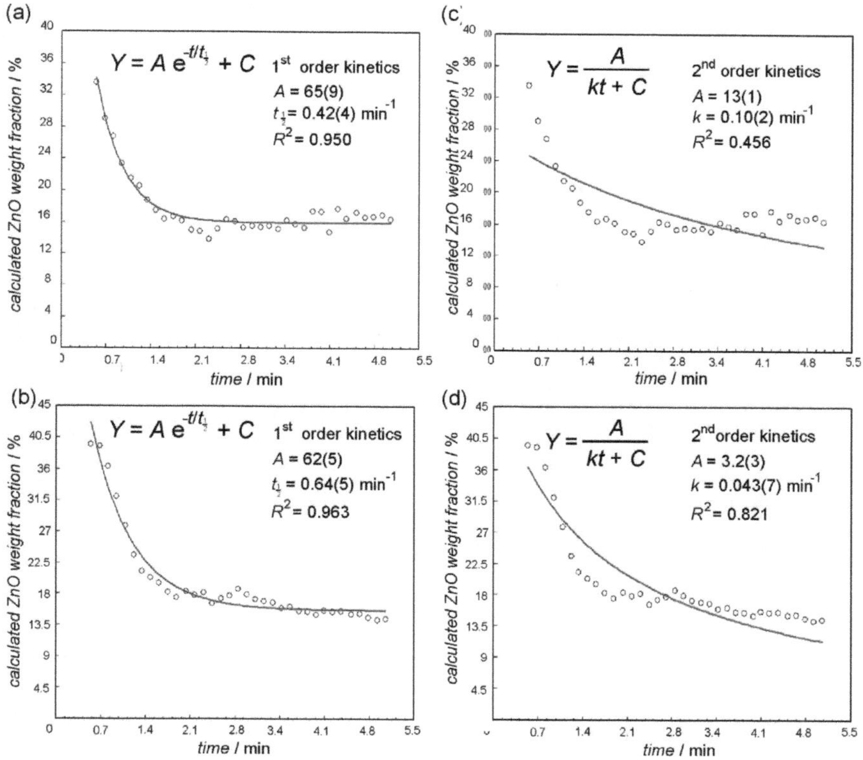

Fig. 8 The non-linear fitting of the time-dependent weight fraction of ZnO, obtained by automated Rietveld refinement of the time-resolved diffraction data, to solution-like reaction kinetics laws. The data for the two identically prepared LAG reactions in (a) and (b) display a superior fit to first-order reaction kinetics as compared to second-order reaction kinetics. The same two sets of data give a visually and numerically poorer fit to second-order kinetics in (c) and (d). For all fitting procedures the data from the first 30 seconds of milling were excluded, in order to minimise the effects of mixture inhomogeneity, and of the initial reactant sample texture.

possible, the possibility to structurally characterise the amorphous phase and its time evolution would arise, allowing for an even deeper insight into the mechanochemical reaction course. These experiments will be the subject of our further studies using the real-time and *in situ* X-ray scattering technique. The current experimental setup should require only one modification, the detector would need to be placed closer to the reaction jar to be able to collect scattering data to higher Q values.

Difficulties due to the presence of a liquid phase

As our next target we wanted to address the reproducibility of the reaction kinetics for a more complex mechanochemical transformation, the reaction of ZnO and 2-ethylimidazole (**HEtIm**) which proceeds *via* open structures with RHO and ANA topologies to form a close-packed qtz-topology ZIF as the final product (Fig. 2a). The corresponding time-resolved diffractograms (Fig. 9) for the LAG reaction conducted at $\eta = 0.08\ \mu L\ mg^{-1}$ demonstrate a noticeable difference between these

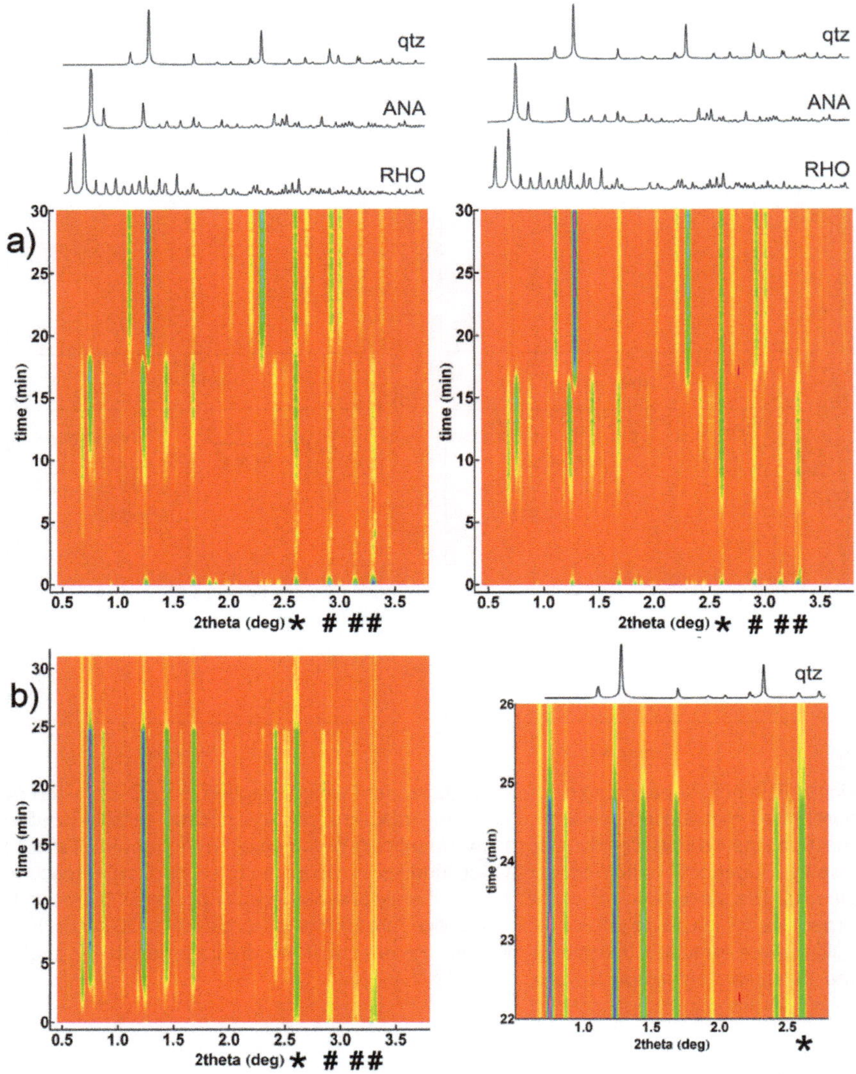

Fig. 9 a) The time-resolved difractograms of the two times repeated mechanochemical ILAG synthesis of porous ZIFs based on 2-ethylimidazole with silicon present (w(Si) = 12%, V(DMF) = 50 μL, η = 0.08 μL mg^{-1}). b) The time-resolved diffractogram of the above ILAG reaction using more liquid (V(DMF) = 150 μL, η = 0.23 μL mg^{-1}) and its zoomed part emphasising the appearance of a weak signal of the qtz-topology product. The calculated diffraction patterns of the ZIFs are given at the top of the figure. The asterisks (*) designate the position of the silicon signal while the hash sign (#) designates the ZnO signal.

two identically prepared reactions, as the reaction products in one of them appear *ca.* 3 minutes earlier. The appearance of the final qtz-topology product also takes place *ca.* 3 minutes earlier in that particular system. The key to this apparent irreproducibility of reaction kinetics lies in the observation that the diffraction signals in both reactions disappear very quickly after the onset of milling. Indeed, the time-resolved diffractogram remains almost completely featureless for almost 5 minutes and 8 minutes in each of the two reaction systems.

Whereas this at first resembles the amorphisation of the reaction mixture, the disappearance of features from the diffraction patterns is complete, also including the reflections of the added crystalline silicon standard. The latter is a clear indication that the loss of the diffraction signal is caused by the lack of sample in the X-ray beam, *i.e.* it is due to an extremely non-uniform sample distribution in the jar. Indeed, LAG reaction mixtures are sometimes seen to strongly adhere to either the walls or the ends of the milling jar, and sometimes even form a hard shell around the milling ball. This also provides a suitable rationalisation for the apparent difference in the reaction kinetics between the two reaction systems, as it is reasonable to assume that the reactivity in a milled, free-flowing powder will be different than for a mostly static sample adhering to the milling vessel or ball. Indeed, such an effect has been noted by Michalchuk and co-workers in the milling cocrystallisation of glycine and malonic acid where two distinct reactivity regimes were observed: (i) the shearing of the sample adhering to the jar wall and (2) the compression of the sample adhering to the jar ends.[38] Whereas these effects of liquid-induced non-uniform sample distribution cannot yet be predicted or resolved, we note that the presence of an internal standard in the mixture allows such events to be identified and distinguished from the reaction mechanism and kinetics. This is well illustrated in the ILAG experiment involving ZnO and 2-ethylimidazole at a higher η value (0.23 μL mg^{-1}). Under such conditions, the abundant liquid additive becomes included as a guest into the open structures of the RHO and ANA intermediates, therefore stabilising them. The stabilisation is evidenced by the persistence of the ANA form up to *ca.* 25 minutes milling (compared to *ca.* 16–19 minutes at $\eta = 0.08$ μL mg^{-1}). Eventually, however, the open structure collapses, as evidenced by the appearance of X-ray reflections of the qtz-framework. Within a couple of minutes, however, the features in the time-resolved diffractogram, including the reflection of the silicon standard, rapidly lose in intensity. Rather than amorphisation, an explanation of such behaviour lies in the release of the grinding liquid that was included in the open ANA-intermediate, resulting in a pasty sample that sticks to the grinding jar walls and is, therefore, effectively removed from the path of the X-ray beam.

Conclusions and outlook

We have demonstrated how the addition of crystalline silicon as an internal X-ray diffraction standard can help circumvent the limitations of the recently introduced technique for *in situ* and real-time reaction monitoring. In particular, the use of an internal standard ameliorated the variation in the diffraction signal intensity caused by the non-uniform distribution of the sample throughout the milling jar and allowed the quantitative analysis of the milled reaction mixtures using the Rietveld method. This allowed, for the first time, the *in situ* detection and evaluation of the amorphous content in a progressing mechanochemical reaction. The results have provided strong evidence that both the liquid-assisted and ion- and liquid-assisted reactions can involve extensive amorphization (up to 35% by weight) of the reaction mixture. As amorphous phases are often proposed as intermediates in mechanochemical processes, we believe that the ability to detect and, indeed, quantify the amorphous content within a milling reaction *in situ* is an important step towards creating an all-encompassing model of mechanochemical reactivity.

References

1 S. L. James, P. Collier, I. Parkin, G. Hyatt, D. Braga, L. Maini, W. Jones, C. Bçlm, A. Krebs, J. Mack, D. Waddell, W. Shearouse, A. G. Orpen, C. J. Adams, T. Friščić, J. W. Steed and K. D. M. Harris, *Chem. Soc. Rev.*, 2012, **41**, 413.

2 A. V. Trask, W. D. S. Motherwell and W. Jones, *Chem. Commun.*, 2004, 890; T. Friščić, D. G. Reid, I. Halasz, R. S. Stein, R. E. Dinnebier and M. J. Duer, *Angew. Chem., Int. Ed.*, 2010, **49**, 712.

3 V. Šepelák, A. Düvel, M. Wilkening, K.-D. Becker and P. Heitjans, *Chem. Soc. Rev.*, 2013, **42**, 7507–7520.

4 (*a*) T. Friščić, *Chem. Soc. Rev.*, 2012, **41**, 3493; (*b*) A. Lazuen-Garay, A. Pichon and S. L. James, *Chem. Soc. Rev.*, 2007, **36**, 846; (*c*) D. Braga, S. L. Giaffreda, F. Grepioni, A. Pettersen, L. Maini, M. Curzi and M. Polito, *Dalton Trans.*, 2006, 1249; (*d*) T. Friščić, I. Halasz, V. Štrukil, M. Eckert-Maksić and R. E. Dinnebier, *Croat. Chem. Acta*, 2012, **85**, 367.

5 (*a*) D. Braga, L. Maini and F. Grepioni, *Chem. Soc. Rev.*, 2013, **42**, 7638; (*b*) A. Bruckmann, A. Krebs and C. Bolm, *Green Chem.*, 2008, **10**, 1131; (*c*) A. Stolle, T. Szuppa, S. E. S. Leonhardt and B. Ondruschka, *Chem. Soc. Rev.*, 2011, **40**, 2317.

6 P. Baláž, M. Achimovičá, M. Baláž, P. Billik, Z. Cherkezova-Zheleva, J. M. Criado, F. Delogu, E. Dutková, E. Gaffet, F. J. Gotor, R. Kumar, I. Mitov, T. Rojac, M. Senna, A. Streletskii and K. Wieczorek-Ciurowa, *Chem. Soc. Rev.*, 2013, **42**, 7571.

7 (*a*) L. Takacs, *Chem. Soc. Rev.*, 2013, **42**, 7649; (*b*) L. Takacs, *J. Mater. Sci.*, 2004, **39**, 4987.

8 C. Suryanarayana, *Prog. Mater. Sci.*, 2001, **46**, 1.

9 T. Friščić, I. Halasz, P. J. Beldon, A. M. Belenguer, F. Adams, S. A. J. Kimber, V. Honkimäki and R. E. Dinnebier, *Nat. Chem.*, 2013, **5**, 66.

10 I. Halasz, S. A. J. Kimber, P. J. Beldon, A. M. Belenguer, F. Adams, V. Honkimäki, R. C. Nightingale, R. E. Dinnebier and T. Friščić, *Nat. Protoc.*, 2013, **8**, 1718.

11 I. Halasz, A. Puškarić, S. A. J. Kimber, P. J. Beldon, A. M. Belenguer, F. Adams, V. Honkimäki, R. E. Dinnebier, B. Patel, W. Jones, V. Štrukil and T. Friščić, *Angew. Chem., Int. Ed.*, 2013, **52**, 11538.

12 E. Boldyreva, *Chem. Soc. Rev.*, 2013, **42**, 7719.

13 T. Friščić and W. Jones, *Cryst. Growth Des.*, 2009, **9**, 1621.

14 (*a*) M. R. Caira, L. R. Nassimbeni and A. F. Wildervanck, *J. Chem. Soc., Perkin Trans. 2*, 1995, 2213; (*b*) V. Štrukil, L. Fábián, D. G. Reid, M. J. Duer, G. J. Jackson, M. Eckert-Maksić and T. Friščić, *Chem. Commun.*, 2010, **46**, 9191.

15 V. Šepelák, S. Bégin-Colin and G. Le Caër, *Dalton Trans.*, 2012, **41**, 11927.

16 J. F. Willart and M. Descamps, *Mol. Pharmaceutics*, 2008, **5**, 905.

17 K. L. Nguyen, T. Friščić, G. M. Day, L. F. Gladden and W. Jones, *Nat. Mater.*, 2007, **6**, 206.

18 A. Jayasankar, A. Somwangthanaroj, Z. J. Shao and N. Rodríguez-Hornedo, *Pharm. Res.*, 2006, **23**, 2381.

19 D. Cinčić, T. Friščić and W. Jones, *J. Am. Chem. Soc.*, 2008, **130**, 7524–7525.

20 S. Karki, T. Friščić and W. Jones, *CrystEngComm*, 2009, **11**, 470–481.

21 L. Fábián and T. Friščić, *CrystEngComm*, 2009, **11**, 743–745.

22 F. C. Strobridge, N. Judaš and T. Friščić, *CrystEngComm*, 2010, **12**, 2409–2412.
23 D. Braga, S. L. Giaffreda, F. Grepioni and M. Polito, *CrystEngComm*, 2004, **6**, 458.
24 G. A. Bowmaker, Effendy, J. V. Hanna, P. C. Healy, S. P. King, C. Pettinari, B. W. Skelton and A. H. White, *Dalton Trans.*, 2011, **40**, 7210.
25 K. Užarević, I. Halasz, I. Đilović, N. Bregović, Mirta Rubčić, D. Matković-Čalogović and V. Tomišić, *Angew. Chem., Int. Ed.*, 2013, **52**, 5504.
26 P. J. Beldon, L. Fábián, R. S. Stein, A. Thirumurugan, A. K. Cheetham and T. Friščić, *Angew. Chem., Int. Ed.*, 2010, **49**, 9640.
27 D. W. Lewis, A. R. Ruiz-Salvador, A. Gomez, L. M. Rodriguez-Albelo, F.-X. Coudert, B. Slater, A. K. Cheetham and C. Mellot-Draznieks, *CrystEngComm*, 2009, **11**, 2272.
28 J. C. Burley, M. J. Duer, R. S. Stein and R. M. Vrcelj, *Eur. J. Pharm. Sci.*, 2007, **31**, 271.
29 A. P. Hammersley, *ESRF Internal Report*, 1997, **ESRF97HA02T**.
30 Topas, version 4.2, Bruker-AXS, Karlsruhe, Germany.
31 Cambridge Structural Database (CSD), v. 5.34 (update Nov 2012).
32 Inorganic Crystal Structure Database, FIZ Karlsruhe, Karlsruhe, Germany.
33 R. J. Hill and C. J. Howard, *J. Appl. Crystallogr.*, 1987, **20**, 467–474.
34 X. Ma, W. Yuan, S. E. J. Bell and S. L. James, *Chem. Commun.*, 2014, **50**, 1585–1587.
35 W. Yuan, T. Friščić, D. Apperley and S. L. James, *Angew. Chem., Int. Ed.*, 2010, **49**, 3916.
36 M. J. Cliffe, C. Mottillo, R. S. Stein, D.-K. Bučar and T. Friščić, *Chem. Sci.*, 2012, **3**, 2495.
37 (*a*) A. Khawam and D. R. Flanagan, *J. Phys. Chem. B*, 2006, **110**, 17315; (*b*) F. L. Cumbrera and F. Sánchez-Bajo, *Thermochim. Acta*, 1995, **266**, 315.
38 A. A. L. Michalchuk, I. A. Tumanov and E. V. Boldyreva, *CrystEngComm*, 2013, **15**, 6403.

Faraday Discussions

PAPER

The scalability in the mechanochemical syntheses of edge functionalized graphene materials and biomass-derived chemicals†

Richard G. Blair,*[a] Katerina Chagoya,[b] Scott Biltek,[c] Steven Jackson,[c] Ashlyn Sinclair,[c] Alexandra Taraboletti[a] and David T. Restrepo[d]

Received 3rd February 2014, Accepted 26th February 2014
DOI: 10.1039/c4fd00007b

Mechanochemical approaches to chemical synthesis offer the promise of improved yields, new reaction pathways and greener syntheses. Scaling these syntheses is a crucial step toward realizing a commercially viable process. Although much work has been performed on laboratory-scale investigations little has been done to move these approaches toward industrially relevant scales. Moving reactions from shaker-type mills and planetary-type mills to scalable solutions can present a challenge. We have investigated scalability through discrete element models, thermal monitoring and reactor design. We have found that impact forces and macroscopic mixing are important factors in implementing a truly scalable process. These observations have allowed us to scale reactions from a few grams to several hundred grams and we have successfully implemented scalable solutions for the mechanocatalytic conversion of cellulose to value-added compounds and the synthesis of edge functionalized graphene.

Introduction

Mechanochemical approaches to chemical syntheses are a growing field. They offer an attractive solution to industries seeking to improve profitability through decreased reaction times and the reduction of solvent waste. Much of the equipment and tools developed for mechanochemical reactions have been adopted from the materials processing community. Attrition technology is focused on the mining and food industries. Mechanical alloying[1–3] and more recently the mechanochemical synthesis of advanced materials[4–6] are becoming more and more accepted as cost-effective tools that also offer the potential for

[a]*University of Central Florida, NanoScience Technology Center, 12424 Research Parkway Suite 400, Orlando, FL, 32826-3250, USA. E-mail: Richard.Blair@ucf.edu; Fax: +1-407-882-2819; Tel: +1-407-823-0639*
[b]*Department of Physics, University of Central Florida, 4000 Central Florida Blvd., Orlando, FL, 32816, USA*
[c]*Department of Chemistry, University of Central Florida, 4000 Central Florida Blvd., Orlando, FL, 32816, USA*
[d]*Garmor, Inc., 6317 McCoy Road. Ste. 100, Orlando, FL, 32822, USA*

† Electronic Supplementary Information (ESI) available: movies of the DEM model showing the compressive forces generated for the various mills. See DOI: 10.1039/c4fd00007b

unique products.[7] The implementation of chemical reactions in a mill has seen a lot of interest.[8–12] However, these studies have had limited scope and do not address routes to industrial implementation. Although quite unique reaction pathways are available mechanochemically, little work has gone into developing pathways to increase scale.[13] In contrast, routes from the bench to full scale in traditional syntheses are well established with known solutions to mixing, heat transfer and material transport issues. Industrial adopters are often concerned with the unknowns faced in implementing a mechanochemical process. This leads to the use of existing tools and the knowledge to develop assumptions in assessing a new process. Often, these assumptions lead to an incorrect understanding and the failure to implement a profitable process.

There are a large array of milling technologies available. Fig. 1 shows some commonly used mills and their attrition modes. The first three mills are often used in academic laboratories for exploratory syntheses. The most popular are the mixer/mill (Fig. 1A) and the planetary mill (Fig. 1B). A mixer/mill is amendable to small exploratory reactions, but does not scale beyond 10 g. A planetary mill can be used for syntheses up to 100 g, but apparatus costs typically limit implementations to the 20–50 g range. Attrition mills represent a linearly scalable technology; however laboratory-scale attritors typically require at least 100 g for efficient operation. By housing these technologies in a small laboratory a range of reaction scales can be investigated (Fig. 2) with an eye toward realizing an industrially relevant solution.

Mixer/mills and planetary mills represent high-energy processing. We have found that reactions for the hydrolysis of cellulose and the oxidation of graphite can be readily moved between the two without the need for excessive process optimization. However, scaling beyond a 50 g reaction requires the utilization of an attrition mill or other scalable approaches. It is our observation that the mechanochemical processes are dominated by two phenomena. Attrition or particle size reduction down to a scale where media impacts can induce meaningful chemical reaction rates, as well as a chemical regime where reactions occur at the point of each impact and are facilitated by the high transient compressive forces generated. Scalability issues were explored through the investigation of two commercially relevant reactions. The solid-catalyzed hydrolysis of cellulose to

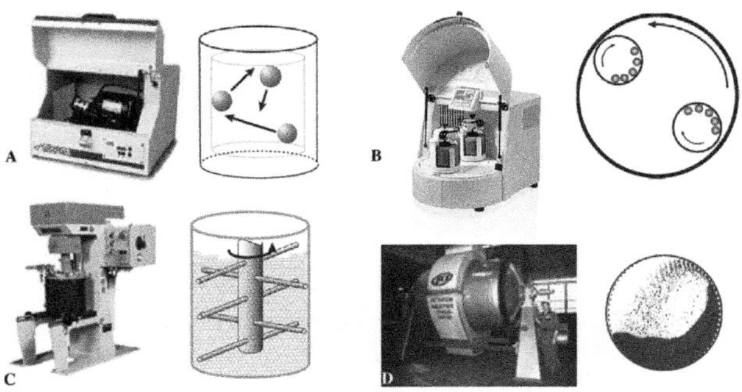

Fig. 1 A small set of milling technologies currently available. The motion of the milling media is indicated in schematics next to each mill. These vary from the small scale (A) SPEX mixer/mill (65 mL),[14] to the medium scale (B) planetary mill (250 mL),[15] (C) attritor (4 L),[16] to the production scale (D) rolling ball mill (hundreds of L).[17,18]

Fig. 2 The reaction scales achievable in a small laboratory. From left to right, a microvial for a SPEX 5100 mixer/mill, a vial for a SPEX 8000 mixer/mill, a vial for a Fritsch Pulverisette 6 planetary mill and a tank for a Union Process 01-HDDM attrition mill.

simple sugars and the oxidation and delamination of flake graphite to produce edge functionalized graphene.

Results and discussion

The scale-up of the mechanocatalytic hydrolysis of cellulose

Biopolymers such as cellulose and hemicellulose are a potentially rich source of industrially significant compounds for use as chemical feedstock or fuel. However current approaches are not feedstock tolerant, require large capital investments and produce significant amounts of waste. A first step in realizing value from these materials is hydrolytic depolymerization to simple sugars. Previous work has shown that cellulose can be readily hydrolyzed by solid acids such as kaolinite[19] and through impregnation with minimal quantities of a strong acid[20–22] combined with mechanical processing. Using basic conditions, lignin can also be mechanocatalytically depolymerized.[23] However this work has been limited to at most 10 g reactions in mixer/mills and planetary mills. In order to realize an industrially useful process, a route to multi-ton implementation must be understood. Our initial studies with toll milling at Union Process[19] indicated a significant induction period (4 hours, Fig. 3) before any appreciable yield was realized. The origin of this induction period was a paradigm difference between the laboratory and the production facility. Typical uses for attrition technology require no chemical change to occur in the ground material. Whereas, mechanocatalytic processes seek to capitalize on the new reaction pathways available under the application of mechanical force. By application of a small laboratory attritor we have been able to develop solutions to the induction period issue.

In Fig. 3, it can be seen that the mixer/mill induces the fastest conversion of cellulose into sugars. The planetary mill is slightly less efficient and attrition milling has significant induction periods. It should be noted that milling for extended periods of time in an attrition mill results in severe dehydration. This is

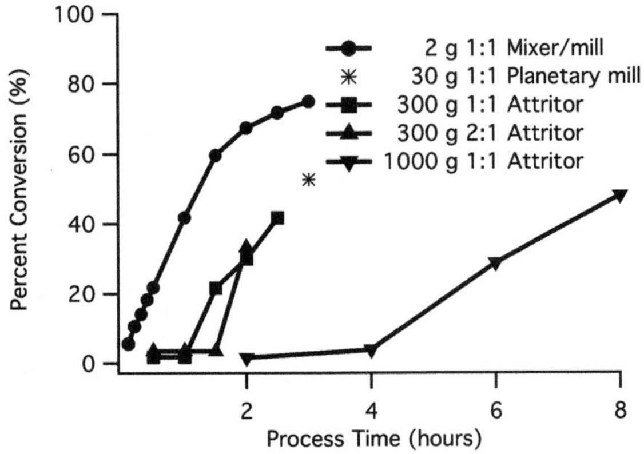

Fig. 3 The conversion of microcrystalline cellulose into soluble saccharides at various scales and cellulose : catalyst ratios.

due to the containment differences between this mill and the others. In the mixer/mill and planetary mill implementation the reaction vessel is sealed and any water loss through frictional heating will be minimal. In an attrition mill, there is significant dead space above the media and water can be transferred from the reaction zone to empty, cooler areas. This effectively reduces the water available for hydrolysis and allows the dehydration reactions to predominate. Upon examining the results in Fig. 3 it becomes apparent that several processes are occurring. Firstly, attrition is reducing the catalyst and cellulose size as well as amorphitizing the cellulose. This process is exemplified by the induction period. Short induction periods indicate effective attrition. Secondly, acid catalyzed hydrolysis is leading to the production of sugars. Thirdly, frictional heating is producing water loss. Fourthly, low water content is leading to the dehydration of the sugars produced. In an attrition mill these issues can be addressed individually. The ideal is a process dominated by hydrolysis. Attrition and amorphitization can be accelerated through the introduction of a solid that is harder than cellulose, but softer than the milling media (440C stainless steel, Mohs hardness = 5.5). We chose dolomite ($CaMg(CO_3)_2$, Mohs hardness = 3.5), a commercially available commodity mineral with a cost per ton similar to or less than that for kaolin.

Previous runs had resulted in slower reaction kinetics when the cellulose : catalyst ratio was increased (Fig. 3). The introduction of the milling aid, dolomite, allowed an increase of the ratio without a reduction in the conversion rates (Fig. 4). Dolomite serves two purposes in this reaction, it improves the attrition of cellulose and dilutes the acidic solid, effectively reducing the dehydration rate. Acid catalyzed hydrolysis only occurs during impacts. By increasing the number of impacts, both the chemical and mechanical processing rates can be improved. This is typically described as the ball to powder ratio (BPR) in mechanical alloying experiments. In a mechanochemical process, increasing the BPR produces more collisions. If the media is free to move then this also keeps the force distribution of the impacts similar. However, if severe confinement is produced through the use of large amounts of reactant or media, then compressive forces will be severely reduced. It can be seen in Fig. 5 that increasing

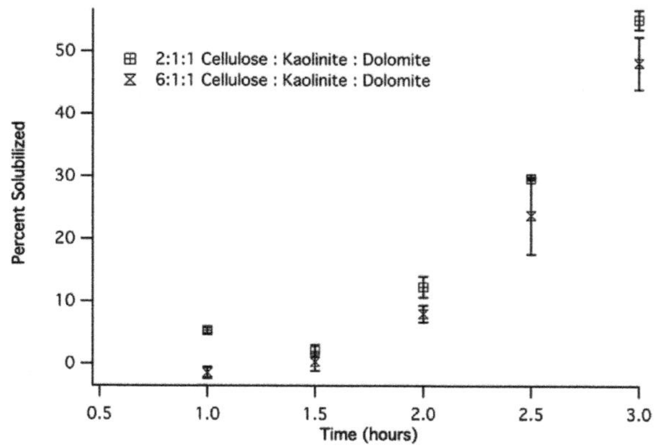

Fig. 4 The enhanced conversion of cellulose to saccharides though the addition of a milling aid.

the BPR through a reduction of the reaction mass resulted in a significant improvement of the reaction kinetics and allowed longer processing without dehydration. The intensity of the collisions can be increased by changing the media size or the milling speed. We explored reactions with 6 mm and 12 mm media at 300 and 700 rpm in the attrition mill. We found that the reaction rate scaled linearly with rotational velocity. The reaction rates also increased with larger media. However, the maximum reaction temperatures though frictional heating (48 °C for 300 rpm/6 mm, 78 °C for 300 rpm/12 mm, 106 °C for 700 rpm/6 mm, and 167 °C for 700 rpm/12 mm) also increased and the dehydration reactions became favored at higher speed/media sizes. Through the application of discrete element modeling (DEM) it is possible to relate the experimental observations to the compressive forces produced in a mill. It was found that the distribution of the compressive forces and the fraction of the impacts that produce non-zero compressive forces are both critical to the reaction rate. In general, higher rotation speeds and a larger media size produce higher collision forces. The complex nature of the media motion in a mill is ideally suited to discrete element modelling (see the movies in the supplemental materials†). The exception is a 28 L attritor where inadequate mixing is evident. In all traditional implementations of the attritor a dead zone forms around the shaft of the tree when the media rises up due to the application of rotation energy. Rotating the tank 90° and using gravity to encourage mixing could eliminate this problem. These examples serve to illustrate the need for whole reactor modelling in scaling mechanochemical reactions. Simple force calculations would not predict inadequate mixing or the formation of dead zones. Favorable conversion kinetics are achieved when the fraction of non-zero force collisions is high (Fig. 6). This represents a large scale confirmation of similar phenomena observed in microscale reactions.[24]

The scale-up of the production of edge functionalized graphene

Graphene has many useful properties including a low coefficient of friction, good electrical conductivity and a high thermal conductivity. However, current

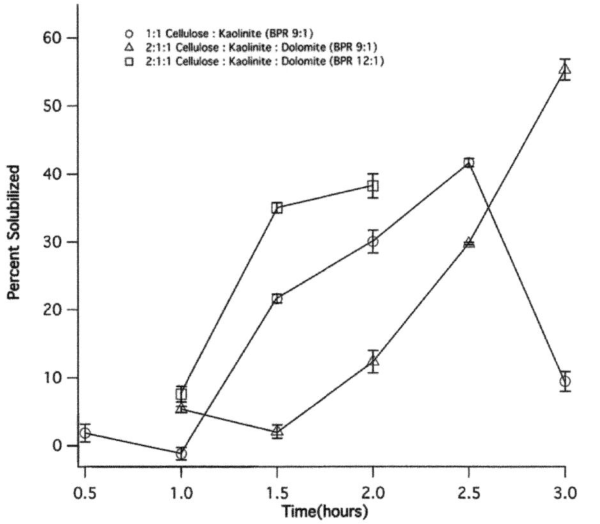

Fig. 5 The improved kinetics are realized with a higher media to reactant ratio. The reactions were performed in a Union Process 01-HDDM with 6 mm media. The reduction of solubilization in the 1 : 1 reaction (circles) is due to the severe dehydration of the reaction products.

approaches for producing graphene are not scalable or produce large quantities of waste. The most popular route to useful quantities of graphene is through the synthesis and exfoliation of graphite oxide. Graphite oxide contains oxygen attached to the layers as both epoxy bridges and hydroxyl groups.[25–27] Its properties vary from graphene in that it is an electrical insulator, thermally decomposes and is significantly more hydrophilic, allowing the complete exfoliation upon suspension in water.[25,27–29] This last property makes graphite oxide a valuable precursor in the mass production of graphene. After exfoliation, the sheets of graphene oxide can be chemically reduced into graphene.[30] The ability to produce graphene oxide-type materials on a commercially relevant scale is very important because of the utility of graphene, produced from it, for developing a number of novel device technologies.[31]

By directly milling graphite powder with a mild oxidizing agent, we can simultaneously oxidize and delaminate graphite producing a product with a few layers suitable for a variety of uses. We found that a urea hydrogen peroxide adduct (UHPA) and 35% hydrogen peroxide can effectively oxidize graphite in a mechanochemical environment. In this application, moving from a mill that relies on inertia for media motion (Mixer/mill and planetary mill) to one that actively stirs the media resulted in improved reaction kinetics and the ability to use a liquid oxidant. Fig. 7 illustrates the problem encountered with UHPA in a planetary mill. The oxidation reaction forms liquid and gaseous products. The liquid products increase the rolling resistance of the media and the gaseous products (eqn 1 to 5) can produce an explosive failure of the vial sealing mechanism. In contrast, a stirred media mill can still move the media through quite viscous mixtures and the milling tank can be designed to allow the release of the reaction products in a controlled manner.

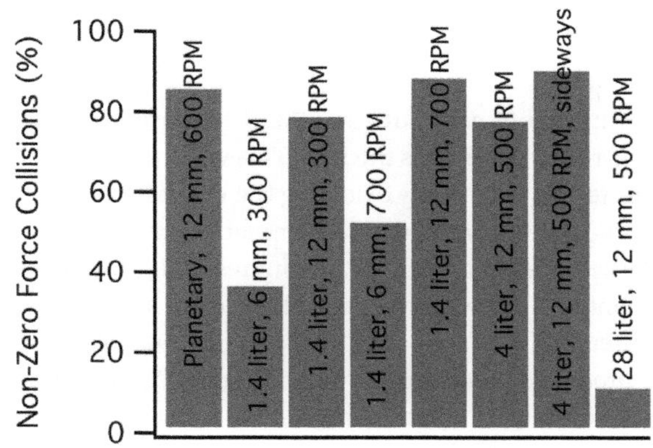

Fig. 6 The best hydrolysis rates are achieved when the fraction of the non-zero force collisions is high. The 4 L and 28 L values have been modelled, but not yet experimentally confirmed.

Fig. 7 The oxidation of graphite in a planetary mill using UHPA results in the formation of a solid mixture that does not tumble.

$$\mathrm{Urea{\cdot}H_2O_2 \rightarrow Urea + H_2O_2} \tag{1}$$

$$\mathrm{H_2O_2 \rightarrow H_2O + O_2} \tag{2}$$

$$\mathrm{nC_{graphite} + O_2 \rightarrow (n-1)C_{graphite} + CO_2} \tag{3}$$

$$\mathrm{C_{graphite} + H_2O_2 \rightarrow C_{graphite}(OH)_2} \tag{4}$$

$$C_{graphite}(OH)_2 + O_2 \rightarrow C_{graphite}(COOH)_2 \quad (5)$$

The scaled synthesis of edge functionalized few-layer graphene represents a reaction where attrition dominates the rate. This was investigated by pre-grinding crystalline graphite and chemically oxidizing the resultant material. 8 to 12 hours are needed to produce finely divided, delaminated graphite. But, high levels of oxidation can be achieved in as little as 30 minutes. This phenomena is illustrated by the oxygen content of the product obtained using several grades of graphite representing different flake sizes and crystallinities. Grade TC306 is a delaminated graphite produced by Asbury carbons whereas TC301 is the same graphite with a lower level of delamination. Grade 146 is a highly crystalline, large flake natural graphite that showed little oxidation without pregrinding. By starting with a delaminated graphite or pregrinding graphite (Table 1) it can be seen that preprocessed materials lead to a product with higher levels of oxygen.

Conclusions

Scaling mechanochemical reaction requires an understanding of the mechanical and chemical processes occurring during the reaction. Through the use of a well-characterized reaction (cellulose hydrolysis) we have developed a measure for the processes in a mill. We have found that the reaction rates and yields are governed by two processes, size reduction and the chemical reaction itself. The profitability of mechanochemical syntheses will only be realized with either high value (edge functionalized graphene) materials or low processing times to reduce energy consumption (cellulose hydrolysis). By relating the results of the cellulose hydrolysis to discrete element models we have shown that the fraction of non-zero force collisions and the distribution of the compressive forces realized in a mill are the critical parameters for realizing an increase in the reaction scale. Models developed for this study will be applied to a 160 L attritor and compared to measured hydrolysis rates.

Experimental

The hydrolysis of cellulose

Small scale milling experiments were performed in 8000M and 8000D SPEX Certiprep mixer/mills. Milling vials were constructed from 440C stainless steel

Table 1 A representative sample of the reactions producing edge functionalized graphene with mass% oxygen

Grade	Mass%, Oxygen	C_nO	Comment
TC306	8.94	13.5	SPEX mill
TC306	12.03	9.80	SPEX mill, Graphite was premilled for 12 hours
TC301	2.70	48.0	SPEX mill
146	20.20	5.27	Planetary mill, 30 g, 12 hours, then 12 hours with UHPA
TC306	13.50	8.54	Attritor®, 50 g with hydrogen peroxide

with an approximate volume of 65 mL. Buna-N quad O-rings were used to maintain a seal during milling. High kinetic energy milling was performed with three 12 mm stainless steel balls weighing approximately 8 g each. Milling was carried out in 30 minute increments, followed by 30 minutes of cooling to reduce wear on the mill motor. Hydrolysis was realized by milling cellulose (Microcrystalline, Alfa Aesar, 1.000 g) with kaolin (Kaopaque 10s, Imerys, 1.000 g).

Large scale milling experiments were performed in a Fritsch Pulverisette 6 planetary ball mill and a Union Process 01-HDDM Attritor®. For planetary milling, vials were constructed of X10CrNiS18-9 stainless steel with an approximate volume of 250 mL. A Viton gasket was used to maintain a seal during milling. Milling was performed with 50 12 mm stainless steel balls (440C) weighing 8 g each. Hydrolysis was realized by milling cellulose (Microcrystalline, Alfa Aesar, 15.000 g) with kaolin (Kaopaque 10s, Imerys, 15.000 g). Milling was carried out in 60 minute intervals with 30 minutes of cooling between each interval for a total of 3 hours of milling.

Attrition milling was performed in a custom 1.4 L tank designed to monitor and hold gas overpressure as well as reduce water loss. Typically, 2.72 kg of 6 mm or 12 mm stainless steel balls (440C) weighing 1 g and 8 g, respectively, were utilized as the milling media. Milling was carried out continuously. Cellulose (Microcrystalline, Alfa Aesar, 150 g) was milled with kaolin (Kaopaque 10s, Imerys, 150.000 g) or a mixture of kaolin and dolomite (Dolowhite, Axner Pottery).

Gravimetric analysis

The extent of the hydrolysis was monitored gravimetrically. The conversion of cellulose to water-soluble oligosaccharides was determined by stirring 0.1 g of the reaction mixture in 30 mL of water. Any oligosaccharide with a degree of polymerization <5 will be solvated. [32] The production of water-soluble products was measured by filtration through a 47 mm diameter Whatman Nuclepore® track etched polycarbonate membrane filter with a pore size of 0.220 μm. The residue was dried in a 60 °C oven for 12 hours and then weighed.

The synthesis of edge functionalized graphene

All of the graphite used in this study was supplied by Asbury Carbons. Small scale milling experiments were performed in 8000M and 8000D SPEX Certiprep mixer/mills. Milling vials were constructed from 440C stainless steel with an approximate volume of 65 mL. Buna-N quad O-rings were used to maintain a seal during milling. High kinetic energy milling was performed with three 12 mm stainless steel balls weighing approximately 8 g each. Milling was carried out in 30 minute increments, followed by 30 minutes of cooling to reduce wear on the mill motor. This was achieved by milling graphite (1.000 g) with a urea hydrogen peroxide adduct (2.902 g). The final product was purified by suspension in water, placed in sealed dialysis tubing and suspended in a 10 L recirculating deionized water bath to remove any remaining urea, the water was changed daily. After three days the sample was removed and dried by evaporation under forced air flow.

Large scale milling experiments were performed in a Fritsch Pulverisette 6 planetary ball mill and a Union Process 01-HDDM Attritor®. For planetary milling, vials were constructed of X10CrNiS18-9 stainless steel with an approximate volume of 250 mL. A Viton gasket was used to maintain a seal during milling. Milling was performed with 50 12 mm stainless steel balls (440C) weighing 8 g each. Milling was carried out in 60 minute intervals with 30 minutes of cooling between each interval for a total of 12 hours of milling. Typically, graphite (TC306, 7.5147 g) was milled in one portion with a urea hydrogen peroxide adduct (30.0284 g). Once all the milling was complete, the reaction mixture was suspended in water and placed in a custom flow dialyzer. The dialyzer consisted of a pump circulating a 10 L bath of deionized water and a peristaltic pump moving 800 mL of a product/water suspension through a dialysis tube suspended in the water bath. The deionized water was changed once a day. After three days, no urea was present in the product. The product suspension was transferred to a glass tray and dried under the air flow of a fume hood.

Attrition milling was performed in a custom 1.4 L tank designed to monitor and hold gas overpressure and 2.72 kg of 6 mm stainless steel balls (440C) weighing 1 g each. Milling was carried out continuously. Typically, graphite (TC306, 50 g) was milled in one portion dry at 700 rpm for 1 hour. Hydrogen peroxide (108 g) was added and the slurry was milled for another hour at 300 rpm. Once all the milling was complete, the reaction mixture was collected and dried under flowing air.

Elemental analysis

For elemental analysis, dry, urea-free samples were analyzed for C, H, N and O content by Micro-Analysis, Inc., Wilmington, DE, www.microana.com. The oxygen content was measured using the Unterzaucher method.

Discrete element modeling

Discrete element models of the milling process were generated using EDEM (DEM Solutions Ltd.).

Acknowledgements

This work was supported in part by funding from Garmor, Inc. (www.garmortech.com) and AMG renewables, LLC (www.amgrenewables.com). We would like to thank Asbury Carbons (Asbury, NJ) for supplying extensive samples of graphite and Imerys, S.A. (Paris, France) for supplying samples of physically delaminated kaolinite. Discrete element modelling *via* EDEM was made possible by the DEM Solutions EDEM Academic Partner Program (EAPP).

References

1 C. Suryanarayana, *Prog. Mater. Sci.*, 2001, **46**, 1–184.
2 V. V. Boldyrev and K. Tkacova, *J. Mater. Synth. Process.*, 2000, **8**, 121–132.
3 L. Takacs, *Hyperfine Interact.*, 1997, **111**, 245–250.
4 P. Balaz, M. Achimovicova, M. Balaz, P. Billik, Z. Cherkezova-Zheleva, J. M. Criado, F. Delogu, E. Dutkova, E. Gaffet, F. J. Gotor, R. Kumar, I. Mitov, T. Rojac, M. Senna, A. Streletskii and K. Wieczorek-Ciurowa, *Chem. Soc. Rev.*, 2013, **42**, 7571–7637.
5 E. Barraud, S. Begin-Colin, G. Le Caer, O. Barres and F. Villieras, *J. Alloys Compd.*, 2008, **456**, 224–233.

6 B. M. Rosen and V. Percec, *Nature*, 2007, **446**, 381–382.
7 D. Restrepo, S. M. Hick, C. Griebel, J. Alarcón, K. Giesler, Y. Chen, N. Orlovskaya and R. G. Blair, *Chem. Commun.*, 2013, **49**, 707–709.
8 S. L. James, C. J. Adams, C. Bolm, D. Braga, P. Collier, T. Friscic, F. Grepioni, K. D. M. Harris, G. Hyett, W. Jones, A. Krebs, J. Mack, L. Maini, A. G. Orpen, I. P. Parkin, W. C. Shearouse, J. W. Steed and D. C. Waddell, *Chem. Soc. Rev.*, 2012, **41**, 413–447.
9 D. Cincic, I. Brekalo and B. Kaitner, *Chem. Commun.*, 2012, **48**, 11683–11685.
10 T. Mitsudome, Y. Mikami, H. Funai, T. Mizugaki, K. Jitsukawa and K. Kaneda, *Angew. Chem.*, 2008, **120**, 144–147.
11 M. Schnuerch, M. Holzweber, M. D. Mihovilovic and P. Stanetty, *Green Chem.*, 2007, **9**, 139–145.
12 B. Rodriguez, A. Bruckmann, T. Rantanen and C. Bolm, *Adv. Synth. Catal.*, 2007, **349**, 2213–2233.
13 A. V. Dushkin, *Chem. Sust. Dev.*, 2004, **12**, 251–273.
14 SPEXCertiprep, in http://www.spexcsp.com/sampleprep/catalog/aid4tid8.html, 2008.
15 Fritsch, in http://www.fritsch.de, 2009.
16 Union Process, in http://www.unionprocess.com/dry_lab.html, 2008.
17 Patterson Industries, in http://www.pattersonindustries.com/7a_pics.html, 2008.
18 K. Palchikoff, in http://ffden-2.phys.uaf.edu/211_fall2002.web.dir/Keith_Palchikoff/sag_physics_2.html, 2009.
19 S. M. Hick, C. Griebel, D. T. Restrepo, J. H. Truitt, E. J. Buker, C. Bylda and R. G. Blair, *Green Chem.*, 2010, **12**, 468–474.
20 Q. Zhang and F. Jérôme, *ChemSusChem*, 2013, **6**, 2042–2044.
21 A. Shrotri, L. K. Lambert, A. Tanksale and J. Beltramini, *Green Chem.*, 2013, **15**, 2761–2768.
22 N. Meine, R. Rinaldi and F. Schüth, *ChemSusChem*, 2012, **5**, 1322–1329.
23 T. Kleine, J. Buendia and C. Bolm, *Green Chem.*, 2013, **15**, 160–166.
24 K. S. McKissic, J. T. Caruso, R. G. Blair and J. Mack, *Green Chem.*, 2014, **16**, 1628–1632.
25 A. Hamwi and V. Marchand, *J. Phys. Chem. Solids*, 1996, **57**, 867–872.
26 H. He, J. Klinowski, M. Forster and A. Lerf, *Chem. Phys. Lett.*, 1998, **287**, 53–56.
27 A. Lerf, H. He, M. Forster and J. Klinowski, *J. Phys. Chem. B*, 1998, **102**, 4477–4482.
28 S. Stankovich, R. D. Piner, X. Chen, N. Wu, S. T. Nguyen and R. S. Ruoff, *J. Mater. Chem.*, 2006, **16**, 155–158.
29 D. A. Dikin, S. Stankovich, E. J. Zimney, R. D. Piner, G. H. B. Dommett, G. Evmenenko, S. T. Nguyen and R. S. Ruoff, *Nature*, 2007, **448**, 457–460.
30 S. Stankovich, D. A. Dikin, R. D. Piner, K. A. Kohlhaas, A. Kleinhammes, Y. W. c. Yuanyuan Jia c, S. T. Nguyen and R. S. Ruoff, *Carbon*, 2007, **45**, 1558–1565.
31 A. K. Geim, *Nat. Mater.*, 2007, **6**, 183.
32 J. B. Taylor, *Trans. Faraday Soc.*, 1957, **53**, 1198–1203.

Faraday Discussions

PAPER

Application of twin screw extrusion to the manufacture of cocrystals: scale-up of AMG 517–sorbic acid cocrystal production

Dominick Daurio, Karthik Nagapudi,* Lan Li, Peter Quan and Fernando-Alvarez Nunez*

Received 21st December 2013, Accepted 11th February 2014
DOI: 10.1039/c3fd00153a

The application of twin screw extrusion (TSE) in the scale-up of cocrystal production was investigated by using AMG 517–sorbic acid as a model system. Extrusion parameters that influenced conversion to the cocrystal such as temperature, feed rate and screw speed were investigated. Extent of conversion to the cocrystal was found to have a strong dependence on temperature and a moderate dependence on feed rate and screw speed. Cocrystals made by the TSE process were found to have superior mechanical properties than solution grown cocrystals. Additionally, moving to a TSE process eliminated the need for solvent.

Introduction

It has long been recognized that understanding and manufacturing the appropriate solid form of active pharmaceutical ingredients (API) is an important part of drug development. To this end, substantial effort is expended in the pharmaceutical industry to understand all possible solid forms of a new chemical entity that is being considered for clinical development. In this arena of solid forms, polymorphs and salts are well known entities while cocrystals have been relatively new entrants. Cocrystals are defined as crystalline materials that are comprised of two or more components that are solids at room temperature (in order to distinguish them from hydrates and solvates) held together by non-covalent forces.[1–3] The distinction between a salt and a cocrystal lies in the fact that there is no proton transfer occurring between the constituents of a cocrystal. There has been an increased interest in cocrystals in the last decade and a number of publications have highlighted the beneficial properties of cocrystals.[4–15] While a lot has been published about how to screen for cocrystals there have been relatively few publications about how to scale up the production of cocrystals. This may be attributed to the fact that cocrystals are only now making their way through clinical pipelines and there is no known example of a marketed cocrystal.

Amgen Inc., Small Molecule Process & Product Development, Thousand Oaks, California, USA

Cocrystals can be prepared by either solvent-based methods or solid state methods. Typical solvent-based techniques involve solution or slurry crystallizations. These techniques are similar to crystallization methods used for the free form or salts of drug substances and are therefore familiar to chemists and engineers. Solution crystallization, in particular, offers good control of chemical purity and material properties. However, the use of solution crystallization for cocrystal scale up does present challenges as knowledge of the ternary phase diagram between the cocrystal constituents and the solvent is necessary and the measurement of such phase diagrams involves a large number of experiments which may be cumbersome to perform.[16–18] In recent years significant progress has been made towards designing simple procedures for using solution crystallization to scale-up the production of cocrystals.[19,20] Nevertheless, the use of solution crystallization for cocrystal scale up is still quite complex. In addition, as with any solution crystallization, drying of solvent from the final product to acceptable levels is required.

Cocrystals have also been traditionally prepared on a small scale using solid state techniques such as grinding and sonication. There are many reports of making cocrystals by neat grinding of the constituents in mills. Recently, liquid-assisted grinding has been developed as a more effective method to make cocrystals.[14,21] Grinding methods avoid challenges associated with kinetic crystallizations and solubility differences between solid starting materials. In addition they can produce phases that cannot be formed by solution crystallization and are environmentally friendly as they avoid the use of solvents. While grinding techniques are attractive methods on a small scale they cannot be used to scale up to large quantities of cocrystals. In recent publications we have introduced twin screw extrusion (TSE) are a viable solid state technique that overcomes the limitations of grinding techniques and can be used for the production of cocrystals.[22,23] Using this process we demonstrated the formation production of cocrystals of caffeine–oxalic acid, AMG 517–sorbic acid, cinnamic acid–nicotinamide, theophylline–citric acid and carbamezapine–saccharin. Dhumal *et al.* has also reported the use of extrusion to make ibuprofen–nicontinamide cocrystals. In this publication, they describe a detailed study of the effect of temperature and processing conditions on cocrystal formation.[24,25] They confirmed that extrusion was an excellent method of producing cocrystals while at the same time being easily amenable to a quality-by-design (QbD) approach. In this paper, the scale up of AMG 517–sorbic acid cocrystals using TSE is described. The solid state data obtained on cocrystals from the TSE approach (labeled TSEC) is compared with that of cocrystals obtained from a standard solution crystallization (labeled SCC) approach. Optimization of the twin screw extrusion process for the scale up of the cocrystal is also described.

Experimental

AMG 517 free base and solution crystallized-AMG 517–sorbic acid cocrystals were provided by the Chemical Process R&D group at Amgen Inc.

All other chemicals used in the study were purchased from Sigma-Aldrich company (St. Louis, MO) and were used as-received for milling and extrusion experiments. All solvents used were of analytical grade.

Ball milling

A Retsch Ball Mill model #MM301 was used to grind cocrystal components in order to obtain reference cocrystal material. In general, about 500 mg of the sample to be milled was placed in a 25-mL chamber with metal grinding balls at room temperature. The vibration frequency of the mill was set at 30 Hz for all experiments. The samples were milled for varying lengths of time between 0 and 60 minutes. Two jars of the same mixture were milled at the same time. About 50 mg of mixture was taken out from each of the 2 jars at 5, 15, 30, 45 and 60 min. Samples after milling were assessed using ^{19}F SSNMR to determine the extent of cocrystallization.

Twin Screw Extrusion (TSE)

A Prism PharmaLab 16mm twin screw extruder (25 : 1 *L*/*D*) was used for all experiments. The extruder has four controllable temperature zones (including the die zone). For these experiments the die zone was not used. The temperature of the extruder barrel was changed from 8 to 135 °C to monitor the effect on cocrystal conversion. Appropriate stoichiometric blends of the starting components were prepared and mixed using a Turbula blender for 10 minutes prior to being charged into the extruder. The charge to extruder was done through a Brabender single screw volumetric feeder that fed directly into the extruder hopper. Screw feeding rates from 5 to 15% were explored in this study. The screw configuration was designed for high mixing capacity and long residence time to enhance the conversion to cocrystal (*c.f.* Fig. 1). The screw design was set up with alternating 16-mm segments of Zone A and Zone B throughout the barrel. Zone A is a purely conveying zone with minimal mixing capacity. Zone A is a single element, 16 mm in length with a pitch of 7.5 mm. Zone B is a high mixing zone with neutral conveying capacity. Zone B is composed of four distributive elements that are 4 mm in length each for a total of 16 mm. The elements in Zone B are offset by 90° between each element (accomplished by combining alternating 0° and 90° paddle elements). All remaining details of the extrusion experiments are provided in the Discussion section. Each experiment processed anywhere between 20–400 grams of material.

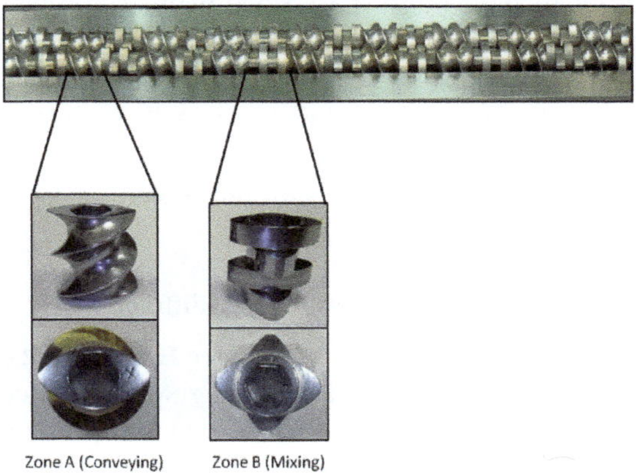

Fig. 1 Screw design employed for TSE showing conveying and mixing elements.

X-Ray powder diffraction (XRPD)

The diffractometer (PANanlytical X'pert, Philips) was equipped with a Cu Kα source ($\lambda = 1.54056$ Å) operating at a tube load of 45 kV and 40 mA. The divergence slit size was 1/4°, while the receiving slit and the detector slit, were 5.0 mm, and 0.1 mm respectively. A small amount of sample was loaded onto the Si 510 zero-background sample holder and scanned between 3 and 40° (2θ) with a step size of 0.008 and a step time of 15.2 s/step in the continuous mode. Data were collected by a high-resolution sealed proportional detector. The Si (111) with a diffraction peak at 28.44° 2θ was used as a standard to calibrate the instrument.

Differential scanning calorimetry (DSC) and thermogravimetric analysis (TGA)

DSC measurements were conducted in crimped aluminum pans using a Q200 (TA Instruments, NewCastle, DE) unit under 50 mL min^{-1} N2 purge. 3 to 5 mg samples were tested each time. Standard DSC scan using a heating rate of 10 °C min^{-1} was used. Indium was used as the calibration standard.

For phase diagram measurements, 5 to 10 mg of the appropriate blend samples were tested using a heating rate of 5 °C min^{-1}. The phase diagram measurements were made in two stages: (a) combining AMG 517 and the cocrystal; and (b) combining sorbic acid and the cocrystal. This method was found to provide more unambiguous results than simply starting with AMG 517 and sorbic acid.

TGA measurements were conducted in Aluminum pans using a Q500 (TA Instruments, NewCastle, DE) unit under 50 mL/min N2 purge. A heating rate of 10 °C/min was used for all TGA runs.

Solid state nuclear magnetic resonance (SSNMR)

All SSNMR measurements were conducted using a Bruker DSX spectrometer operating at a ^{1}H resonance frequency of 600 MHz. A Bruker 2.5 mm or 4-mm double resonance magic angle spinning (MAS) probe head was used to record all NMR data. A sample spinning speed of 14 to 35 kHZ was used. The presence of a trifluoromethyl moiety in the AMG 517 structure, allows for the investigation of samples by ^{19}F SSNMR. As the ^{19}F nucleus has high natural abundance and sensitivity, traditional single pulse NMR experiments can provide rapid quantitative information about the presence of different phases in the sample. A ^{1}H 90° pulse length of 2.5 μs and a cross polarization contact time of 2 ms were employed. ^{1}H decoupling was achieved with a spinal 64 sequence using a pulse length of 5 μs. A recycle delay of 10 seconds was used. Teflon with chemical shift of −122 ppm was used as the chemical shift reference.

Particle size distribution by dry dispersion laser diffraction

Particle-size analysis was performed on a Sympatec Helos H1782 equipped with a Rodos/M dry disperser and Aspiros Micro Dosing System feeder. The disperser injector diameter was 4 mm and feeding rate was 30 mm s^{-1}. Primary dispersing pressure of 1.0 bar and a R3 lens (0.9 to 175 μm) was used. The start/stop trigger threshold was set at a minimum optical concentration of <0.1%.

BET surface area

BET surface area measurement was performed on a Micromeritics Tristar II Surface Area and Porosity Analyzer 3020 using Krypton absorbate. Adsorption volumes were determined at multiple points from 5 to 25% p/p0 pressure range at 77.5 Kelvin.

Bulk and tap density

Tap density was determined using a Varian P/N 50-1200 tap density tester following USP method. API, sieved through #30 mesh and was dispensed into the 100mL graduated cylinder. The volume and weight were recorded for bulk density analysis. The cylinder was tapped and the tap density calculated.

Scanning electron microscopy

SEM images were acquired from an EVO® MA15 with LaB6 source Carl Zeiss Microscopy system. Sample was loaded on a carbon tape adhered to aluminum pin stub and sputter coated with gold and palladium.

Flow function

The flow function was determined using the RTS-XS Ring Shear Cell with an XSMV4 cell. The drug substance was passed through a 30 mesh screen prior to testing to ensure that the material was delumped. Measurements were performed at various consolidation pressures (1, 2, 3, and 6 kPa), and flow functions were recorded. The flow function coefficient (FFC) is calculated from these experiments as the ratio of the consolidation stress to the unconfined yield strength.

HPLC

A reversed-phase HPLC gradient analysis was performed on AMG 517 free base, SCC, and TSEC samples. The analysis was run on an Agilent HP1100 series HPLC system using a Phenomenex 150 × 4.6 mm, 3 micron Luna C18(2) HPLC column. The method was validated for SCC, and it is stability-indicating and capable of separating most of major impurities. The method is capable of accurately quantifying AMG 517 in SCC and TSEC samples. To demonstrate the stability of the TSEC sample, it was stressed at 60 °C closed conditions and 40 °C/75%RH open conditions for 4 weeks and analyzed *via* HPLC to assay for chemical purity and XRD to assay for physical stability.

Results and discussion

AMG 517 was developed as a potent and selective antagonist of VR1 indicated in the treatment of acute and chronic pain. The structure of AMG 517 is shown in Fig. 2. Based on superior *in-vivo* performance following oral administration in male Sprague Dawley rats, the sorbic acid cocrystal of AMG 517 was selected for development over AMG 517 free base.[4] About 15 kg of the cocrystal was synthesized *via* solution crystallization (SCC) to support development. The XRPD and DSC data of SCC are shown in Fig. 3. The XRPD data of the sample is in good agreement with the powder pattern calculated from the single crystal structure. The singe crystal structure of the material has been solved and the details are

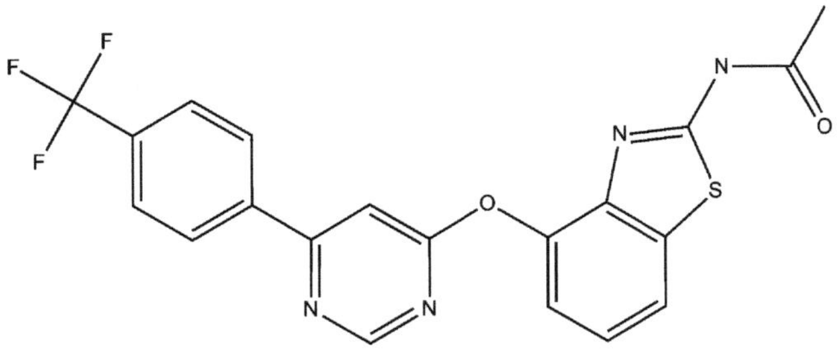

Fig. 2 Chemical structure of AMG 517.

reported in ref. 4. The first endotherm with an onset temperature of 204 .8 °C in the DSC thermogram represents the melting of the cocrystal while the second endotherm with an onset of 223.3 °C represents the melting of the residual free acid. The data show that the cocrystal is an anhydrous crystalline solid.

I. Cocrystal synthesis using ball milling

Prior to extrusion experiments, ball milling was conducted on a 500 mg scale to determine if mechanical action could lead to cocrystal formation. Ball milling was conducted at 30 Hz, the highest available frequency in the model used. Samples removed from the ball mill at different times were analyzed by ^{19}F SSNMR. By the 5-minute milling time, a conversion of >80% was seen. After ball milling for 30 minutes, complete conversion to cocrystal was observed. Based on the data obtained from ball milling it is clear that the AMG 517–sorbic acid cocrystal can be easily generated by mechanical action and as such it was decided to use the AMG 517–sorbic acid cocrystal system as an example case to illustrate the scalability of cocrystal production using TSE. Since an abundance of high quality SCC was also available, it would be possible to make a head-to-head comparison of TSEC with SCC material in order to gauge the effectiveness of TSE as an alternative to solution based processes.

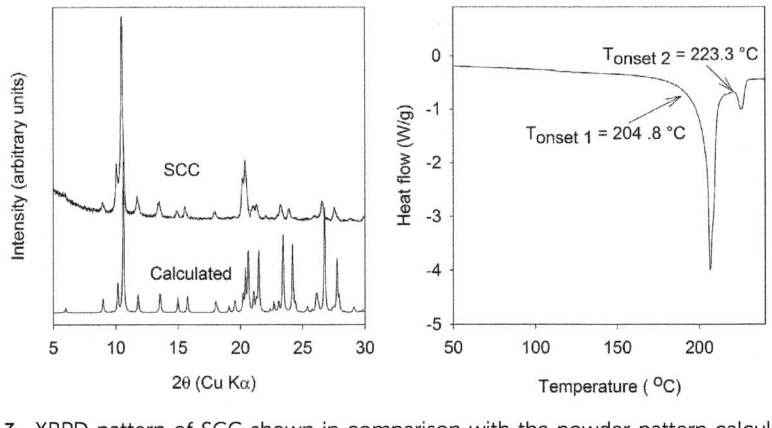

Fig. 3 XRPD pattern of SCC shown in comparison with the powder pattern calculated from the single crystal structure. (b) DSC thermogram of SCC.

II. Optimization of experimental parameters for TSE

Screw design, temperature, feed rate and the RPM of the screws are the primary experimental parameters that can affect cocrystal conversion in the extruder. In previous studies with other cocrystal systems, the effect of screw design on cocrystal conversion was examined by using several designs intended to provide a varying degree of mixing and conveying.[22] From these studies, it was determined that for mechanochemical applications a screw configuration that provides for intense mixing where new surfaces are continuously generated was crucial to achieve high conversion. Based on these prior observations, the screw design that provides for high mixing was used for all experiments in this study.

The most important factor that can influence cocrystal formation is the temperature in the extruder barrel. The effect of temperature on cocrystal conversion was studied by varying the temperature in the barrel between 10 and 135 °C. For these experiments, the feed rate was kept at 5% and the screw RPM was maintained at 25. ^{19}F SSNMR analysis of samples from each of these runs is shown in Fig. 4. The preblend sample prior to extrusion, containing stoichiometric equivalents of the two starting materials, did not show any conversion to the cocrystal. This shows that the mechanical action provided by TSE is critical for cocrystal formation. The data also shows that even at the lowest temperature of 10 °C, a high conversion to the cocrystal is achieved. However, complete conversion to the cocrystal was only achieved at 115 °C, indicating that increasing the temperature only improves the kinetics of the process. All temperatures used in this study up to 115 °C are less than the individual melting points of AMG 517 (226.9 °C) and sorbic acid (133.5 °C). As the cocrystal formation occurs for all temperatures used it indicates that the mechanism of cocrystal formation may not be mediated by a liquid phase formation due to the eutectic temperature. Based on previously published studies, it was found that cocrystal conversion may or may not be mediated by eutectic formation.[22] For example, eutectic formation was clearly shown to mediate cocrystal formation for the nicotinamide–cinnamic acid system. However, mechanism based on eutectic formation was found to be unlikely for the caffeine–oxalic system. While it seems unlikely that eutectic formation leads to cocrystal production in the case of AMG 517–sorbic acid system, it was decided to investigate this further by developing a complete phase diagram for the system.

A phase diagram for the AMG 517–sorbic acid system is shown in Fig. 5. The phase diagram is typical of what is expected of a binary system with compound formation. The solid lines represented fits to the Schroeder–van Laar equation which relates the temperature at which crystallization occurs (T) to the mole fraction of the solute (χ):

$$T = \left(\frac{1}{T_m^*} - \frac{R\ln\chi}{\Delta H_{fus}}\right)$$

Where T_m^* and ΔH_{fus} are the melting and the heat of fusion of the pure component and R is the universal gas constant. The equation fits for the cocrystal-free base part of the phase diagram well while the fit is not good for sorbic acid–cocrystal part of the phase diagram. This indicates that ideal behavior assumed by the Schroeder–van Laar equation is not applicable for the sorbic acid–cocrystal system.

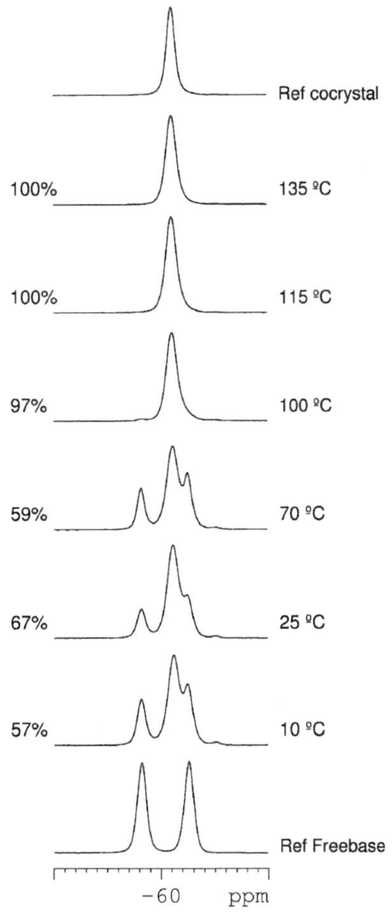

Fig. 4 ^{19}F SSNMR data for extruded AMG 517–sorbic acid samples as a function of temperature. The numbers shown in percentages refer to the percentage conversion to the cocrystal.

Two eutectic points are evident from the data: (a) $T_{eu1} \sim 404$ °K (131 °C) between sorbic acid and AMG 517–sorbic acid cocrystal and (b) $T_{eu2} \sim 467$ °K (194 °C) between AMG 517–sorbic acid cocrystal and AMG 517. Both these eutectic points are higher than most of the temperatures used in the extrusion study. Thus formation of the cocrystal in the extruder is not mediated by eutectic formation. This still leaves the question of the mechanism responsible for cocrystal formation open. To explain the mechanism of mechano-chemical reactions, usually an intermediate phase is invoked which can be (a) liquid state, (b) vapor state or (c) amorphous state. The existence of an intermediate liquid state in the formation of an AMG 517–sorbic acid cocrystal has been ruled out through phase diagram measurements. The vapor pressure of these compounds does not lend them to sublimation and therefore the presence of an intermediate vapor phase in mediating cocrystal formation is also unlikely. Nair *et al.*[26] and Friscic *et al.*[27] have proposed that an amorphous intermediate state could mediate cocrystal formation. Surface amorphization of materials through the high shear provided in the

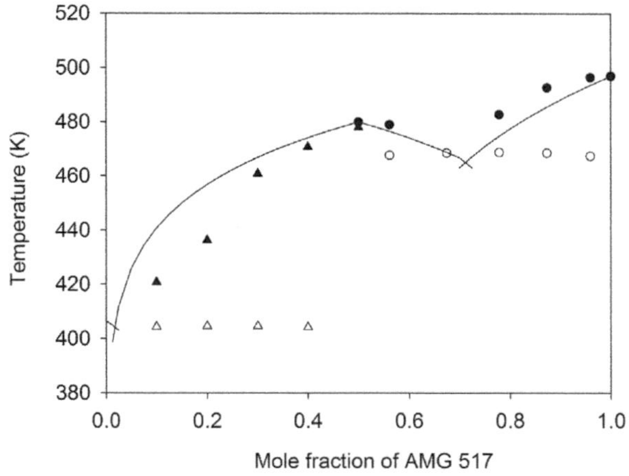

Fig. 5 Solid–liquid phase diagram for a binary system comprised of AMG 517 and sorbic acid. (Δ) T_{eu1}, between sorbic acid and cocrystal, (▲) points on the liquid curve for sorbic acid and cocrystal, (O) T_{eu2}, between cocrystal and AMG 517, and (●) points on the liquid curve for cocrystal and AMG 517.

mixing zones of the extruder is possible. This amorphous phase, owing to enhanced mobility, could then mediate cocrystal formation. Additionally, new surfaces are continuously being generated in the TSE process which could facilitate the propagation of cocrystal formation through an amorphous intermediate state. While there is no direct evidence of surface amorphous formation, this seems to be the most likely mechanism driving cocrystal formation in the case of the AMG 517–sorbic acid system.

The effect of feed rate and RPM on cocrystal conversion was investigated at 100 °C. The feed rate was varied from 5 to 15% while keeping the RPM constant at 100. Based on XRPD data it was clear that there is residual starting material in all cases with the lowest amount of starting material seen for the 5% feed rate. These results show that a high RPM of 100, even the lowest feed rate would not provide complete conversion to the cocrystal. Similarly screw speeds of 25, 50 and 100 RPM were investigated at a feed rate of 5% to understand the effect of residence time on conversion. It is assumed that at the lower the RPM there is a longer residence time in the extruder. In line with the previous observations, the lower RPMs of 25 and 50 provided near-complete conversion while the 100 RPM shows the presence of unreacted starting material. The XRPD data for screw speed and RPM variation is shown in Fig. 6. These data indicate that the extent of conversion to the cocrystal shows a strong dependence on temperature while showing moderate dependence on feed rate and screw speed. For the final scale up, the following conditions were chosen: 115 °C, 5% feed rate, and 50 RPM. The 115 °C temperature and a feed rate of 5% were chosen to provide complete conversion to cocrystal. The screw speed was set in the middle of the range to 50 RPM to ensure adequate residence time for full conversion while at the same time providing for a reasonable process duration.

III. Production of AMG 517–sorbic acid cocrystal by TSE

Stoichiometric quantities of AMG 517 and sorbic acid were mixed together in a bottle and blended by hand. The pre-blended material was fed through a hopper

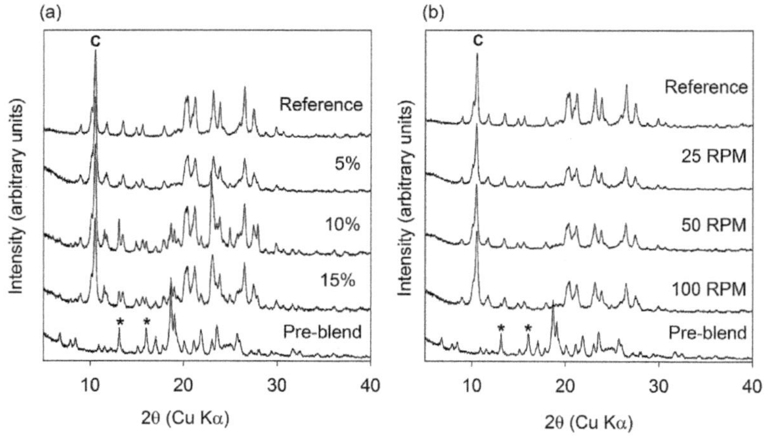

Fig. 6 XRPD data showing the influence of in-process parameters on the extent of conversion to the cocrystal at 100 °C: (a) effect of variation of feed rate and (b) effect of variation of screw speed. C: Peak characteristic of cocrystal, *: Peak characteristic of AMG 517.

into the extruder at a feed rate of 5% while the barrel was maintained at 115 °C. The extruder was run for a total time of 4 hours and 16 minutes. Samples were periodically collected from the extruder as time-point composites. Frequent sampling in 5 minute intervals was done for the first 15 minutes and the sampling frequency was reduced to an hourly basis from one through four hours. These samples were analyzed by XRPD and ^{19}F SSNMR to check for the extent of conversion to the cocrystal. At all time points complete conversion to the cocrystal was observed (data not shown), indicating the scale-up process performs uniformly over time. Total converted cocrystal collected was approximately 140 grams. As this is a continuous process, the extrusion process can simply be run for a longer duration to generate more material. Furthermore, a higher throughput can be achieved by moving to a larger diameter twin screw extruder (*e.g.* moving from a 16 mm extruder 27 mm extruder). The essential TSE parameters for the run are summarized in Table 1. The RPM, temperature, and feed rate are tightly controlled during the extrusion process. The only observation during the scale up run was that the output rate decreased from about 0.63 g min^{-1} to 0.49 g min^{-1}. This decrease in output over time is due to the feeder getting caked with pre-blend which in turn decreases the conveying efficiency. The feeder deficiency can be easily mitigated by moving from a volumetric based feeder to a gravimetric based feeder. Nevertheless, the experiment demonstrated that cocrystal production can be successfully scaled using TSE.

Table 1 Experimental TSE parameter variation during scale up

TSE parameter	Mean (RSD)	Min–max
Torque/N m	6.9 (2.3)	3–19
Temperature/°C	115 (0)	N/A
RPM	49.5 (2.1)	43–57
Feed rate (%)	4.96 (0.27)	2–6

IV. Comparison of TSEC and SCC

Solid state properties. Fig. 7a shows XRPD patterns of TSEC and SCC in comparison with the powder pattern calculated from the single crystal structure. The experimental XRPD patterns of the TSEC and SCC materials are in good agreement with the calculated powder pattern indicating that the cocrystal is formed in both processes. Using the quantitative HPLC analysis the conversion of AMG 517 free base to the AMG 517–sorbic acid cocrystal in the TSEC was calculated to be higher than 99%. The percent conversion of the free base to the cocrystal was found to comparable between the TSEC and the SCC samples. DSC thermograms of the TSEC in comparison with the free base and sorbic acid are shown in Fig. 7b. The peak melting temperature of the cocrystal was found to be in-between that of the starting components. Thus, thermal data provide further confirmation of the cocrystal formation by the TSE process. Based on these data, twin screw extrusion was found to be a viable option for producing AMG 517–sorbic acid cocrystals.

Chemical stability of TSEC and SCC. As the TSE process involves the use of high temperature to force complete conversion to the cocrystal, the question regarding the chemical stability of the final material has to be addressed. The melting point of the cocrystal was found to be 207 °C and therefore the thermal excursion of the material to 115 °C for a short period (<6 minutes) in the extruder is unlikely to cause any thermal degradation. Nevertheless, a comprehensive head-to-head stability analysis of the TSEC and SCC was initiated at accelerated storage conditions to assess the chemical liability of the materials. The stability data are summarized in Table 2. The data shown in Table 2 are on a %-relative-area basis. The TSEC sample is of comparable purity to the SCC sample. However, the SCC sample shows a lesser percentage of total impurities than the TSEC sample. This is due to the fact that crystallization in solution is able to reject impurities while TSE based crystallization is unable to do so. This is one of the drawbacks of the solid state method of producing cocrystals in that chemical purification is not possible. Additionally the TSEC sample stressed under

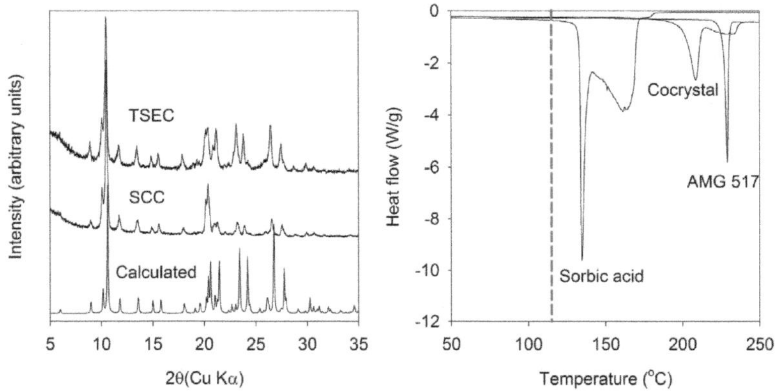

Fig. 7 (a) XRPD patterns of TSEC and SCC in comparison with the powder pattern calculated from the single crystal structure. (b) DSC thermogram of the cocrystal shown in comparison to that of the starting components. The dotted line in the Figure refers to the temperature of extrusion.

Table 2 Chemical stability of TSEC under accelerated storage conditions

	Free base ($T = 0$)	TSEC ($T = 0$)	TSEC 40C/75%RH	TSEC 60C/2 weeks	SCC ($T = 0$)
AMG 517	98.90	98.92	98.8	98.86	99.50
% total impurities	0.90	0.90	0.94	0.93	0.34

accelerated conditions shows excellent stability. The data presented show that TSE can be used to produce high quality cocrystals provided that starting AMG 517 free base is clean. The high temperature used in the TSE process does not produce any additional impurities in the sample.

Mechanical properties. A comparison of the mechanical properties of TSEC and SCC are provided in Table 3. The SEM micrographs of particles are shown in Fig. 8 and the particle size distribution as measured by dry dispersion laser diffraction is shown in Fig. 9. SCC particles have a rod-shaped morphology with significant particle aggregation while the TSEC particles have irregular morphology with significant particle aggregation. SEM data also show that while SCC crystals have smooth surfaces, TSEC crystals have corrugated surfaces. The morphological and textural differences between the samples are in line with expectations based on the different processes that were used to generate these crystals. In order to break the aggregation, 1 bar dispersing pressure was used in the PSD measurements. The PSD measurements show that TSEC has a broader distribution than SCC. The PSD metrics for the two samples are shown in Table 3. The D90 for the TSEC sample is nearly 1.8 times that of the SCC sample, indicating presence of bigger particles for the TSEC sample. The measured surface area of the samples is contrary to expectations based on PSD measurement. It would be expected that the SCC sample would have the higher surface area as it has the lower particle size. However, the data show that the TSEC sample has 3× the surface area of the SCC sample. This result can be rationalized through the surface difference between the two samples as seen in the SEM images. The TSEC sample that presents a highly corrugated surface and hence a larger surface area than the SCC samples, which have smooth surfaces. This is one of the advantages of the TSE process, in that it can readily produce material with a higher surface area due to mechanical action. The use of the TSE process may eliminate the need for secondary processing, such as milling, to improve the surface area.

Table 3 Comparison of the mechanical properties of TSEC and SCC

Property	Solution co-crystal	TSE co-crystal
PSD	$D_{10} = 1.7$ μm $D_{50} = 6.9$ μm $D_{90} = 30.5$ μm $M_v = 12.4$ μm	$D_{10} = 1.4$ μm $D_{50} = 9.8$ μm $D_{90} = 53.4$ μm $M_v = 19.5$ μm
Surface area	0.9 $m^2 g^{-1}$	2.9 $m^2 g^{-1}$
FFC @ 2 kPa	1.5	4.4
Bulk density	0.21 g cm^{-3}	0.49 g cm^{-3}
Tap density	0.38 g cm^{-3}	0.85 g cm^{-3}

Fig. 8 SEM micrograph of (a) SCC and (b) TSEC.

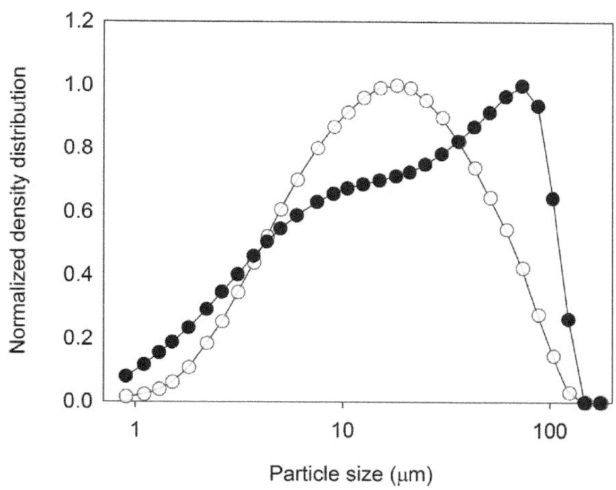

Fig. 9 PSD distribution of (●) SCC and (O) TSEC.

In addition to the surface area improvement, TSEC material also shows improved bulk density and flow function coefficient. The flow function data for SCC and TSEC are shown in Fig. 10. The flow of the TSEC sample is classified as *easy flowing* when compared SCC material which is classified as *very cohesive*. The improvements achieved for the TSEC can again be rationalized through the densification that can be achieved when using the TSE process. The improvements in bulk density and flow characteristics of the active ingredient can lead to improved performance in the drug product. For example, the improvements achieved by using the TSE process can be advantageous when making high strength tablets with increased drug loads using dry granulation. There a few reports in the literature of cocrystallization improving the mechanical performance of the pharmaceutically active ingredient. In this report it is demonstrated that not only cocrystallization but also the method of cocrystallization is important when considering improvements in mechanical properties.

Finally, the use of solvents in making the cocrystal has been completely eliminated by the use of the TSE process. For example, the current process of making the SCC uses nearly a liter of solvent to make 100 grams of material. The

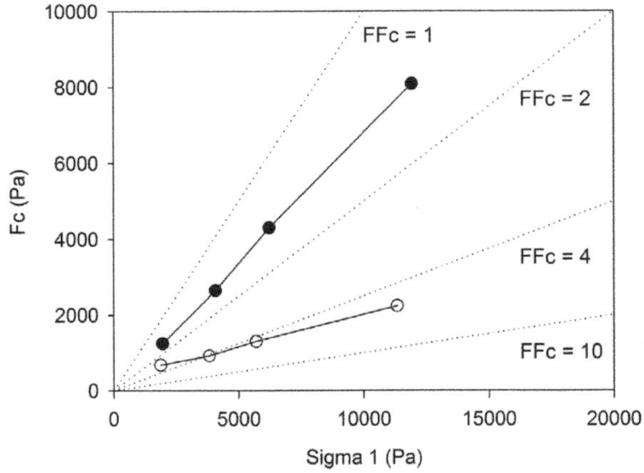

Fig. 10 Flow function of SCC (●) and TSEC (○). Typical classification of flowability: FFc < 1 is *non-flowing*, 1 < FFc < 2 is *very cohesive*, 2 < FFc < 4 is *cohesive*, 4 < FFc < 10 is *easy-flowing*, 10 < FFc is *free-flowing*.

entire need for the solvent has been obviated by moving to a continuous, mechano-chemically scalable process. Therefore in addition to the aforementioned advantages, TSE is a green chemistry process.

Conclusions

The scale up of AMG 517–sorbic acid cocrystal was demonstrated using TSE. A high mixing screw design was used to produce the cocrystals, as mixing in the extruder was demonstrated to be critical for cocrystal formation. The mechanism of cocrystal formation in the extruder was found not to be mediated by eutectic formation. However, temperatures greater than 100 °C were required to ensure complete conversion to the cocrystal. Feed rate and screw speed were found to have only a moderate influence on cocrystal conversion. TSEC were found to have improved surface area, bulk density, and flow properties when compared to SCC. In addition, the entire need for the solvent has been obviated by moving to TSE as a production method. TSE can be considered to be a continuous, efficient, scalable, and an environmentally friendly process for the production of cocrystals.

References

1 A. D. Bond, *CrystEngComm*, 2007, **9**, 833–834.
2 G. R. Desiraju, *CrystEngComm*, 2003, **5**, 466–467.
3 J. D. Dunitz, *CrystEngComm*, 2003, **5**, 506.
4 A. Bak, A. Gore, E. Yanez, M. Stanton, S. Tufekcic, R. Syed, A. Akrami, M. Rose, S. Surapaneni, T. Bostick, A. King, S. Neervannan, D. Ostovic and A. Koparkar, *J. Pharm. Sci.*, 2008, **97**, 3942–3956.
5 R. Banerjee, P. M. Bhatt, N. V. Ravindra and G. R. Desiraju, *Cryst. Growth Des.*, 2005, **5**, 2299–2309.
6 A. M. Chen, M. E. Ellison, A. Peresypkin, R. M. Wenslow, N. Variankaval, C. G. Savarin, T. K. Natishan, D. J. Mathre, P. G. Dormer, D. H. Euler, R. G. Ball, Z. Ye, Y. Wang and I. Santos, *Chem. Commun.*, 2007, 419–421.

7 S. L. Childs, L. J. Chyall, J. T. Dunlap, V. N. Smolenskaya, B. C. Stahly and G. P. Stahly, *J. Am. Chem. Soc.*, 2004, **126**, 13335–13342.
8 D. J. Good and N. Rodriguez-Hornedo, *Cryst. Growth Des.*, 2009, **9**, 2252–2264.
9 D. P. McNamara, S. L. Childs, J. Giordano, A. Iarriccio, J. Cassidy, M. S. Shet, R. Mannion, E. O'Donnell and A. Park, *Pharm. Res.*, 2006, **23**, 1888–1897.
10 J. F. Remenar, S. L. Morissette, M. L. Peterson, B. Moulton, J. M. MacPhee, H. R. Guzman and O. Almarsson, *J. Am. Chem. Soc.*, 2003, **125**, 8456–8457.
11 N. Variankaval, R. Wenslow, J. Murry, R. Hartman, R. Helmy, E. Kwong, S.-D. Clas, C. Dalton and I. Santos, *Cryst. Growth Des.*, 2006, **6**, 690–700.
12 S. Karki, T. Friscic, L. Fabian, P. R. Laity, G. M. Day and W. Jones, *Adv. Mater.*, 2009, **21**, 3905–3909.
13 C. C. Sun and H. Hou, *Cryst. Growth Des.*, 2008, **8**, 1575–1579.
14 A. V. Trask, W. D. S. Motherwell and W. Jones, *Chem. Commun.*, 2004, 890–891.
15 A. V. Trask, W. D. S. Motherwell and W. Jones, *Cryst. Growth Des.*, 2005, **5**, 1013–1021.
16 R. A. Chiarella, R. J. Davey and M. L. Peterson, *Cryst. Growth Des.*, 2007, **7**, 1223–1226.
17 S. J. Nehm, B. Rodriguez-Spong and N. Rodriguez-Hornedo, *Cryst. Growth Des.*, 2006, **6**, 592–600.
18 N. Rodriguez-Hornedo, S. J. Nehm, K. F. Seefeldt, Y. Pagan-Torres and C. J. Falkiewicz, *Mol. Pharmaceutics*, 2006, **3**, 362–367.
19 A. Ainouz, J.-R. Authelin, P. Billot and H. Lieberman, *Int. J. Pharm.*, 2009, **374**, 82–89.
20 A. Y. Sheikh, S. Abd. Rahim, R. B. Hammond and K. J. Roberts, *CrystEngComm*, 2009, **11**, 501–509.
21 T. Friscic, A. V. Trask, W. Jones and W. D. S. Motherwell, *Angew. Chem., Int. Ed.*, 2006, **45**, 7546–7550.
22 D. Daurio, C. Medina, R. Saw, K. Nagapudi and F. Alvarez-Nunez, *Pharmaceutics*, 2011, **3**, 582–600.
23 C. Medina, D. Daurio, K. Nagapudi and F. Alvarez-Nunez, *J. Pharm. Sci.*, 2010, **99**, 1693–1696.
24 R. S. Dhumal, A. L. Kelly and A. R. Paradkar, *J. Pharm. and Pharmacol.*, 2010, **62**, 1418–1419.
25 R. S. Dhumal, A. L. Kelly, P. York, P. D. Coates and A. Paradkar, *Pharm. Res.*, 2010, **27**, 2725–2733.
26 A. Jayasankar, A. Somwangthanaroj, Z. J. Shao and N. Rodriguez-Hornedo, *Pharm. Res.*, 2006, **23**, 2381–2392.
27 I. Halasz, A. Puškarić, A. J. S. Kimber, P. J. Beldon, A. M. Belenguer, F. Adams, V. Honkimäki, R. E. Dinnebier, P. Bhavanita, W. Jones, V. Štrukil and T. Friscic, *Angew. Chem.*, 2013, **125**, 11752–11755.

Faraday Discussions

PAPER

Self-sustaining reactions as a tool to study mechanochemical activation

L. Takacs

Received 13th December 2013, Accepted 8th January 2014
DOI: 10.1039/c3fd00133d

Ever since Faraday reported on the reduction of AgCl with Zn and Sn the "dry way", the study of Mechanically induced Self-propagating Reactions (MSRs) has been an important area within mechanochemistry. An interesting phenomenon is the mutual suppression of ignition in some mixed metal–chalcogen systems, such as in $(1 - x)$(Sn + Se) + x(Zn + Se) powders. Here both the Sn + Se and Zn + Se mixtures react in a self-sustaining manner after some activation time, but when they are combined, the reaction is gradual in the middle of the concentration scale. Mechanically induced metal–chalcogen combination reactions were studied by Chakurov *et al.* in the 1980s, using a low-energy vibratory mill. Similar measurements were carried out in our laboratory using the more energetic SPEX 8000 shaker mill. The results show qualitative similarities, but the details are different. It is suggested that the loss of MSR in mixed systems is the consequence of the very different properties of the binary systems, so that either one of the components (Zn) or a product formed gradually without ignition (*e.g.* SnSe) can act as an inert component relative to the rest of the system. Several examples are presented and the effect of the milling conditions is discussed. Finding new systems with similar behaviour and detailed studies of the activated state are needed to understand MSR in these systems.

Introduction

If a sufficiently exothermic powder mixture (*e.g.* Al–Fe_2O_3 or Ti–B) is processed in a ball mill, a self-sustaining reaction can be induced after some activation time, typically a few minutes to several hours long. This process is called a Mechanically induced Self-sustaining Reaction (MSR). Most of the reaction takes place within a few seconds after ignition and the released reaction heat causes a sudden increase of the mill temperature.

Whether a reaction is exothermic enough to ignite and propagate a self-sustaining process depends mainly on its adiabatic temperature, often simplified to the ratio of the reaction enthalpy to the room temperature heat capacity of the products ($\Delta H/C$). But whether an MSR takes place or not depends also on many

Department of Physics, University of Maryland Baltimore County, Baltimore, MD 21250, USA. E-mail: takacs@umbc.edu

other factors, like the heat conductivity, activation energy, and the mechanical properties of the powders, as well as the type of the mill and the parameters of operation such as the ball-to-powder mass ratio.

MSRs occupy a special place in the history of mechanochemistry. Michael Faraday made one of the earliest remarks on a chemical reaction initiated by grinding in 1820.[1] His objective was to decide whether AgCl could be reduced with hydrogen gas. Reacting AgCl with Zn directly was a control experiment. Heating the mixture caused a violently exothermic reaction that produced silver. Faraday also initiated the reaction by grinding. He wrote, "...if dry chloride of silver in powder be triturated in a mortar with zinc filings, the two bodies immediately act, and a heat above that of boiling water is produced." This statement suggests MSR and a recent attempt to produce the same effect in a SPEX 8000 Mixer Mill indeed resulted in the temperature jump characteristic of ignition.[2] Faraday repeated his experiment with tin, causing an even more powerful effect and achieved some reduction of AgCl also with copper and iron.

The first systematic investigation of MSRs was carried out in the 1980s by Chakurov and his co-workers in Bulgaria. They studied combination reactions in dry powder mixtures of a metal, such as Zn, Cd or Sn with the chalcogen elements S, Se or Te affected by grinding in a vibratory mill. Their results were published in a series of five papers that still constitute one of the richest systematic investigations of an MSR system.[3–7] The first paper[3] provides the earliest example of using the mill temperature to detect the ignition of a self-sustaining reaction. It contains ignition time results for 18 systems, sufficient to look for systematic trends. They observed that inert additives increase the ignition time and in larger quantities make the ignition of MSR impossible.[5]

In some mixed systems, such as (Sn–Zn) + S, a remarkable change of behaviour was observed as a function of composition.[7] When an equimolar mixture of Zn and S powders was milled, MSR was initiated after some activation time. The same happened when a mixture of Sn and S powders was milled. Surprisingly, however, the ignition time increased if the two mixtures were mixed, and the (Sn–Zn) + S blends reacted gradually without the ignition of a self-sustaining process in an intermediate concentration interval. Similar behaviour was observed in other (Zn, Sn, Cd) – (S, Se, Te) mixtures. The qualitative explanation is based on the fact that Sn reacts with S much more easily than Zn does, but its reaction heat is much lower. Thus on the Sn-rich side of the composition scale, the relatively inert Zn retards the ignition of the Sn + S reaction. On the Zn-rich side, some Sn sulphide forms early, but the corresponding reaction heat is not sufficient to ignite the self-sustaining reaction. At a later stage of the process, the SnS (or other tin sulphide) product acts as an inert additive and suppresses ignition. In the middle of the concentration scale neither reaction escalates into a self-sustaining process and the reaction remains gradual. The current paper will look at this phenomenon again, adding new experimental results and looking for more general explanations.

Chakurov's 1987 paper[5] has an interesting historical significance. Mechanical alloying was invented in 1966 without any knowledge of the rest of mechanochemistry[8] and the two research areas developed independently for over twenty years. The first cross-reference was made by Schaffer and McCormick,[9] who discovered MSR in CuO–metal systems. Although they called the process "combustion synthesis by mechanical alloying" while it was "explosive mechanochemical synthesis" according to Chakurov's terminology, they realized that the two phenomena were essentially the same and cited the work of Chakurov.[5] This reference is the first connection that broke the separation of the two areas.

The discovery of MSRs in oxide–metal and metal–metalloid systems resulted in significant activity in the 1990s.[10] The study of activation and ignition contributed substantially to our understanding of the mechanism of mechanochemical reactions. Several applications are related to MSRs as well, for example the preparation of some refractory compounds[11] and the synthesis of nanocomposites and free nanoparticles.[12] Knowing how to avoid ignition is necessary to ensure the safe activation of energetic materials such as thermites.[13] Mechanical activation and ignition are very complex processes. They depend on the chemical and mechanical properties of the powders as well as on the processing conditions. Many open questions remain that keep the investigation of MSRs an active research area.

In this paper, the unique behaviour of the Sn–Zn–S and similar metal–chalcogen systems will be revisited. Some existing results will be summarized and critically analyzed and new measurements will be added to develop a better understanding of the phenomenon.

The kinetics of a typical MSR process

Ball milling can turn a highly exothermic reaction into a self-sustaining process in cases like the reduction of a metal oxide with a more reactive metal (*e.g.* CuO–Al, Fe_3O_4–Zr), the formation of refractory compounds from a reactive metal and a metalloid (*e.g.* Mo–Si, Ti–C), and the combination reaction between a metal and a chalcogen element or a pnictide (*e.g.* Zn–S, Cd–Te, Ni–P). But independent of the chemical nature of the system, the reaction consists of the same three steps. (i) Activation, during which size reduction and mixing take place, chemically active defect sites are created, but little product is formed. (ii) Ignition, that occurs when a critical degree of activation is reached. At this point, the impact of a milling ball initiates a self-sustaining thermal process that propagates through the powder charge as a Self-propagating High-temperature Synthesis (SHS) reaction.[14] The release of the reaction heat results in an abrupt increase of the mill temperature. (iii) Continued milling brings the reaction to completion and results in a uniform product.

Ignition can be detected by measuring the temperature of the milling vial. The recording shown in Fig. 1 was obtained while milling a $(Sn_{30}Zn_{70})$ + Se mixture with a SPEX 8000 Mixer Mill, using a round-ended hardened steel vial and 20 small (diameter 6.35 mm) steel balls. The two curves were recorded by thermocouples attached to the opposite end surfaces of the vial. While differences in heat transfer result in minor differences between the two curves, the main features are identical: There is an initial temperature increase due to the dissipation of the mechanical energy of milling. As an equilibrium is approached between the energy input and the heat loss to the environment, the temperature levels off. At a well-defined moment, in this case 2440 s after starting the mill, an abrupt temperature increase is observed due to the large energy release during the self-sustaining reaction. Observing this sudden jump of the temperature is the clearest indication that an MSR happened. Shortly after the jump, the temperature begins to cool toward a steady-state value and no further release of chemical

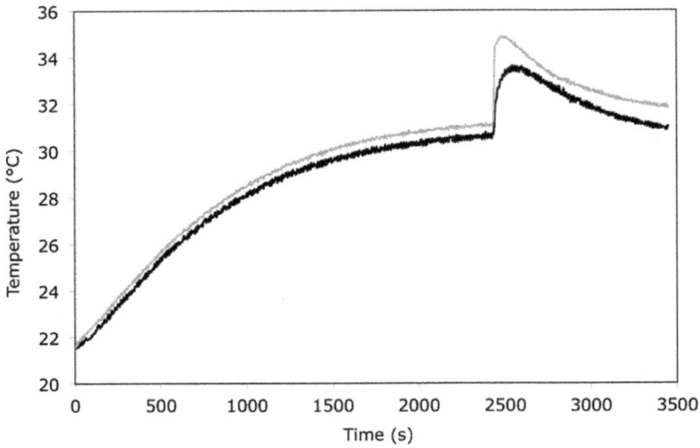

Fig. 1 Temperature of the milling vial while processing a $(Sn_{30}Zn_{70})$ + Se mixture using a SPEX 8000 mill. The two curves were measured at the opposite end plates of the milling vial.

energy is observed. When Rusanov and Chakurov studied the same reaction using a single-ball vibration mill, they observed ignition after about 6 hours of milling.[7] The difference is a consequence of the much lower intensity of their mill. Many similar measurements have been published in the literature; the detailed shape of the temperature–time curve and the ignition time depend both on the reaction and the milling conditions.[3,10,15]

The effects of inert additives

Adding an inert material to a reactive mixture delays the ignition of MSR and above a certain concentration suppresses ignition entirely. As the behaviour of the quasi-binary metal–chalcogen systems – the main subject of this paper – will be explained using clues from the effects of inert additives, it is necessary to first investigate MSR in mixtures containing additives.

Inert additives, such as an oxide or a polymer, change the conditions in many ways: They make the reaction less exothermic as they contribute to the heat capacity but not to the reaction heat, thereby decreasing $\Delta H/C$. They change the heat conductivity. The inert particles get between the reactants and reduce the contact area. The additives change the mechanical properties of the powder charge; hard particles can accelerate particle size reduction while soft additives reduce the efficiency of milling.

According to Chakurov's model[5] of mechanochemical reactions, the activation process begins with size reduction, mixing, and the accumulation of lattice defects, followed by the formation of composite aggregates, typically a few tens of microns in diameter. If an agglomerate grows large enough, self-heating by an incipient reaction can overwhelm heat loss at the surface and the reaction extends to its entire volume. Heat transfer to the neighbouring aggregates will ignite further reaction and lead to the propagation of MSR across the powder charge. The critical degree of activation is characterized by reaching a certain aggregate size. According to this model, the main effect of an inert additive is to inhibit the formation of large aggregates by forming layers that separate one block of the reactant mixture from the other. This is essentially a geometric percolation effect, thus it is natural to characterize it in terms of a critical volume fraction that ensures sufficient fragmentation to prevent ignition. Chakurov determined the critical volume fraction for eleven metal–chalcogen mixtures, using SiO_2 as the inert component.[5] The formation of ZnS was the most exothermic reaction and it required 50 volume % of SiO_2 to prevent its ignition.

In order to establish the volume fraction as the most relevant parameter, Chakurov studied the suppression of ignition in equimolar Sn–S mixtures using a polymer, SiO_2, rhyolite, and Fe as inert additives,[5] but the number of experiments was not sufficient to clearly verify a general trend. Also, Fe may react with S rather than behave as an inert additive; in fact, FeS can form from Fe and S powders *via* MSR in a high-energy mill.[16] Similar experiments have been carried out in a SPEX 8000 Mixer Mill using 5-g powder batches milled in a round-ended hardened steel vial with 5 balls, 12.7 mm in diameter. Both Sn + S and Sn + 2S compositions were used for the reactive mixture and Al_2O_3, NaCl, and graphite were added as inert materials (Fig. 2). The addition of a small amount of Al_2O_3 left the ignition time practically unchanged, suggesting that the hard oxide particles increased the efficiency of milling, and this effect was sufficient to balance the reduction of $\Delta H/C$ up to a point. If more than a few volume percent of Al_2O_3 was added, the reaction was not exothermic enough any more and the reaction became gradual. NaCl could participate in the reaction in some way, as it left the ignition time unchanged or even decreased it with increasing concentration, until ignition became impossible. Only graphite behaved the way usually expected from an inert additive. Similar results have been obtained for the effect of additives on Zn + S mixtures.[17]

In summary, treating the effect of inert additives as a geometric percolation effect is an oversimplification. Nevertheless, it remains true that additives

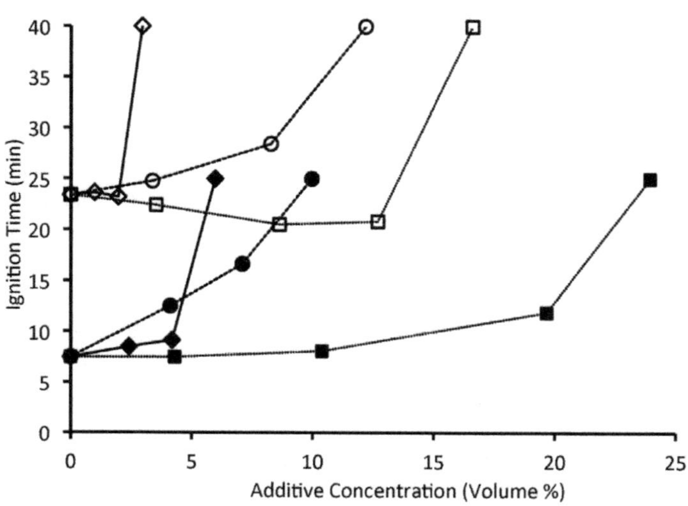

Fig. 2 The effect of additives on the ignition time of Sn–S mixtures. Closed symbols represent Sn + S, open symbols Sn + 2S stoichiometry. Diamonds, squares and circles correspond to Al_2O_3, NaCl, and graphite as the additive.

interfere with the ignition of MSR. At low concentration they tend to increase the ignition time and beyond a critical amount they make ignition impossible. In a high-energy mill, the energetic impacts cause significant forced reaction and perhaps local melting, thus ignition can start in the powder caught between the colliding surfaces. The percolation model may apply better to a low-energy mill where the individual impacts transfer relatively little energy to the powder, thus processing causes activation and agglomeration, but the MSR begins with the self-ignition of a free aggregate.

Suppression of MSR in quasi-binary metal chalcogen systems

In a few cases, a surprising phenomenon happens when two reactive powder mixtures are blended together. It is possible that MSR occurs in either of the two mixtures, but when they are mixed together in different proportions, the ignition time increases and at some compositions MSR is not ignited at all. This phenomenon was first observed by Rusanov and Chakurov[7] in some $(M_1–M_2) + Ch$ systems, where M_1 and M_2 represent Zn, Cd, or Sn and Ch is a chalcogen element. A broad composition interval with no MSR was found in the (Zn–Cd) + Te and (Sn–Zn) + Te systems, ignition was suppressed only in a narrow composition range in (Sn–Zn) + S and (Sn–Zn) + Se, and only an increase of the ignition time relative to the binary systems and a cusp in the concentration dependence were observed in (Zn–Cd) + S.

The explanation given by Chakurov considered this behaviour a percolation phenomenon, similar to the effect of inert additives, but here the main components of the powder charge were assumed to play the role of the additives.[7] A key feature of these systems is that the reactivity of the two metals is always very different, as indicated by the very different ignition times of the $M_1 + Ch$ and $M_2 + Ch$ mixtures. Thus one reaction occurs much more easily than the other. When the more reactive metal (the one with shorter ignition time) is the main component, the less reactive metal acts as an inert additive. When the less reactive metal is the main component, some chalcogenide of the more reactive metal forms gradually before ignition and acts as an inert additive later. The presence of an inert component holds back the reaction either way, thus the ignition time increases at both ends of the concentration scale. Also, there are critical amounts of the inert components, above which the reactions become gradual. Depending on the relative position of the two critical compositions, a region with no MSR or a cusp in the concentration dependence of the ignition time can be observed.

Investigations on the same systems using a SPEX 8000 Mixer Mill showed that although some qualitative features are the same, the details are very different. Fig. 3 shows the concentration dependence of the ignition time for the (Sn–Zn) + S system. The two vertical dashed lines show the postulated critical concentrations. While the results obtained with the vibratory mill clearly show the "percolation phenomenon," the ignition time obtained with the SPEX mill varies smoothly, not even a cusp is visible. Nevertheless, the ignition time increases from either end of the concentration scale, thus there remains some similarity between the two behaviours.

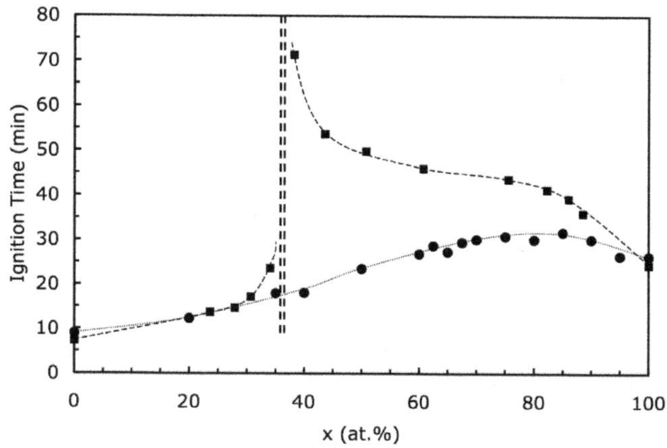

Fig. 3 Concentration dependence of the ignition time in (1 − x)Sn + xZn + S mixtures as measured in a low-energy vibratory mill[7] (squares) and in a SPEX 8000 shaker mill (5 g powder, five 12.7-mm balls; circles). The data obtained with a vibratory mill are multiplied by 15 to aid comparison.

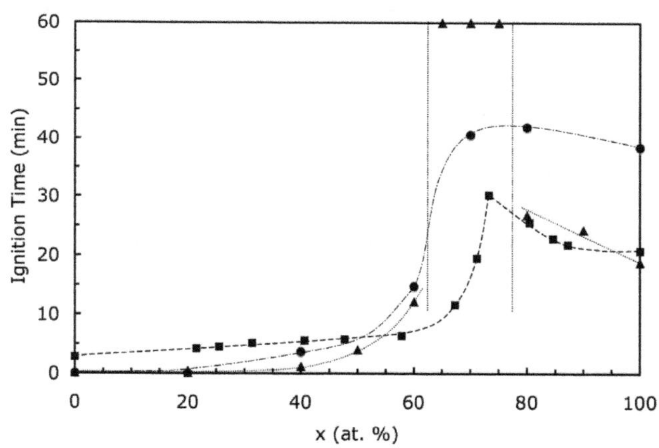

Fig. 4 Concentration dependence of the ignition time in (1 − x)Sn + xZn + Se mixtures as measured in a low-energy vibratory mill[7] (squares) and in a SPEX 8000 shaker mill (5 g powder, twenty 6.35-mm balls, circles and 5 g powder, five 12.7-mm balls, triangles). The vibratory mill data are multiplied by 15 to aid comparison.

The (Sn–Zn) + Se system shows the opposite relationship, shown by the ignition times presented in Fig. 4. The system is more reactive – the ignition times are about a factor of 2 shorter – thus larger critical concentrations are expected; in fact, they are expected to cross over so that only a cusp should remain in the concentration dependence, but no MSR-free concentration range should exist. The results obtained using a low-energy mill exhibit such behaviour.[7] If similar mixtures are processed in a SPEX 8000 mill using mild conditions (10 small balls, 6.35 mm diameter each) the ignition time varies in a similar way, although a rounded maximum is obtained instead of a sharp cusp. But if milling is carried out with five 12.7-mm balls, providing a higher impact energy and larger ball-to-powder mass ratio, a rather broad concentration range, from at least 65 at.% to 75 at.%, is found where no MSR is ignited. The results show that the ignition time always increases compared to the binary systems, but the suppression of ignition depends on the milling conditions as well. (The reaction in the binary Sn + Se mixture happens as soon as the mill is started and only the substitution of some Zn for Sn makes the ignition time finite.)

Beside the monosulphide and -selenide discussed above, Sn can also form disulfide (SnS_2) and diselenide ($SnSe_2$). (In fact, mixed-valence compounds such as Sn_2S_3 exist also, but we did not investigate reactions with the corresponding stoichiometry.) The ignition times for $(1-x)(Sn + 2S) + x(Zn + S)$[18] mixtures and their Se analogues are shown in Fig. 5. The ignition time increases starting from either end of the concentration scale and broad MSR-free concentration interval is observed in both cases. (Notice that the reactions with selenium were initiated at relatively mild milling conditions, using 20 balls, 6.35 mm in diameter. MSR is ignited over the entire composition range, if milling is carried out with 5 balls of 12.7 mm. These systems were not investigated by Chakurov's group.)

The cessation of ignition is very different when approaching the MSR-free middle region from the two ends of the concentration scale. On the Sn-rich side, the ignition time increases moderately, to about 34 and 20 min in the sulphur- and selenium-containing systems, respectively, but if the Zn concentration is increased further, ignition is lost abruptly. On the Zn-rich side (x decreasing from 100%), the ignition time increases much farther, to at least 60 min, before the reaction turns gradual. The difference may be explained the following way: it seems that the activation process is not held back by the increasing Zn content on the Sn-rich side, but at some point the reaction is no longer exothermic enough to ignite the MSR. (Zn is not involved, as it is not activated enough at that point.) On the other side, the increasing Sn content results in an increasing amount of gradually formed tin sulphide and the milling efficiency decreases due to the presence of soft Sn, but the reaction remains quite exothermic. Thus the ignition

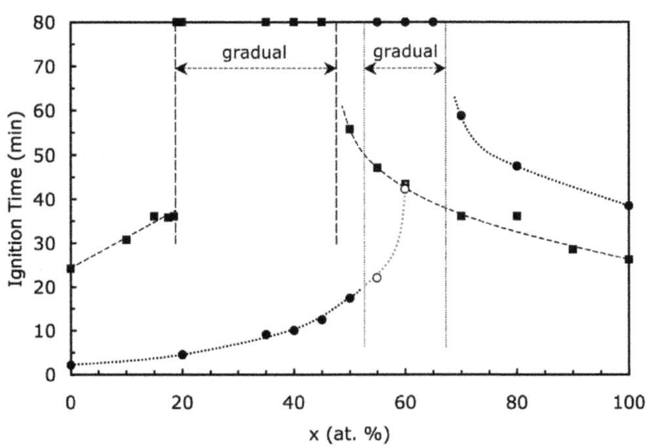

Fig. 5 Concentration dependence of the ignition time in powder mixtures $(1-x)(Sn + 2S) + x(Zn + S)$ (prepared using 5 g powder and five 12.7-mm balls, squares) and $(1-x)(Sn + 2Se) + x(Zn + Se)$ (5 g powder and twenty 6.35-mm balls, circles) measured in a SPEX 8000 shaker mill. The open circles show cases with no ignition but a noticeable increase of temperature (*cf.* Fig 6.)

time increases substantially before the ignition of MSR becomes impossible. The effect of inert additives shows similar differences, *e.g.* little increase of the ignition time is observed before the loss of ignition, if the additive is a hard oxide, but the ignition time increases substantially before the reaction turns gradual with the addition of graphite (Fig. 2). Similar asymmetry is observed in off-stoichiometric Sn + S and Zn + S mixtures between the S-rich and S-lean compositions.[19] The ability to describe such fine details is a sensitive test of the validity of any future model of the ignition process.

When using a SPEX mill to process (Zn–Cd) + S and (Zn–Cd) + Se powder mixtures, MSR was observed at any composition. The ignition time was a smooth monotonous function of concentration, without any cusp. It is possible that gradual reaction or a transition between MSR and gradual kinetics could be obtained by using substantially different milling conditions. The use of more energetic milling conditions, such as larger milling balls as in the case of (Sn–Zn) + Se (Fig. 4) or a small amount of powder like in (Sn–Zn) + S,[20] may change the kinetics to gradual for all compositions or just in a restricted concentration interval. The reason may be that the reaction is initiated between the colliding surfaces before the conditions become favourable for the propagation of the reaction across the entire powder charge. It may be self-sustaining in the small volume of powder involved in the collision, but it seems gradual on the macroscopic scale. It is also possible that milder conditions result in gradual reaction; after all, a cusp in the concentration dependence of the ignition time was observed in this system when it was processed in a low-energy mill.[7]

Qualitative explanation of the observed behaviour

Whether the effect of inert additives is indeed a simple percolation phenomenon or more complex changes, such as the reduction of the adiabatic temperature and the change of the milling efficiency are at work, it remains true that an additive suppresses the ignition of MSR above a critical amount, and this is usually preceded by an increase of the ignition time. Thus the main question is whether Zn on one side and gradually produced Sn (or Cd) sulphide or selenide on the other can indeed be considered an inert component.

The most important property of a self-sustaining reaction is its adiabatic temperature, or its simplified form, the ratio of the reaction enthalpy to the room temperature heat capacity of the products ($\Delta H/C$).[10,21] For MSR (and SHS) this

Table 1 Reaction heat ($\Delta H = -\Delta H_{298}$) and "simplified adiabatic temperature" ($\Delta H/C$, where C is the heat capacity of the products) for the formation of some metal–chalcogen compounds using data from Ref. 22

Compound	ΔH/kJ	$\Delta H/C$/K
ZnS	205.2	4516
SnS	107.9	2191
SnS_2	153.6	2189
ZnSe	163.2	3144
SnSe	94.1	1788
SnSe2	124.7	1743

quantity is the primary measure of the exothermicity of the reaction. A large enough value means that the degree of self-heating is sufficient to propagate the reaction. In an oxide–metal or a metal–metalloid mixture, MSR is usually possible if $\Delta H/C$ is at least 2000 K;[21] smaller values can be sufficient for metal–chalcogen and metal–halide systems. The relevant parameters of the reactions considered in this paper are listed in Table 1. Notice that the formation of the Zn compounds is much more exothermic than the formation of Sn compounds. Based on this property alone, by far the shortest ignition times would be expected for the reactions with Zn. The fact that this is not the case suggests that other properties, such as activation energy and the efficiency of ball milling, must also be considered.

An important factor is the formation of product compounds before the ignition of MSR by a gradual mechanochemical reaction. XRD results suggest that the formation of Sn compounds happens much more readily than the formation of ZnS or ZnSe.[17,18] Therefore, Sn compounds are indeed present, thus they can act as inert additives in the quasi-binary systems. The situation is made more complicated by the existence of at least two mixed-valence sulphides, Sn_2S_3 and Sn_3S_4 and the analogous selenides. On the microscopic scale, the mechanochemical reaction takes place by diffusion between grains, just like any solid-state reaction. Thus independent of the overall composition of a powder mixture, at least four different Sn-S or Sn-Se compounds can form. Little ZnS or ZnSe is produced before ignition and no mixed Sn-Zn-S or Sn-Zn-Se compound was observed.

The ignition of MSRs is investigated partly to gain insight on the mechanism of mechanochemical activation in general. Depending on the energy balance between the reaction heat and the heat loss to the environment, MSR may or may not be ignited, but the activation process is essentially the same. The main advantage of working with an MSR is the existence of a reference state, defined by the point of ignition, while in a gradual mechanochemical reaction the phase

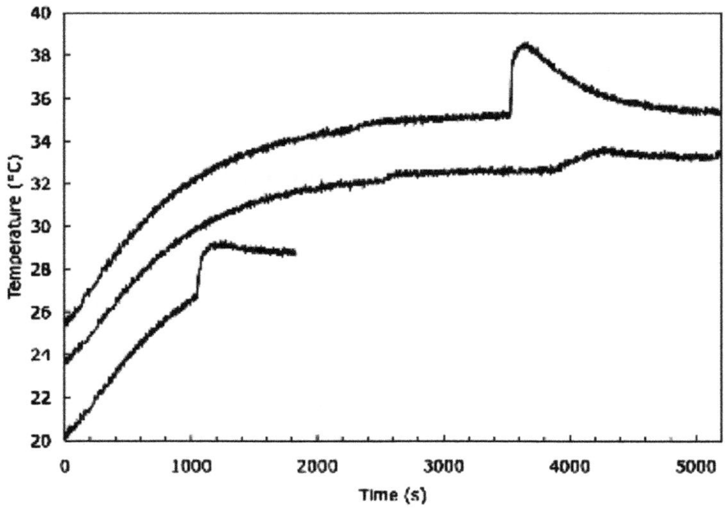

Fig. 6 Temperature of the milling vial as a function of time while milling $(1-x)(Sn + 2Se) + x(Zn + Se)$ mixtures using a SPEX 8000 mill with 5 g of powder and twenty 6.35-mm balls. From bottom to top, $x = 0.5$, 0.6, and 0.7.

composition and properties change smoothly. If the composition of the powder charge or the milling conditions is changed slightly, it is reflected by a change of the accurately observable ignition time of an MSR. It is more difficult to detect minor changes of gradual kinetics.

Interestingly, even if no MSR is ignited in a milling experiment, the first step is still activation with little chemical change, followed by a period of relatively fast reaction. Fig. 6 shows the temperature–time curves for three $(1 - x)(\mathrm{Sn} + 2\mathrm{Se}) + x(\mathrm{Zn} + \mathrm{Se})$ mixtures. (The ignition times of the same powder mixtures were presented in Fig. 5.) At $x = 0.5$, MSR was ignited after 1045 s (17.4 min.) At $x = 0.7$, on the Zn-rich side of the interval with no MSR, the ignition time was 3530 s (58.8 min.) The characteristic abrupt temperature increase was observed in both cases. No MSR was observed at $x = 0.6$. Nevertheless, there is a period of relatively fast temperature increase, caused by fast, although not self-sustaining, reaction after 2525 s. That is about the time when ignition would be anticipated by extrapolating the Sn-rich side of the ignition time–concentration curve. Thus the reaction rate increased after 2525 s of activation, even though it did not escalate into an MSR process. Some selenides – mostly of Sn – were produced during this period. They acted as inert additives, thus it took much additional milling, up to about 5160 s of total milling time, before another period of fast reaction started. Some relatively fast temperature increases can be noticed again, although not the sudden jump typical of an MSR. These results support the assumption that the activation process must be quite similar in related systems, even if one reacts according to a gradual kinetics and the other propagates as a self-sustaining reaction. Accordingly, the information obtained from the study of MSR is applicable to similar systems, even if they react gradually.

Searching for similar phenomena in different reactions

The fact that one can take two powder mixtures that are both capable of reacting with an "explosive" self-sustaining kinetics and combine them to end up with a mixture that reacts slowly according to a gradual kinetics is very surprising. The opposite behaviour would be much more plausible: If the reaction begins in one component, it transfers heat to the rest of the powder igniting the other component also. The two reactive mixtures support each other rather than act as inert additives for each other. The behaviour of the metal–chalcogen systems is quite unique and follows from the very different properties of the binary mixtures. In order to better understand the details of the phenomenon, it would be beneficial to find other, preferably very different, systems that exhibit similar phenomena.

We investigated two possible combinations, both without success. The first involved the reduction of CuO with Si ($\Delta H/C = 6278$ K) and Zn ($\Delta H/C = 2913$ K). Although the reaction with Si is the more exothermic, the hard Si particles do not break up and mix easily, thus the contact area between the different grains is small and does not increase quickly with activation. Thus ignition takes place only after a relatively long activation period. On the other hand, the reaction with Zn is much less exothermic, but the relative softness and pliability of Zn results in larger contact area and easier reaction, resulting in similar, in fact, shorter,

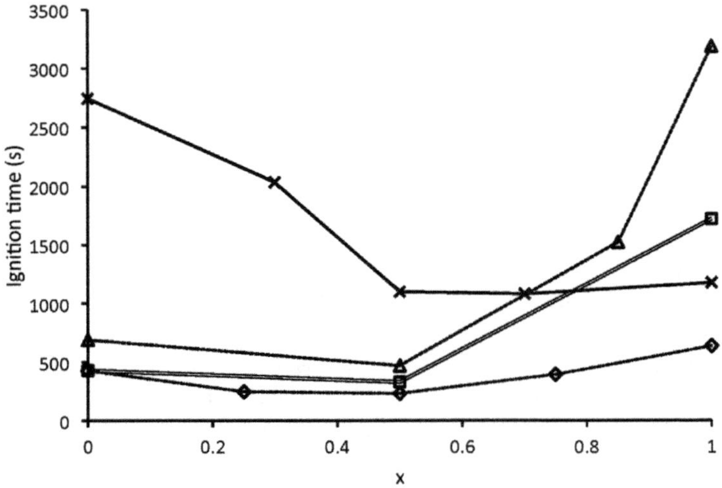

Fig. 7 Ignition time as a function of composition in mixtures activated in a SPEX 8000 Mixer Mill with round-ended hardened steel vial. Data for CuO + (1 − *x*)Zn + (*x*/2)Si (3 g powder with 3 balls, triangles; 1.5 g powder with 3 balls, squares; 3 g powder with 6 balls, diamonds; each with 12.7-mm balls) and (1 − *x*)$FeCl_3$ + (*x*/2)Fe_2O_3 + Al (3 g powder milled with five 9.525-mm balls.)

ignition times. Based on the asymmetric properties, it was reasonable to expect anomalous concentration dependence similar to the one observed in the metal–chalcogen systems. Here Si could behave as the inert component on the Zn rich side of the concentration scale and Zn could react early to form ZnO that acts as an inert additive in Si-rich samples. Nevertheless, as shown in Fig. 7, the ignition time decreases at both ends of the concentration scale and has a flat minimum in the middle, rather than an increase at the ends and a maximum or an ignition free range. As the behaviour can depend on the milling conditions, three different combinations of powder mass and number of balls were tried. Each curve shows a minimum, although the ignition time of the Si-rich compositions varied more with the milling conditions.

Displacement reactions involving chlorides are less exothermic than the similar reactions with oxides, but the chlorides are softer and consequently they react more easily with a reducing metal. As a result, similar ignition times are observed in spite of the very different properties. We investigated the reduction of ferric chloride–hematite mixtures with Al. The $\Delta H/C$ values for the binary $FeCl_3$ + Al and the Fe_2O_3 + 2Al reactions are 2633 K and 6586 K, respectively. According to the data in Fig.7, the ignition time of the mixtures has a shallow minimum.

While these attempts were unsuccessful, it is still possible that some other system may exhibit anomalous concentration dependence. Metal–chalcogen combination reactions with transition metals such as Mn could be of interest. The formation of phosphides may show similar behaviour. More complex reactions, such as the displacement reactions discussed above, would be more interesting. In any case finding systems that show the mutual suppression of ignition in systems that are very different from the metal–chalcogen mixtures could add substantially to our understanding of the ignition of MSR and more broadly of mechanochemical activation in general.

A simple model for reactions in mixed systems

The unexpected increase of the ignition time and the occasional loss of ignition toward the middle of the concentration scale were explained qualitatively based on the very different physical and chemical properties of the component systems. In the following, a simple probabilistic model is introduced to explain the concentration dependence of the probability of ignition. A cluster of reactive grains is considered. If this cluster is caught between the colliding milling tools, its temperature increases due to the dissipation of mechanical energy. This temperature increase may induce reaction in one of the grains. If that happens, the reaction heat is released, increasing the temperature of the entire cluster that can lead to reaction in another grain. As this will result in further temperature increase, the reaction will be transferred to additional grains and eventually escalate into an MSR. Thus it is reasonable to assume that the probability that the reaction extends to a second grain during a collision determines the probability of ignition. A larger probability results in shorter ignition time, while a lower probability leads to longer ignition time and eventually to the loss of ignition.

Let's consider a cubic cluster of $3 \times 3 \times 3$ grains, equal in size, each either an equimolar powder mixture of metal M_1 and chalcogen Ch, capable of the reaction $M_1 + Ch \rightarrow M_1Ch$, or metal M_2 and the same chalcogen Ch, capable of the reaction $M_2 + Ch \rightarrow M_2Ch$. The mean composition of the sample is $xM_1 + (1 - x)M_2 + Ch$. If this cluster is compressed between two colliding surfaces during ball milling, its temperature is increased to T_1. The probability that the central grain is an $M_1 + Ch$ mixture is x, and that it reacts due to the temperature increase is proportional to

$$P_1 = x\alpha \exp(-E_1/T_1), \quad (1)$$

where k_BE_1 is the activation energy and α is the frequency factor of the reaction. If all the heat is used to increase the temperature of the cluster, the temperature increases to

$$T_{21} = T_1 + Q_1/27C, \quad (2)$$

where Q_1 is the reaction heat and C is the heat capacity of a single grain, assumed to be the same for every grain for sake of simplicity. Similarly, the probability that the central grain contains M_2 and Ch and it reacts to form M_2Ch due to the initial temperature increase is proportional to

$$P_2 = (1 - x)\beta \exp(-E_2/T_1), \quad (3)$$

and it raises the temperature of the cluster to

$$T_{22} = T_1 + Q_2/27C. \quad (4)$$

Let's assume that the probabilities are small, so that the probability of initiating a reaction in more than one grain at a time is negligible.

The temperature increase to T_{21} or T_{22} increases the probability of further reaction in a neighbouring grain. For example, if the initial reaction took place in an $M_1 + Ch$ mixture than the probability of a second, thermally ignited reaction in a grain that happens to consist of $M_2 + Ch$ is

$$P_{12} = (1 - x)\beta \exp(-E_2/T_{21}). \quad (5)$$

Similar probabilities can be assigned to the other possible cases.

Combining the above expressions, the probability of "ignition," defined as an initial reaction in the centre of a cluster followed by further reaction in one of the neighbouring grains due to the initial heat release is proportional to

$$P(x) = P_1 (P_{11} + P_{12}) + P_2 (P_{21} + P_{22}). \quad (6)$$

For most reasonable combinations of the parameters, the function $P(x)$ is close to a straight line or has a maximum at an intermediate concentration. But if the properties of the two reactions are very different, it is possible that the variation has a minimum. The curve shown in Fig. 8 was obtained using parameters resembling the properties of the Zn–Sn–S system: Q_1 = 205.2 kJ mol^{-1} for the formation of ZnS, Q_2 = 107.9 kJ mol^{-1} for the formation of SnS; the average of the heat capacities of ZnS and SnS is C = 47.35 J mol^{-1} K^{-1} (the individual values differ only by 10%). It is more difficult to initiate the formation of ZnS, thus its activation energy was chosen to be E_1 = 25 000 K while for the formation of SnS E_2 = 12 500 K. The pre-exponential factors $\alpha = 10^{16}$ and $\beta = 2 \times 10^8$ were chosen to make the probabilities of the binary reactions in Zn + S and Sn + S about equal.

Although this model is far too simple to address most issues related to the ignition of self-sustaining reactions, it shows that asymmetry of the physico-chemical properties of the two reacting mixtures can lead to decreasing probability and consequently to increased ignition time or loss of ignition, when the mixtures are combined.

Conclusions

The mutual suppression of ignition, this far only observed in a few mixed metal–chalcogen systems, is a unique and complex phenomenon that can serve as a vehicle to gain more insight on the mechanism of the ignition of self-sustaining reactions

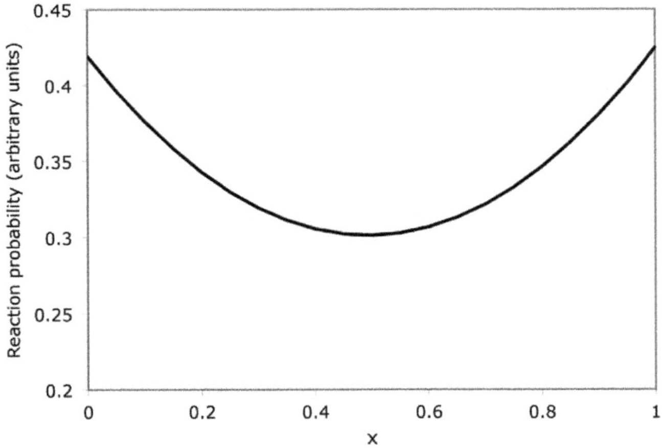

Fig. 8 Calculated probability $P(x)$ of a self-sustaining reaction in a mixture with composition $xM_1 + (1 - x)M_2$ + Ch.

and of mechanical activation in general. Although the main features, namely the increase of the ignition time when one reactive mixture is blended with another and the possible loss of MSR in an intermediate concentration range, are the same in every case, the details are sensitive to the composition and the milling conditions.

This interesting phenomenon calls for further investigations. Finding new systems that exhibit similar behaviour would be of interest, especially if reactions other than the combination of metals with a chalcogen element would be involved. Using a broader variety of mills and milling conditions could provide new insight also. Only ignition time data were presented in this paper. However, detailed investigations of the activated states using XRD to obtain phase composition, grain size and strain, microscopy to study changes of the microstructure, calorimetry to test reactivity, *etc.* are also essential for learning about the actual processes during activation and ignition. Some results are available in the literature,[10,15] but more investigations are needed to understand this phenomenon and to utilize it to better understand mechanochemical processes in general.

Notes and references

1 M. Faraday, *Quart. J. Sci. Lit. Art*, 1820, **8**, 374.
2 L. Takacs, *J. Therm. Anal. Calorim.*, 2007, **90**, 81.
3 Chr. G. Tschakarov, G. G. Gospodinov and Z. Bontschev, *J. Solid State Chem.*, 1982, **41**, 244.
4 Chr. G. Tschakarov, V. Rusanov and G. Gospodinov, *J. Solid State Chem.*, 1985, **59**, 265.
5 Chr. Chakurov, V. Rusanov and J. Koichev, *J. Solid State Chem.*, 1987, **71**, 522.
6 V. Rusanov and Chr. Chakurov, *J. Solid State Chem.*, 1989, **79**, 181.
7 V. Rusanov and Chr. Chakurov, *J. Solid State Chem.*, 1990, **89**, 1.
8 J. S. Benjamin, in *New Materials by Mechanical Alloying Techniques*, ed. E. Artz and L. Schultz, DGM Informationsgesellschaft, Oberursel, 1989, p. 3.
9 G. B. Schaffer and P. G. McCormick, *Metall. Trans. A*, 1990, **21**, 2789.
10 L. Takacs, *Prog. Mater. Sci.*, 2002, **47**, 355.
11 E. Gaffet, F. Charlot, D. Klein, F. Bernard and J.-C. Niepce, *Mater. Sci. Forum*, 1998, **269–272**, 379.
12 J. Ding, T. Tsuzuki, P. G. McCormick and R. Street, *J. Phys. D: Appl. Phys.*, 1996, **29**, 2365.
13 D. Stamatis and E. L. Dreizin, *Combust. Flame*, 2011, **158**, 1631.
14 J. J. Moore and H. J. Feng, *Prog. Mater. Sci.*, 1995, **39**, 243.
15 L. Takacs, in Experimental and Theoretical Studies in Modern Mechanochemistry, ed. F. Delogu and G. Mulas, *Transworld Research Network*, Trivandrum, Kerala, India, 2010, p. 153.
16 L. Takacs and V. K. Garg, in *Mössbauer Spectroscopy in Materials Science*, ed. M. Miglierini and D. Petridis, Kluwer Academic Publishers, 1999, p. 143.
17 L. Takacs, *J. Mater. Synth. Process.*, 2000, **8**, 181.
18 L. Takacs and M. A. Susol, *J. Solid State Chem.*, 1996, **121**, 394.
19 L. Takacs and M. A. Susol, *Mater. Sci. Forum*, 1996, **225–227**, 559.
20 Bakhshai, V. Soika, M. A. Susol and L. Takacs, *J. Solid State Chem.*, 2000, **153**, 371.
21 Z. A. Munir and U. Anselmi-Tamburini, *Mater. Sci. Rep.*, 1989, **3**, 277.
22 O. Kubaschewski, C. B. Alcock and P. J. Spencer, *Materials Thermochemistry*, 6th edn Pergamon Press, Oxford, 1993.

Faraday Discussions

PAPER

Scale-up of organic reactions in ball mills: process intensification with regard to energy efficiency and economy of scale†

Achim Stolle,* Robert Schmidt and Katharina Jacob

Received 16th December 2013, Accepted 15th January 2014
DOI: 10.1039/c3fd00144j

The scale-up of the Knoevenagel-condensation between vanillin and barbituric acid carried out in planetary ball mills is investigated from an engineering perspective. Generally, the reaction proceeded in the solid state without intermediate melting and afforded selectively only one product. The reaction has been used as a model to analyze the influence and relationship of different parameters related to operation in planetary ball mills. From the viewpoint of technological parameters the milling ball diameter, d_{MB}, the filling degree with respect to the milling balls' packing, $\Phi_{MB,packing}$, and the filling degree of the substrates with respect to the void volume of the milling balls' packing, Φ_{GS}, have been investigated at different reaction scales. It was found that milling balls with small d_{MB} lead to higher yields within shorter reaction time, $t_{reaction}$, or lower rotation frequency, rpm. Thus, the lower limit is set considering the technology which is available for the separation of the milling balls from the product after the reaction. Regarding $\Phi_{MB,packing}$, results indicate that the optimal value is roughly 50% of the total milling beakers' volume, $V_{B,total}$, independent of the reaction scale or reaction conditions. Thus, 30% of $V_{B,total}$ are taken by the milling balls. Increase of the initial batch sizes changes Φ_{GS} significantly. However, within the investigated parameter range no negative influence on the yield was observed. Up to 50% of $V_{B,total}$ can be taken over by the substrates in addition to 30% for the total milling ball volume. Scale-up factors of 15 and 11 were realized considering the amount of substrates and the reactor volume, respectively. Beside technological parameters, variables which influence the process itself, $t_{reaction}$ and rpm, were investigated also. Variation of those allowed to fine-tune the reaction conditions in order to maximize the yield and minimize the energy intensity.

Institute for Technical Chemistry and Environmental Chemistry (ITUC), Friedrich-Schiller University Jena, Lessingstr. 12, D-07743 Jena, Germany. E-mail: Achim.Stolle@uni-jena.de; Fax: +49 3641 948 402; Tel: +49 3641 948 413

† Electronic Supplementary Information (ESI) available: Further experimental details. See DOI: 10.1039/c3fd00144j

Introduction

Mechano-chemical reactions in general and specifically those carried out in ball mills have attracted serious attention throughout many fields of inorganic, organometallic, and organic synthesis as well as materials science.[1–8] Mechano-chemical reactions are characterized by a specific stress load which initiates directly (or indirectly) the breaking and making of chemical bonds.[9] If the mechanical stress is to be employed to initiate chemical reactions in solid materials by the release of frictional energy, two principal methods exist: manual grinding using a mortar-and-pestle; and ball milling using laboratory equipment like a planetary ball mill (PBM).[1] Manual grinding is the easiest and thus the most cost-effective way to initiate reactivity between solids on the mechanical pathway. However, it has been shown that those reaction protocols lack the necessary robustness, reliability and process controllability.[10] This is due to the fact that the applied force is directly linked to the physical characteristics of the experimenter, which becomes critical if the reaction time is long. Thus, accomplishment of the reactions by different persons may lead to different results, except for reactions with a low activation barrier or a high thermodynamic driving force. From a chemical engineering point of view ball mills should be employed since these offer the possibility to control the reaction time, energy density, and other parameters.[1,11] Two different types of ball mills are applicable for organic synthesis at the laboratory scale: PBMs and vibration or mixer ball mills (MBMs). In case of PBMs the grinding media (in the present case milling balls) are accelerated due to rotation of the milling chamber which itself is mounted on a sun disc rotating in the opposite direction. The resulting trajectories of the milling balls result in frictional forces (gliding) and impact processes between the balls, the chamber, and of course the mill charge between them. The kinetic energy is transformed and dissipated as thermal energy and allows the initiation of chemical reactions resulting in phenomena like *reactive milling*, *mechano-chemical reactions*, or *mechanically-assisted reactions*.[1,2,12] In case of MBMs the milling chambers are accelerated horizontally or in the horizontal and vertical plane following a circular arc or the surface of a spherical cap, respectively. Independently from the mill type it is possible to control the energy entry by control of the process conditions.[13]

Despite the advantages of lab-scale PBMs (and also MBMs) over the utilization of a mortar-and-pestle there are some serious disadvantages or knowledge-gaps in comparison to traditional methods of energy entry, like thermal heating or advanced technologies like microwave-assisted synthesis. Lab-scale ball mills generally have no possibility for cooling or heating, which makes energy management and heat recycling complicated or even impossible. From an engineering viewpoint the lack of controllability (regarding the input power) is critical: the regulation by torque sensors mounted to the rotation axis is impossible due to constructive limitations. Thus, important parameters which are necessary for scale-up are not as easy to achieve as in stirred media ball mills (SBMs). Only a few scaled-up examples in the field of organic synthesis in ball mills have been published to date.[14,15] Those that have been published are rather an alignment of different reactions showing the general possibility that scale-up is possible, than a systematic investigation of scale-up parameters and process conditions.

Within the present study the authors want to show not only that a scale-up including a technology switch from PBM to SBM is possible, but they also want to understand the mechanisms and principles in order to establish a scale-up protocol with suitable parameter sets. The study is dedicated to multi-ball milling systems with the same principle of milling ball movement: rotation. Thus, the present study does not include scale-up in MBMs and the present report will only consider the first step of the program: a scale-up within the system of PBMs. Those experiments would not be possible with MBMs due to the volumetric restrictions of the milling beakers' volume in the case of lab-scale apparatus.

Generally, processes in ball mills are characterized by different parameters, regardless of the type of process, *i.e.* a mechano-chemical or a mechanically-assisted reaction or particle comminution, and the type of ball mill. In contrast to particle refinement solely, the balance volume for reactions in ball mills has to be enlarged taking into account the chemical parameters of the specific reactions.[1,11,16] The large number of influencing variables can be categorized into:

(i) Chemical parameters,
(ii) Technological parameters, and
(iii) Process parameters.

Chemical parameters summarize all those variables which are directly linked to the chemical transformations taking place in the chamber of the ball mill, which include: (i) the type of chemical reaction; (ii) the presence of catalysts or additives; (iii) the reagent ratio; or (iv) the presence of low amounts of liquid(s) in the case of liquid-assisted grinding, LAG.[17–22]

With respect to **technological variables**, the importance of (i) the type of ball mill, (ii) the milling material, (iii) the number and (iv) size of milling balls as well as (v) the filling degree of the milling beakers for the accomplishment of chemical reactions should be considered. This set of parameters describes the technology which is used for comminution.

Process parameters are useful for controlling the energy entry during the process. Thus, (i) operating frequency rpm, (ii) reaction time $t_{reaction}$, and (iii) temperature T are the main values which should be considered. Whereas rpm is directly correlated to the kinetic energy of the grinding media and thus, to the energy which can be dissipated as heat to the mill charge, $t_{reaction}$ and T assign for the progression of a chemical reaction and are linked to the reaction kinetics.

In order to scale-up reactions in ball mills all those parameters have to be considered. In the following chapters the influence of the main parameters on the outcome of a model reaction will be discussed.

Experimental

General information

All chemicals were purchased from Sigma-Aldrich or Alfa Aesar and used as received. The reactions were accomplished in a Fritsch Pulverisette P7 premium line and a Fritsch Pulverisette P6 classic line planetary ball mill (Fritsch GmbH, Idar-Oberstein, Germany) using 45 ml (**P7-45**) and 250 (**P6-250**) or 500 ml milling beakers (**P6-500**) made from stainless or tempered steel, respectively (Table 1). The milling balls were made from magnesia-stabilized zirconia at varied diameter, d_{MB}, and milling ball number, n_{MB}. Both d_{MB} and n_{MB} determine the milling ball filling degree, $\Phi_{MB,packing}$. This parameter is defined as the volumetric ratio of

Table 1 Experimental parameters (for the relationship between individual parameters see Electronic Supporting Information)

	Symbol [Unit]	**P7-45-**	**P6-250-**	**P6-500-**
Revolution Speed of disk	rpm [min^{-1}]	750–850	450–650	450–650
Gear ratio	k [—]	−2	−1.82	−1.82
Rotational speed of beaker	2rpm [min^{-1}]	1500–1700	819–1183	819–1183
Milling ball diameter	d_{MB} [mm]	2, 5, 10	10, 20, 30	10, 20, 30
Total volume of milling beaker[a]	$V_{B,total}$ [ml]	45	250	500
Milling ball filling degree	$\Phi_{MB,packing}$ [—]	0.3–0.8	0.18–0.8	0.3–0.8
Amount of starting material i[b]	$n_{initial,i}$ [mmol]	**20**	**100**–200	**200**, 300

[a] Material for $V_{Btotal} = 45$ and 500 ml: stainless steel. Material for $V_{Btotal} = 250$ ml: tempered steel. [b] Bold-face numbers indicate the standard reaction conditions.

the milling balls' packing volume (including the void volume between the milling balls) to the total volume of the milling beaker, $V_{B,total}$ (eqn (1)).

$$\Phi_{MB,packing} = \frac{V_{MB,packing}}{V_{B,total}} \tag{1}$$

Experimental procedure

The milling beakers were equipped with the respective number of milling balls. Afterwards equimolar amounts of vanillin and barbituric acid were added in the given order. Milling was accomplished at the respective frequency, rpm, and milling time, $t_{reaction}$, according to the parameter values shown in Table 1 and assigned in the Figures and Table of the following sections. The nomenclature of the respective experiments is as follows: type of ball mill-beaker volume ($V_{B,total}$)-batch size ($n_{initial,i}$). For instance, **P7-45-20** is related to experiments in the PBM P7 classic line with 45 ml beakers and a batch size of 20 mmol related to one starting material. The relationship between $\Phi_{MB,packing}$, d_{MB}, and n_{MB} for the individual experimental conditions can be found in the Electronic Supporting Information.†

Analytical procedure

After the milling process, samples of the crude reaction product (100 mg) were placed on top of a thin layer of silica gel and were extracted with ethyl acetate (5 ml). The organic extract was analyzed with GC-FID using a 7890A-GC (Agilent Technologies) with the following measurement conditions: HP5 column (30 m lengths, 0.32 mm diameter, 0.25 mm film thickness); H_2 (12 psi); temperature program: 80 °C (hold for 1 min), 15 K min^{-1} up to 200 °C, 30 K min^{-1} up to 280 °C (hold for 1 min); injector temperature: 280 °C; detector temperature: 300 °C.

Results and discussion

One long-term goal of the project is to close the gap between lab-scale applications of ball mills, *i.e.* planetary ball mills (PBMs), and stirred media ball mills (SBMs) for utilization at a technical scale. Thus, a volumetric scale-up factor of 220 should be realized starting with $V_{B,total} = 0.045$ (PBM) to 10 l (SBM) whereas in the present case experiments in two different PBMs will be reported carried out in milling beakers with $V_{B,total} = 45$ (**P7**) to 500 ml (**P6**; volumetric scale-up factor = 11). This affords some experimental prerequisites which narrow down the number of model reactions to a few examples. The following criteria should be fulfilled by the model reaction.

(i) The starting material(s) should be available in the necessary quantity at reasonable costs and suitable purity.

(ii) The application of hazardous substances and the formation of gases during processing should be avoided.

(iii) The materials should be air-stable allowing processing and storage under a standard atmosphere.

(iv) To avoid three-phase systems with possible changes in viscosity and plasticity of the mill charge, dry media milling should be accomplished.

(v) The melting points of the starting material(s), the product(s) and intermediate(s) should be ≥ 70 °C to avoid formation of eutectics which would influence the morphology of the mill charge.

(vi) The reaction should have a high selectivity and quantitative conversion ($\leq 97\%$) should be reachable within $10 \leq t_{reaction} \leq 45$ min (using a PBM with $V_{B,total} = 45$ ml).

(vii) In order to control and monitor the reaction progress fast and robust analytical methods should be applied. Any promoting effect during work-up or sample preparation should be excluded.

(viii) The grinding chambers or its interior should be made from stainless steel.

(ix) To avoid chemical interactions with the substrate(s), milling balls should be made from magnesia-stabilized zirconia (MSZ).

Decision criteria (i)–(vi) relate to the class of chemical parameters. Especially the availability considering costs and purity as well as possible hazards during handling and operation influence the suitability of existing protocols for organic, ball mill-assisted reactions. Several candidates, for instance metal-catalyzed, cycloaddition, or condensation reactions, were evaluated as model reactions.[1,2,7] Kaupp published several examples which may be considered as model reactions, although there are often limitations regarding the melting points and/or the utilization/liberation of gaseous reactants.[14,15] Both limitations exclude an application of PBMs as reactors, thus an internal scale-up prior to technological switch to SBM is not possible.

However, two suitable candidates were short-listed: (i) imine formation starting from anilines and benzaldehydes; and (ii) Knoevenagel condensation of benzaldehydes with CH-acidic compounds.[23–27] Both fulfil most of the criteria, whereas it is well known that many condensation reactions yielding imines proceed already upon co-crystallization, which is an indicator for a low activation barrier and high reaction rate even in the solid state.[14,28–31] Furthermore, analysis is complicated due to a high solid state reactivity and spontaneous product formation in solution. Thus the Knoevenagel reaction between vanillin (**1**) and barbituric acid (**2**) yielding 5-(4-hydroxy-3-methoxybenzylidene)pyrimidine-2,4,6(1*H*,3*H*,5*H*)-trione **3** (Scheme 1) was chosen as the two-component model reaction.[24,32–35] All criteria with respect to chemical parameters are fulfilled and processing in both types of intended ball mills – PBM and SBM – will be possible.

Evaluation of a rapid, robust and reliable analytical method in order to control and monitor the reaction progress is a second prerequisite for the investigation of

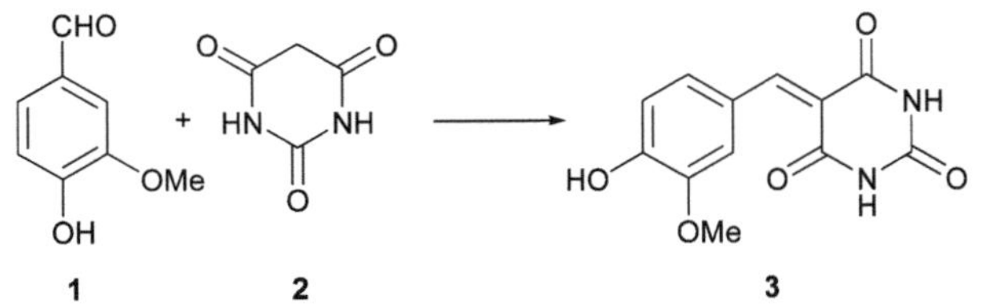

Scheme 1 Model reaction for systematic scale-up studies.

other parameters. With respect to applicability, solid-state analytic techniques like ATR-FTIR, solid-state NMR, or (P)XRD are the methods of choice, since work-up incorporating solvents can be omitted. Unfortunately the methods lack of possibilities for quantification of the substrates and/or the product. For instance, with ATR-FTIR only semi-quantitative information can be obtained from signals assigning the presence of the CHO-group in **1**. Solution-based methods like ^{1}H and ^{13}C NMR confirm the identity of the product and the absence of any side products or isomers. A post reaction in deuterated solvents like DMSO-D_6 or $CDCl_3$ was not observed. However, high throughput analysis as well as quantification with the required precision is difficult. Since the product formation is accompanied by coloration of the solid product mixture UV-VIS measurements in DMF as solvent were performed. Those allow precise quantification, but a high amount of DMF (100 ml per sample) is required for sample preparation. Solid–liquid extraction of a small representative sample with 5 ml EtOAc allowed the selective separation of unconverted **1** from **2** and **3**, which was analyzed chromatographically (GC-FID). Since no side and consecutive products were detectable (>99% selectivity) conversion and yield were calculated from those measurements based on a calibration for **1**.

General considerations

According to the criteria summarized above, restrictions follow for some technological parameters in order to reduce the parameter space. The type of ball mill is a PBM in every case (Table 1).[13,36] The basic effect leading to the specific trajectories of the milling balls is the rotation around their central axis and rotation in the opposite direction around the central axis of the sun disk on which the beakers are mounted. Thus the movement type, *i.e.* rotation, is in principle identical to the conditions in an SBM. The PBMs used herein are different in size, mainly determined by the distance of the two axes and the gear ratio of the two resulting rotations (Table 1). Secondly, stainless ($V_{B,total}$ = 45 and 500 ml)/tempered steel ($V_{B,total}$ = 250 ml) was chosen as material for the milling beakers and milling balls were made from MSZ. All experiments were carried out at room temperature and milling was accomplished without cycling or reverse operation mode. The following parameters were varied and their influence will be discussed:

- rotation frequency, rpm [min^{-1}] (process parameter),
- milling ball diameter, d_{MB} [mm] (technological parameter),
- milling ball filling degree, $\Phi_{MB,packing}$ [—] (eqn (1); technological parameter),
- milling beaker volume, $V_{B,total}$ [ml] (technological parameter), and

• the grinding stock filling degree, Φ_{GS} [—] (eqn (6)–(8); technological parameter).

The last two parameters, $V_{B,total}$ and Φ_{GS}, relate also to the batch size, $n_{initial,i}$, as a chemical parameter. The parameter spaces for all variables are summarized in Table 1 and for the relationship between d_{MB}, n_{MB}, and $\Phi_{MB,packing}$ the reader is referred to the Electronic Supporting Information.† In most cases the dependency of those parameters to the reaction time $t_{reaction}$ is also discussed. Furthermore, the net energy input will be considered in the investigation.

Influence of the rotation frequency

As first parameter the influence of the rotation frequency, rpm, on the yield of the model reaction was investigated for two different PBMs. resulting in three different basic experimental conditions (Table 1). The results are summarized in Fig. 1 and are arranged such that $n_{initial,i}$ increases from top to bottom. In all cases sigmoidal curves were found, whereas the time at which quantitative yield, $t_{97\%}$, is reached depends on rpm and the batch size, $n_{initial,i}$. Generally, the higher the rpm the earlier a yield $\geq$ 97% is reached, independently of the type of ball mill. Despite the increasing value for $n_{initial,i}$, from top to bottom the reactions can be completed in less than 15 min. Especially, in the bottom case (**P6-500-200**) $t_{reaction} \leq 10$ min is sufficient for quantitative conversion operating at rpm = 650 min^{-1}. The promoting effect of the rotation frequency has been demonstrated for many reactions in ball mills and is related to an increase of the system's kinetic energy which is proportional to rpm^2.[1,37] Beside $t_{reaction}$, it has been shown that the influence of rpm on a chemical reaction in ball mills is the most important one.[38] Several examples are described in literature indicating this mathematical correlation between, *e.g.*, yield or conversion and the rpm at which a ball mill is operated. Generally a higher rpm results in an increase of those variables. Examples regarding organic synthesis are very widespread resembling many fields of synthesis chemistry: for instance cross-coupling;[10,38] redox chemistry;[39–42] organocatalytic reactions;[43] condensation reactions;[34,44–46] polymer grafting;[47] or reactions of carbohydrates.[48]

Although Fig. 1 indicates that the correlation between rpm and yield can be generalized and seems to be independent of the batch size and the size of the utilized ball mill, there remains one challenge. Based on rpm alone it is not possible to compare the results directly. Specifically the energetic conditions in the milling beaker depend on rpm and on the angular velocity, ω, which itself depends on rpm and the distance between the two rotating axes (comparison between **P7** and **P6**). Both ball mills have also different rpm_{max}, which are 1100 and 650 min^{-1}, respectively. Additionally, rpm_{max} depends also on d_{MB}: the higher d_{MB} the lower is rpm_{max}, in order to avoid abrasion and process instabilities due to the high energy impact of the milling balls (**P7**: $rpm_{max} = 850$ min^{-1} for $d_{MB} = 10$ mm). Furthermore, the milling beaker size should be considered (comparison between **P6-250** and **P6-500**). Increasing $V_{B,total}$ results in an increasing inner diameter of the milling beaker, d_B, since its height is limited due to constructional restrictions. A higher d_B influences the centrifugal force applied to the moving milling balls and therefore their kinetic energy is different. Those considerations allow to construe the findings regarding the striking independency of $t_{97\%}$ from $n_{initial,i}$. Despite the lower rpm in case of **P6** the higher distance

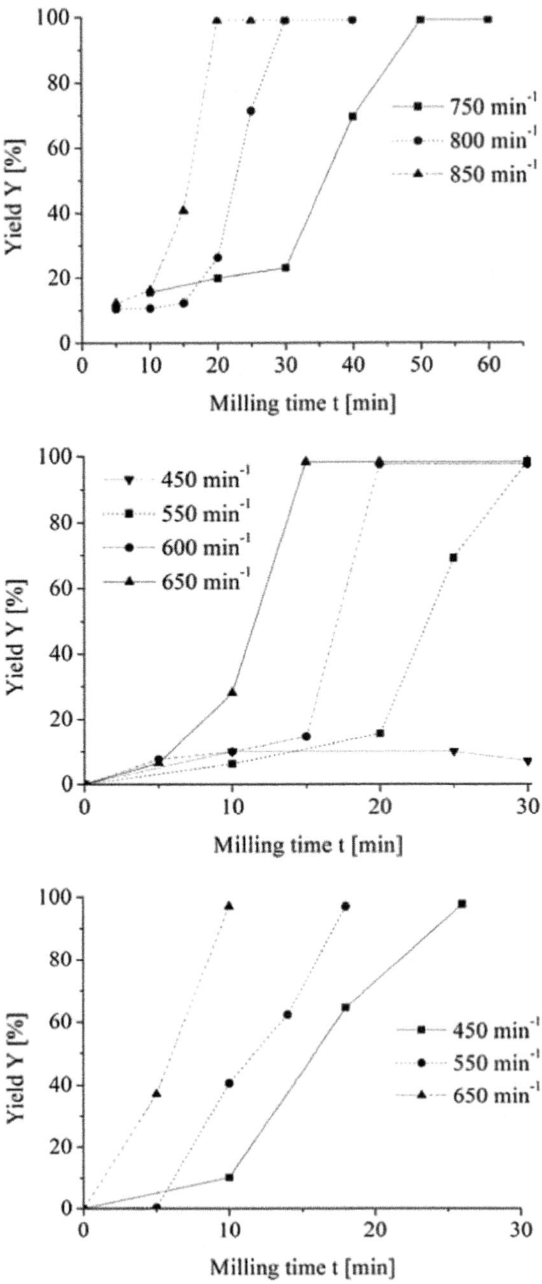

Fig. 1 Influence of milling time, $t_{reaction}$, and rotation frequency, rpm, on the yield of the model reaction, Y (Scheme 1), for different ball mills and experimental conditions: **P7-45-20** (top; $d_{MB} = 10$ mm, $\Phi_{MB,packing} = 0.35$), **P6-250-100** (middle; $d_{MB} = 20$ mm, $\Phi_{MB,packing} = 0.45$), and **P6-500-200** (bottom; $d_{MB} = 10$ mm, $\Phi_{MB,packing} = 0.50$). For the experimental conditions see Table 1 in the Experimental section and the explanations provided.

between the rotational axes results in similar energetic conditions (compared to **P7**) allowing conversion of an increased amount of starting materials in nearly the same time.

Influence of the milling ball diameter

In order to interpret the findings regarding the influence of d_{MB} on $t_{97\%}$, some preliminary considerations are necessary. From experiments and theoretical considerations in the field of wet comminution in SBMs, a cubic proportionality between d_{MB} and the stress energy is indicated (eqn (2)).[37,49,50]

$$E_{\text{stress,feed}} = d_{\text{MB}}^3 \cdot v_p^2 \cdot \rho_{\text{MB}} \cdot K^{-1} \text{ with } K = f(E_{Y,i}) \quad (2)$$

Thus, one might expect, the higher the d_{MB} the higher is the stress energy transferred to the mill feed. However, this implication considers only one aspect: the energy which is transferable during one single collision. Most significant differences in the present case are of the mill type, the type of milling (wet *vs.* dry milling), and the total filling degree of the milling reactor, Φ_{total} (eqn (7); see later), and the fact that a chemical reaction takes place. Thus, a more general consideration is required regarding the influence of d_{MB} on the stress energy E. In general, the stress energy is considered as an energy distribution summing up all stress energies, SE, from all single stress events (eqn (3)).

$$E = \sum_{i=1}^{\text{SN}} \text{SE}_i = \text{SF} \cdot \overline{\text{SE}} \cdot t_{\text{reaction}} \quad (3)$$

The number of stress events, SN, itself is the product of $t_{reaction}$ and the stress frequency, SF.[51] SF and SN correlate with n_{MB} which is inversely proportional to d_{MB}, assuming that the overall mass of the milling balls, $m_{MB,total}$, is constant. Thus, both parameters, d_{MB} and n_{MB}, influence the outcome of the individual comminution process or chemical reaction in ball mills. Since these parameters also influence the median of the stress energy, the distribution variation of both values may result in different yields, even if following the above mentioned prerequisite: $m_{MB,total} \approx$ constant.[40]

Fig. 2 provides an overview of the time-dependent yield, changing d_{MB}, the experimental setup and, linked to it, $n_{initial,i}$. Experiments have been carried out under the premise of nearly constant $m_{MB,total}$, thus if d_{MB} was changed, n_{MB} had to be adopted. The correlation between these three parameters is provided within the Electronic Supporting Information.† A general trend, which is significant for all cases, is the fact that smaller milling balls result in a lower $t_{97\%}$, independent of the chosen ball mill and $n_{initial,I}$, which increases from top to bottom. This effect is highlighted for the results at the lowest scale (**P7-45-20**). With 10 mm balls after 40 min a yield of 40% was found, $t_{97\%}$ for 5 and 2 mm ball were 30 and 20 min, respectively. Thus, it seemed to be obvious that SF (eqn (3)) dominates the energy transfer to the substrates and the single stress energy (eqn (2)) is less important for the present case. Keeping $m_{MB,total}$ constant while decreasing d_{MB} (and increasing n_{MB}) results in an increase of the active surface area of the milling balls which is proportional to $(d_{MB})^2$. Since the active surface area is important for energy transfer to the substrates, beside particle refinement these findings are comprehensible. Besides this striking evidence of the influence of d_{MB}, it should

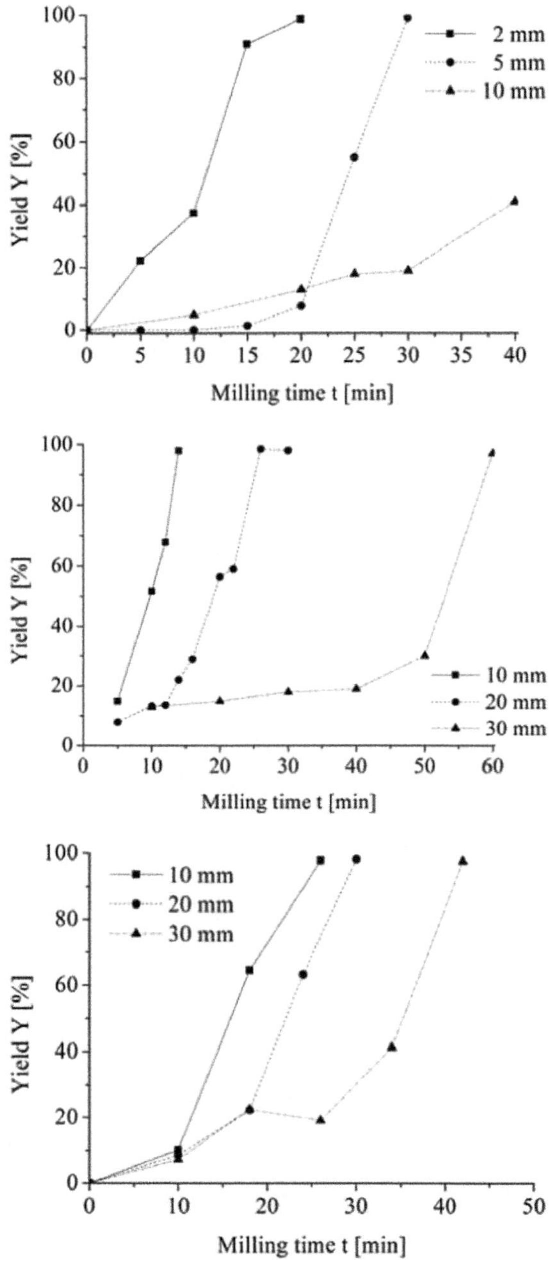

Fig. 2 Influence of milling time, $t_{reaction}$, and milling ball diameter, d_{MB}, on the yield of the model reaction, Y (Scheme 1), for different ball mills and experimental conditions: **P7-45-20** (top; rpm = 800 min^{-1}, $\Phi_{MB,packing} = 0.45$ for $d_{MB} = 2$ and 5 mm, $\Phi_{MB,packing} = 0.50$ for $d_{MB} = 10$ mm), **P6-250-100** (middle; rpm = 650 min^{-1}, $\Phi_{MB,packing} = 0.28$–0.30), and **P6-500-200** (bottom; rpm = 450 min^{-1}, $\Phi_{MB,packing} = 0.50$). For the experimental conditions see Table 1 in the Experimental section and the explanations provided.

be noted that the literature reports the opposite effect: if $m_{MB,total}$ is kept constant d_{MB} has no influence.[1,40] However, it has to be mentioned that the authors did not investigate the time dependency. Literally speaking, if the observed yields are in the top part of the sigmoidal curves in every case, such conclusions become evident. Furthermore, the different complexity of the reactions themselves, as well $n_{initial,i}$, should be taken into account. Fig. 2 reveals, that $t_{reaction}$ also has a tremendous effect, and thus has to be considered in the interpretation of those experiments.

Analysis of the data from Fig. 2 for the experiments at low scale (**P7**) allows for some additional insights. First of all, the question might arise why, at the large scale experiments (**P6**), no milling balls with $d_{MB} < 10$ mm were applied? Even during experiments at the smaller scale (**P7**), it was observed that working with $d_{MB} = 2$ mm results in an enclosure of the milling balls in the substrate–product mixture and no further movement was possible (Fig. 3). In particular, at high rpm such observations were made. To avoid such effects, which are more critical at a larger scale, it was decided not to use those milling balls, although in future experiments this should be reconsidered, since it is well known that SBMs are often operated at $d_{MB} < 1$ mm. On the other hand, the application of 20 or 30 mm balls in the low-scale PBM (**P7**; like in **P6**) is critical, since in this case the PBM in combination with 45 ml beakers had to be operated with $n_{MB} < 3$. This would result in severe abrasion (high impact energies) and PBMs should be avoided in contrast to MBMs. These may be operated with one or two milling balls only.

Comparison of the data in Fig. 2 with $d_{MB} = 10$ mm at different scales leads to variations of the values for $t_{97\%}$. With **P7-45-20**, even after 40 min milling, no quantitative conversion was observed, $t_{97\%}$ were 12 and 26 min using **P6-250-100** and **P6-500-200** as experimental conditions, respectively. With 20 mm balls, values of 26 and 30 min were found. Thus, at lower scale the effect of decreasing d_{MB} is more pronounced, at first sight. A more detailed analysis revealed that two other important parameters were changed beside the scale-up from 100 to 200 mmol for $n_{initial,i}$, respectively. One was rpm, an increase of rpm for **P6-500-200** to 650 min^{-1} resulted in $d_{MB} = 10$ mm with a $t_{97\%}$-value of 9 min (Fig. 1). The second difference becomes obvious when calculating the milling ball filling degree (with respect to ball packing), $\Phi_{MB,packing}$ (eqn (1)), which was different and ranges from 0.28–0.30 (**P6-250-100**) to 0.50 (**P6-500-200**). Thus, this parameter also influences the results, which will be discussed in the following chapter.

Fig. 3 Hindered milling ball movement by enclosure of the milling balls in the substrate–product mixture. Reaction conditions: P7-45-20, rpm = 1000 min^{-1}, $d_{MB} = 2$ mm, $\Phi_{MB,packing} = 0.46$.

Influence of the degree of milling ball filling

Fig. 4 visualizes the influence of $\Phi_{MB,packing}$ on the yield of the model reaction for different d_{MB}. Experiments have been conducted in the large PBM in 250 ml milling beakers at a constant reaction time of 12 min and rpm = 650 min^{-1} (**P5-250-100**), which are the optimal conditions already outlined for $\Phi_{MB,packing}$ = 0.28–0.30 in Fig. 2. The results indicate strikingly the importance of this parameter, since independent from d_{MB} the curves pass through a maximum, which is in the range of $0.43 \leq \Phi_{MB,packing} \leq 0.50$.

The reason for this behavior relates to the trajectories of the milling balls in their milling beakers. According to eqn (2) and (3) the overall stress energy depends on SE and SF. If $\Phi_{MB,packing}$ is < 0.4 the number of stress events, SN, and therefore also SF and the medium SE are reduced which leads to the observed effect regarding the yield. However, the movement of the milling balls inside the beakers is not distorted. For an efficient energy transfer the grinding bodies need space for acceleration and movement through the milling beaker. In case of $\Phi_{MB,packing}$ > 0.5 the free space for the trajectories is restricted or completely blocked and the milling balls cannot be accelerated, make their way through the vessel, and cause impacts (Fig. 4).[11,36,52,53] Thus, SF is decreased (see eqn (3)). Beside hampered acceleration, inefficient energy transfer, and decreased SF; centrifugal transport of the milling balls along the milling beaker wall (see Fig. 3)[54] can cause increased abrasion of the milling beaker.[11,52]

Manufacturers of PBMs generally advise their customers to leave at least 33% of $V_{B,total}$ for the trajectories of the milling balls and the geometrical volume, $V_{MB,material}$, of the milling balls should not exceed 33%.[52] However, $\Phi_{MB,packing}$ (eqn (1)) considers the packing volume of the milling balls. This is reasonable from an engineering point of view since the filling process of a milling beaker with milling balls results in ball packing, whereas the void volume, V_{void}, depends on the space filling degree, $1 - \varepsilon$ (eqn (4)).

$$V_{MB,packing} = V_{MB,material} + V_{void} = \frac{V_{MB,material}}{1 - \varepsilon} \tag{4}$$

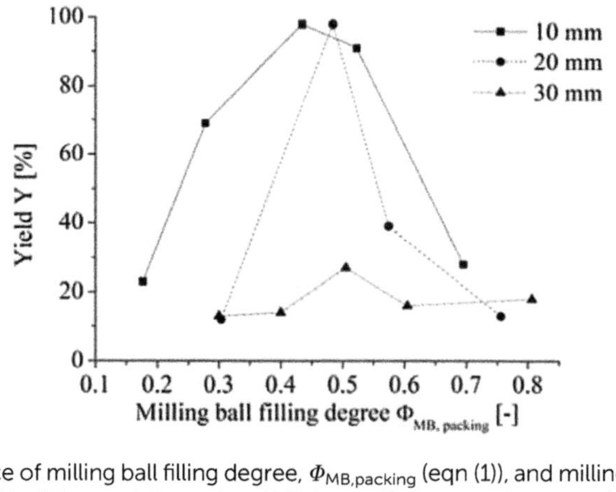

Fig. 4 Influence of milling ball filling degree, $\Phi_{MB,packing}$ (eqn (1)), and milling ball diameter, d_{MB}, on the yield of the model reaction, Y (Scheme 1), for experiments using **P6-250-100** (rpm = 650 min^{-1}, $t_{reaction}$ = 12 min). For an indication of the experimental conditions see Table 1 in the experimental section and the explanations provided.

The space filling degree can be calculated from the porosity, ε, of the packing, which can be measured or calculated similar to powders. The geometric volume of the milling balls, $V_{\mathrm{MB,material}}$, as well as the correlated geometric filling degree, $\Phi_{\mathrm{MB,material}}$, can be calculated according to the following equation.

$$\Phi_{\mathrm{MB,material}} = \frac{V_{\mathrm{MB,material}}}{V_{\mathrm{B,total}}} = \frac{\pi \cdot d_{MB}^{3} \cdot n_{MB}}{6 \cdot V_{\mathrm{B,total}}} = \frac{m_{\mathrm{MB,total}}}{V_{\mathrm{B,total}} \cdot \rho_{\mathrm{MSZ}}} \cong \Phi_{\mathrm{MB,packing}}(1-\varepsilon) \tag{5}$$

Based on parameters like d_{MB} and n_{MB} or $m_{\mathrm{MB,total}}$ and the material density of the milling balls, ρ_{MSZ}, the value of $\Phi_{\mathrm{MB,material}}$ can be estimated. According to eqn (4) and the assumption that $\rho_{\mathrm{MSZ}} \gg \rho_{\mathrm{air}}$ (which allows to neglect the mass of the void volume) it is possible to correlate $\Phi_{\mathrm{MB,material}}$ and $\Phi_{\mathrm{MB,packing}}$ by consideration of ε. Recalculation of the values for $\Phi_{\mathrm{MB,packing}}$ from Fig. 4 results in $\Phi_{\mathrm{MB,material}}$ values of 0.26, 0.27, and 0.28 for $d_{\mathrm{MB}} = 10$, 20, and 30 mm, respectively (see the Electronic Supporting Information†). This is roughly in accordance with the manufacturers' advice. Actually, the optimal values for $\Phi_{\mathrm{MB,packing}}$ (Fig. 4) and thus those for $\Phi_{\mathrm{MB,material}}$ are slightly lower, which means n_{MB} could be reduced, improving the overall process costs.

Fig. 4 indicates that for $d_{\mathrm{MB}} = 30$ mm within 12 min the yield did not increase over a value of roughly 25% (see also Fig. 2, middle). In order to evaluate those data additional experiments have been conducted allowing comparison of the time required to reach quantitative yield, $t_{97\%}$, and $\Phi_{\mathrm{MB,packing}}$. Results are presented in Fig. 5 for the general conditions **P6-250-100** (top) and at enlarged scale **P6-500-200** (bottom). Although, the values slightly differ, it is obvious for the small-scale experiments that, independently from d_{MB}, the curves pass through a minimum, whose location regarding the values for $\Phi_{\mathrm{MB,packing}}$ is almost identical to those shown in Fig. 4. At an optimized filling degree $t_{97\%}$ ranges from 12–18 min. Enlarging $V_{\mathrm{B,total}}$ and $n_{\mathrm{initial,i}}$ narrows down the optimum value for $\Phi_{\mathrm{MB,packing}}$ to exactly 0.50 for 20 and 30 mm milling balls. For the Knoevenagel condensation with $d_{\mathrm{MB}} = 10$ mm (**P6-500-200**) no minimum was detected.

Influence of the substrate filling degree

During the experiments conducted up to now the batch size, $n_{\mathrm{initial,i}}$, was not varied in the ball milling systems utilized (**P7-45-20:** $n_{\mathrm{initial,i}} = 20$ mmol, **P6-250-100:** $n_{\mathrm{initial,i}} = 100$ mmol, **P6-500-200:** $n_{\mathrm{initial,i}} = 200$ mmol). Changing this value in order to increase the scale of reaction would consequently result in different conditions regarding the ratio of substrate amount (GS) and milling balls. One possibility to measure this ratio is the so called "ball-to-powder-ratio" which is often applied to ascribe experimental conditions in the field of inorganic synthesis in ball mills or dry comminution.[11,55] This value is often expressed as a mass ratio, which makes this parameter inapplicable to the description of general relationships from an engineering perspective. The large discrepancy with respect to material or bulk densities makes a comparison of different systems impossible. Although the weighed-in mass is important in general, in order to describe the correlation to other values like $\Phi_{\mathrm{MB,packing}}$ or $V_{\mathrm{B,total}}$ a volume-based parameter has to be applied. Since the substrates and the product are free-flowing powders,

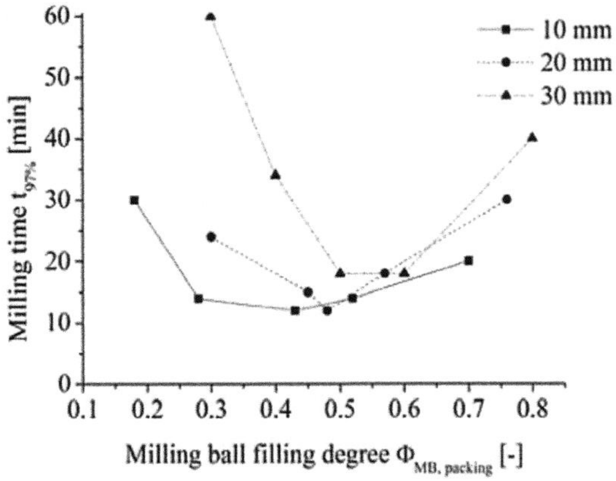

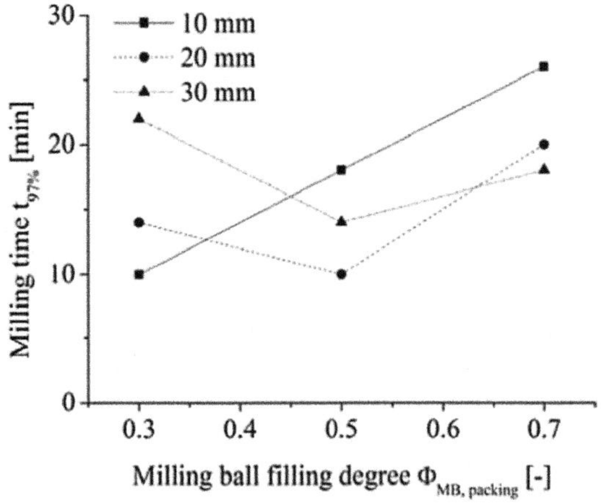

Fig. 5 Influence of milling ball filling degree, $\Phi_{MB,packing}$ (eqn (1)), and milling ball diameter, d_{MB}, on the time to reach quantitative yield, $t_{97\%}$, for different basic experimental conditions **P6-250-100** (top; rpm = 650 min^{-1}), and **P6-500-200** (bottom; rpm = 550 min^{-1}). For an indication of the experimental conditions see Table 1 in the Experimental section and the explanations provided.

their bulk volume, $V_{GS,packing}$, can be calculated according to eqn (6) considering the individual masses, $m_{GS,i}$, and bulk densities, $\rho_{GS,i}$.

$$\Phi_{GS,packing} = \frac{V_{GS,packing}}{V_{B,total}} \cong \frac{\sum_i^n \frac{m_{GS,i}}{\rho_{GS,i}}}{V_{B,total}} \tag{6}$$

Correlation to $V_{B,total}$ leads to the filling degree of the substrates, $\Phi_{GS,packing}$, which is limited by the total beaker volume and the geometric filling degree of the milling balls: $0 \leq \Phi_{GS,packing} \leq 1 - \Phi_{MB,material}$. Due to the low porosity and the high space filling degree of the substrate packing with respect to that of the milling balls, the void volume is neglected in the first approximation and the total volumetric filling degree of the milling beaker, Φ_{total}, can be calculated as shown in eqn (7).

$$\Phi_{total} = \Phi_{MB,material} + \Phi_{GS,packing} \tag{7}$$

This parameter ranges from $0 \leq \Phi_{total} \leq 1$, which resemble the empty and the completely filled milling beaker. With respect to wet and dry comminution processes another parameter is introduced in eqn (8) describing the filling degree, Φ_{GS}, in relation to the void volume of the milling balls' packing.

$$\Phi_{GS} = \frac{V_{GS,packing}}{V_{MB,packing} - V_{MB,material}} = \begin{cases} < 1 \\ = 1 \\ > 1 \end{cases} \tag{8}$$

In order to illustrate the meaning of Φ_{GS}, three cases are exemplified. If Φ_{GS} is < 1, the substrates are located in the free space of the packed milling balls. The upper limit is exemplified by $\Phi_{GS} = 1$ which means that the void volume in ball packing is completely filled with the substrate(s). The third case, $\Phi_{GS} > 1$, indicates a substrate volume which is higher than V_{void}. Practically this implies the substrates cover over the packed balls. The upper limit is restricted by the milling beaker volume.

Table 2 summarizes the results for two different $V_{B,total}$ and varying $n_{initial,i}$. The calculation of Φ_{GS} leads to values of >1 indicating that the substrate volume is significantly higher than the V_{void} of the ball packing. The total filling degree, Φ_{total}, ranges from 0.55–0.85 and 0.58–0.73 for the 250 and the 500 ml beakers, respectively. The most interesting effect can be found in the last column of Table 2. Despite the high values for Φ_{total}, the times to reach quantitative yield were 30 and 10 min, depending on $V_{B,total}$. The variations found were negligible, which leads to the conclusion that the degree of substrate filling has no significant influence in the investigated parameter range. Even a residual free volume for milling ball movement of 15% ($1-\Phi_{total} = 1-0.85$) of the original beaker volume is apparently enough to achieve high conversion. However, it should be taken into account that organic substrates are light-weight materials with a low Young's modulus. (Reactive) treatment of harder (inorganic) materials might lead to other results. However, since conditions with roughly $\Phi_{total} \geq 75\%$ are realizable for

Table 2 Experimental results for variation of the batch size, $n_{initial,i}$, and the effect on the filling degrees, Φ_i [−],[a] and the time to reach quantitative yield, $t_{97\%}$, for the model reaction in Scheme 1

$n_{initial,i}$/mmol	$\Phi_{MB,material}$	$\Phi_{MB,packing}$	$\Phi_{GS,packing}$	Φ_{GS}	Φ_{total}	$t_{97\%}$/min
P6-250 ($V_{B,total}$ = 250 ml, d_{MB} = 20 mm, n_{MB} = 15, rpm = 550 min^{-1})						
100	0.25	0.45	0.30	1.42	0.55	30
125	0.25	0.45	0.38	1.78	0.63	28
150	0.25	0.45	0.45	2.13	0.70	30
175	0.25	0.45	0.53	2.49	0.78	29
200	0.25	0.45	0.60	2.84	0.85	28
P6-500 ($V_{B,total}$ = 500 ml, d_{MB} = 20 mm, n_{MB} = 33, rpm = 650 min^{-1})						
200	0.28	0.50	0.30	1.28	0.58	10
300	0.28	0.50	0.45	1.92	0.73	10

[a] Calculation according to the equations given in parentheses: $\Phi_{MB,material}$ (eqn (5)), $\Phi_{MB,packing}$ (eqn (1)), $\Phi_{GS,packing}$ (eqn (6)), Φ_{GS} (eqn (8)), and Φ_{total} (eqn (7)).

organic synthesis in ball mills, a scale-up and technological change to dry-operated SBM seems to be possible.

Influence of the milling beaker volume

As Table 2 indicates, the initial beaker volume, $V_{B,total}$, is of importance for the efficiency of the reaction in terms of $t_{97\%}$. If the volume is increased, generally the inner diameter, d_B, of the milling beaker increases, since its height is limited due to constructional prerequisites (the clamping device of the milling beaker on the sun disk). The SE (Eqs. 2 and 3) distributed by impacts of the milling balls depend on the centrifugal force which is proportional to the rotational axis of the system. Thus, enlargement of the beaker volume should result in higher medium SE and the same results in terms of yield should be observed in shorter reaction time or at lower frequency. The results shown in Fig. 6 revealed that at rpm = 650 min^{-1} the $t_{97\%}$-values are in the range of 10–20 min depending on $\Phi_{GS,packing}$ and $V_{B,total}$. By a decrease in rpm the gap between the $t_{97\%}$-values for the 250 and 500 ml beaker become larger, whereas reactions at a larger scale required generally longer reaction times. However it should be considered that the absolute reaction rate is higher since $n_{initial,i}$ is doubled. Thus, the considerations were confirmed by experiment.

Fig. 7 compares the previous results energy-wise. Interestingly, experiments at the largest scale (**P6-500-200**) revealed an energy minimum at rpm = 550 min^{-1} with respect to the overall energy input, E_{brutto}, as well as to the molar energy consumption, E_m (eqn 9).[10] Thus, the higher the reaction scale the lower is the specific energy demand, at least at medium rpm. Increase of the revolution speed to its maximum revealed similar results energy-wise for reactions in 250 and 500 ml beakers. It should be pointed out that neither E_{brutto} nor E_m are identical to the (medium) stress energy outlined in Eqs. 2 and 3.

$$E_m = \frac{E_{brutto}}{n_{initial,i} \cdot Y} \tag{9}$$

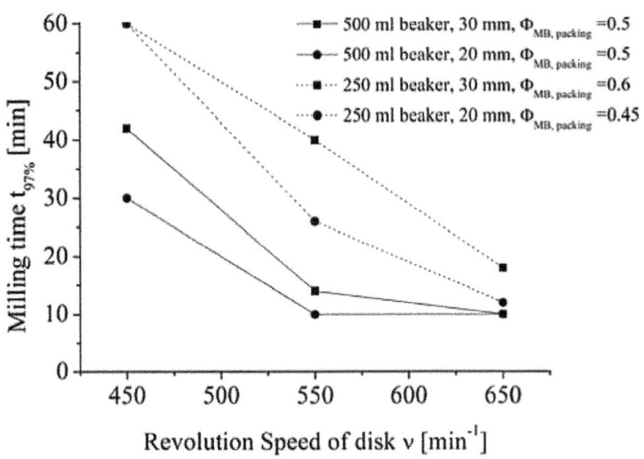

Fig. 6 Influence of beaker size, $V_{B,total}$, and revolution speed of the sun disk, rpm, on the time to reach quantitative yield, $t_{97\%}$, for experiments in the large-scale ball mill (**P6-250-100**, **P6-500-200**). For an indication of the experimental conditions see Table 1 in the Experimental section and the explanations provided.

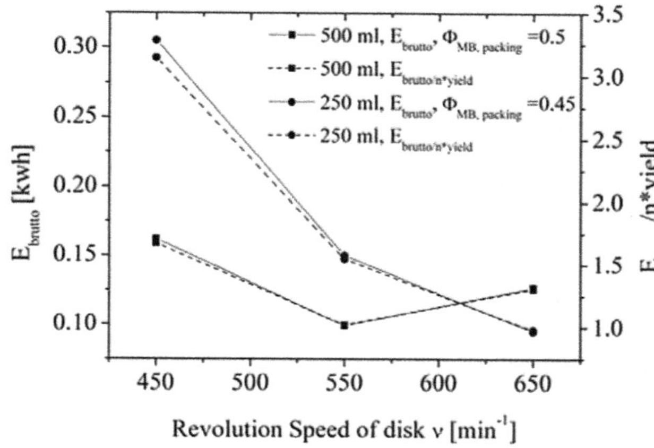

Fig. 7 Influence of beaker size, $V_{B,total}$, and revolution speed of the sun disk, rpm, on the energy consumed for experiments in the large-scale ball mill (**P6-250-100**, **P6-500-200**). For an indication of the experimental conditions see Table 1 in the experimental section and the explanations provided.

Conclusion

The aim of the present study was to investigate the influence of technological and process parameters relevant to the scale-up of an organic model reaction carried out in planetary ball mills. As a model reaction, the Knoevenagel condensation of vanillin and barbituric acid was chosen, since the substrates are available in high quantity at reasonable costs and sufficient purity, the reaction is selective and no other reagents or additives were required, and the reaction progress can be followed with analytical methods which avoid possible post reactions in solution.

Several parameters have been investigated. Generally, it was demonstrated that the combination of at least two parameters leads to interesting insights to their relationship. It was possible to scale up the model reaction from 20 to the 300 mmol-scale which resulted in scale-up factors of 15 and 11 with respect to scale-up of the batch size and the reactor volume, respectively. The experiments demonstrated that the increase of the batch size as well as the reactor volume did not necessarily lead to an increase of the reaction time. Especially, the variation of the milling ball diameter, d_{MB}, the rotation frequency, rpm, and the filling degree of the milling beakers considering the ball packing, $\Phi_{MB,packing}$, and the substrates, Φ_{GS}, allowed fine-tuning of the reaction conditions. With respect to d_{MB} it was found that small milling balls are advantageous, since the number of stress events is considerably increased. Contrary to comminution processes, the stress energy seemed to be less important. This allows us to conclude that the reaction is rather thermally-initiated than real mechano-chemical initiation. However, the influence of rpm exemplifies the importance of the mechanical processes in order to increases the system's kinetic energy, *i.e.* the system's temperature, and the homogeneity of the heterogeneous solid–solid mixture. Experiments regarding the influence of the filling degree revealed, that $\Phi_{MB,packing}$, should be in the range of 0.5, which is identical to an overall milling ball volume of roughly 25% of the total beaker volume. With respect to Φ_{GS} it was found that up to 50% of the beaker's total volume can be taken up by the substrates without a disadvantageous effect on the reaction yield.

Acknowledgements

This study was carried out within the project RESPEKT (Resource efficient chemical synthesis – process development of solvent-free reactions in ball mills) funded by the Deutsche Bundesstiftung Umwelt (DBU; AZ 29622-31). We would like to thank our project partners – Fritsch GmbH, C. F. Burmeister, S. Breitung-Faes and A. Kwade from IPAT (Technical University Braunschweig) and Zoz Group – for fruitful discussions.

Notes and references

1 A. Stolle, T. Szuppa, S. E. S. Leonhardt and B. Ondruschka, *Chem. Soc. Rev.*, 2011, **40**, 2317–2329.
2 S. L. James, C. J. Adams, C. Bolm, D. Braga, P. Collier, T. Friščić, F. Grepioni, K. D. M. Harris, G. Hyett, W. Jones, A. Krebs, J. Mack, L. Maini, A. G. Orpen, I. P. Parkin, W. C. Shearouse, J. W. Steed and D. C. Waddell, *Chem. Soc. Rev.*, 2012, **41**, 413–447.
3 P. A. May and J. S. Moore, *Chem. Soc. Rev.*, 2013, **42**, 7497–7506.
4 V. Sepelak, A. Duvel, M. Wilkening, K.-D. Becker and P. Heitjans, *Chem. Soc. Rev.*, 2013, **42**, 7507–7520.
5 S.-E. Zhu, F. Li and G.-W. Wang, *Chem. Soc. Rev.*, 2013, **42**, 7535–7570.
6 D. Braga, L. Maini and F. Grepioni, *Chem. Soc. Rev.*, 2013, **42**, 7638–7648.
7 G.-W. Wang, *Chem. Soc. Rev.*, 2013, **42**, 7668–7700.
8 E. Boldyreva, *Chem. Soc. Rev.*, 2013, **42**, 7719–7738.
9 In the *IUPAC Compendium of Chemical Technology, 2nd edn* (the "*Gold Book*"), ed. A. D. McNaught and A. Wilkinson, Blackwell Scientific Publications, Oxford, 1997.
10 F. Schneider, T. Szuppa, A. Stolle, B. Ondruschka and H. Hopf, *Green Chem.*, 2009, **11**, 1894–1899.
11 C. Suryanarayana, *Prog. Mater. Sci.*, 2001, **46**, 1–184.
12 B. Rodríguez, A. Bruckmann, T. Rantanen and C. Bolm, *Adv. Synth. Catal.*, 2007, **349**, 2213–2233.
13 C. F. Burmeister and A. Kwade, *Chem. Soc. Rev.*, 2013, **42**, 7660–7667.
14 G. Kaupp, J. Schmeyers, M. R. Naimi-Jamal, H. Zoz and H. Ren, *Chem. Eng. Sci.*, 2002, **57**, 763–765.
15 G. Kaupp, *CrystEngComm*, 2006, **8**, 794–804.
16 C. Maheshwari, A. Jayasankar, N. A. Khan, G. E. Amidon and N. Rodríguez-Hornedo, *CrystEngComm*, 2009, **11**, 493–500.
17 T. Friščić, A. V. Trask, W. Jones and W. D. S. Motherwell, *Angew. Chem., Int. Ed.*, 2006, **45**, 7546–7550.
18 T. Friščić and L. Fábián, *CrystEngComm*, 2009, **11**, 743–745.
19 T. Friščić, S. L. Childs, S. A. A. Rizvi and W. Jones, *CrystEngComm*, 2009, **11**, 418–426.
20 T. Friščić, D. G. Reid, I. Halasz, S. R. S., R. E. Dinnebier and M. J. Duer, *Angew. Chem., Int. Ed.*, 2010, **49**, 712–715.

21 F. C. Strobridge, N. Judăs and T. Friščić, *CrystEngComm*, 2010, **12**, 2409–2418.
22 F. Bernhardt, R. Trotzki, T. Szuppa, A. Stolle and B. Ondruschka, *Beilstein J. Org. Chem.*, 2010, **6**, 7.
23 A. McCluskey, P. J. Robinson, T. Hill, J. L. Scott and J. K. Edwards, *Tetrahedron Lett.*, 2002, **43**, 3117–3120.
24 G. Kaupp, M. Reza Naimi-Jamal and J. Schmeyers, *Tetrahedron*, 2003, **59**, 3753–3760.
25 G. Kakade, B. Madje, M. Ware, R. Balaskar and M. S. Shingare, *Org. Chem.: Indian J.*, 2007, **3**, 104–106.
26 K. Mogilaiah, H. S. Babu, K. Vidya and K. S. Kumar, *Indian J. Chem., Sect. B: Org. Chem. Incl. Med. Chem.*, 2010, **49B**, 390–393.
27 M. A. Pasha and K. Manjula, *J. Saudi Chem. Soc.*, 2011, **15**, 283–286.
28 J. Schmeyers, F. Toda, J. Boy and G. Kaupp, *J. Chem. Soc., Perkin Trans. 2*, 1998, 989–994.
29 G. Rothenberg, A. P. Downie, C. L. Raston and J. L. Scott, *J. Am. Chem. Soc.*, 2001, **123**, 8701–8708.
30 G. Kaupp, J. Schmeyers and J. Boy, *Chemosphere*, 2001, **43**, 55–61.
31 A. Naqvi, M. Shahnawaaz, A. V. Rao, D. S. Seth and N. K. Sharma, *E-J. Chem.*, 2009, **6**, S75–S78.
32 J. T. Li, H. G. Dai, D. Liu and T. S. Li, *Synth. Commun.*, 2006, **36**, 789–794.
33 C. S. Reddy, A. Nagaraj and P. Jalapathi, *Indian J. Chem.*, 2007, **46B**, 660–663.
34 R. Trotzki, M. M. Hoffmann and B. Ondruschka, *Green Chem.*, 2008, **10**, 767–772.
35 A. Ilangovan, S. Muralidharan and S. Maruthamuthu, *J. Korean Chem. Soc.*, 2011, **55**, 1000–1006.
36 S. Rosenkranz, S. Breitung-Faes and A. Kwade, *Powder Technol.*, 2011, **212**, 224–230.
37 M. Becker, A. Kwade and J. Schwedes, *Int. J. Miner. Process.*, 2001, **61**, 189–208.
38 F. Schneider, A. Stolle, B. Ondruschka and H. Hopf, *Org. Process Res. Dev.*, 2009, **13**, 44–48.
39 W. Qiu and T. Hirotsu, *Macromol. Chem. Phys.*, 2005, **206**, 2470–2482.
40 T. Szuppa, A. Stolle, B. Ondruschka and W. Hopfe, *Green Chem.*, 2010, **12**, 1288–1294.
41 T. Szuppa, A. Stolle, B. Ondruschka and W. Hopfe, *ChemSusChem*, 2010, **3**, 1181–1191.
42 R. Schmidt, A. Stolle and B. Ondruschka, *Green Chem.*, 2012, **14**, 1673–1679.
43 B. Rodríguez, A. Bruckmann and C. Bolm, *Chem.–Eur. J.*, 2007, **13**, 4710–4722.
44 S. Wada and H. Suzuki, *Tetrahedron Lett.*, 2003, **44**, 399–401.
45 Z. Zhang, Y.-W. Dong, G.-W. Wang and K. Komatsu, *Synlett*, 2004, 61–64.
46 E. M. C. Gérad, H. Sahin, A. Encinas and S. Bräse, *Synlett*, 2008, 7202–7204.
47 H. Watanabe, E. Matsui, Y. Ishiyama and M. Senna, *Tetrahedron Lett.*, 2007, **48**, 8132–8137.
48 P. R. Patil and K. P. R. Kartha, *J. Carbohydr. Chem.*, 2008, **27**, 279–293.
49 A. Kwade, *Int. J. Miner. Process.*, 2004, **74**, S93–S101.
50 S. Breitung-Faes and A. Kwade, *Powder Technol.*, 2011, **212**, 383–389.
51 G. Kakuk, I. Zsoldos, Á. Csanády and I. Oldal, *Rev. Adv. Mater. Sci.*, 2009, **22**, 21–38.

52 P. Baláz, in *Mechanochemistry in Nanoscience and Minerals Engineering*, Springer-Verlag, Berlin, 2008, pp. 103–132.
53 P. Baláz and E. Dutkova, *Miner. Eng.*, 2009, **22**, 681–694.
54 S. Bernotat and K. Schönert, in *Ullmann's Encyclopedia of Industrial Chemistry*, VCH, Weinheim, 1998, vol. B2, pp. 1–39.
55 L. Takacs, *Prog. Mater. Sci.*, 2002, **47**, 355–414.

Faraday Discussions

DISCUSSIONS

Mechanistic understanding, catalysis and scaling up of mechanochemistry: General discussion

DOI: 10.1039/C4FD90007C

Dr Maini opened the discussion of the paper by Ivan Halasz: Thank you for your pioneering work on *in situ* and real time characterization of mechanochemistry reactions. Regarding the instrumentation set-up, is it possible to focus the X-ray beam only on one side to avoid the splitting of the profile patterns?

Dr Halasz replied: I don't believe focusing of the X-ray beam is possible. What could be possible is to pass the X-rays at the very bottom of the internal chamber of the milling vessel in order to avoid diffraction from two sample parts sitting on the opposite walls of the vessel. However, since we have been able to model the split peaks at higher angles, we considered this approach to be sufficiently good. On the practical side, if the X-rays were passed through the very bottom of the internal chamber of the vessel, we would need more time for precise positioning of the vessel relative to the X-ray beam in consecutive experiments.

Dr Maini asked: The quantification of the amorphous is quite tough also under normal conditions and it is an indirect measurement. In your diffraction pattern you describe only the patterns of ZnO, silicon and ligand (which is almost zero). How do you take into account the contribution of the salt? Can the movement of the jar create problems in the description of the peaks and the final quantification of the amorphous state?

Dr Halasz responded: We did not see any diffraction signal from the added salt. This may be due to the salt becoming amorphous, dissolving in added ethanol, reacting with ZnO and/or imidazole. We note also that the amount of the added salt was less then 20 mg and it could have been insufficient to yield an interpretable diffraction signal after it had been distributed throughout the reaction vessel. In general, the quality of the diffraction images is not particularly good. Jar movement is certainly something that does not help the quality of the data but this effect is minor. More important is the size of the sample and the fact that it was distributed on opposite walls of the reaction vessel giving rise to split peaks at higher angles. To answer the question, I don't think the jar movement causes problems for peak-shape description and quantification.

Professor Takacs asked: Have you considered using several small balls rather than one or two relatively large ones? They may result in more uniform powder inside the mill and in shorter blanking of the X-ray beam.

Dr Halasz responded: This is a very good idea. We haven't tried this yet but we will in the future. Even if the blanking of the X-ray beam might not reduce with many small balls, we may achieve a more uniform blanking. This is well worth trying we believe.

Professor Takacs remarked: The values of the amorphous fraction are rather large, but only obtained indirectly. It would be useful to validate the results by conventional *ex situ* XRD results.

Dr Halasz replied: The weight fraction of the amorphous phase at around 30%, as obtained from *in situ* measurements, does seem to be rather large. We have reproduced the experiments and performed quantification on data collected *ex situ* on a laboratory diffractometer. The liquid assisted grinding (LAG) experiment showed 12% of amorphous phase (see figure) while the ILAG experiment was inconclusive since there appeared another crystalline phase which we were not able to identify (see Figure). However, this shows how incomparable the results from *in situ* and *ex situ* measurements are. For the *ex situ* analysis, the sample was dried from ethanol and was necessarily aged between milling and data collection. The mixture composition could thus have changed due to ethanol evaporation (what was dissolved in ethanol could have crystallized) and crystallization of what could have been the amorphous ZIF. We therefore feel that the discrepancy between *in situ* and *ex situ* quantification currently does not undermine the validity of *in situ* obtained data. But yes, the validity of *in situ* data needs to be further validated before it could be considered correct.

Professor James asked: I don't understand the method for determining the amount of amorphous material; can you quantify the amount of amorphous zinc oxide present?

Dr Halasz answered: The method where an internal diffraction standard is used (here it was silicon) determines the amount (weight fraction) of the total amorphous material in the mixture. It does not determine any specifics of the composition of the amorphous phase. The amount of amorphous zinc oxide could possibly be determined indirectly from the data about the amount of remaining crystalline zinc oxide and the amount of formed ZIF but this was not attempted in this work.

Professor James asked: In our recent work on Raman monitoring we used *ex situ* analysis; which is necessarily a stop/start process so it is important to ensure that the reaction doesn't continue when milling has stopped. The overall finding was that the kinetics were second order which differs from your results. Does your reaction continue when the milling is stopped?

Dr Halasz responded: Our reaction seems to continue after milling is stopped. When we have tried to perform quantification of the reaction mixture by collecting PXRD data on a laboratory diffractometer, we observed different compositions. This strongly suggests that the reaction mixture had changed by aging before the patterns were collected. If for no other reason, ethanol that was used as the liquid additive during liquid-assisted grinding must have evaporated, and what was dissolved in it could have crystallized.

Besides this, we don't have a clear explanation for the difference in kinetic behaviour between your and our work but we must bear in mind that the reaction conditions in these two studies were different (different liquids and different ratios of the liquid volume and mass of solid reactants).

Professor James commented: If under your conditions the reaction continues when the milling is stopped then it seems to be inherently kinetically different to ours; perhaps because a larger amount of solvent is present so thermal diffusion may play a greater role. In our system temperature had no effect on the rate, which was puzzling at first but is understandable since the rate of collisions between particles is independent of temperature, it depends mainly on the speed of the milling.

Dr Belenguer addressed Dr Friščić and Dr Halasz: Do you agree that we are not talking about amorphous *vs.* crystal? Amorphous is defined as anything that does not diffract. On the other hand a "pre-crystalline" nucleus that brings the structural arrangement of the product with itself may also not diffract. The amorphous-to-diffraction content includes anything not-crystalline-to-diffraction in the milling process, which means the amorphous-to-diffraction reactant content, the amorphous-to-diffraction intermediate state, and the amorphous-to-diffraction product content, and the solvent (in the case of LAG). Do you think all these amorphous contents can be defined as one single phase? Or are they rather several different phases, all amorphous-to-diffraction?

In our work (solid-state dynamic covalent chemistry) the sample does not lose the information of the crystal structure during the solid state reaction even if it is quite plausible that our reaction also undergoes a transition where the sample would not diffract, and so can be labelled as an amorphous phase. Amorphous is defined as anything that does not diffract and it encompasses so many things that cannot be seen by XRD. There must be a moment/stage in a grinding reaction that allows for complete "phase information formatting". I mean: there is a moment in which the molecule/component separates from its original reactant crystal "x" (or crystal surface "x") and loses its information about it, so that it can contribute to the formation of product crystal "y" (or crystal nucleus "y", or crystal surface "y"). Whether the amorphous content detected in Halasz and coauthors' work represents this not-anymore-"x"-but-not-yet-"y" intermediate state is doubtful. Can you comment?

Dr Friščić replied: Understanding of the nature of the amorphous phase, specifically its structure, can be accomplished by X-ray diffraction using the pairwise distribution function (PDF) analysis. Work related to structure analysis of an amorphous metal–organic framework after milling has been reported by Cao, Bennett, Keen, Goodwin and Cheetham (*Chem. Commun.*, 2012, **48**, 7805) and, in principle, I imagine similar procedures could be applied for all sufficiently simple mechanochemically obtained amorphous systems.

Professor Jones asked: It would be interesting to see how the patterns you obtain at each stage of the experiments compare with the standard intensity *vs.* two-theta plots that one normally sees in similar papers. I say this because it would be useful to know what impact a 35% amorphous component might have on the appearance of the pattern.

Dr Halasz responded: We have repeated the experiments in the laboratory in order to collect the standard laboratory X-ray diffraction patterns. This is elaborated in the answer to Question 413 and the corresponding PXRD patterns have been attached. The ILAG synthesis was inconclusive since an unidentified crystalline phase appeared but the LAG experiment did show *ca.* 12% of an amorphous content. The LAG pattern does not exhibit the obvious "amorphous hump" recognisable by the naked eye. Smaller weight fraction of the amorphous phase in the *ex situ* obtained pattern may be due to crystallisation of the amorphous phase upon standing between sampling and data collection and is surely due to evaporation of ethanol which was used as the liquid additive. However, the pattern not exhibiting a characteristic "amorphous hump" indicates that the amorphous phase may be ubiquitous in milling experiments where quantification by PXRD was not performed by the internal standard method. We believe it is very difficult and tricky to compare *ex situ* and *in situ* data due to the inability to establish the extent the sample may be changed between sampling and *ex situ* analysis.

Dr Stolle said: Please comment on the general applicability of the method to other reactions under ball milling conditions.

Dr Halasz answered: The applicability of *in situ* monitoring by diffraction is not limited to specific systems or specific types of material. One might assume that one would need atoms of heavier elements to obtain an interpretable scattering signal. Of course, having, *e.g.*, metals in the system should help with diffraction but we have been able to measure also purely organic reactions with reactants having oxygen as the heaviest atom. Another issue is the reaction mixture amorphization. In this case, organic materials may be more difficult to monitor because they might more readily become amorphous. But this is a general limitation of diffraction rather then the limitation of the presented *in situ* diffraction. A possible way around would be to monitor evolution of the pair distribution function with the time of milling. This could possibly provide insight into changes in the amorphous phase. We plan to explore this possibility and perform these investigations in the future.

Ms Tröbs asked: How stable are your milling jars? How many milling cycles are you using one milling jar for? Since you are analysing your reactions quantitatively, how can you make sure that the grinding balls do not influence your measurements too much?

Dr Halasz replied: We have been using the same jar in typically 10 to 20 milling experiments each lasting 30 minutes. After that, wear on the inside of the jar could be noticed by the naked eye in the sense that the jars would become blurry. Regarding the influence of the balls, this is an open question. Balls will absorb incident radiation but this will not effect the final measured pattern because this will only decrease the radiation intensity impinging on the sample. However, should the balls absorb already scattered radiation, this could affect relative intensities at different 2theta directions. This effect is difficult to estimate, we believe it is negligible since our 2theta range is very small (form 0.5 to 4 degrees). This means that, should the scattered radiation pass through the ball, its attenuation would be very similar for all directions. Therefore, the assumption that the balls do not alter relative intensities, which is essential for quantification, most likely holds. Thus far, we haven't devised any means to avoid balls in the beam.

Dr Maini remarked: Have you validated your results on the quantification of the amorphous phase with other measurements or technique also *ex situ*?

Dr Halasz replied: This is a difficult question because *ex situ* analysis is not directly comparable to *in situ* measured data. Namely, the sample can change between sampling and *ex situ* analysis. If for no other reason, the material which may have been dissolved in ethanol (which was here used as the grinding liquid), will likely crystallise upon ethanol evaporation. We have performed *ex situ* analysis by X-ray diffraction on LAG (using ethanol) and ILAG (using ethanol and ammonium nitrate) reactions (see figures). The LAG reaction showed a different outcome as compared to *in situ* data: the ZnO was completely consumed as opposed to *ca.* 1/3 of it remaining during *in situ* measurements. The *ex situ* analysis showed *ca.* 12% of amorphous phase present as opposed to *ca.* 30% form *in situ* data. It is impossible to firmly state whether the difference is caused entirely due to the sample being changed after milling was stopped and *ex situ* data have been collected or if the *in situ* collected diffraction patterns do not yield accurate information. We believe the former to be more likely.

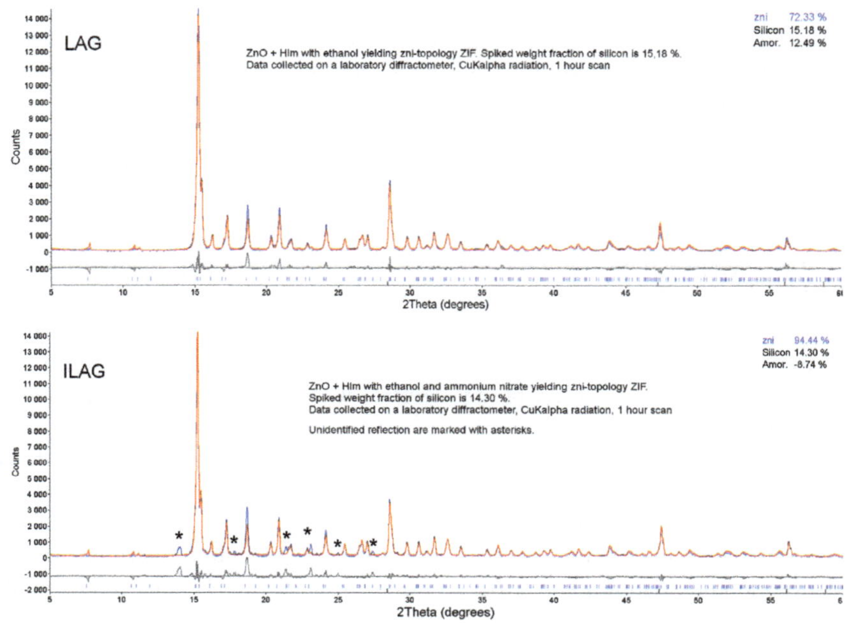

Fig. 1

Ex situ analysis of ILAG reaction was inconclusive since an unknown crystalline phase appeared not allowing for quantification by powder diffraction (see Fig. 1).

To validate the diffraction data, we would need another independent *in situ* technique capable of quantification of the amorphous content.

Dr Maini asked: The measurements start some time after the milling starts. This stand by time is needed to homogenise the reacting mixture. Have you tried to homogenise your starting mixture before introducing it in the vessel?

Dr Halasz answered: This is a very good idea that would certainly help in obtaining a homogeneous reaction mixture sooner after the milling process is initiated. Thus far, we wanted to avoid any reaction before the milling was started. In the future, we shall attempt experiments with reactants premixed as well.

Dr Belenguer addressed Dr Friščić and Dr Halasz: In Figure 5 and 7 the product reaches a plateau between 3 and 4 minutes into the reaction. Do your *in situ* measurements support that ball mill grinding reactions are driven to equilibrium and hence governed by thermodynamics? In your paper the milling kinetic curves of Fig. 4, 5 and 7 reach a plateau in the end. These plateaus have been seen in other systems such as mine. Is this a thermodynamic equilibrium? If not, why and what is it?

Dr Friščić answered: There are few known and well-studied cases of mechanochemical equilibrium reported in the literature. Some of these include the classical work on calcite–aragonite equilibration addressed by a number of papers (*e.g.*, Schrader and Hoffmann, *Z. Anorg. Allg. Chem.*, 1969, **369**, 41; Isobe and Senna, *Faraday Trans.*, 1, 1988, **84**, 1199), reversible Diels–Alder reactions of fullerenes (Murata, Kato, Fujiwara and Komatsu, *J. Org. Chem.*, 1999, **64**, 3483) and base-assisted equilibration of disulfide bonds (Belenguer, Friščić, Day and Sanders, *Chem. Sci.*, 2011, **2**, 696). In the systems addressed herein I do not believe thermodynamic equilibration plays a major role as these are highly driven reactions involving the neutralization of an oxide to form water.

Dr Halasz added: I don't believe we have enough information at present to speculate on thermodynamic or kinetic product formation in this case.

Dr Belenguer then communicated: In the previous related qualitative paper (Friščić *et al.*, *Nat. Chem.*, 2013, **5**, 66), the *in situ* kinetic studies demonstrated that the ball mill grinding process transformed the microsize crystals of the starting materials to nanosized crystals of the product at equilibrium. The equilibrium domain size value at the end of the experiment is reported to be in the order of tens of nm.

Nanosize crystallites have higher enthalpies and free energies than large crystals because of a positive surface energy. Furthermore in contact with a solvent as in LAG, the particle surfaces are likely to be solvated. At this nanosize level the surface/volume ratio will be high. Thus, while thermodynamics conventionally assumes the surface effects to be negligible (*i.e.*, infinite bulk structures), this is not the case in nanosized particles such as continuously mechanically ground systems. The free energy contribution of the surface solvated structure to the stability of a certain polymorph of the product is significant and can determine which polymorph is formed, as in the system I studied. These thermodynamic aspects are general, and should apply to any milling system independent of the mechanisms involved in the chemical reactions.

Professor Boldyreva also communicated: My congratulations with this very interesting and promising work. I have a minor comment and a proposal. The comment is that in some cases it was also possible to monitor similar trends in accumulation of the products of mechanochemical reactions *ex situ*, if the treatment was interrupted step after step (see I. A. Tumanov, A. F. Achkasov, E. V. Boldyreva, and V. V. Boldyrev, Following the products of mechanochemical synthesis step by step, *CrystEngComm*, 2011, **13**(7), 2213–2216; E. A. Losev and E. V. Boldyreva, The role of a liquid in "dry" co-grinding: A case study of the effect of water on mechanochemical synthesis in a "L-serine–oxalic acid" system, *CrystEngComm*, 2014, **16**(19), 3857–3866). A proposal is to try to use the possibility of focusing synchrotron beam at different parts of the milling jar in order to follow the formation of different products at these different parts (see the contribution by A. Michalchuk to this Discussion). I suggest that we try to do this as a joint project.

Dr Blair opened the discussion of the paper by Richard Blair: Movies of the DEM models discussed are available in the online supplementary materials at: http://pubs.rsc.org/en/content/articlelanding/2014/fd/c4fd00007b#!divAbstract

Professor Galembeck asked: You determine the transformation of cellulose into sugars by measuring the amount of dissolved material. Do you have information on the composition of the solutes, are they just glucose and its oligomers?

Dr Blair responded: There are interesting compositional differences depending on the milling mode. When we use a SPEX mixer mill we analyzed the material through silanization and GC/MS analysis. No oligomers greater than $n = 2$ were observed. Over milling can favor dehydration and retro-aldol products such as levoglucosan, levoglucosenone, 5-hydroxymethylfurfural, and furfural. Schuth *et al.* used a small planetary mill and observed a wider range of oligomers.

N. Meine, R. Rinaldi, and F. Schüth, Solvent-Free catalytic depolymerization of cellulose to water-soluble oligosaccharides, *ChemSusChem*, 2012, **5**(8), 1322.

Professor Suslick commented: From your data, the way in which you avoid the long induction time is to scale down the reaction dramatically. Presumably a higher loading means fewer collisions within the mill.

Dr Blair responded: Actually, we moved to a smaller mill. We kept the media : reactant loading ratio the same for all scales. The "induction" period with the larger attritor was reduced by running the mill using operational parameters different from the manufacturer's recommendation. Additionally, the use of a milling aid like magnesium carbonate reduces the physical attrition time that I think is responsible for the induction time. There are more collisions when larger

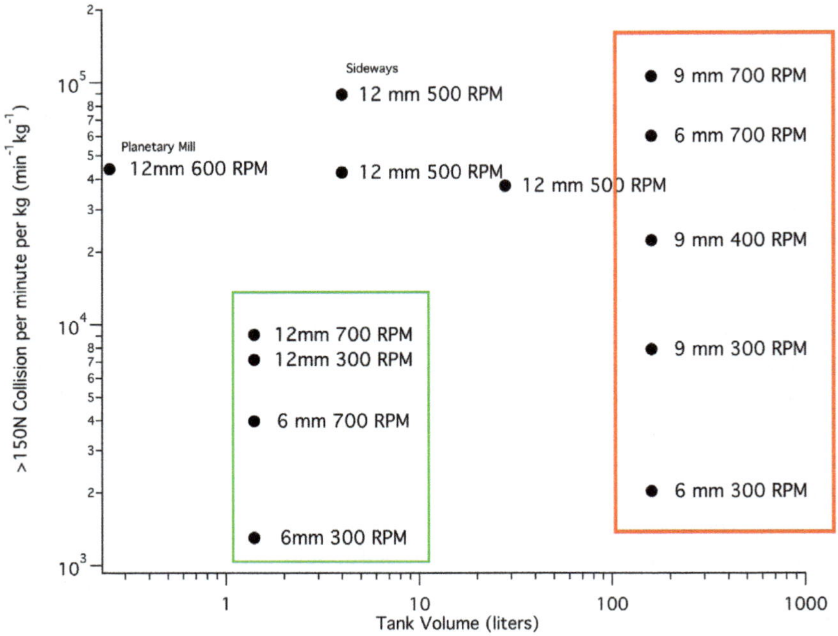

Fig. 2

mills are used. A more accurate parameter is collisions kg^{-1} of reactant. High energy collisions are needed to induced hydrolysis. See Fig. 2 that shows expected values. You can see the 160 L tank should provide energetics and conversion efficiencies similar to that seen in a planetary mill.

Professor Suslick remarked: Energy efficiency is rather critical here because cellulose is a low value added material. Conversion of cellulose has to be extremely cheap: you don't get to use many kWh.

Dr Blair replied: Profitability can be realized in several ways. Biomass will hydrolyze to hexoses, pentoses, and lignins. If the hexoses can be produced and sold at 100/MT there is a market for them. The pentoses (xylose currently can be had for 1000/ton) are higher profit items. Of course, a large cheap source may exceed demand. In that case, there are organisms that will ferment pentose in the absence of hexoses. We have identified a catalyst that allows depolymerization of lignin as well. Of course, the market for these compounds has not been fully exploited and the old adage "you can make everything from lignin, but money" is relevant. Finally, we are working on a new process to take hexoses directly to commodity chemicals. Currently, we have had success in producing isopropanol from glucose. Other compounds such as dihydroyacetone and glyceraldehyde offer the chance for improved profitibility. The question that needs to be answered is how energetic must the media collisions be to induce hydrolysis. Additionally, process optimization can be used to reduce milling time. There is a cost/profit sweet spot that can be obtained by running the hydrolyis to partial completion. It appears that hydrolysis itself is rapid. If we can decouple hydrolysis from attrition (perhaps separate mills as was discussed in the meeting) we may be able to realize significant energy savings. Finally, use of attrition technology is the first step toward at scale operation. As I stated, there are other technologies that purport high impact energies with much lower energy consumption.

Mr Margoutidis commented: It has been recently demonstrated by researchers working in the field of cellulose depolymerization that glucose yield and selectivity can be controlled by altering the microwave power density/distribution (*J. Am. Chem. Soc.*, 2013, **135**, 11728). Considering the interest for increasing the energy yield of mechanochemical transformations (paper 21), do you think that studying the rpm parameter (for selected types of milling) might reveal an energy minimum for the maximum percent of soluble saccharides obtained in your experiments? If yes, which additional parameter (solid acid catalysts, force of collisions manipulation by ball diameter, packing degree of vial, *etc.*) plays the most critical role in selectivity of the products obtained by mechanical processing of polysaccharides? Could the weakening of the intermolecular hydrogen bond network by just grinding (no additive) reveal a possible role of CH_2OH groups in the cleavage of the chains and hopefully approach a more random (rather than end only) depolymerization?

Dr Blair responded: We have found that conversion rates are directly proportional to the compressive force and number of collisions per unit time per kilogram of material (Fig. 3). In general, the faster the milling the faster the kinetics. I am sure at very high milling speeds dehydration reactions would dominate and the desired product (saccharides) would diminish. For a given mill the media size

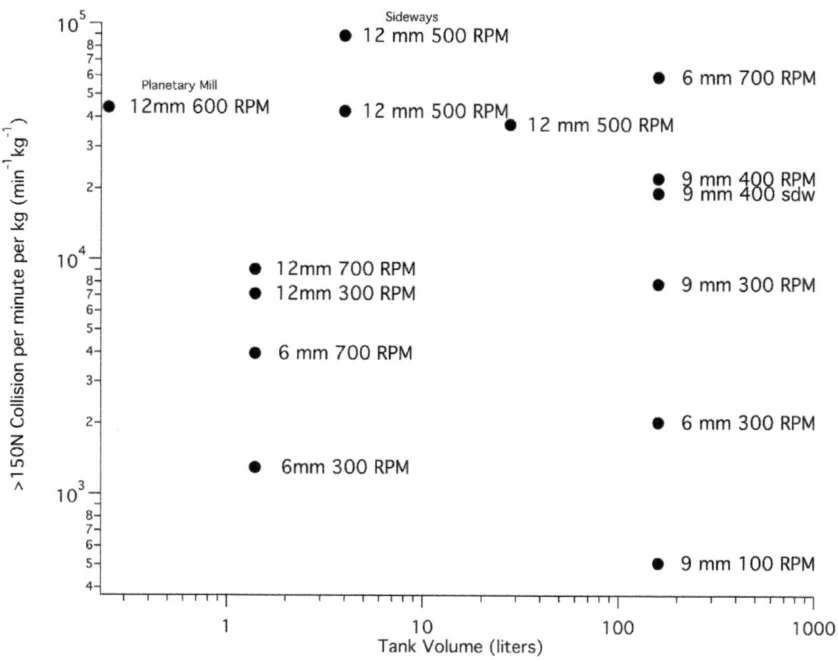

Fig. 3

plays the most important role in selectivity. Larger sizes produce increased compressive forces which can lead to unwanted dehydration if not properly monitored. Solid acid catalysts are important as well, as very high surface acidities will also favor dehydration over hydrolysis.

Grinding alone seems to disrupt the crystallinity of the feedstock, but no measurable hydrolysis was observed. The acidity of the CH_2OH proton is low and this is consistent without observations that a certain level of surface acidity is needed for the reaction to proceed.

Dr Stolle commented: How have mechanical properties of the mill feed been integrated in the DEM-models?

Dr Blair responded: If you mean introduction of material into the mill as in a continuous or flow process, it is possible to include this in the model, but we have limited our current models to batch systems.

If you mean the amount of reagent/feed/material in the mill during the milling then we have approached this in two ways:

1) Solid material has been modeled as solid spheres much smaller than the milling media. The model can be improved by using a set of spheres bound by a force model in place of each single sphere of material. We have not implemented that yet. This approach is computationally intensive.

2) Liquid material is more easily modeled by changing the coefficient of restitution of the material impacts (steel/steel, WC/steel, *etc.*). There is a relationship between the viscosity of the liquid and the coefficient of restitution.

G. G. Joseph, R. Zenit, M. L. Hunt and A. M. Rosenwinkel, Particle–wall collisions in a viscous fluid, *J. Fluid Mech.*, **433**, 329.

Dr Baláž asked: In your presentation, you showed a plot, in which a different number of collisions in different kinds of mills was significant. Could you please comment more precisely in which kind of mills the number of collisions is the highest and the other way round?

Dr Blair replied: Its a balance between number of collisions and forces generated by each collision. Large mills with small media will produce a lot of collisions, but the impact forces will be quite low. Larger media will increase the collision force, but decrease the total number of collisions. A good measure is to think about collisions > certain force per minute per mass of process material. This takes into account ball to powder ratio, impact forces, and number of collisions. With constant size media, collisions scale with reactor volume (see attached figure). Across mill types, using collision min^{-1} kg^{-1} its SPEX > planetary > or = attritor (in certain configurations) > pebble mill.

Professor Takacs remarked: There is a series of high-energy horizontal attritors, called the "Simoloyer" manufactured by Zoz GmbH that is scalable from 0.5 to 900 liters. Much was also published on the working of this mill. A good place to start exploring this equipment is the file:

http://www.zoz-group.de/zoz.main/pdf_content/publications/v31.pdf

Dr Blair replied: Thank you, I have heard of them. Their website is a bit odd. They may well have a solution to the energy/scale demands of out process.

Professor Mack addressed Dr Blair: When looking at your DEM models what is the best way to increase the compressive force of a Spex mill?

Dr Blair replied: Increasing the density of the balls is the simplest. Tungsten carbide has almost twice the density and will immediately double the impact forces without changing the motion. Increasing ball size is the next easiest, but mixing will be affected by the increased volume of the media. Increasing milling intensity works to an extent, but there is not a linear relationship between mill speed and impact forces. Doubling the mill speed does not double the impact forces (more like 1.5x).

Professor Baláž asked: The presented mills are of a batch mode. Is it possible to utilize the mills also in a continuous mode?

Dr Blair responded: Yes, attrition mills can be run in continuous mode.
http://www.unionprocess.com/tech_papers/dry_milling.pdf
However, we have not explored continuous processes yet. We want to fully understand the process in batch mode before adding the complication of a continuous process.

Professor Jones asked: In one of your slides you showed a number of large attrition mills. I wonder if you could comment on the feasibility of connecting such a bank of mills in series so that the feed of one is fed into the next and so on. If such is possible do you know if, for example, it has been possible to deliberately aim to have gradual and controlled particle size reduction or indeed physical and/or chemical conversion as the material is passed from mill to mill.

Dr Blair responded: It is indeed possible to feed one mill with the product of another. Stirred media mills have exit ports, material can be conveyed through this to another mill. In principle it should work and I am sure this has been implemented by the mining industry. It is a very good idea that warrants further study.

Mr Jörres commented: How do you get out material/product in the case of multi-kilogram scale reactions (vessels were shown in your talk) and how do you work up these reactions.

Dr Blair answered: Product is removed in one of two ways. The attritor tree can be raised out of the tank and the tank can be tilted *via* a gear mechanism (see Fig. 4) to dump everything. The processed material is a fine power and can be sieved to recover from the media. This can be time consuming. Alternatively, there is a pneumatic gate valve (green circle in image) on the bottom of the tank with a screen. The valve can be opened and the tree turned at a low speed (50–100 rpm). The product will sift out and the media will remain behind. Soluble sugars are removed by aqueous extraction. The catalyst and unreacted material is recovered by centrifugation such as a Sharples centrifuge or similar technology.

Fig. 4

The aqueous fraction can continue for C5 and C6 separation, or fermentation. The solid residue can be dried and returned to the reactor with additional feedstock. At a certain point lignin content will be too high for cost effective recovery of sugars. We are currently investigating three strategies for lignin removal. This is in conjunction with a licensee so I cannot go into detail on that step.

Professor Mack opened the discussion of the paper by Dominick Daurio: Since the twin screw extruder is so large in scale, how do you conduct the R&D of a system to know if the twin screw extruder would be a good match?

Dr Nagapudi replied: For R&D puposes, something like a minilab would be ideal. However, scaling from the minilab is not straightforward. It is only good for proof of concept.

Professor Suslick addressed Dr Nagapudi: The mechanochemical forces needed to make molecular co-crystals is relatively modest. Can you use the same extruder for harder materials?

Dr Nagapudi replied: This may be challenging for harder materials. However we have not attempted any studies with these types of materials in our labs.

Professor Suslick commented: Can you give us a sense of the order of magnitude cost for a small twin-screw extruder unit?

Dr Nagapudi answered: Lab scale extruders start at around 75 000. However typical units used in this study can run up to 200–300 000.

Professor Galembeck said: Reactive processing in the plastics industry has been a great success and extruders are fitted with injection and venting ports as well and co-extrusion is also common. Do extruders used in the pharma area have similar features, which ones are most important?

Dr Nagapudi said: Yes, extruders in the pharma area are the same ones as that being used in the plastics world.

Professor Galembeck commented: You mentioned problems related to abrasion and wear. I believe that extruders used in the pharma area are not operated at very high temperatures, as happens with polymers and other materials. For this reason, the barrel and screw can be built using materials that are not commonly used in other areas. Can you comment on this?

Dr Nagapudi replied: I agree. If materials of construction are properly selected, then abrasion and ware will not be a major issue.

Professor Beyer asked: The examples of mechanochemical processes by twin-screw extrusion shown in the paper all involve non-covalent interactions. What is the potential of the method for covalent modifications?

Dr Nagapudi replied: While the paper discusses only examples of non-covalent systems, we have had success with bond forming reactions as well. These results have not yet been made public yet. In our opinion if an organic reaction can be performed by ball milling then there is a very good chance that it can be scaled using twin screw extrusion. We are in the process of writing two papers that discuss scale-up of reactions using the extruder.

Dr Maini commented: It is clear that the temperature is the key parameter to improve your conversion and reach the 100%. The temperature set is really close to the melting point of the ascorbic acid. Can you exclude the effect of melting ?

Dr Nagapudi answered: Effect of melting in driving the conversion cannot be excluded. However, eutectic melting is most likely not the mechanism of co-crystal formation.

Dr Maini remarked: How important is the homogenization of the powder before extruding? How a bad homogenization can affect the result?

Dr Nagapudi responded: Homogenization is important. However, it is not necessary to homogenize the starting materials as this can be accomplished in the extruder.

Professor Jones said: The technique of TSE is an extremely promising method and the work from Amgen has demonstrated how successful the approach can be with regard to co-crystal formation. Could you comment on the likelihood of being able to control the stoichiometry of the resulting co-crystal product since not only polymorphism and stoichiometric variation are possible in such systems?

Dr Nagapudi responded: From a single experiment that we tried with suberic acid and nicotinamide, we were not successful in controlling the stoichiometry of the co-crystals.

Professor Jones said: Could you comment on progress which has been made, either by yourself or others, on the preparation of co-crystals within polymer loaded-TSE experiments either with co-crystals as the feed or indeed co-crystal formation within the extruder in the presence of polymer?

Dr Nagapudi answered: We have not directly attempted to do the kind of experiment you are referring to. However, people have shown that one can form co-crystals in the presence of excipients (some of which are polymeric).

Professor Jones asked: I assume that we are generally considering the equivalent of 'neat grinding' in the method you have described. Would it be possible to add small amounts of liquids at various stages of the process to control, for example, polymorphic outcome?

Dr Nagapudi replied: Yes this is possible. There are injection ports in the extruder that can be used to control solvent addition.

Mr Jörres asked: Did you optimize the shape of the mixing elements of the screw in your aparatus?

Dr Nagapudi responded: This is one of the key optimization studies that have to be done.

Professor Baláž remarked: What about wear of the materials used for working parts of extruders when abrasive materials are being processed?

Dr Nagapudi responded: One has to deal with wear and screws will need resurfacing from time to time or one will have to completely replace them.

Ms Brekalo remarked: You mentioned that you can use LAG in the twin screw extrusion. Is the extruder a closed system, or will the liquid that is put inside evaporate after a while. If it does evaporate, you do not have LAG the entire time, so the mechanism and products might be different than in a closed system.

Dr Nagapudi answered: The extruder is not a closed system and evaporation of solvent is possible. In this context, the products may be different from conventional LAG using ball milling.

Ms Brekalo asked: If we envision a two step process, and each step can be done mechanochemically, would it be possible to create a sequence of two extruders where the product of one extruder becomes the substrate for another, along with the addition of other reactants, liquids, *etc.*?

Dr Nagapudi replied: Yes this is theoretically possible.

Dr Lamaty asked: You mention UNCA and succinic anhydride of amino acids; have you tried both derivatives?

Dr Nagapudi answered: Yes both have been tried.

Dr Lamaty commented: Was CO_2 release a problem in the process?

Dr Nagapudi responded: It is not a problem as long as appropriate venting is done.

Dr Lamaty asked: The yield is lower in the continuous process: is this because of difficulties of recovering the product?

Dr Nagapudi answered: You are right. At the small scale yield loss is due to recovery. However on a larger scale this should get better because machine losses remain more or less constant as once increases scale.

Dr Lamaty said: Can you perform a multi-step synthesis by connecting more extruders?

Dr Nagapudi answered: Yes this should be feasible.

Dr Blair opened the discussion of the paper by Laszlo Takacs: Mechanical activation seems to describe the apparent storage of energy in a reaction mixture. This does not seem physically possible. At the extreme one could "store" energy by simply milling a reaction to 99% of its activation energy and stopping. The "activated" material could be packaged and the energy stored released when a minor amount of addition energy is added to overcome the activation energy. I will agree that this is essentially what's happening if one grinds a metal such as magnesium to a fine powder. It is easier to ignite. But, this it not from stored energy – its just an increase in surface area. The activation energy of the reaction $Mg + O \rightarrow MgO$ is still the same. While this may be observed for self-propagating high temperature synthesis reactions (SHS) and solid-state metathesis reactions (SSM). It doesn't seem applicable to reactions with lower activation energies. For instance, both the solid-state acid catalyzed hydrolysis of cellulose and the synthesis of alkaline earth carbides from the elements proceed in a continuous manner. If mechanical activation was as described then I would expect any mill to induce reactions if milling was applied long enough. However, this is not the case. We can mill cellulose or carbon/metal mixtures for weeks in a low energy mill with no observable reaction (S. M. Hick, C. Griebel, D. T. Restrepo, J. H. Truitt, E. J. Buker, C. Bylda, and R.G. Blair, Mechanocatalysis for biomass-derived chemicals and fuels, *Green Chem.*, 2010. **12**, 468; S. M. Hick, C. Griebel, and R.G. Blair, Mechanochemical synthesis of alkaline earth carbides and intercalation compounds, *Inorg. Chem.*, 2009, **48**(5), 2333).

Wouldn't it be more accurate to describe this phenomena in terms already in general use; defect formation, size reduction, increase in surface energy, *etc.* This would then allow a relation between milling energetics and reaction rates.

The lower energies reported for reactions after mechanical activation could be ascribed to size reduction. It is well known that nanosized materials have different thermal properties such as reduced melting points. It would be reasonable to infer that the reactivities would be enhanced when smaller crystalline domains are present.

Dr Blair communicated: Should we as a community choose a more descriptive set of terms for the observed reduction in reaction onset, increased reaction rates, improved reactivity, instead of the blanket term mechanical activation? For example, increased surface energy is more descriptive. Add increased defect concentrations, reduction in crystallinity, and other phenomena and one has a set of phenomena that can be used to describe mechanochemical reactions. This choice of language allows these phenomena to be conceptually decoupled and investigated as factors on their own. For example, some reaction may require amorphitization before any appreciable reaction is observed. Milling technology can then be chosen and optimized for specific phenomena.

Dr Blair said: Materials transport in the vials is not uniform. This use of reaction delay for energy characterization may not be the complete story. It has been shown that these types of reactions initiate in random areas and then self-propagate through the vial. The onset of ignition seems more complex than just an energy dose argument. This can be seen in work by Cocco *et al.* who filmed the reaction 5Ti + 3Si.

M. Monagheddu, S. Doppiu, C. Deidda, and G. Cocco, The self-combustion of structurally co-deformed powder mixtures: a direct view of the process, *J. Phys. D: Appl. Phys.*, 2003, **36**(15), 1917.

Professor Takacs replied: Declaring that ignition occurs when a critical degree of activation is reached, meaning a certain amount of mechanical dose has been received by the charge, is certainly not the last word in understanding the initiation of a self-sustaining process. Nevertheless, it is true in a rather broad range of parameters, not only for the Zn+S reaction, but also for thermites and metal-metalloid reactions like 5Ti+3Si or Zr+C. Ignition of the cited 5Ti+3Si reaction took place 232 min after starting the mill. That corresponds to about half a million collisions between the ball and the end plates of the vial. Thus although the material transport could have been uneven, the charge was probably quite uniformly activated by the time of ignition. A strong argument for the uniformity of activation can be derived from how ignition happens. After 232 min without any sign of ignition, a spark appeared and extinguished; soon enough another appeared and disappeared, then the third one initiated a reaction that travelled through the entire volume. This sequence of events took only 250 ms. Therefore, the powder had to be ready to ignite at several points of the vial simultaneously, suggesting quite uniform activation. Exactly where the final spark formed was random, depending on the fluctuation of impact energy and material distribution. The Ti+C system behaved in a very similar way, as shown by C. Deidda *et al.*, *J. Metastable Nanocryst. Mater.*, 2003, **15–16**, 215.

Thus the powder immediately before ignition must have been in a uniformly activated critical state.

Of course, many questions remain open: what is the exact nature of this critical state, what processes are taking place during activation on both the macro and micro scales, what is the exact mechanism of ignition, *etc.* Those questions require further investigation.

Dr Baláž communicated: Self-sustaining reactions are mainly discussed in the field of inorganic reactions. Do you have any evidence of such reactions taking place in case of organic reactions?

Professor Takacs replied: I am not aware of any case where both components had been organic. Most organic reactions are not exothermic enough or release gas during the reaction that makes the experiment dangerous. There may be examples in the literature on the sensitivity of explosives. Mulas investigated the reaction of hexachlorobenzene and calcium hydride some time ago; the ignition times varied from a few minutes to about an hour, depending on the composition (G. Mulas *et al.*, *J. Sol. State Chem.*, 1997, **129**, 263).

Professor Jones communicated: Could you comment please on any observed relationship between particle size and the ignition time? I refer specifically to Fig. 2 of your paper and wonder how the curves would be influenced by the reactant particle sizes?

Professor Takacs replied: There are two very different particle size scales involved in the typical ball milling experiments. The free particles or agglomerates are typically on the order of 10 μm, while the coherently scattering grains decrease toward 10 nm or so during milling. Experiments show decreasing grain size as the powder approaches ignition, but mixing, defect density, heat transfer and other factors are also important. The particle size of the starting powder seems less important, as long as typical commercial powders are used. The particles get deformed, shattered, and agglomerated very quickly and the original particle size becomes irrelevant. It is probable though, that using much smaller particles, approaching the nanometer range, would make a significant difference in the ignition time. I am not aware of any systematic investigation in that direction. Research on nanothermites shows much-increased sensitivity as the particle size is reduced.

Professor Suslick communicated: As a critically important control, what happens when you take zinc and mill it by itself for 15 min (*i.e.*, just before the ignition time), then do the same with selenium (mill it by itself for slightly less than the ignition time), and then finally mill the pre-milled powders together. How long does it take before ignition of the pre-milled powders?

Professor Takacs replied: This is an interesting experiment that needs to be done. But there are complications. For example, milling causes agglomeration, even cold welding, and it is much stronger in an element than in a mixture. Thus milling ductile Zn alone results in large particles or even chunks of metal 'hammered' to the wall of the vial. Only the grains inside will be refined to the nanometer scale. Selenium may behave a little better. Mixing the separately milled components can only happen on the size scale of the agglomerates, micrometers or larger, while co-milling mixes on the size scale of the grains. Perhaps a different system with brittle components that do not agglomerate

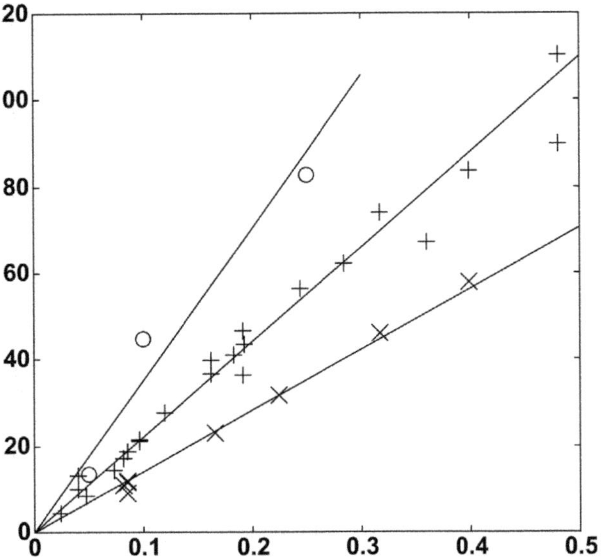

Fig. 5 Ignition time as a function of the charge ratio for the MSR between Zn and S. Results obtained with a SPEX 8000 Mixer Mill and flat- and round-ended vials and with a Fritsch P-6 mill at 550 rpm are shown by "x", "+", and "o", respectively.

strongly during milling would be better for this experiment. But in spite of the challenges, understanding the roles of grinding and mixing separately would be certainly very beneficial.

Professor Suslick communicated: I am very impressed with the reproducibility of the ignition times of the self-sustaining reactions. Since it has always been very difficult to compare one mechanochemical apparatus or technique with another, could one use the ignition time as a calibration tool?!

Professor Takacs replied: Yes. If the impact velocity of the balls is the same for a mill independent of the charge, the mechanical dose "absorbed energy per powder mass" is proportional to the ball-to-powder ratio. If ignition happens when a critical total dose has been reached, the ignition time is proportional to the powder-to-ball mass ratio for a given mill as shown in the figure. The slope is characteristic to the milling intensity. Thus for the three cases shown, the SPEX mill with flat-ended vial is the most intense, using a round-ended vial decreases the intensity 1.57 fold, and the P-6 mill is less intense by a factor of 2.5 (Fig. 5). These kinds of data can be used to scale experimental kinetics obtained with different mills. The interesting questions arise, when the differences are beyond the different time scale.

L. Takacs and V. Šepelák, *J. Mater. Sci.*, 2004, **39**, 5487.

Professor Jones noted: The intense milling has been seen to cause the surface layers of a crystal to become amorphous. Such images are similar to the concept in polishing and lubrication studies of a Beilby layer (Hopkins, *Phil Mag.*, 1936,

21, 820) which is amorphous. Similarly grinding minerals usually result in the formation of defect centers as seen in triboluminescence and electron spin resonance (*Clays and Clay Minerals*, 1982, **30**(1), 73). Presumably such centres could be the source of radical species that catalyse reactions.

Professor Galembeck replied: Yes. Moreover, carbon free radicals react with atmospheric oxygen forming peroxide free radicals and they can also participate from other reactions, *e.g.*, electron transfer producing ionic species.

Professor James asked: To what degree does the reproducibility of your results depend on the source material; how tolerant is it of starting materials from different batches for example, which may differ in terms of particle size or surface passivation?

Professor Takacs replied: The formation of zinc sulfide from a mixture of elemental powders is a very forgiving reaction. When the powder gets compressed between the colliding milling tools the first time, it is drastically deformed immediately, thus the original particle size and surface oxidation have little effect. There are systems that are much more sensitive to the properties of the starting powders and need to be handled more carefully. For example titanium is oxygen sensitive and hard, thus obtaining reproducible ignition times for the formation of TiC takes meticulous care and a consistent source of starting powders. Magnesium-based thermite reactions are not very reproducible due to the stickiness of magnesium.

Dr Irikura commented: The excellent reproducibility of the ignition delays is impressive. Prof. Takacs described an initial rapid attrition in particle size, presumably followed by a slower attrition until some critical particle size or agglomeration state is reached. Has anyone tried using non-spherical grinding balls that would fill space better? With small void spaces, maybe particle attrition would be faster.

Professor Takacs replied: I am not aware of any experiment using non-spherical grinding pieces for the activation and ignition of a self-sustaining reaction. The closest is a mixture of balls with different sizes. In that case the inverse proportionality between ball-to-powder ratio and ignition time remains valid. Filling space well is probably not an important requirement as the milling container itself is filled only to a fraction of its capacity to allow room for the balls and the powder charge to move around.

Dr Friščić addressed Professor Takacs: Is there a general consensus whether mechanochemical activation is based on effects at the level of crystal structure (*e.g.*, amorphisation, structural defects) or at the level of individual molecules (*e.g.*, placing molecules into conformations of high energy, distorted HOMO-LUMO levels).

Professor Takacs replied: I do not think there is a simple general answer to this question for bulk materials. The formation of lattice defects results in shifting energy levels thus causing increased reactivity. Changes on different length sales are connected. The situation is simpler in single-molecule studies.

Dr Friščić continued: On the mechanism of mechanochemical activation: I would appreciate any information that Prof. Takacs or the other candidates might have on structural changes in the material induced by milling before ignition. For example, have there been any recent studies of the X-ray diffraction patterns of the materials ground for different amounts of time before the mixture ignited? Can we assume that the period of mechanochemical activation is based on a combination of amorphisation and improved particle mixing – or are there any other factors to consider?

Professor Takacs responded: This is a very difficult problem. Although it has been demonstrated by XRD that milling decreases the average grain size and increases random strain due to the introduction of defects, it is hard to make a direct connection between this information and the ignition time. Heat transfer is another important factor, both within the powder and between the powder and the environment. Ignition is a threshold phenomenon, but it is hard to predict what combination of grain size, strain, and other parameters would result in ignition under certain milling conditions. It is possible to identify correlations, but hard to develop functional relationships.

Dr Stolle communicated: Although most of the milling processes used for synthetic purposes but also for particle refinement and mechanical activation employ spheres as grinding media other types of grinding media are available in general (regular cylindrical geometry). Additionally, in the case of autogeneous milling irregular shaped feed material function as grinding media. Despite the advantages of grinding spheres for specific processes other types may be applicable.

Dr Blair said: Your ignition delays seem inversely proportional to the bulk modulus of the metals used. Have you looked at this? It would provide an additional parameter that must be considered when comparing reactions.

Professor Takacs responded: Mechanical activation is related to the creation of lattice defects, such as dislocations and interfaces. Those relate to plastic processes, thus correlation is expected between the ignition time and parameters describing strength. Indeed, there is some correlation between hardness and ignition time in the case of metal–chalcogen reactions (Avvakumov and Kosova, 1991). The bulk modulus is an elastic property, thus it is probably less important. In general, activation and ignition depend on many parameters, thus it is difficult to find meaningful correlations.

Dr Khripin communicated: Regarding the use of ignition to calibrate a ball mill, it was pointed out during the comment session that this may not be practical due to variations in powders from different suppliers. This issue could be addressed by involving a standards organisation, such as NIST located in Gaithersburg, MD. NIST routinely makes and releases standard reference materials.

For example, they could make and keep sealed ampoules of a standard ignition powder, selling them as needed for mill calibration.

Professor Takacs communicated in reply: The accuracy desired for the comparison of different mills or the same mill under different operating conditions is a few percents at best. On that level, small amounts of impurity and somewhat varying particle size are not serious concerns. Thus commercial materials are acceptable, as long as the materials are not air-sensitive, mix easily, are not very hard, do not agglomerate or cold-weld into large pieces, and the ignition times are conveniently between a few minutes and a few hours. Especially if a larger mill is evaluated, it is also important that the adiabatic temperature is not very high, otherwise the reaction may damage the mill. The formation of ZnS from elemental powders turned out to be a good choice. On the other hand, reactions like the formation of ZrC or an iron oxide-Al thermite would have obvious shortcomings.

Professor Mack opened the discussion of the paper by Achim Stolle by communicating: Since you are doing large scale reactions that produce a lot of water, what do you do to remove the water from the reaction?

Dr Stolle answered: Right now, we are not 100%-sure where the water goes during the reaction. Generally there are two options, either it is chemically stored in the crystal lattice (there are some hypotheses in the literature) or it is evaporated during elevated process temperature. Right now we are trying to address both questions by a) PXRD studies and b) *in situ* humidity sensors for large scale equipment. At lower scale *in situ* pressure measurement is an option but the pressure increase is dominated by the temperature increase rather then the evaporation of water.

To answer your question: you can run your reaction under slight vacuum conditions at a large scale. Drying of the product at slightly elevated temperatures (>120 °C) leads to decomposition of the product. The way, which we follow right now, is that we did not remove the water at all. The presence of water seemed not to interfere with the product morphology since a powder is formed in most cases with identical properties compared to commercial samples.

The general question regarding the necessity of work-up which very much depends on the utilization of the product on a large scale. If the water does not interfere with up-stream reactions, there is simply no need to remove the water. If the water needs to be removed (either for further reactions or in case of product properties) you can use vacuum drying at large scale (even during milling).

Mr Jörres asked: In your presentation you mentioned that optimized filling degrees are around 0.5 but even with higher degrees good results were achieved. Did you also try lower filling degrees and did you observe any different results? Did a lower filling degree lead to a lower conversion, or is it just the loss of energy that makes these reactions less effective?

Dr Stolle replied: We tried filling degrees ranging from 0.15 to 0.8 and recognized an optimal in the range of 0.45 value as demonstrated in the manuscript (Fig. 4). It might be important to mention that this specific filling degree considers the milling ball packing ($F_{MB,packing}$). With respect to the sole volume this optimal value represents approximately 0.25–0.3 ($F_{MB,material}$).

For the calculation of the different filling degrees see the manuscript and the Supporting Information.

My comment regarding the possibility to run the planetary ball mill at a filling degree of up to 0.8 regards the total filling degree (F_{total}), thus the combination of milling balls and grinding stock. The filling degree regarding the milling balls has been kept constant as demonstrated in Table 2 of the manuscript.

Dr Hamilton communicated: You identify important parameters for the scale-up of organic reactions. You draw conclusions for optimal parameters using one model; how general do you expect your findings will be, especially in regards to optimal filling ratios, ball size, *etc.*? Won't this depend on the specific energetics of each reaction?

Dr Stolle replied: It is correct that the specific reaction conditions depend on the energetic requirements of the tested model reaction regarding the activation barrier. Thus, generalization to the complete field of mechanochemistry in ball mills (inorganic, organic) will not be possible. If you consider the chosen model reaction as an example for the processing of soft matter (organic reactions) in ball mills general conclusions can be drawn for this type of chemistry. Especially regarding the fact that smaller milling balls and a medium filling degree seemed optimal and in agreement with distributers' advice for particle refinement.

The general question (which cannot be answered by scale-up studies only) is the question regarding the location of the reaction. For reactions involving inorganics it seemed clear that the single stress energy is important which might lead to the conclusion that the reactions occur in the collision zones. In contrast for reactions of organic matter the frequency of the single stress events seemed to be the determining factor. Thus, those reactions can take place in the bulk, if the particle size is small hence the surface for interparticle contacts is increased. Under this premise the reaction is controlled by the global energy present in the system and not by the stress energy only.

Professor Beyer commented: From gas phase studies it is well established that condensation or hydrolysis reactions require several, typically at least three, water molecules to lower the barrier for proton transfer. This would mean that a minimum water content is required for the reaction to proceed efficiently. Is this effect, at least in part, responsible for the observed induction period? Do you plan to investigate the effect of water concentration in the reaction vessel on the rate?

Dr Stolle replied: The reactions have been carried out under standard atmosphere conditions. For scaled-up syntheses in horizontal stirred media mills we considered to operate under inert gas or standard atmosphere and equip the machine with a humidity sensor to trace the water generated during the reaction. This might answer the question of where the water is situated after the reaction (either as crystalline water or in the gas phase).

We are right now starting to investigate the temperature effect on the reaction in the ball mill which may allow us an insight into the kinetics and if necessary we might also consider to investigate the humidity influence (*e.g.*, by running the reaction in a dry inert gas atmosphere).

Mr Zhang addressed Dr Stolle: There is just some information for the scale-up ball mills. Actually in the area of environmental remediation, very huge high energy ball milling machines have been used for the mechanochemical destruction of persistent organic pollutants (POPs). One Japanese company (Radical Planet, http://www.radical-planet.com/) built a planetary ball mill machine (250 L or 500 L for each pot) and used it to destroy the high concentration POPs wastes such as DDT. And a New Zealand company (Environmental Decontamination Limited, http://www.edl-asia.com/) built a series of stirring mills (265 L for each unit) for the remediation of contaminated soil.

Mr Zhang then commented: What do you think is the biggest problem for the scale-up of a ball milling machine?

Professor Baláž said: Today, the ball mills with a capacity around 10 000 L are being applied.

Dr Blair added: A big challenge, especially in larger mills is combining chemical needs with mechanical motion. This represents a paradigm shift from standard application of mills for size reduction. Great efforts are taken to eliminate chemical reactions, we want them to occur. This means that the overall design of a mill must be assessed in terms of applicability to mechanochemical reactions. For example, stirred media mills can produce quite severe segregation of material and media. This is great for size reduction because it reduces frictional heating by decreasing the time the material spends in contact with the media. For chemical reactions this can lead to slow reaction rates.

Dr Stolle concluded: The only comment I can provide is: Yes, you are absolutely right. This is actually part of our present research. However, what might be a benefit for mechanochemical reaction in such devices is the temperature increase due to frictional and impact processes. Normally stirred media mills are cooled to get rid of this dissipative heat. Contrarily the kinetics of chemical reactions is often positively influenced by higher temperatures (in the case of Arrhenius behaviour, the absence of intermediate melts, and the exclusion of side reactions). Thus operating a mill at its temperature equilibrium might be advantegeous with respect to the energy balance compared to particle refinement only, since the energy demand for cooling is reduced. Surely, this will depend on the mill design, the process parameters and of course the individual mechanochemical process?

Professor Jones asked: Could you comment briefly on the manner in which the material is mixed in the horizontal drums? Is it by use of baffles and if so are there any known important design features which influence, for example, the kinetics of conversion?

Dr Stolle responded: The mixing behavior of substances in a horizontal drum depends on the drum size, it's design and the rotation speed. If mixing is the only purpose baffles can be used to mix and induce transport of the grinding stock in the horizontal direction from one end of the drum to the other. If particle size reduction is of importance grinding media has to be added also. With respect to high kinetic processing in stirred media mills or high kinetic ball mills mixing is initiated by the grinding media trajectories inside of the drum.

To the best of our knowledge there are no studies investigating the effect of baffle design and position on a mechanochemical process by means of kinetics and/or conversion.

Professor Jones asked: Your paper considers factors such as ball diameter, rotation frequency, and degree of filling. Would you be able to comment on the influence of temperature and how practical it might be to run the reaction at controlled higher (or indeed lower) temperatures?

Dr Stolle replied: Like in every chemical process temperature is an important parameter. Regrettably planetary ball mills did not offer the possibility for temperature regulation contrarily to horizontal stirred media mills or attritors.

Investigation of the temperature influence (not only control but regulation) are in the focus of the present research. With respect to processes in ball mills, our aim is to take advantage of the dissipative temperature increase due to friction and impact processes and run the mill at temperature equilibrium (the temperature which is reached in the milling vessel under certain conditions without external regulation) or slightly increased temperature. Thus the energy demand for cooling or heating can be reduced to minimum. The equilibrium temperature strongly depends on similar parameters as outlined before (filling degree, milling ball mass, milling ball diameter). Operating temperatures below the equilibrium temperature would strongly increase the energy demand for cooling and reaction rate would decrease.

Professor Jones asked: Is there a simple correlation between the structures you observe and/or anticipate with other known complexes for which crystal structures are known?

Dr Stolle answered: The presence of different morphologies has not been investigated yet. To evaluate the process, the chemical conversion of the substrates to the product was measured.

Faraday Discussions

PAPER

Advances in elucidating mechanochemical complexities *via* implementation of a simple organic system

Adam A. L. Michalchuk,*[ab] Ivan A. Tumanov,[bc] Valeri A. Drebushchak[bd] and Elena V. Boldyreva*[bc]

Received 20th December 2013, Accepted 6th February 2014
DOI: 10.1039/c3fd00150d

Mechanochemistry is becoming increasingly popular amongst both the academic and industrial communities as an alternative method for inducing physical and chemical reactions. Despite its rapidly expanding application, little is understood of its mechanisms, greatly limiting its capacity. In the present work the application of specialty devices allowed submission of the simple organic system, α-glycine + β-malonic acid, to isolated shearing and impact treatment. In doing so, unique products were observed to result from each of these major mechanical actions; shear inducing formation of the known salt, glycinium semi-malonate (GSM), and impact yielding formation of a novel phase. Correlation of these isolated treatments with a more common ball mill indicated two unique regions within the milling jar, each characterised by varying ratios of shear and impact, leading to different products being observed. It is widely accepted that, particularly when considering organic systems, mechanical treatment often acts by inducing increases in local temperature, leading to volatilisation or melting. A combination of DSC and TGA were used to investigate the role of temperature on the system in question. Invariably, heating induced formation of GSM, with evidence supporting a eutectic melt, rather than a gas-phase reaction. Shear heating alone is unable to describe formation of the novel phase obtained through impact treatment. By considering the formation and character of mechanically produced tablets, a model is described that may account for formation of this novel phase. This system and methodology for mechanochemical study offers intriguing opportunities for continued study of this widely used and exciting field.

1 Introduction

Mechanochemistry is often believed to be simply an alternative method to induce thermal reactions, in addition to its milling capabilities. The first evidence in

[a]*School of Chemistry, The University of Edinburgh, Edinburgh, UK. E-mail: adam.michalchuk@gmail.com*
[b]*Novosibirsk State University, Novosibirsk, Russian Federation*
[c]*Institute of Solid State Chemistry and Mechanochemistry SB RAS, Novosibirsk, Russian Federation. E-mail: eboldyreva@yahoo.com*
[d]*Institute of Geology and Mineralogy SB RAS, Novosibirsk, Russian Federation*

opposition to this simplicity can be found in the works by M. C. Lea[1] demonstrating the unique decomposition products on heating and mechanical treatment of silver halides. Later, a detailed comparative study of the products of thermal and mechanochemical decomposition of a series of inorganic salts demonstrated that the products formed on crystal cleaving depend on the rate of fracture propagation.[2–6] Slow propagation rates resulted in decomposition products coinciding with those of thermal decomposition, while rapid fracture propagation yielded products corresponding to radiolysis.[7,8] Much effort was therefore made through the 20th century in an attempt to understand the mechanisms that underlay mechanochemical reactions, especially in the event that mechanical heating offered little by means of an explanation.[9,10] While substantial progress has been made to understanding mechanochemical mechanisms,[11–13] there remain more questions than answers. With current trends of mechanochemistry suggesting immense potential for such areas as co-crystal screening,[14–18] green chemistry,[12,19–22] materials processing and tuning,[23–26] amongst others, and substantial application in industry – in particular the pharmaceutical industry – it is of growing importance to obtain an understanding of the fundamentals of mechanochemistry. As a particularly pressing example, one might consider the effects of mechanical treatment on the polymorphism of pharmaceutical compounds, which may result from explicit treatment or as a consequence of technological processes.[27,28] In the latter, unexpected transformations may occur, having drastic consequences on both the tableting (*e.g.* paracetamol[29]) and biological performance[30] of the compound, *e.g.* by altering dissolution profiles.[31] Of equal importance is the growing field of "dry synthesis" of solid pharmaceuticals, their salts and co-crystals.[32–36]

Efforts in understanding mechanochemical transformations have seen the development of methods for monitoring mechanochemical reactions on co-grinding at intermediate intervals either *ex situ*[3,39–42] or *in situ*,[3,43,44] with the latter now possible at very short time intervals without the need to stop mechanical treatment.[32,45] Here an accumulative approach to mechanistic mechanochemistry is presented in which the complex problem is initially segregated into somewhat scientifically well-developed components, ultimately combining to yield a more complete understanding of the process. In this approach, a mechanochemical reaction can be thought of as being composed of two parts: (1) an interaction between the mechanoreactor and the system; and (2) an interaction between the system components.

The present work aims to address the mechanochemistry of a model system, α-glycine + β-malonic acid, by means of this approach, expanding on previous works that have attempted to separate the major types of mechanical action, impact[40] and shear.[41] In section 3.1 we present a summary of an earlier paper[39] in which an initial understanding of the workings of a common ball mill were presented. Section 3.2 then extends to the second component of this methodology, discussing work to elucidate the intra-system interactions. Finally, section 3.3 presents a model to extend section 3.2 to otherwise unexplainable trends.

The selected system was originally identified by Losev *et al.* as a system with immense mechanochemical complexity.[37,38] The mechanochemical product of α-glycine + β-malonic acid was found to be tunable under treatment in an SPEX-8000 mill by means of controlling the initial mixture composition. Small quantities of β-malonic acid led to an α → γ-glycine polymorphic transition, while mechanical treatment of near-equimolar compositions of the system induced formation of a known salt, glycinium semi-malonate (GSM).[46] This complexity was further ameliorated when studies[39] demonstrated product selectivity through controlling the type of mechanochemical treatment applied, Fig. 1. With current interest in both amino acids and small dicarboxylic acids, this relatively simple, yet mechanochemically complex system offered an intriguing opportunity to study mechanistic mechanochemistry on a system of current scientific and biological interest.

2 Experimental

2.1 Materials

All materials used were commercially available and were used without further purification. The chemicals used were α-glycine (Reactiv, Russia) and β-malonic acid (Fluka, grade ≥ 99.0%).

2.2 Experimental protocols

2.2.1 GSM salt growth. Pure GSM salt was obtained by slow evaporation from an aqueous solution of equimolar quantities of the reagents. Purity was verified by XRPD and DSC.

2.2.2 Milling. Milling was performed using a Retsch Cryomill ball mill. 0.5 g mixture of equimolar quantities of α-glycine and β-malonic acid were prepared in milling jars (stainless steel: 7 mm internal diameter × 44.6 mm total internal length). To each jar, two milling balls (stainless steel: 0.691 g, 5.6 mm diameter) were added. Three variations of experimental protocol were then observed. In the first, samples were milled at 24 Hz for the appropriate duration of time. Alternatively, samples were milled at 24 Hz for 20 min, followed by a pause of 3 min, after which a second bout of 20 min milling was endured. This process of pause and mill was repeated a second time for a total milling time of 60 min. Finally, the third protocol entailed milling samples at 24 Hz for 20 min, followed by manual mixing of the sample (during which any pellet formed was carefully broken) and resubmitting the sample for 20 min of milling. This process of mixing and milling

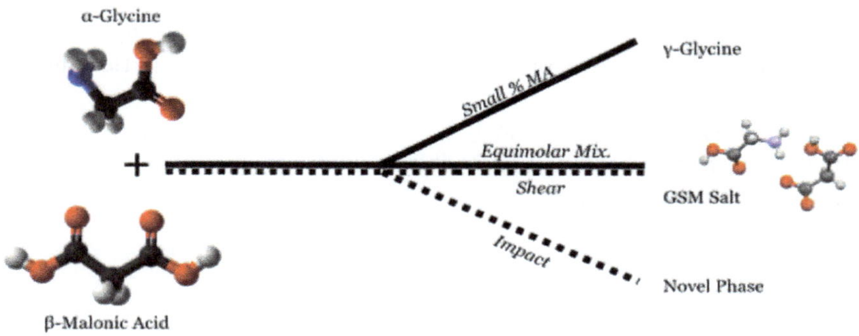

Fig. 1 Schematic outlining the complexities of α-glycine + β-malonic acid mechanochemical reactivity. This encompasses studies by Losev *et al.*[37,38] (solid line) and Michalchuk *et al.*[39] (dashed line).

was again repeated a second time for a total milling time of 60 min. In all cases, samples were immediately characterized by XRPD.

2.2.3 Impact treatment. Impact treatment was applied using a purpose-built device.[40] Treatment was performed at 2.3 Hz with *ca.* 10 mJ per impact. All powders were sieved to control particle size, p: 100 mm < p < 200 mm. 0.125 g samples were prepared as 75 mol% α-glycine/25 mol% β-malonic acid and manually rotated in a glass vial to ensure sufficient mixing prior to impact treatment. Impact treatment was then performed for the desired period of time, after which the sample was immediately tested by XRPD.

2.2.4 Shear treatment. Shear treatment was applied using a purpose-built device.[41] Treatment was performed at 33 rpm under 1 kg loaded mass, or approximately 4 kPa. All powders were sieved to control particle size, p: 100 mm < p < 200 mm. 0.5 g samples were prepared as 75 mol% α-glycine/25 mol% β-malonic acid and manually rotated in a glass vial to ensure sufficient mixing prior to impact treatment.

2.2.5 Vibratory treatment. Vibratory treatment was performed using a NARVA Vibrator DDR-GM9458 (30 W, 50 Hz). 0.5 g samples were treated in the presence of one milling ball (stainless steel: 4.049 g, 10.0 mm diameter) for sufficient time to cause a noticeable change in consistency (powder to sticky) of the entire mixture. This change occurred between 8 and 20 min. The samples were then allowed to stand for 1 h to allow solidification of the sticky sample. Phases were verified by XRPD.

2.2.6 Liquid assisted grinding. Liquid assisted grinding (LAG) was performed using a range of solvents. Following addition of 0.5 g sample to the milling vessel, a single drop (20 μL) liquid was added prior to milling: $\eta = 0.04$.[47] Continuous milling (machine as described above) was performed for varying lengths of time. The sample was characterised by XRPD and quantitative phase analysis performed as outlined below.

2.3 Characterisation

2.3.1 X-Ray powder diffraction. All samples were characterised by XRPD analysis using a Bruker GADDS diffractometer: Cu-Kα radiation ($\lambda = 1.5418$ Å, $5.0° \leq 2\theta \leq 46.0°$, with an operating potential of 40 keV and a current of 40 mA. All data were obtained in reflection mode with scanning time of 180 s per frame.

2.3.2 Differential scanning calorimetry. DSC was carried out using a DSC-204 Netzsch machine with standard aluminium crucibles. Samples of approximately 3–4 mg powder were used with particle size, p: 100 μm < p < 200 μm. Heating rate was 6.0 K min^{-1}.

2.3.3 Thermal gravimetric analysis. TGA was carried out using a NETZSCH TG 209 with standard aluminium crucibles. Samples of approximately 14–16 mg powder were used with particle size, p: 100 μm < p < 200 μm. Heating rate was 6.0 K min^{-1}.

2.4 Programmes and literature crystallographic data

For all programmes, literature crystallographic data were obtained from the CCDC database: α-glycine (GLYCIN29), β-malonic acid (MALNAC06), GSM (AWIHIY). Void surface was calculated using Crystal Explorer 3.0.[48,49] Quantitiative phase analysis was performed using Powder Cell V2.3[50] and errors were estimated at 5% in accordance with previous works on this system.[38]

3 Results and discussion

3.1 Complexities of mechanochemical processes

Studies of the mechanochemical reactivity of the α-glycine + β-malonic acid system were performed on individual reactant components. Under no conditions were there any observable effects on the consistency of the product mixture, nor did mechanical treatment induce any noticeable effect on the phase of individual reactant components. In contrast, mechanical treatment on mixtures of these components led to formation of a range of sticky product mixtures. Except in the case of shear treatment, in which apparent 100% reaction yields were often obtainable, all reaction mixtures contained residual reactant along with one or both obtainable products, GSM salt and an unidentifiable, novel phase. While GSM has previously been shown to be formed through various crystallisation techniques,[37–39] the novel phase was not obtainable but through mechanical treatment as outlined below.

3.1.1 Isolated treatment - impact. The use of equimolar α-glycine + β-malonic acid mixtures proved impractical due to both product formation resulting from mixing of initial samples and the inability to produce product mixtures containing both products, each in isolation. Instead, 3 : 1 (*i.e.* 75 mol% α-glycine + 25 mol% β-malonic acid) mixtures were used, overcoming these issues and allowing observation of an initial induction period. A series of independent experiments indicate that impact treatment of the current system results in formation of both GSM and the novel phase, Fig. 2. Interestingly, these two products appear to occur in succession. GSM first appears following less than 10 min of treatment (*ca.* 13.8 J of mechanical energy) and, by 60 min of treatment (*ca.* 82.8 J), a product mixture containing only the novel phase is observed. Within this interval, at approximately 45 min (*ca.* 62.1 J) both product phases are observed. Of further note is the

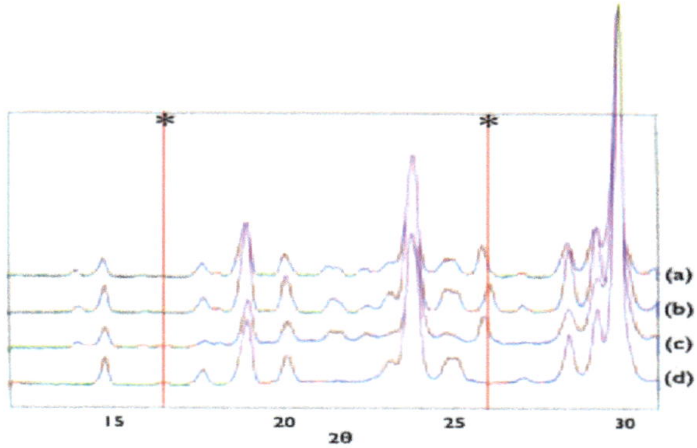

Fig. 2 XRPD of a series of impact experiments conducted on 3 : 1 α − glycine + β-malonic acid mixtures. For clarity, only the major peaks of the product phases are indicated (*) at 16.5° (novel phase) and 26.0° (GSM). Treatment duration was (a) 20 min, (b) 30 min, (c) 45 min and (d) 60 min. These data demonstrate the conversion of GSM to novel phase upon sufficient impact treatment.

fact that, regardless of the product phase, only minimal amounts can be produced; a phenomenon that will be discussed in Section 3.3. It must be mentioned that as each sample was prepared independently, sample mixing may differ and as such the quoted energy values may only be indicative of the general process. Regardless, the immense time and energy gaps between these notable events in the evolution of the system suggest that minor discrepancies in the mixing will have a negligible effect on the outcome.

A further caution must be mentioned with respect to the working of the impact device. While the impact device induces mechanical action through impact motion, it is critical to acknowledge that energy is not necessarily translated in an identical fashion through the sample mixture, as will be discussed in section 3.3.

3.1.2 Isolated treatment – shear. Similar to impact studies, 3 : 1 mixtures of α-glycine + β-malonic acid were used for shear treatment studies. A sample was submitted for 3 min of shearing treatment, resulting in a sticky substance. Quickly solidifying, XRPD of the sample indicated complete conversion of all available malonic acid to GSM, with no indications of the novel phase being present, Fig. 3. This sample was subsequently resubmitted for a further 20 min of shear treatment, with no observable change. Of note here was the fact that the product mixture of the second treatment was not noticeably sticky, suggesting the formation of a liquid phase intermediate, perhaps through a eutectic melt, between initial reactants for the formation of GSM. Subsequent treatment of GSM product therefore does not yield this sticky phase. It is interesting to compare the product yield obtained through shear treatment with that of impact treatment. The largely improved yield of the shearing device is likely the result of continued particle mixing throughout the treatment and inhibition of tablet formation.

Curious to observe the effect of shear treatment on the novel phase, a product mixture containing noticeable quantities of the novel phase was produced by vibratory treatment and submitted to successive bouts of 30 min shearing, Fig. 4. Analysis of the resulting product mixtures clearly indicates that shear treatment greatly favours formation of GSM, while leading to loss of the novel phase, which is no longer observable following 90 min of shear treatment.

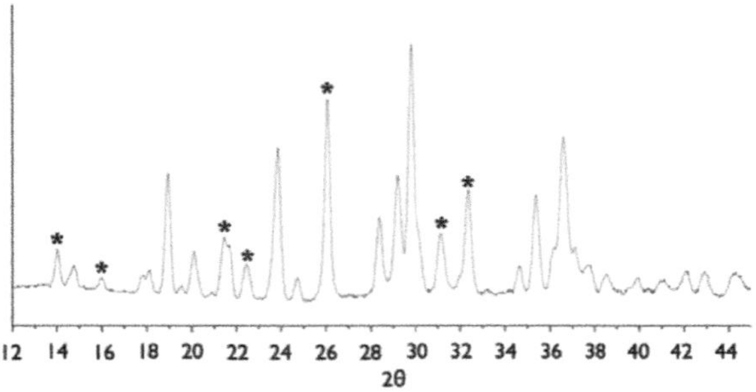

Fig. 3 XRPD resulting from shear treatment of mixtures of 75 mol% α-glycine + 25% β-malonic acid. Sample was treated for 3 min. Labelled peaks (*) indicate GSM, with remaining peaks belonging to α-glycine. Small peaks at 27° and 33–34° may indicate traces of β-malonic acid. It appears that shear treatment leads to a nearly complete conversion of available reactants to the product phase.

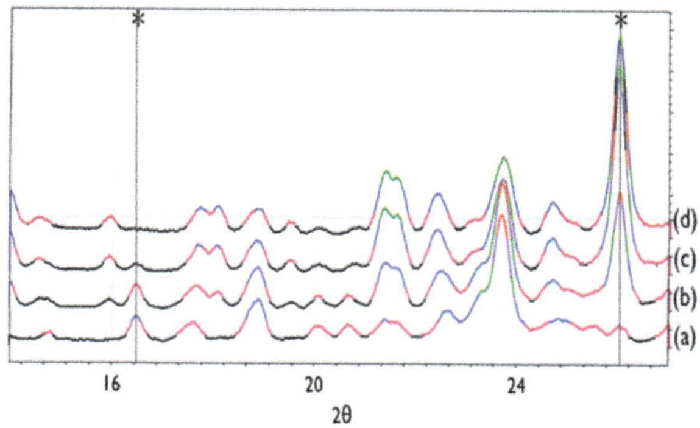

Fig. 4 XRPD of a sample prepared by vibratory treatment, subsequently treated by successive bouts of shear treatment. For clarity, the product phases can be monitored by their most prominent peaks, indicated (*) at 16.5° (novel phase) and 26° (GSM). (a) Initial mixture, (b) 30 min shear, (c) 60 min shear and (d) 90 min shear. It is clear that shear treatment favours GSM formation, leading to the loss of the novel phase.

3.1.3 Milling studies. A comparison of the effects of isolated treatments on the α-glycine + β-malonic acid system to the observed reaction under ball milling was particularly intriguing. The trend of product formation upon milling appears to be most similar to that observed under impact treatment, *viz.* initial production of GSM followed by conversion to the novel phase, Fig. 5. Interestingly, however, the rate of this conversion differs drastically. While the GSM → novel phase transition occurs rather slowly under impact treatment – with an intermediate composition containing both GSM and the novel phase – this process occurs very abruptly under milling treatment. This is perhaps not surprising considering the intensity and frequency associated with each treatment. However, the question remains as to whether milling treatment can then be thought of as acting in the same, or a similar, way as impact treatment.

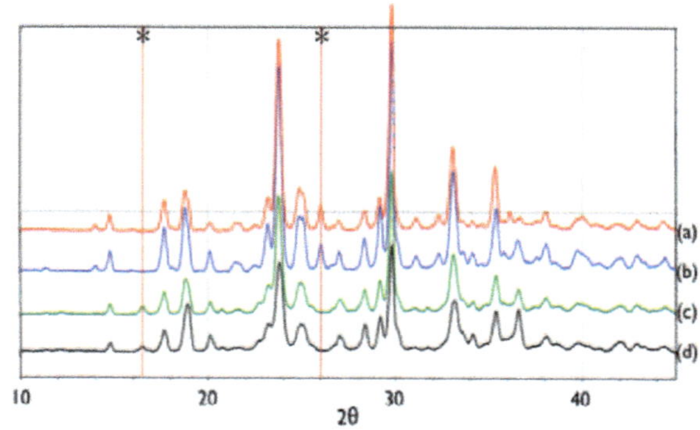

Fig. 5 Milling (24 Hz) α-glycine + β-malonic acid mixture of equimolar composition. For clarity, the two product phases can be monitored by their most intense peaks at 16.5° (novel phase) and 26.0° (GSM salt), indicated by (*). (a) Milling 30 min, (b) milling 45 min, (c) milling 60 min and (d) milling 90 min. Milling treatment initially leads to GSM formation, transitioning to novel phase production after sufficient treatment.

3.1.3.1 Reactive regions in a ball mill. In an attempt to understand the relationship between impact, shear and mill treatments it proved interesting to consider the fact that general mechanoreactors act upon their samples in a variety of ways within the reaction vessel. In the case of the Retsch cryo-mill ball mill used in these studies, two unique reaction zones were in fact observed, Fig. 6. Sampling from the walls of the milling vessel indicates the presence of GSM, with no signs of the novel phase. In contrast, sampling from the milling jar ends (a tablet), indicates signs of the novel phase, with minimal amounts of GSM. The presence of two unique zones of reactivity within the milling jar is clear, with the walls being consistent with the previously demonstrated shearing product and the ends being consistent with the previously demonstrated impact product. The existence of these two regions may prove of grave importance to both academia and industry, in which, over time, *e.g.* on storage, small traces of undesired product may induce a complete conversion of desired product into this undesired phase.

It is important to note here that different ball mills will act differently on their samples, depending largely on the orientation of the milling jar to the displacement axis and the displacement path followed by the milling jar. In addition, factors including the ratio of sample to milling bodies and regime of treatment may also affect the outcome.

3.2 Thermal studies – a mechanism for GSM formation

Despite having demonstrated the selective effects of shearing and impact treatment, little can so far be said regarding the mechanisms involved in the apparent "solid-state" reaction. Substantial work has been done attempting to understand the mechanisms by which solids react.[51–54] At present a number of common mechanisms have been proposed. These consist of: (1) liquid phase intermediates, including from liquids present in the initial mixture (*e.g.* moisture or residual solvent)[15,34,55–60] or by formation of a eutectic melt;[55,56,61–65] (2) gas phase intermediates of a single or multiple reactant components;[36,55,65–68] and (3)

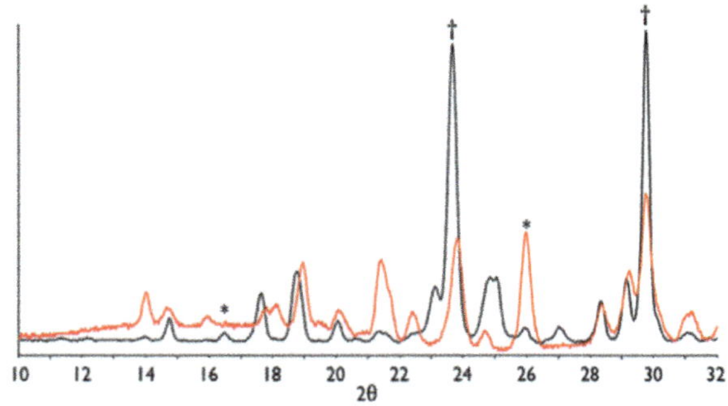

Fig. 6 XRPD patterns of milling equimolar α-glycine + β-malonic acid 24 Hz, 3 × 20 min, with no intermittent mixing. Sample was tested from the milling jar ends (black) and milling jar walls (red). For clarity, the products (*) can be identified by their major peaks at 16.5 (novel phase) and 26° (GSM). Similarly, the relative amounts of reactant (†) can be monitored by their major peaks at 30° (glycine) and 23.8° (malonic acid); their intensities coincide with the observed product formation. It is evident that zones with distinct reactivity exist within a single milling jar.

solid–solid mass transfer. It is not difficult to comprehend mechanisms involving either (1) or (2), however, the feasibility of (3) remains highly debated, in particular when considering non-spherical, bulky organic molecules. At present, arguments in favour of (3) include highly mobile intermediate phases, such as amorphous solids,[69] which can potentially ease solid-phase diffusion processes. It is widely accepted that mechanical treatment induces a certain level of heating, whether this be the result of shear heating or so-called "hot spots." Therefore, the low melting point, eutectic and volatile nature of organic crystals makes them prime systems to expect these heating effects to be of potential consequence, with mechanisms (1) and (2) being very easy to imagine possible.

3.2.1 Identifying thermal character of reactant compounds. DSC of β-malonic acid identifies two thermal events, the first occurring at *ca.* 97 °C, with the second at *ca.* 132 °C. The latter is consistent with both literature and TGA data (shown below), corresponding to melting. The former event, however, offers an interesting query. Previous experiments have suggested this phase transition to occur at 79 °C[70] and 87 °C,[71] both measured by DSC. The experimental protocol for ref. 70 indicates that prior to DSC measurements malonic acid was re-crystallised from aqueous solution. It is possible that this process may have introduced contaminants or moisture to the sample, which would account for a largely decreased temperature for the phase transition as compared to the current and other literature data. One might also consider the effects of variations in particle size between experiments, a factor known to have strong effects on DSC data.[72–75] Consistent with previous works[75,76] on α-glycine, DSC displayed no thermal events below its decomposition temperature, Fig. 7.

Throughout the mechanochemical experiments, under no conditions did the mechanoreactors become warm to the touch. As such, it is difficult to imagine that temperatures corresponding to the melting temperatures would be reached at a global level within any of the tested systems. It was therefore interesting to investigate the possibility of eutectic melting at inter-particle contacts for the system of study. As an initial test system, 3 : 1 mixtures consistent with those used in mechanochemical studies were used for DSC experiment, Fig. 8. The initial endothermic peak, taken to correspond to melting of the system, is seen to occur

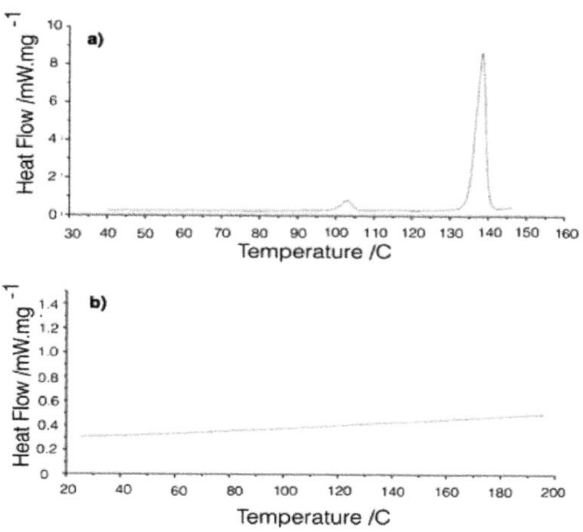

Fig. 7 DSC data of reactant compounds, (a) β-malonic acid, (b) α-glycine.

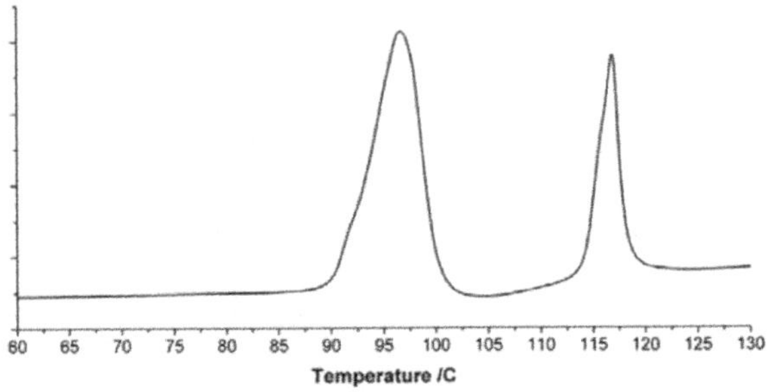

Fig. 8 DSC curve for 25 mol% β-malonic acid + 75 mol% α-glycine mixture. The initial endothermic peak corresponds to reactant melting, followed by an exothermic recrystallisation step. The second endothermic peak is consistent with GSM melting.

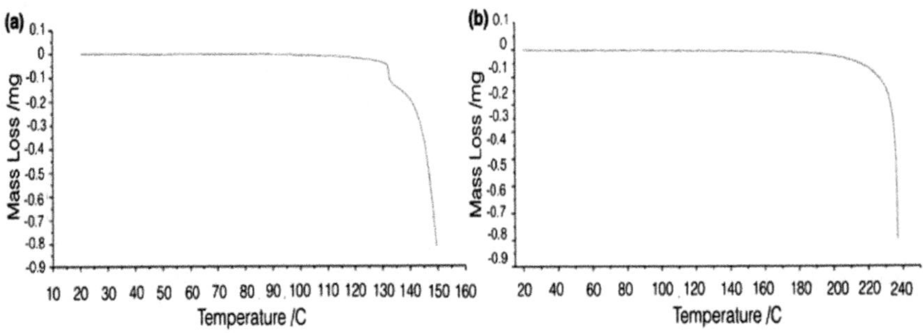

Fig. 9 TGA analysis of reactant components. (a) β-malonic acid system; (b) α-glycine. It does not appear that volatilization of reactant components occurs at sufficiently low temperatures to support a gas-phase mechanism for GSM formation, which itself melts at *ca.* 112 °C.

at *ca.* 88 °C, well below the melting temperature of either reactant component. This peak is followed by a shallow exothermic event, likely recrystallisation of GSM salt. The second endothermic event occurring at *ca.* 112 °C is consistent with the melting temperature of GSM.[77] Testing systems of equimolar composition show this melting temperature to drop to as low as *ca.* 60 °C. It is therefore clear that the system expresses substantial eutectic character. While these temperatures are still well above what would be expected for a reactor that does not feel warm to the touch, it is possible that they may be indicative of the energy equivalent attainable at mechanochemical hot spots. This combination of thermal data with mechanochemical studies offers an intriguing opportunity to study the character of hot spots in depth.

3.2.2 The nature of the intermediate fluid phase: gas or liquid? Observing that heat is sufficient to induce formation of GSM from an initial α-glycine + β-malonic acid mixture the question remains as to the nature of this phase.

TGA analysis of β-malonic acid, Fig. 9a, suggests two thermal events, the first at *ca.* 132 °C, with the second event being characterized by a gradual onset, with a maximum gradient following *ca.* 142 °C. The former is consistent with the melting temperature of β-malonic acid, at which point there is obvious volatilisation (mass loss > 0.05 mg), with the latter being taken as the evaporation of the substance. In contrast, TGA of α-glycine, Fig. 9b, displays only a single thermal event, taken to correspond to the sublimation of the substance with loss (>0.05 mg) occurring from 214 °C. The decomposition point is clearly visible, occurring at (232 °C) at which point rate of mass loss drastically hastens. This result is consistent with previously reported temperatures for the decomposition of glycine.[78]

By comparison of these results, it follows that for a gas-phase intermediate to have any notable impact on the mechanochemical reaction, energies within the mechanoreactor would be required to reach the equivalent of *ca.* 132 °C. From DSC it is known that GSM is formed well below 132 °C and thus, a gas-phase intermediate does not appear to be a plausible mechanism. It is important to note here that, while it is obvious a gas-phase intermediate is not *required* for the formation of GSM, its role in the overall mechanism of GSM formation cannot be completely discounted when considering mechanical conditions. It is possible that hot spots may in fact induce temperatures required for volatilisation, or perhaps impact may lead to a "loosening" of surface molecules, thus favouring volatilisation at reduced temperatures.

It is important to reiterate that, under no circumstances throughout the mechanochemical studies did any mechanoreactors become warm to the touch, indicating, at least at a global level, these volatilisation temperatures were not obtained.

3.3 Insights into mechanisms of novel phase synthesis: the physical basis of consecutive products

Having discussed the effects of impact and shear on the α-glycine + β-malonic acid system, as well as having shed some light on the processes occurring within a simple ball mill, it becomes interesting to consider the time-dependence of the observed reaction pathway, *i.e.* the existence of consecutive products. It is clear from the aforementioned experiments that GSM results from shear treatment – which, in accordance with section 3.2 is likely the result of shear heating – with the novel phase obtainable only *via* impact treatment. This thus demands answers to: (1) why the novel phase does not appear immediately upon submission of a sample to impact treatment; (2) why the novel phase is favoured by impact treatment; and (3) the ultimate fate of GSM. In considering this we present here an extension to the mechanochemical model of Section 3.2, presenting initial evidence supporting a route to the novel phase that requires the existence of the mechanochemically produced tablet and its corresponding conditions.

3.3.1 Tableting: A 'switch' to novel phase synthesis. From section 3.1 it appears that all treatment initially favours GSM formation, while the novel phase can only be obtained after sufficient impact treatment. An explanation for this may be found in considering the processes involved under impact treatment, a treatment that ultimately yields formation of a solid powder tablet. Initial stages of this tableting process are characterised by loose particles being thrust past one another by the impacting body. With particles thrusting past one another there is undoubtedly substantial shearing between particles. As vacant space within the tablet is reduced, the intensity and length of shearing across particles decreases, ultimately yielding a densely packed tablet consisting of particles incapable of further motion. At this point one could imagine that nearly all incident impact force would be transferred through the tablet as particle–particle impact and minimal shearing, Fig. 10. Therefore, one could imagine implementing such a system as α-glycine + β-malonic acid to measure tableting rates and the forces and energies required. For example, 3 : 1 mixtures of the present system appear to undergo this transition between 45 and 60 min of treatment – *i.e.* between 62.1 J and 82.8 J energy.

It is of interest here to point out that a mechanism involving tablet formation for novel phase production would offer a convincing explanation as to the minute maximum quantities of obtainable product. It is important to note that mechanical treatment of GSM does not form solid tablets; it is likely that mechanically induced melting of the initial reactant components leads to tableting.

3.3.2 **The importance of tableting on reactivity.** To investigate the role of tablet formation on the mechanochemical product, alternative milling protocols were implemented, Fig. 11. Experiment 1, in which the tablet was manually destroyed at intermittent points in the milling cycle, yielded substantial quantities of GSM, with no notable indications of the novel phase. In contrast, the product mixture obtained from Experiment 2, in which the tablet was left intact, indicated no signs of GSM, with observable quantities of the novel phase. It is therefore clear that the existence of this tablet plays a crucial role in specifying the resulting product and may offer an explanation for the observed selectivity of shear and impact treatment; both of which may induce melting, but only the latter of which does so within the presence of a tablet framework, yielding nucleation under constrained conditions.

Alternatively, one may suggest that Experiment 1 is characterised by substantial amounts of shear treatment as a result of the aforementioned tableting process being repeated. In contrast, Experiment 2 allows full tablet formation, at which point the relative quantities of shear and impact transferred through the tablet by the mechanoreactor differ substantially.

3.3.3 **Consecutive mechanochemical products.** This concept of tablet formation as a requirement for novel phase growth presents an intriguing starting point for an underlying mechanism of formation. There remains, however, a rather pressing issue: what happens to GSM? From experiment it is obvious that the initial stages of treatment lead to formation of GSM salt, and one might expect that this product would persist, irrespective of continued treatment. This, however, is not the case. Initially it could be suggested that continued treatment may induce GSM degradation (either to the novel phase or reversion to reactants),

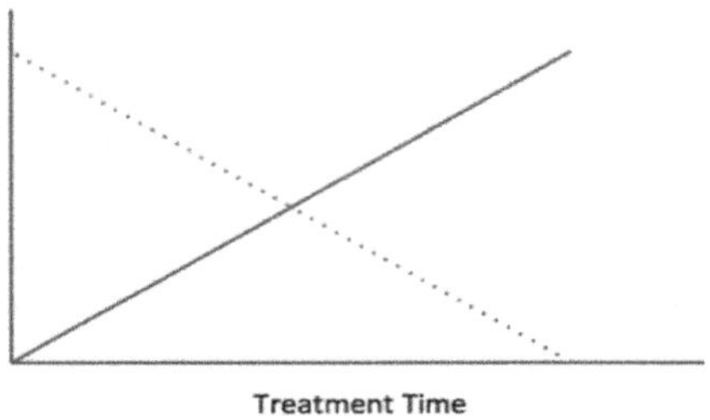

Fig. 10 Diagramatic representation of the change in tablet free space (dashed) and % impact (solid) as a function of treatment time.

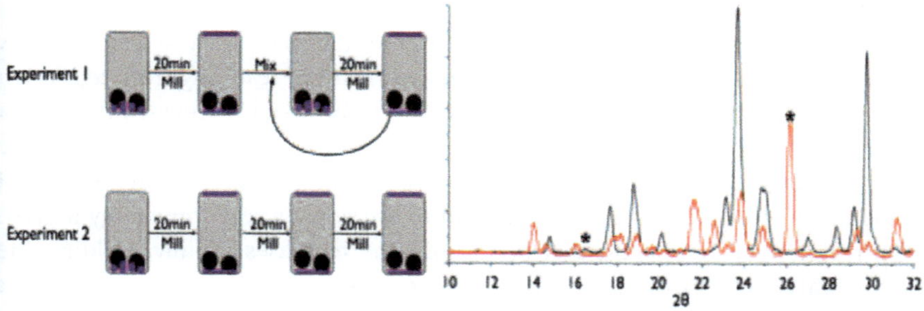

Fig. 11 (Left) Schematic of experimental protocol implemented. (Right) PXRD of resulting products from experiment 1 (red) and experiment 2 (black).(*) Indicates the most intense peak of each product phase: Novel phase (16.5°) and GSM (26°). It is clear that these two very similar protocols yield drastically different products as a result of mechanical treatment.

amorphisation or melting, all of which would lead to the loss of the XRPD pattern corresponding to GSM. In other works,[39,77] however, evidence was presented against these possibilities as routes to the novel phase, with further evidence discounting the possibility of mechanically-induced GSM phase transition as the underlying culprit.

Here we suggest an alternative explanation, rooted in a theory of the mechanical energy available in the system.

In mechanochemical studies, the total excess energy of a system (E) can be defined as the sum of the current energy of the system (E_i) plus some input energy from the mechanical pulse (ΔE), accounting for the loss of energy due to relaxation of the system (ΔE_R) eqn (1),

$$E = E_i + \Delta E - \Delta E_\mathrm{R} \tag{1}$$

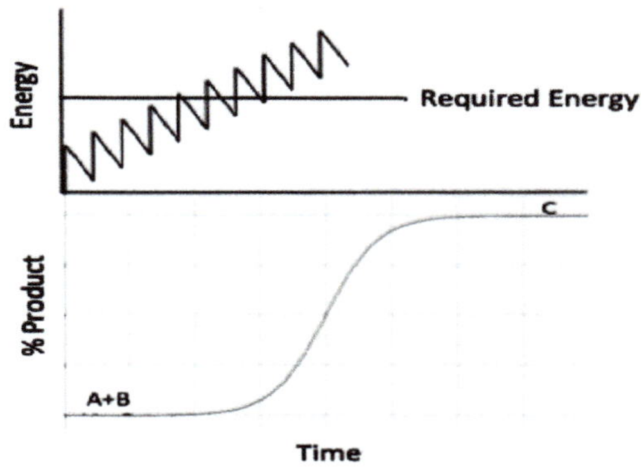

Fig. 12 Diagramatic representation of a system's energy under mechanical treatment (top) and the corresponding reactant (A + B) → product (C) conversion (bottom). The repetitive increase and decrease of the energy of the system is caused by the interplay between mechanical action and the relaxation of the system. The time between successive mechanical action (increasing energy) can be controlled, thus allowing tuning of the rate of energy accumulation. The rate of relaxation (decreasing energy) is a property of the system of study.

One might therefore expect the rate of product accumulation to follow a sigmoidal trend, Fig. 12. In such a system, the interplay between mechanically induced increases in energy and the decrease in energy resulting from relaxation leads to a gradual increase in the total energy of the system, provided the rate of successive bouts of mechanical stressing exceeds the rate of relaxation. As the overall energy of the system increases, there exist, initially, momentary periods of time in which sufficient energy is present to overcome a reaction energy barrier. With prolonged treatment, this period of time increases rapidly, yielding the sigmoidal trend. The observed rate of GSM to novel phase transition is consistent with such a model, with an initial decline in GSM followed by its rapid disappearance, Fig. 13. In addition, this model accounts for the extended time required for full conversion of products observed upon impact treatment as compared to milling treatment; weaker and slower impulses lead to a slower increase in energy, Section 3.1.3.

This model demands two major requirements:

1. The rate of treatment must exceed that of the relaxation process.

2. The rate of treatment must exceed that of nucleation or growth.

This model therefore would offer an exciting opportunity to measure experimentally the rate of relaxation processes and crystal nucleation and growth. By extension of this model, two mechanisms can be suggested to explain consecutive mechanochemical products:

1. The products are directly competitive at the nucleation stage and nuclei can be converted with sufficient energy, *i.e.* one nucleation process to both products.

2. The products are competitive in that they require the same reactants, but nucleation is controlled by the type of energy applied and/or the environment in which nucleation occurs. Loss of GSM results not from conversion to the novel phase, but from loss of crystallinity of the product. *i.e.* two different nucleation processes take place, each to a unique product.

Much work has been done to study relaxation processes in inorganic crystals,[79] however, at present, similar work remains to be performed on organic crystals. Of particular interest is the lifetime of metastable states in crystals, estimated at 10^{-2}–10^{-3} s.[79] Given that inorganic crystals are based on strong interatomic

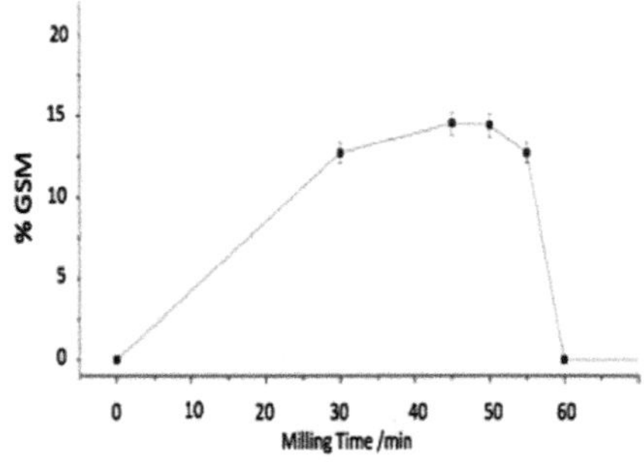

Fig. 13 Percent of glycinium semi-malonate observed in the product mixture following neat milling over a range of durations of milling treatment. The loss of GSM appears to occur initially between 50–55 min, accelerating greatly between 55–60 min, by which point GSM is no longer detectable. This trend is consistent with a sigmoidal increase in mechanically induced energy or pressure in the reacting system.

interactions, while organic crystals are stabilised by weak intermolecular forces, it is feasible to assume that similar processes may be longer-lived – *i.e.* relax at slower rates – in organic systems. Assuming all blows strike the same sample, the impact device used in these studies applies mechanical stress every 0.435 s, with the milling device doing the same every 0.0417 s. It is thus not infeasible, in particular in sight of tablet frameworks discussed below, to suggest that such an interplay of energetic process may be occurring in the α-glycine + β-malonic acid system.

As demonstrated in Section 3.2, GSM appears to be the result of shear-induced melting and re-crystallisation in the absence of a tablet. Initial stages of re-crystallisation may be characterised by a metastable product, as suggested for the nucleation of reactant glycine crystals,[80–82] which can be inhibited from full relaxation – for example by hindering self-assembly processes – to the GSM product nuclei whilst treatment is continued. If sufficient treatment is provided, an energy barrier can be overcome, causing conversion of this hypothetical metastable phase (which may simply be characterised by nanometer metastable nuclei[82]) to an alternative phase, either the novel or an amorphous GSM state. Studies demonstrating the increased ductility of organic substances under tabletting conditions – provided a low-temperature phase transition exists, such as in GSM[77] – lend support to the theory of an amorphous state,[83] with additional evidence for the amorphisation of GSM from experiments with shear treatment.

Provided sufficient, continuous shear treatment is applied, shearing of pure reactants leads to formation of an apparent partially amorphous GSM product, Fig. 14. It is therefore plausible that continued treatment may convert a metastable product to an amorphous state, thus leading to a disappearance of GSM product under continued treatment. Such an effect may be similar to the observation of mechanical amorphisation of sulphathiazole reported previously.[84] It is interesting here to reiterate that conducting any form of mechanical treatment on mixtures initially containing GSM has, to date, proved incapable of producing this effect of disappearing products, and may be the result of providing reaction mixtures with initially stable GSM nuclei. Consistent with the above model, this

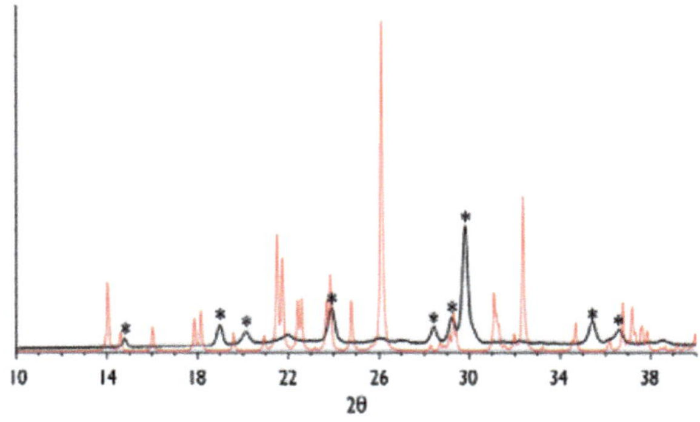

Fig. 14 XRPD resulting from shear treatment of mixtures of 75 mol% α-glycine + 25% β-malonic acid. Sample was subjected to 25 min of continuous shearing treatment. The experimental pattern (black) is compared against the theoretical pattern for GSM (red). Labelled peaks (*) indicate α-glycine, with no obvious signs of malonic acid. Minor, broad features suggest amorphisation of GSM as the resulting product.

suggests that once nucleation has relaxed through an initial metastable state to the stable GSM product it is no longer susceptible to mechanical amorphisation under these experimental conditions, therefore further offering an explanation as to why mechanical treatment of pure GSM yields no effect.[39]

3.3.4 Tablet conditions to novel phase formation. In a mill we can assume realistically that reactants will only react at jar walls and ends, *i.e.* where a three-body collision between the reactants and milling body occurs with any substantial force. When this takes place at jar walls, the shearing product, GSM, is produced, with no indication of the ability to yield either the novel phase or amorphisation, as observed in the shearing device. In contrast, when this collision occurs at the end of a milling jar it compresses and begins to form a tablet. We have demonstrated how truly continuous (*i.e.* with the shearing device) treatment can lead to GSM amorphisation and here extend this concept to discontinuous treatments such as impact and milling by considering the concept of a tablet.

A comparison of milling series conducted across 3 : 1 and equimolar mixtures, Fig. 15 offers interesting evidence towards the role of a tablet. While the product transition for equimolar mixtures occurs between 55 and 60 min of treatment, that for 3: 1 mixtures occurs instead between 40 and 45 min, a marked discrepancy with a minimum difference of 10 min. It is known that, within a solid, the rate – and potentially feasibility – at which a product forms is, to some extent, dependent on the stress this transformation has on the surrounding environment.[52] In its simplest form a pellet can be considered as an extended solid structure and thus it is a not great stretch to propose similar effects within a tablet.

Adopting a measure for crystal density, ρ, as a ratio of the crystal void space (CVS) and unit cell volume (UCV), eqn (2), a simple comparison, Table 1, shows

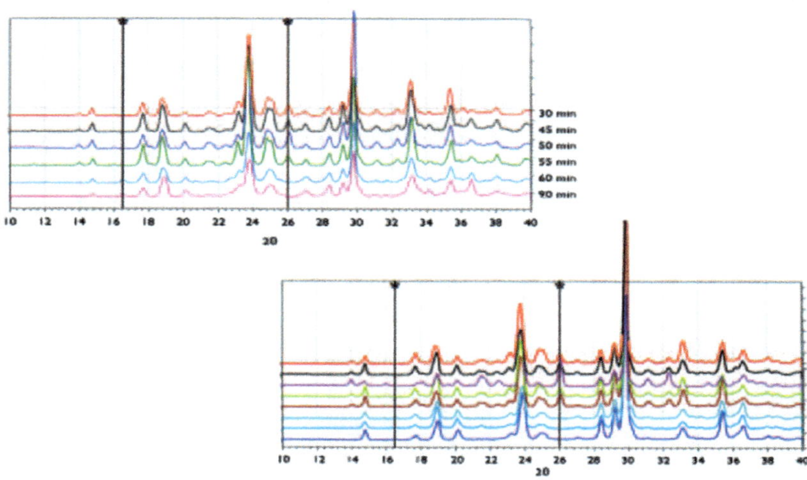

Fig. 15 XRPD resulting from milling at 24 Hz over a series of times. Similar series were conducted for equimolar mixtures (top) and 3 : 1 mixtures (bottom) of α-glycine + β-malonic acid. For clarity, only the major peak of each product phase is indicated (*) at 16.5° (novel phase) and 26.0° (GSM). Both compositions exhibit the same trend of GSM to novel phase conversion upon sufficient treatment. However, this transition occurs under notably less treatment for 3 : 1 mixtures as compared to equimolar mixtures.

Table 1 Unit cell volume (UCV) and crystal void space (CVS) dimensions of reactant and products involved in mechanochemical reactions

Substance	UCV/Å^3	CVS/Å^3	CVS/UCV
α-Glycine	310.122	5.25	1.70%
β-Malonic acid	213.836	19.15	8.96%
GSM	751.201	62.47	8.32%

that, while the free space available within malonic acid and the product phase, GSM, are nearly identical, that available in the glycine unit cell is substantially less. Thus, with increasing quantity of glycine, the amount of free space available in the tablet decreases and the stress of relaxation to produce the product phase likely also increases. In addition, decreasing available void space would also decrease the space available for molecular reorganisation required for re-crystallisation.

It should be noted here that when extending this measure of void space from a crystal to a tablet, an additional term must be added to account for inter-grain void spaces within the tablet itself. Given that in the current system the tablet appears sufficiently strong to induce its effect even under the low frequency, low intensity impact treatment on both 3 : 1 and equimolar mixtures, it can be approximated that, for this system, composition has negligible effects on tablet-ability. As such, this additional inter-grain void space term can be taken as a constant and therefore can be disregarded for the purposes of this model. It is critical to emphasise that this approximation will not hold for all systems, as it is known that excipients can have drastic consequence on compressibility [85]

$$\rho = \frac{\mathrm{CVS}}{\mathrm{UCV}} \times 100 \tag{2}$$

Continued mechanical treatment will therefore strengthen/compress the tablet framework and thus work in opposition to the relaxation, expansion and formation of the GSM phase, thereby offering a mechanism by which the afore-mentioned metastable phase could be maintained. This further reduces the effective value gap between the rate of successive mechanical action and the life-time of metastable states, discussed above, by elongating the lifetime of the metastable state. Further, it follows that extra resistance against relaxation offered by the tablet in such a scheme will likely increase the relaxation rates of mechanically induced stresses. Therefore, for a system in which this resistance is larger, *e.g.* by means of a denser pellet – such as the 3 : 1 mixture – the rate of increase of the energy in the system will hasten, therefore decreasing the time required to induce GSM amorphisation. This is consistent with experiment, Fig. 15.

Under the present model, access to the novel phase is obtained through a unique nucleation event occurring within the constrained, elevated-pressure conditions of the tablet framework, and is not a product of an energetic conver-sion of nuclei, but instead an alternative nucleation event directed by this framework. This can be easily visualised by altering the axes of Fig. 12, bottom, such that the vertical axis represents the fraction of tablet environments at or exceeding the required transformation pressure. Therefore, the initial production of the novel phase would not necessarily be indicative of completion of the tableting process, but instead of the point at which sufficient stress exists in the tablet to induce a change of crystallisation product. Relaxation processes within the tablet would largely be composed of the relaxation of pressure, upon which nucleation can then occur.

An additional interpretation for the effect that altering reactant mixture composition may have on the rate of the GSM → novel phase transition may be proposed based on the stoichiometries of the reaction products. While it is known that altering mixture composition under mechanochemical conditions can induce control of the outcome of product stoichiometry,[86] this effect would simply be expected to alter the relative proportions of product phases obtainable. This is largely due to the fact that the novel phase can only be formed within a tablet, an environment in which particle mixing is no longer a factor. It therefore follows that the composition will affect the amount of product attainable by altering the relative amount of heterogeneous particle interfaces capable of reaction. As outlined in Section 3.2, a global melt is not attainable within the mechanoreactors and as such any reaction will occur solely at these heterogeneous interfaces. The total mixture composition will therefore have minimal effect on the *effective* concentration of the reaction.

While one might alternatively propose a concept borrowed from fluid phase chemistry as an alternative to the above, attempting to explain why alterations in reactant mixture composition may influence the rate of the reaction by means of increasing reactant concentrations, this argument is rendered invalid once tableting is considered.

With this completed model in mind, one could imagine that both altering the rate of tablet formation and allowing tablet relaxation at intermittent points throughout a mechanochemical experiment would alter the amount and type of product obtained. In the event that a tablet is formed slowly, the majority of nuclei would be expected to form as GSM-type and therefore continued treatment would simply result in conversion to an amorphous state. This is simply to say that the vast majority of nuclei will be formed during the tableting process, not within the tablet, with *some* nuclei forming as the novel phase-type. In contrast, rapid formation of tablets would ensure that a larger proportion of nucleation processes occur *within* a tablet and will therefore be of the novel phase-type.

To test this, two experiments were performed: (1) milling experiments with altered intermittent relaxation processes, testing GSM formation; and (2) a comparison of novel phase formation with tableting rate.

In the first set of experiments a sample was milled for three consecutive 20-minute bouts of treatment, altering relaxation time between successive bouts, Fig. 16. Noting that 60 min of continuous treatment yields no signs of GSM formation, these data demonstrate an interesting trend. Indeed, it appears that allowing elongated relaxation times between treatments does increase GSM formation. Unfortunately, due to the immobility of tableted particles, this experimental protocol is only capable of producing small quantities of GSM. As such, elongation beyond 6 min showed no notable increase in GSM production. The observed trends are consistent with the model proposed here, suggesting that the tablet framework is capable of suspending nucleation processes in the, perhaps, the proposed metastable state.

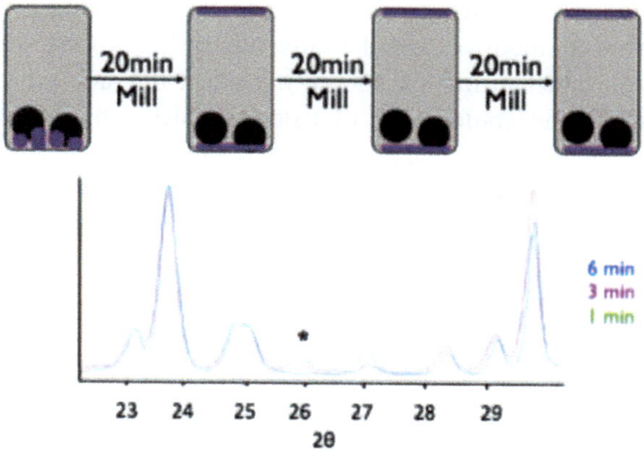

Fig. 16 (Top) Milling protocol used. (Bottom) XRPD of resulting mixtures, showing an increase in the amount of GSM present with increasing wait times between successive 20-minute milling cycles. The major peak of GSM is indicated (*). It appears that increasing tablet relaxation time at intermittent intervals leads to a detectable increase in GSM growth.

Choosing two forms of mechanical treatment at either end of the range of tableting rate, *i.e.* impact (2.1 Hz) and vibratory treatment (50 Hz), a second set of experiments comparing tableting rate and product formation was performed, Fig. 17. Interestingly, consistent with the proposed model, increasing the tableting rate drastically increases the observed amount of novel phase. An alternative explanation for these results may reside in the number of tablets observed (noting that vibrational treatment creates many small tablets distributed about the milling body) however this logic would imply that a larger sample, such as that used in milling studies, would also yield substantially larger quantities of the novel phase; this is not observed over milling studies on mixtures of equivalent composition to impact and vibrational studies.

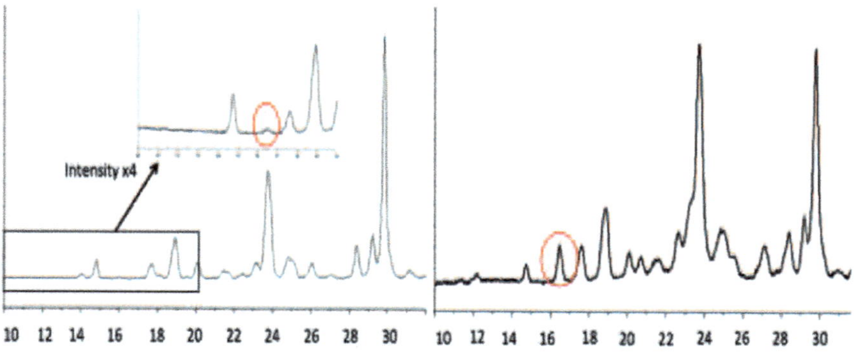

Fig. 17 XRPD patterns resulting from (left) 60 min impact treatment and (right) 8 min vibratory treatment of a 3 : 1 α-glycine + β-malonic acid system. Highlighted peak corresponds to the novel phase. It can be suggested that the intensity of treatment, and thereby the rate of energy or pressure increase in the system, has a positive correlation with the amount of the novel phase attainable.

These results further suggest that GSM and the novel phase cannot be two products that are competitive from the same nucleation process – if this were the case, the rate of tablet formation should not affect the amount of the product observed. Instead, one would expect that slower tablet formation, such as in the mill or impact device, would yield a mixture of the two observable product phases, GSM and the novel phase.

Further investigation into the processes that underpin the mechanochemical reactivity of the α-glycine + β-malonic acid system may allow exciting opportunities to study the kinetics and energetics of mechanical processes in relation to organic systems.

4 Liquid assisted grinding

Liquid assisted grinding (LAG) is becoming an increasingly popular variation of mechanical treatment, offering in many cases substantial improvement and increased control over traditional approaches.[13,14,87] In fact, studies have demonstrated the ability of liquids to have an impressive level of control of mechanochemical processes, in some cases specific solvents favouring a particular product, with other cases showing that a solvent is crucial for the reaction to occur at all.[11,15,32,60,87–90] With these potential benefits, the applications and mechanisms governing LAG have been the focus of numerous studies, however, at present there remains no substantiated theory describing the full workings of this mechanochemical variation. Current theories have suggested a number of roles for the liquid under LAG conditions, including acting as a lubricant to ameliorate component mixing, a medium to facilitate tableting or affecting dielectric permeability, polarizing or even ionizing components, or acting as a solvent, dissolving one or more components thereby allowing the reaction to occur in solution.[11] Current trends in mechanistic LAG studies involve comparison of identical reactions in the presence of a range of solvents. In doing so one can often begin eliminating possible interpretations for the effect of a liquid on the mechanochemical reaction of choice. For example, if liquids that do not dissolve

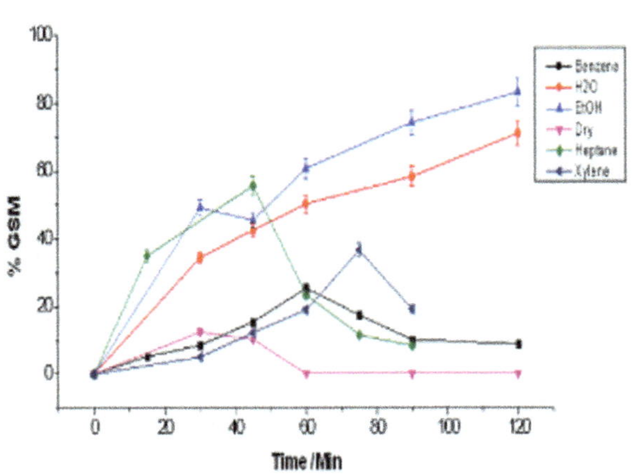

Fig. 18 Rate of formation of GSM under neat and LAG milling experiments over a range of solvents.

any reactant component still exhibit an effect on co-grinding, one can obviously exclude transition to a solution phase reaction.[60]

Following this methodology, a series of solvents were tested for the α-glycine + β-malonic acid system, Fig. 18. The formation of GSM appears to be facilitated by polar liquids, which dissolve the reactant components to a great extent. This is consistent with the aforementioned hypothesis that the formation of this salt proceeds more easily through a fluid state, either a eutectic or, in the case of LAG, through solution at the inter-particle contacts. Examining the effects of non-polar solvents, however, also exhibits an intriguing effect. In the early stages of treatment, these solvents appear to increase (albeit to a much lesser amount as compared to polar solvents) the quantity of GSM produced, as compared to neat grinding. In addition, the GSM → novel phase transition point is also elongated. Following from section 3.3, one might therefore suggest that non-polar solvents, which do not dissolve reactant components, may instead induce their effects by affecting tablet formation. Inhibiting tablet densification would greatly reduce the stabilising capabilities of the tablet framework on the hypothetical metastable state, thereby hastening the stabilisation of GSM nuclei and subsequent growth: *i.e.* increased quantities of GSM. In addition, longer treatment times would be required to achieve sufficient conditions within the tablet to yield the novel phase, thereby explaining the elongation of this GSM → novel phase transition point.

Further studies are required to continue to gain deeper insights into this very complex problem. However, such studies will be dramatically simplified by first obtaining a detailed understanding of mechanochemical mechanisms of the system in question.

5 Conclusion

The α-glycine + β-malonic acid system has to date displayed an immense level of mechanochemical complexity. Under treatment in an SPEX-8000 mill[37,38] the resulting product of mechanical treatment is selective to control of the initial reactant mixture composition, demonstrating an α → γ-glycine phase transition or formation of GSM salt low mol% malonic acid and equimolar mixtures, respectively. We add to this complexity by demonstrating a second divergent reaction pathway, dependent on the type of mechanical treatment applied to the system of study. Speciality devices allowed for segregation of basic mechanical actions, impact and shearing. Impact treatment was observed to ultimately yield formation of a novel phase, at present unobtainable by any other crystallisation methods, whereas shear treatment induced GSM salt formation. This selectivity was independent of the reactant mixture composition. Combination of these observations to mechanical treatment in a common ball mill demonstrated the existence of two unique reaction regions; sampling from the milling jar ends showed novel phase synthesis, while sampling from milling jar walls showed GSM formation. These results could be of drastic consequence to academia and industry, with particular importance to the pharmaceutical industry where small quantities of undesired product may induce a transformation following packing, thereby having consequence on such factors as bioavailability or patent protection.

The importance of tableting for the formation of the novel phase, *i.e.* as a mechanism for the effect of impact treatment, was explored. It was demonstrated that the existence of a tablet is crucial for novel phase synthesis, with further studies suggesting the role of a tablet is to alter crystallisation conditions. We propose a model in which the two products, GSM and the novel phase, diverge at the initial nucleation stage. Under continuous treatment *e.g.* shearing, GSM nuclei are inhibited from relaxing to a stable crystalline state, eventually adopting a non-crystalline state. In contrast, whereas the tablet framework is evidenced to maintain this GSM metastable state, initial crystallisation within this framework leads to novel phase-type nucleation, ultimately responsible for novel phase formation. Considering shearing studies, in which shear treatment (likely shear melting and recrystallisation) under ambient conditions demonstrated a clear novel phase → GSM transition, it can be suggested that, even with novel phase seeds present, GSM is the favoured crystal product. However, whether this is thermodynamic or kinetic, as in the case of the kinetic preference for α-glycine under ambient conditions,[75] remains to be studied.

Acknowledgements

This study was supported by The School of Chemistry, The University of Edinburgh (A. M.) and grants 11-03-00684-a, 13-03-00795 and 13-03-92704 from RFBR (V. D./E. B.). The authors would also like to express their gratitude to Mr E. A. Losev for his assistance with the use of equipment. Thanks are also given to Dr S. G. D. Henderson and Prof. C. R. Pulham, as well as to the newly established Year Abroad agreement between The School of Chemistry (UoE) and Novosibirsk State University, without which this study would not have been possible. Finally, the authors would like to thank Prof. V. V. Boldyrev for his on-going support and fruitful discussions.

References

1 M. C. Lea, *Am. J. Sci.*, 1894, **147**, 377.
2 J. Gilman, *MRS Online Proc. Libr.*, 1999, **539**, 145–151.
3 V. V. Boldyrev, *Treatise on Materials Science and Technology*, 1983, pp. 186–223.
4 F. Urakaev, O. Pozdnyakov, Yu. P. Savintsev and V. V. Boldyrev, *Kinet. Catal.*, 1978, **19**, 1442–1447.
5 F. Urakaev, O. Pozdnyakov, Yu. P. Savintsev and V. V. Boldyrev, *Kinet. Catal.*, 1977, **18**, 350–358.
6 V. Regel, O. Pozdnyakov, F. Urakaev, B. Bylskii and V. V. Boldyrev, *Dokl. Phys. Chem.*, 1975, **221**, 634–637.
7 V. Boldyrev, J. Zarko and A. Deribas, *Chem. High Energy*, 1967, **1**, 117.
8 V. Boldyrev, E. Awwakumav, H. Harenz, G. Heinicke and L. Strugova, *Z. Anorg. Allg. Chem.*, 1972, **393**, 79.
9 S. Yamamura, K. Terada and Y. Momose, *Drug Stability*, 1997, **1**, 209–216.
10 V. Boldyrev, E. Awwakumov, H. Harenz, G. Heinicke and L. Strugova, *Z. Anorg. Allg. Chem.*, 1972, **393**, 152.
11 E. V. Boldyreva, *Chem. Soc. Rev.*, 2013, **42**, 7719–7738.
12 E. V. Boldyreva, *Frontiers in Mechanochemistry and Mechanocal Alloying*, CSIR-National Matallurgical Laboratory, Jamshedpur-831007, India, 2011, pp. 17–30.
13 T. Friščić and W. Jones, *Cryst. Growth Des.*, 2009, **9**, 1621–1637.

14 M. Eddleston, S. Sivachelvam and W. Jones, *CrystEngComm*, 2013, **15**, 175–181.
15 K. Fucke, S. A. Myz, T. P. Shakhtshneider and E. V. Boldyreva, *New J. Chem.*, 2012, **36**, 1969.
16 K. Yamamoto, S. Tsutsumi and Y. Ikeda, *Int. J. Pharm.*, 2012, **437**, 162–171.
17 C. B. Aakeröy, A. B. Grommet and J. Desper, *Pharmaceutics*, 2011, **3**, 601–614.
18 A. V. Trask, D. A. Haynes, W. D. S. Motherwell and W. Jones, *Chem. Commun.*, 2006, 51–53.
19 D. Rajguru, B. Keshwal and S. Jain, *Med. Chem. Res.*, 2013, **22**, 5934–5939.
20 L. Reddy, T. Reddy, R. Mohan, A. Mahesh, Y. Lingappa and N. Reddy, *Chem. Pharm. Bull.*, 2013, **61**, 1114–1120.
21 D. Talukdar and A. Talukdar, *Green Chem. Lett. Rev.*, 2013, **6**, 55–61.
22 V. Štrukil, M. D. Igrc, L. Fábián, M. Eckert-Maksić, S. L. Childs, D. G. Reid, M. J. Duer, I. Halasz, C. Mottillo and T. Friščić&cacute, *Green Chem.*, 2012, **14**, 2462–2473.
23 Y. Sagara, T. Komatsu, T. Ueno, K. Hanaoka, T. Kato and T. Nagano, *Adv. Funct. Mater.*, 2013, **23**, 5277–5284.
24 H. Honbo, K. Takei, Y. Ishii and T. Nishida, *J. Power Sources*, 2009, **189**, 337–343.
25 K. Ichimura, K. Aoki, H. Akiyama, S. Horiuchi and S. Horie, *J. Mater. Chem.*, 2010, **20**, 4784–4791.
26 Y. Nakai, E. Fukuoka, S. Nakajima and K. Yamamoto, *Chem. Pharm. Bull.*, 1977, **25**, 3340–3346.
27 T. P. Shakhtshneider, *Solid State Ionics*, 1997, **101–103**, 851.
28 H. K. Chan and E. Doelker, *Drug Dev. Ind. Pharm.*, 1985, **11**, 315.
29 P. D. Martino, A.-M. Guyot-Hermann, P. Conflant, M. Drache and J.-C. Guyot, *Int. J. Pharm.*, 1996, **128**, 1–8.
30 M. Otsuka and N. Kaneniwa, *Int. J. Pharm.*, 1990, **124**, 307.
31 J. F. Remenar, S. L. Morissette, M. L. Peterson, B. Moulton, J. M. MacPhee, H. R. Guzmán and Ö. Almarsson, *J. Am. Chem. Soc.*, 2003, **125**, 8456–8457.
32 I. Halasz, A. Puškarić, S. A. Kimber, P. J. Beldon, A. M. Belenguer, F. Adams, V. Honkimäki, R. E. Dinnebier, B. Patel, W. Jones, V. Štrukil and T. Friščić&cacute, *Angew. Chem., Int. Ed.*, 2013, **52**, 11538–11541.
33 A. Delori, T. Friščić and W. Jones, *CrystEngComm*, 2012, **14**, 2350.
34 S. A. Myz, T. P. Shakhtshneider, K. Fucke, A. P. Fedotov, E. V. Boldyreva, V. V. Boldyrev and N. I. Kuleshova, *Mendeleev Commun.*, 2009, **19**, 272–274.
35 S. Childs and K. Hardcastle, *Cryst. Growth Des.*, 2007, **7**, 1291–1304.
36 M. A. Mikhailenko, T. P. Shakhtshneider and V. V. Boldyrev, *J. Mater. Sci.*, 2004, **39**, 5435–5439.
37 E. A. Losev, M. A. Mikhailenko and E. V. Boldyreva, *Dokl. Phys. Chem.*, 2011, **439**, 153–156.
38 E. A. Losev, M. A. Mikhailenko, A. F. Achkasov and E. V. Boldyreva, *New J. Chem.*, 2013, **37**, 1973–1981.
39 A. A. Michalchuk, I. A. Tumanov and E. V. Boldyreva, *CrystEngComm*, 2013, **15**, 6403–6412.
40 I. A. Tumanov, A. F. Achkasov, E. V. Boldyreva and V. V. Boldyrev, *CrystEngComm*, 2011, **13**, 2213.
41 I. A. Tumanov, A. F. Achkasov, S. A. Myz, E. V. Boldyreva and V. V. Boldyrev, *Dokl. Phys. Chem.*, 2014, under review.
42 I. A. Tumanov, A. F. Achkasov, E. V. Boldyreva and V. V. Boldyrev, *Russ. J. Phys. Chem. A*, 2012, **86**, 1014–1017.

43 K. L. Nguyen, T. Friščić, G. M. Day, L. F. Gladden and W. Jones, *Nat. Mater.*, 2007, **6**, 206–209.

44 H. Lin, G. Zhang, P. Hsu and S. Lin, *Microchem. J.*, 2013, **110**, 15–20.

45 T. Friscic, I. Halasz, P. J. Beldon, A. M. Belenguer, F. Adams, S. A. Kimber, V. Honkimaki and R. E. Dinnebier, *Nat. Chem.*, 2012, **5**, 66–73.

46 E. A. Losev, B. A. Zakharov, T. N. Drebushchak and E. V. Boldyreva, *Acta Crystallogr., Sect. C: Cryst. Struct. Commun.*, 2011, **67**, 297–300.

47 T. Friščić, S. Childs, S. A. A. Rizvi and W. Jones, *CrystEngComm*, 2009, **11**, 418–426.

48 S. K. Wolff, D. J. Grimwood, J. J. McKinnon, M. J. Turner, D. Jayatilaka and M. A. Spackman, *Cryst. Explorer*, 2012.

49 M. J. Turner, J. J. McKinnon, D. Jayatilaka and M. A. Spackman, *CrystEngComm*, 2011, **13**, 1804–1813.

50 W. Kraus and G. Nolze, *Powder Cell.*

51 D. Braga and F. Grepioni, *Angew. Chem., Int. Ed.*, 2004, **43**, 4002–4011.

52 V. V. Boldyrev, M. Bulens and B. Delmon, *Control of the Reactivity of Solids*, Elsevier, 1979.

53 *Reactivity of Molecular Solids*, ed. E. V. Boldyreva and V. V. Boldyrev, Wiley and Sons, Ltd., 1999.

54 V. V. Boldyrev, *J. Therm. Anal.*, 1993, **40**, 1041–1062.

55 *Pharmaceutical Salts and Co-crystals*, ed. J. Wouters and L. Quéré, RSC, Cambridge, 2011.

56 G. Rothenberg, A. P. Downie, C. L. Raston and J. L. Scott, *J. Am. Chem. Soc.*, 2001, **123**, 8701–8708.

57 V. V. Boldyrev, *Powder Technol.*, 2002, **122**, 247–254.

58 D. Balarew, *Z. Anorg. Allg. Chem.*, 1924, **136**, 216–220.

59 L. M. Kulberg, L. S. Mustafin and A. I. Cherkasov, *Ukr. Chem. J.*, 1959, **18**, 547–552.

60 S. A. Myz, T. P. Shakhtshneider, N. A. Tumanov and E. V. Boldyreva, *Russ. Chem. Bull.*, 2012, **9**, 1782–1793.

61 D. D. Saratovkin and P. A. Savintcev, *Proc. Acad. Sci. U. S. S. R.*, 1941, **33**, 303.

62 R. P. Rastogi and P. S. Bassi, *J. Phys. Chem.*, 1964, **68**, 2398–2406.

63 K. B. Gerasimov and V. V. Boldyrev, *Mater. Res. Bull.*, 1996, **31**, 1297–1305.

64 H. Yamashita, Y. Hirakura, M. Yuda, T. Teramura and K. Terada, *Pharm. Res.*, 2012, **30**, 70–80.

65 K. Chadwick, R. Davey and W. Cross, *CrystEngComm*, 2007, **9**, 732.

66 T. P. Shakhtshneider and V. V. Boldyrev, in *Reactivity of Molecular Solids*, ed. E. V. Boldyreva and V. V. Boldyrev, Wiley and Sons, Ltd., 1999, pp. 271–311.

67 R. P. Rastogi, P. S. Bassi and S. L. Chadha, *J. Phys. Chem.*, 1963, **67**, 2569–2573.

68 R. Kuroda, Y. Imai and N. Tajima, *Chem. Commun.*, 2002, 2848–2849.

69 K. Seefeldt, J. Miller, F. Alvarez-Nunez and N. Rodriguez-Hornedo, *J. Pharm. Sci.*, 2007, **96**, 1147–1158.

70 M. Fukai, T. Matsuo and H. Suga, *Thermochim. Acta*, 1991, **183**, 215–243.

71 S. Ganguly, J. R. Fernandes, G. R. Desiraju and C. N. R. Rao, *Chem. Phys. Lett.*, 1980, **69**, 227.

72 M. Sorescu and T. Xu, *J. Therm. Anal. Calorim.*, 2011, **107**, 463–469.

73 V. Minkov, V. Drebushchak, A. Ogienko and E. V. Boldyreva, *CrystEngComm*, 2011, **13**, 4417–4426.

74 K. Yukimitu, R. Oliveira, E. Araújo, J. Moraes and L. Avanci, *Thermochim. Acta*, 2005, **426**, 157–161.

75 E. V. Boldyreva, V. A. Drebushchak, T. N. Drebushchak, I. E. Paukov, Y. A. Kovalevskaya and E. S. Shutova, *J. Therm. Anal. Calorim.*, 2003, **73**, 419–428.
76 P. Langan, S. A. Mason, D. Myles and B. P. Schoenborn, *Acta Crystallogr., Sect. B: Struct. Sci.*, 2002, **58**, 728–733.
77 A. A. Michalchuk, V. A. Drebushchak and E. V. Boldyreva, *CrystEngComm*, 2014, under review.
78 V. Y. Yablokov, I. L. Smel'tsova, I. A. Zelyaev and S. V. Mitrofanova, *Russ. J. Gen. Chem.*, 2009, **79**, 1704–1706.
79 G. Heinicke, *Tribochemistry*, Akademic-Verlag, 1985.
80 V. Drebushchak, A. Ogienko and E. V. Boldyreva, *J. Therm. Anal. Calorim.*, 2012, **111**, 2187–2194.
81 N. Surovtsev, S. Adichtchev, V. Malinovsky, A. Ogienko, V. Drebushchak, A. Manakov, A. Ancharov, A. Yunoshev and E. V. Boldyreva, *J. Chem. Phys.*, 2012, **137**, 065103.
82 B. Hamilton, M. Hillmyer and M. W. and, *Cryst. Growth Des.*, 2008, **8**, 3368–3375.
83 M. Cespi, G. Bonacucina, L. Casettari, S. Ronchi and G. F. Palmieri, *Int. J. Pharm.*, 2013, **448**, 320–326.
84 T. P. Shakhtshneider and V. V. Boldyrev, *Drug Dev. Ind. Pharm.*, 1993, **19**, 2055.
85 A. O. Onyekweli, O. O. Kunle and E. I. Okoye, *Res. J. Pharm. Technol.*, 2013, **6**, 1019–1031.
86 S. Karki, T. Friščić and W. Jones, *CrystEngComm*, 2009, **11**, 470–481.
87 G. Bowmaker, *Chem. Commun.*, 2013, **49**, 334–348.
88 L. Loots, H. Wahl, L. V. D. Westhuizen, D. a. Haynes and T. L. Roex, *Chem. Commun.*, 2012, **48**, 11507–11509.
89 D. Braga, S. Giaffreda, F. Grepioni, M. Chierotti, R. Gobetto, G. Palladino and M. Polito, *CrystEngComm*, 2007, **9**, 879–881.
90 A. Trask, W. Motherwell and W. Jones, *Chem. Commun.*, 2004, 890–891.

Faraday Discussions

PAPER

Enhancing covalent mechanochemistry in bulk polymers using electrospun ABA triblock copolymers†

A. L. Black Ramirez,[a] A. K. Schmitt,[b] M. K. Mahanthappa[b] and S. L. Craig[a]

Received 8th January 2014, Accepted 6th February 2014
DOI: 10.1039/c4fd00001c

The mechanochemical activation of covalent bonds in bulk polymers is often characterized by low conversions. Here we report that the activation of *gem*-dibromocyclopropane (*g*DBC) mechanophores embedded in a poly(1,4-butadiene) (PB) is enhanced when a central *g*DBC-PB block is flanked by two polystyrene (PS) end blocks in an ABA-type triblock architecture. Electrospinning the PS-(*g*DBC)PB-PS leads to even greater activation in aligned fiber mats under tension.

Covalent polymer mechanochemistry[1] has in recent years provided access to new chemical reactions,[2–8] mechanistic insights,[1,4,8–17] and polymer transformations.[1,2,18] In bulk polymers, mechanochemical activation has been used as the basis for new classes of responsive polymers that demonstrate stress/strain sensing,[15,16,19–23] molecular level remodeling and stress-strengthening,[4,23–30] and the release of small molecules that are potentially capable of triggering further chemical or material response.[31,32] These demonstrations of mechanochemical reactivity in the bulk, however, are often limited by low levels of mechanophore activation, particularly in cases where the reactions require large (~nN) forces. For example, we recently reported that *gem*-dibromocyclopropane (*g*DBC) mechanophores[33] embedded in the main chain of poly(1,4-butadiene) (PB) can be activated in the bulk to undergo a ring opening reaction to a 2,3-dibromoalkene product under unconstrained uniaxial compression, but only with very low levels of activation (approximately 0.3%) accompanying dramatic, irreversible deformation of the bulk *g*DBC-PB.[6] Uniaxial tension is even less effective, with no mechanophore activation detected by ^{1}H NMR in films stretched to failure.[6]

The lack of mechanical activity was attributed to the relatively modest stresses achieved during tension, at which point the material presumably fails through the disentanglement of the polymer chains. General approaches that can enhance the

[a]*Department of Chemistry, Duke University, Durham, NC, 27708-0346, USA*
[b]*Department of Chemistry, University of Wisconsin-Madison, 1101 University Avenue, Madison, WI 53706, USA*

† Electronic Supplementary Information (ESI) available: synthesis, characterization and additional SAXS data. See DOI: 10.1039/c4fd00001c

force-induced reactivity of a given mechanophore, therefore, are highly desirable. One approach is to devise and optimize molecular-level solutions, such as the recently reported "lever arm effect" that are controlled at the level of the polymer backbone structure.[8,34] Here, we consider an alternative strategy, in which the mechanophore reactivity is influenced by changes in the polymer morphology and macroscopic material structure.

We hypothesized that a cross-linked material would allow for higher stresses and strains to be achieved, improving the magnitude of the forces transduced to the individual polymer chains. Because the insolubility of covalently cross-linked systems would challenge quantitative analysis, and because of discouraging exploratory results in covalently cross-linked polymer gels, we turned to triblock thermoplastic elastomers as an attractive strategy for enhanced mechanical properties. The microphase separation between the blocks could lead to physical cross-linking in samples that could still be readily dissolved in an appropriate solvent. Because PB has proven to be a useful platform for the *g*DBC mechanophores, we focused our attention on poly(styrene-*b*-1,4-butadiene-*b*-styrene) (SBS) triblocks (Fig. 1). The SBS microphase separates into glassy PS and amorphous PB regions, in which the stiff glassy regions act as non-covalent crosslinks that provide high strength and the soft amorphous regions accumulate the large stresses that might trigger mechanochemical activation. We envisioned that incorporating the *g*DBC moiety into the amorphous block would allow for greater force transduction to the mechanophore, relative to that found in pure *g*DBC-PB.

Phase transfer catalyzed dibromocarbene addition (Fig. 1) to the alkenes in the PB block of SBS (140 kDa, 30 wt% styrene, Sigma Aldrich) produced a poly(styrene-*b*-(*g*DBC-*ran*-1,4-butadiene)-*b*-styrene) polymer **1** with 50 mol% overall mechanophore content (by ^{1}H NMR), all of which is randomly distributed in the central soft block, and a total molecular weight of 220 kDa as determined by gel permeation chromatography with multi-angle light scattering detection. Films were cast from a 6.5 wt% solution in toluene, from which the solvent was allowed to slowly evaporate under ambient conditions in a fume hood over 7 days, followed by drying under vacuum at ambient temperature for 24 h. Small-angle x-ray scattering (SAXS) was used to characterize the morphologies of the films. Films of **1** adopted a cylindrical phase with a domain spacing of 21.7 nm, as compared to a spherical morphology, $d = 33.1$ nm, for the SBS precursor (Fig. 2). Uniaxial tensile tests were performed at a constant crosshead velocity of 0.2 mm s^{-1} (initial strain rate of ~0.03 s^{-1}). Two types of loading were investigated: single, constant strain to failure, or cyclic loading in which the films is stretched to 700% strain on the initial loading cycle, relaxed to 0% strain, and stretched in increasing increments of approximately 100% in subsequent cycles (800% strain on the second, 900% on the third, *etc*...).

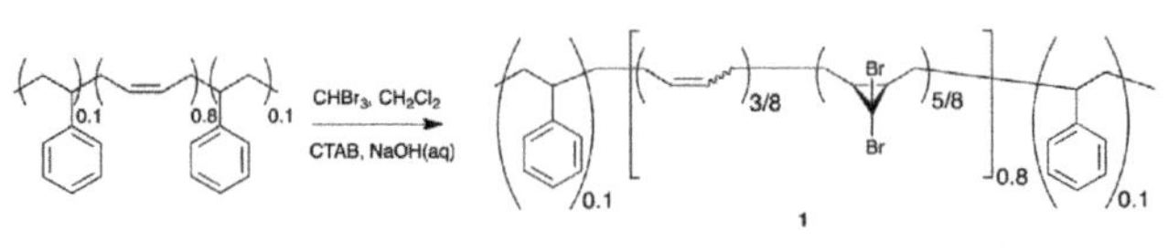

Fig. 1 Phase transfer catalysed carbene addition to generate mechanophore-rich SBS **1**. Note that the central butadiene-*g*DBC block is random.

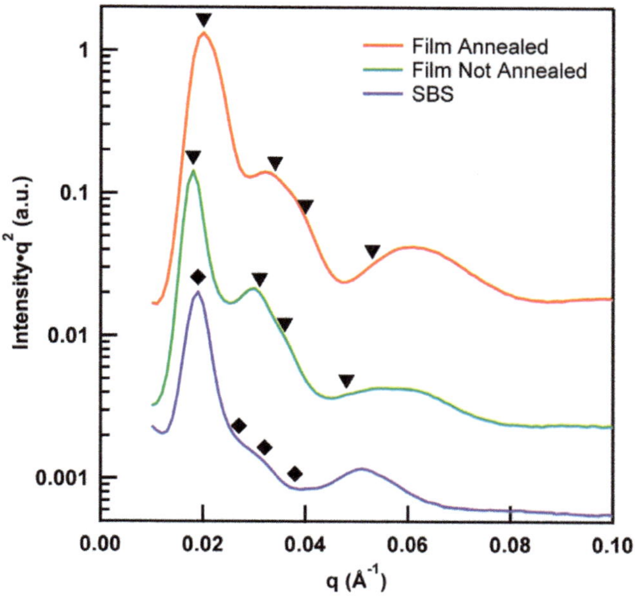

Fig. 2 A small-angle X-ray scattering Kratky plot of $q^2I(q)$ *vs.* q of the annealed film of modified SBS, the non-annealed film of modified SBS and the film of the as-received SBS. The film of the as-received SBS (bottom) exhibits a spherical morphology ($d = 33.1$ nm) with observed scattering peaks (q^*, $\sqrt{2}q^*$, $\sqrt{3}q^*$ and $\sqrt{4}q^*$) denoted with filled diamonds and a broad, spherical form factor scattering peak centered at $q = 0.051$ Å^{-1}. The modified SBS films exhibit a cylindrical morphology with a domain spacing of 21.7 nm for the annealed film (20.9 nm for unannealed) as evidenced by the observed scattering peaks (q^*, $\sqrt{3}q^*$, $\sqrt{4}q^*$ and $\sqrt{7}q^*$) (filled triangles) and a broad, cylindrical form factor scattering peak centered at $q = 0.061$ Å^{-1} ($q = 0.055$ Å^{-1}).

In contrast to those made from pure *g*DBC-PB, the films made from polymer **1** undergo measurable (but still sparse) activation. For both the single strain and cyclic loading cycles, the ^{1}H NMR spectra reveal that approximately 0.2% of the *g*DBC mechanophores are opened to their 2,3-dibromoalkene products (Fig. 3). This activation, while very small, is not observed in unstretched films and represents an enhancement from the undetectable levels of activation observed previously in pure *g*DBC-PB.[6]

Lee *et al.* previously demonstrated that chain alignment in the direction of strain increases the activation of spiropyran mechanophores embedded in a polyurethane.[35] We wondered whether electrospinning aligned fibers might influence the transduction of force and the associated mechanophore activation in **1**. Polymer **1** was electrospun from a 16 wt% solution in $CHCl_3$ to produce ~2 μm fibers that were collected into aligned mats (Fig. 4). These mats, approximately 30 μm thick, were subsequently annealed for 2 h at 70 °C under vacuum.

The mats were again placed in a microstrain analyzer and subjected to uniaxial strain velocities of 0.2 mm s^{-1}, mimicking the conditions employed with the solvent cast films. Whether single strain or cyclic loading cycles were employed, enhanced activation of 1.5% ring opening was observed, which is roughly a factor of seven times greater than that in the films (Fig. 5). Importantly, the analysis of the fibers following electrospinning and annealing, but in the absence of applied

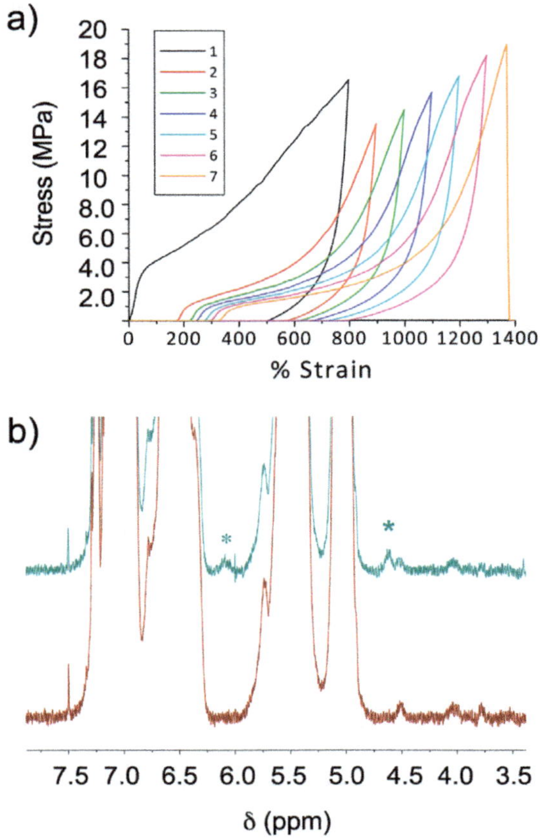

Fig. 3 The cyclic loading at increasing strains (a) results in a modest but detectable formation of the 2,3-dibromoalkene product of *g*DBC ring opening, indicated by ^{1}H NMR resonances denoted by asterisks at $\delta = 6.09$ and 4.60 ppm (b; red spectrum is of film held in absence of tension, teal spectrum is of film loaded in (a)).

tension, showed no detectable *g*DBC ring opening during the fabrication of the mats. A similar lack of activation was observed in mats placed in the microstrain analyzer for **1** h but without an applied strain (Fig. 5).

Although covalent mechanochemistry has been demonstrated previously in triblock polymer architectures,[36] to the best of our knowledge this is the first quantitative comparison of activity in a homopolymer relative to that same polymer as a block in a triblock copolymer. In addition, this is the first report of incorporating mechanophores into electrospun fibers, and the substantially enhanced activation observed as a result both points to and raises interesting questions about the underlying mechanism. The physics at play are undoubtedly complex and beyond the scope of this manuscript, but we consider and briefly evaluate some possible contributions.

Firstly, we note that the mats withstand greater stresses than the films (36 *vs.* 20 MPa), and loading is typically correlated with activity. When unannealed fiber mats are employed, however, the stress at failure (15 MPa) was comparable to or slightly less than that of the solvent cast films, and the enhanced mechanophore

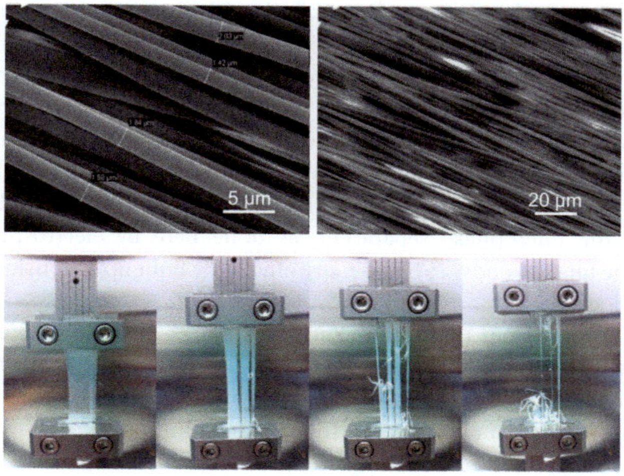

Fig. 4 SEM images (top) of the electrospun fibers of 1. The aligned fiber mats (bottom) were subjected to uniaxial tension and the fibers from the gauge section were collected just prior to or after failure for analysis by ^{1}H NMR.

activation (relative to the solvent cast films) of 1.2% was observed. The difference in activation is apparently due to factors beyond the actual stress at break. Secondly, the stress concentration in the mats is undoubtedly different than that in the films, as evidenced by the distribution of the failed fibers under tension (Fig. 4, bottom). We note, however, that if the strain is stopped as soon as the first fiber fails, total activation across the gauge section is still 1.0%, which is well above that observed in the solvent cast films after complete failure of the film.

Another, initially counterintuitive, influence might be the extent of ordering in the bulk. We were initially surprised to observe that, in contrast to the solvent cast films, no ordered microphase segregation was observed by SAXS in the fibers (see ESI†); there is only a weak correlation hole scattering peak that is symptomatic of

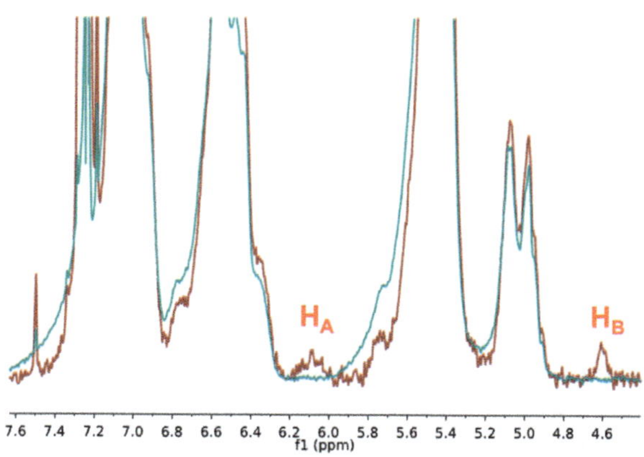

Fig. 5 The ^{1}H NMR overlay of the dissolved fibers after electrospinning and annealing (blue) and then straining to failure at 0.2 mm s^{-1} (red). The resonances indicative of mechanophore activation appear at $\delta = 6.09$ and 4.60 ppm.

the length scale of the block copolymer radius of gyration, and no higher order peaks. The observed lack of ordering with the increased activation is contradictory to our original hypothesis and motivation, but there is reason to believe that it might actually be beneficial to have less order. Prior studies of microphase separated copolymers suggest that the weak point in such materials is often the glassy domain.[37] That is, the glassy polystyrene breaks before substantial strain is built up in the amorphous butadiene regions. It is therefore possible that by disrupting the microphase separation, as done here by electrospinning, the macroscopic tension ends up being more efficiently built up as tension along the *g*DBC-PB block.

Finally, we point out that electrospinning induces chain alignment with the fiber axis, which is in turn aligned with the tension applied to the fiber mats.[38] As the chain and mechanophore alignment are correlated with mechanochemical activation,[15,35] this component of the electrospinning process might also contribute to the enhanced reactivity in the fiber mats.

Conclusions

Incorporating mechanophore-rich polybutadiene as the central block of an SBS-based triblock thermoplastic elastomer yields films with better mechanical properties and enhanced mechanochemical responses relative to mechanophore-embedded polybutadiene alone. The enhanced activation suggests some potential advantages for the design of mechanochemically active bulk polymers. A much greater enhancement (a factor of seven beyond that observed in the films), however, is observed when the functionalized triblocks are electrospun into aligned fiber mats. Somewhat surprisingly, an enhanced mechanical activity occurs despite the absence of ordered microphase separation, which is consistent with earlier reports in which the glassy styrene domains provide a weak point for fracture that might dissipate energy that might otherwise be channelled into mechanophore activation. It is difficult to draw mechanistic conclusions at this time, because a direct comparison between the electrospun mats and solvent cast films is compromised by the numerous structural factors that change simultaneously. Nevertheless, the levels of activation observed in these systems motivate further exploration of mechanophore-rich triblock[18] thermoplastics and electrospun fibers as platforms for bulk polymers with covalent mechanochemical activity. Polymer microstructure provides another handle through which to modulate mechanochemical activity, complementing previously noted "lever arm" effects at the single molecule level.[8]

Acknowledgements

This material is based upon work supported by the U.S. Army Research Laboratory and the U.S. Army Research Office under grant number W911NF-11-1-0071. A. L. B. R. acknowledges a DoD SMART Fellowship. The SAXS data were acquired at the Advanced Photon Source (APS) (Sector 12-ID-B), and we thank Dominic V. Perroni for access to synchrotron beam time through General User Proposal No. 31075 Use of the APS, an Office of Science User Facility operated for the U.S. Department of Energy (DOE) Office of Science by Argonne National Laboratory (Argonne, IL), was supported by the U.S. DOE under Contract No. DE-AC02-06CH11357.

Notes and references

1 M. M. Caruso, D. A. Davis, Q. Shen, S. A. Odom, N. R. Sottos, S. R. White and J. S. Moore, *Chem. Rev.*, 2009, **109**, 5755.

2 C. R. Hickenboth, J. S. Moore, S. R. White, N. R. Sottos, J. Baudry and S. R. Wilson, *Nature*, 2007, **446**, 423.

3 J. M. Lenhardt, M. T. Ong, R. Choe, C. R. Evenhuis, T. J. Martinez and S. L. Craig, *Science*, 2010, **329**, 1057.

4 H. M. Klukovich, Z. S. Kean, S. T. Iacono and S. L. Craig, *J. Am. Chem. Soc.*, 2011, **133**, 17882.

5 J. N. Brantley, K. M. Wiggins and C. W. Bielawski, *Science*, 2011, **333**, 1606.

6 J. M. Lenhardt, A. L. Black, B. A. Beiermann, B. D. Steinberg, F. Rahman, T. Samborski, J. Elsakr, J. S. Moore, N. R. Sottos and S. L. Craig, *J. Mater. Chem.*, 2011, **21**, 8454.

7 K. M. Wiggins and C. W. Bielawski, *Angew. Chem., Int. Ed.*, 2012, **51**, 1640.

8 H. M. Klukovich, T. B. Kouznetsova, Z. S. Kean, J. M. Lenhardt and S. L. Craig, *Nat. Chem.*, 2013, **5**, 110.

9 S. S. Konda, J. N. Brantley, B. T. Varghese, K. M. Wiggins, C. W. Bielawski and D. E. Makarov, *J. Am. Chem. Soc.*, 2013, **135**, 12722.

10 S. Akbulatov, Y. Tian and R. Boulatov, *J. Am. Chem. Soc.*, 2012, **134**, 7620.

11 R. Boulatov, *Nat. Chem.*, 2013, **5**, 84.

12 M. J. Kryger, A. M. Munaretto and J. S. Moore, *J. Am. Chem. Soc.*, 2011, **133**, 18992.

13 Z. S. Kean, Z. Niu, G. B. Hewage, A. L. Rheingold and S. L. Craig, *J. Am. Chem. Soc.*, 2013, **135**, 13598.

14 J. Ribas-Arino and D. Marx, *Chem. Rev.*, 2012, **112**, 5412.

15 B. A. Beiermann, S. L. B. Kramer, J. S. Moore, S. R. White and N. R. Sottos, *ACS Macro Lett.*, 2012, **1**, 163.

16 C. K. Lee, B. A. Beiermann, M. N. Silberstein, J. Wang, J. S. Moore, N. R. Sottos and P. V. Braun, *Macromolecules*, 2013, **46**, 3746.

17 R. Groote, B. M. Szyja, E. A. Pidko, E. J. M. Hensen and R. P. Sijbesma, *Macromolecules*, 2011, **44**, 9187.

18 Z. S. Kean, A. L. Black Ramirez and S. L. Craig, *J. Polym. Sci., Part A: Polym. Chem.*, 2012, **50**, 3481.

19 C. M. Kingsbury, P. A. May, D. A. Davis, S. R. White, J. S. Moore and N. R. Sottos, *J. Mater. Chem.*, 2011, **21**, 8381.

20 B. A. Beiermann, D. A. Davis, S. L. B. Kramer, J. S. Moore, N. R. Sottos and S. R. White, *J. Mater. Chem.*, 2011, **21**, 8443.

21 D. A. Davis, A. Hamilton, J. Yang, L. D. Cremar, D. Van Gough, S. L. Potisek, M. T. Ong, P. V. Braun, T. J. Martinez, S. R. White, J. S. Moore and N. R. Sottos, *Nature*, 2009, **459**, 68.

22 Y. Chen, A. J. Spiering, S. Karthikeyan, G. W. Peters, E. W. Meijer and R. P. Sijbesma, *Nat. Chem.*, 2012, **4**, 559.

23 S. Karthikeyan and R. P. Sijbesma, *Macromolecules*, 2009, **42**, 5175.

24 B. R. Crenshaw and C. Weder, *Macromolecules*, 2006, **39**, 9581.

25 A. L. Black, J. A. Orlicki and S. L. Craig, *J. Mater. Chem.*, 2011, **21**, 8460.

26 A. L. Ramirez, Z. S. Kean, J. A. Orlicki, M. Champhekar, S. M. Elsakr, W. E. Krause and S. L. Craig, *Nat. Chem.*, 2013, **5**, 757.

27 B. R. Crenshaw, M. Burnworth, D. Khariwala, A. Hiltner, P. T. Mather, R. Simha and C. Weder, *Macromolecules*, 2007, **40**, 2400.
28 N. Bruns, K. Pustelny, L. M. Bergeron, T. A. Whitehead and D. S. Clark, *Angew. Chem., Int. Ed.*, 2009, **48**, 5666.
29 S. Y. Cho, J. G. Kim and C. M. Chung, *Sens. Actuators, B*, 2008, **134**, 822.
30 Y. K. Song, K. H. Lee, W. S. Hong, S. Y. Cho, H. C. Yu and C. M. Chung, *J. Mater. Chem.*, 2012, **22**, 1380.
31 C. E. Diesendruck, B. D. Steinberg, N. Sugai, M. N. Silberstein, N. R. Sottos, S. R. White, P. V. Braun and J. S. Moore, *J. Am. Chem. Soc.*, 2012, **134**, 12446.
32 M. B. Larsen and A. J. Boydston, *J. Am. Chem. Soc.*, 2013, **135**, 8189.
33 D. Wu, J. M. Lenhardt, A. L. Black, B. B. Akhremitchev and S. L. Craig, *J. Am. Chem. Soc.*, 2010, **132**, 15936.
34 H. M. Klukovich, Z. S. Kean, A. L. Black Ramirez, J. M. Lenhardt, J. Lin, X. Hu and S. L. Craig, *J. Am. Chem. Soc.*, 2012, **134**, 9577.
35 C. K. Lee, D. A. Davis, S. R. White, J. S. Moore, N. R. Sottos and P. V. Braun, *J. Am. Chem. Soc.*, 2010, **132**, 16107.
36 S. Jiang, L. Zhang, T. Xie, Y. Lin, H. Zhang, Y. Xu, W. Weng and L. Dai, *ACS Macro Lett.*, 2013, **2**, 705.
37 M. K. Mahanthappa, M. A. Hillmyer and F. S. Bates, *Macromolecules*, 2008, **41**, 1341.
38 H. Fong and D. H. Reneker, *J. Polym. Sci., Part B: Polym. Phys.*, 1999, **37**, 3488.

Faraday Discussions

PAPER

Increasing the energy yield of mechanochemical transformations: selected case studies

Anatoly Politov*[ab] and Olga Golyazimova[a]

Received 16th December 2013, Accepted 25th February 2014
DOI: 10.1039/c3fd00143a

The products of mechanical treatment are surface atoms or molecules, substances with a crystal structure different from their initial one (another polymorph, amorphous), point or linear defects, radicals and new chemical substances. It is often assumed, that to increase the yield of the products of a mechanical treatment, it is necessary to increase the treatment time and the mechanical power input. In view of the low energy yield of many mechanochemical transformations, this leads to high power consumption and contamination of the matter under treatment with the wear products of the material of a mill or reactor, in which the mechanical treatment is carried out. As a result, the technological attractiveness of mechanochemical processes is reduced, so that many mechanochemical transformations that have been discovered recently do not reach the stage of commercialization. In the present paper we describe different examples of increasing successfully the energy yield of mechanochemical processes, by a factor of several times to several orders of magnitude, for inorganic and organic substances. An increase in the energy yield of mechanochemical transformations opens new possibilities for their practical usage. In particular, the methods of preliminary treatment and the modes of conducting enzymatic processes that may find application in the production of second-generation biofuels are discussed using lignocellulose materials as examples.

Introduction

Despite the great diversity of mechanochemical reactions, their practical applications still require improvement. In particular, the efficiency of the transformation of mechanical energy into chemical reaction remains low. A substantial fraction of mechanical energy is consumed for plastic deformation and dissipates in the form of heat to increase the temperature of the substance under treatment and the working volume of the devices. Mechanical initiation of

[a]*Institute of Solid State Chemistry and Mechanochemistry, Novosibirsk, Russia*
[b]*Novosibirsk State University, Novosibirsk, Russia. E-mail: anpolitov@yahoo.com; Fax: +7 383 3321550; Tel: +7 903 9981819*

chemical reactions relates to non-thermal methods of increasing reactivity of substances. For the qualitative description of these processes, similarly to the cases of photolysis and radiolysis, it was proposed to describe the efficiency of mechanochemical transformations with the help of the energy-related yield of products[1]

$$G_M = M/D\ , \tag{1}$$

where M is the amount of product obtained as a result of mechanical treatment (moles); D is the dose of mechanical energy input (MJ); and G_M is the energy yield (efficiency) of the mechanochemical transformation (mol MJ^{-1}). The energy yield of various processes is 10^{-4}–10^2 mol MJ^{-1}.

Eqn (1) can be expressed through the mechanical power input per unit mass of the matter under treatment, and treatment time. The amount of the product will be equal to

$$M = G_M W\tau \tag{2}$$

where W is the mechanical powder input (g), and τ is treatment time (s).

The energy yield G_M is a characteristic value for a given mechanochemical transformation. According to eqn (2), there are two methods to increase the yield of products: to increase either the time of treatment, or the input power. However, both methods cause an increase in energy consumption and result in the contamination of the products. This can be compared with the case of a photochemical reaction: the quantum yield being constant, it is possible to increase the product yield by increasing either the exposure, which in its turn may be increased by increasing the irradiation time or the intensity. For photochemical reactions, in some cases, one may succeed in enhancing the quantum yield by choosing an appropriate sensitizing agent. For many thermal reactions the mechanisms of which have already been studied, catalysts were proposed; they cause an increase in reaction rates and provide the formation of target products at lower temperature and pressure.

The energy yield of a reaction is determined by the physicochemical and mechanical properties of the substances under treatment and seems to be unchangeable. However, in some cases the knowledge of the mechanism of a mechanochemical reaction helps one to find the way to increase the G_M, *i.e.* to obtain the necessary products with essentially lower power consumption. In the present work we consider selected mechanochemical transformations in inorganic, organic substances, as well as a case study of the enzymatic hydrolysis of biopolymers. We demonstrate how the knowledge of the mechanisms and of the elementary processes allows one to increase the energy yields of these processes. An increase in the efficiency of the mechano-enzymatic hydrolysis of natural polysaccharides is important for such practical applications as obtaining the second-generation biofuel.

The use of mechanical methods to process lignocellulose materials, in order to obtain the second generation biofuel

Second generation biofuels are manufactured from fermentable saccharides obtained from natural biopolymers through enzymatic rather than acid hydrolysis. Lignocellulose, not a food material, serves as the source of biopolymers. Viewed as a solid-state reaction, the formation of saccharides from lignocellulose is a heterogeneous process of depolymerization of natural polysaccharides resulting from the hydrolysis that takes place at moderate temperature (55 °C) due to the action of a complex of cellulose enzymes. It is important to emphasize that enzymatic hydrolysis is a solid state reaction, and its course is mainly diffusion controlled. There is a problem related to the delivery of enzymes into the reaction zone. So, first of all, it is necessary to make the substrate surface maximally available for enzymes. Second, the reaction products (saccharides), that inhibit the enzymes and are likely to limit their mobility, are accumulated in the reaction zone. The two problems can be solved using mechanical treatment. It is well-known that grinding of the initial raw material can help to increase the reaction area. However, a wide application of this method is limited by high energy consumption required to obtain fine particles. To solve the second problem, mechanical treatment should be carried out *in situ*, in order to remove the products that get accumulated in the reaction zone. However, as was shown in ref. 2 & 3 even mere mixing the solution of reactants in a laminar reactor inactivates the dissolved enzymes. Decreasing energy consumption and increasing the energy yield of the two mechanochemical processes is a challenge.

An increase in the energy yield of grinding lignocellulose materials

The surface-active additives (surfactants) are well-known to have a pronounced effect on decreasing the energy consumption of grinding solids. For example, water is a surfactant for sand, so that the comminution of sand and quartz proceeds with higher efficiency (that is, with lower power consumption) in the presence of water. Quite on the contrary, in the case of plant materials, water is a plasticizer for natural biopolymers that are present in plants, and prevents grinding by directing the mechanical energy to plastic deformation. Plants can be considered as a composite material in which sequential hierarchic destruction of components can be achieved. At first, the decomposition of a plant into tissues occurs. At this stage of destruction, the particle size becomes comparable with the size of a plant cell and equals tens of micrometers. In the second stage, the cells are destroyed. One cannot fully separate the stages of tissue and cell destruction; long-term treatment involves simultaneous destruction of tissue and cells. To simplify primary destruction, it is necessary to remove the substances that hold tissues together. For fine grinding, it is necessary to affect the cell walls of a plant. Treatment in diluted alkalis or acids destroys the bonds between tissues and simplifies the subsequent destruction of the plant material[4,5] However, treatment with acids and alkalis also changes the pH of the plant material and does not constitute an ecologically safe technology. Having studied the structure of the plant cells,[6] we have chosen a complex of enzymes to treat the plant material. After this treatment, the destruction of the plant material was even more efficient than after the treatment with acids and alkalis. On the other hand, the treatment with enzymes is an ecologically safe procedure and allows one to use the material for subsequent enzymatic hydrolysis of natural polysaccharides incorporated in the material. The results of this approach are presented in Tables 1 and 2 and in Fig. 1 and 2.

Preliminary treatment of plant materials with the enzyme was carried out at 30–50 °C for 24 h. A so-called empty fruit bunch (EFB) was used as the lignocellulose material in this study. This is a large-scale waste product from the production of palm oil in equatorial countries.

Grinding of dry EFB was performed in the laboratory mill Fritsch pulverisette 5 (Fritsch, Germany) in agate cylinders with agate balls. Rice husk is ground with a laboratory disintegrator DESI 11 (AS Desintegrator, Estonian Republic). The granulometric composition of ground lignocellulose material before and after enzymatic treatment is shown in Table 1.

The efficiency of grinding before and after preliminary treatment of the raw material can be estimated, and the energy yield calculated by a quantitative comparison of the granulometric composition of ground material. The enzymatic treatment increases the efficiency of grinding by a factor of 3–4 (Table 2). For instance, after grinding EFB after enzymatic treatment for 10 minutes in the laboratory mill Fritsch pulverisette 5, the amount of the fine fraction is 4 times larger than after grinding the initial material with the same moisture content.

The amount of saccharides that passed into solution as a result of preliminary enzymatic treatment was determined. The analysis of the solution showed that the extent of carbohydrate hydrolysis during enzymatic treatment was about 1–2%. So, an insignificant change of the composition causes substantial changes of the mechanical properties of the raw material, and this, in turn, causes an increase in grinding efficiency by a factor of 3–4.5 (depending on the particle size).

This effect can be explained in the following way. First of all, the bonding between the tissues is destroyed, which causes an increase in the content of the fraction with particle size less than 125 micrometers. On the other hand, the hydrolysis causes a partial destruction of the cellulose fibers. These defect sites become the concentrators of strain, similar to the case when a small crack or a cut in the material causes a substantial decrease in its strength.

Table 1 Grinding of lignocellulosic material EFB before and after special pretreatment

Size of particles, micron	Mass fraction of particles (%)					
	EFB initial feed stock	EFB before treatment	EFB after treatment	EFB before treatment	EFB after treatment	EFB before treatment
Time of grinding/min	0	10	10	70	70	120
Large 500	69	67	29	26	—	—
300–500	13	18	21	10	1	2
200–300	8	6	15	19	2	5
125–200	5	4	16	13	9	12
80–125	2	2	7	8	15	12
Less 80	3	3	12	24	73	69

Table 2 Increase of energy yield of grinded lignocellulosic materials

Feed stock	Time of grinding, type of mill	$D < 125$ μm, mass % before treatment	$D < 125$ μm, mass % after treatment	G^a_M/G^b_M	$D < 50$ μm, mass % before treatment	$D < 50$ μm, mass % after treatment	G^a_M/G^b_M
EFB	10 min, Fritsch	5	20	4	—	—	—
EFB	70 min, Fritsch	32	88	2.75	11	48	4.4
Rice husk	Disintegrator DESI 11	42	90	2.1	—	—	—

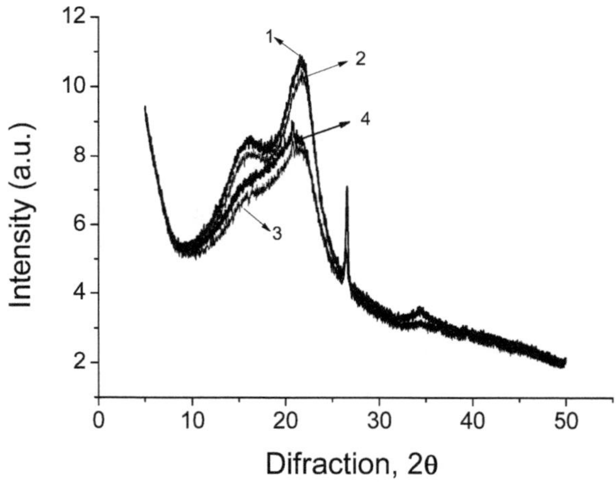

Fig. 1 X-Ray patterns of EFB ground before and after enzymatic treatment. Patterns 1 and 4 correspond to samples that were ground for 120 minutes, patterns 2 and 3 to those ground for 70 minutes.

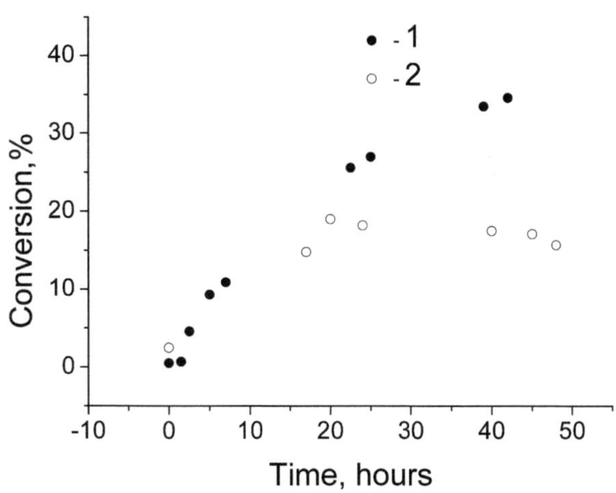

Fig. 2 Influence of mechanical pre-treatment on the hydrolysis of EFB mechanically treated in two different ways, with the mechanical energy being the same: 1 – EFB ground after enzymatic pretreatment, 2 – EFB ground before enzymatic treatment.

An increase in grinding efficiency is observed for all kinds of plant material and with different types of mills. It is shown in Table 2 that grinding of rice shells in the disintegrator after preliminary enzymatic treatment occurs is more than 2 times more efficient. The same amount of input energy causes different results. On the other hand, the same size is achieved on treatment of preliminarily processed raw material with lower energy consumption. The particle size distributions in the powders shown in the two last columns of Table 1 are approximately the same.

However, the energy consumption required to grind the material after preliminary enzymatic treatment was 1.7 times lower.

The rate of substrate hydrolysis depends not only on the particle size, but also on the amorphization degree of cellulose. Efficient grinding (let us call it "smart grinding") not only destroys the material efficiently, but it also promotes cellulose amorphization.

The diffraction patterns 1 and 2 recorded from the particles 71–40 micrometers in size, and diffraction patterns 3 and 4, recorded from the particles smaller than 40 micrometers are plotted in Fig. 1. Samples 1 and 4 were obtained after EFB grinding for 120 minutes, while samples 2 and 3 were obtained after EFB grinding for 70 minutes. The crystallinity index of cellulose, calculated from the intensities of diffraction reflections at 17 and 22 angular degrees, is equal to 43 for the 71–40 μm fraction and 29 for the fraction less than 40 micrometers. The same degree of cellulose amorphization is almost twice as fast if achieved after smart grinding.

It was discovered that the fine fraction of lignocellulose material is hydrolyzed to higher transformation degree than the coarse fraction. The results of the enzymatic hydrolysis of the cellulose incorporated in EFB are shown at Fig. 2. The extent of hydrolysis was estimated as the amount of saccharides that passed into the solution. The size of particles in sample 2 was 200–300 micrometers. Sample 1 was a fine powder obtained as a result of smart grinding, with a particle size less than 80 micrometers. One can see, that the hydrolysis of the raw material composed of the particles of 200–300 micrometers stops at the conversion degree of 15–17%, while the hydrolysis of fine particles proceeds to a higher conversion degree and reaches 35% after 40–45 hours.

An increase in the efficiency of the mechano-enzymatic treatment of the reaction mixture *in situ*

The hydrolysis stops before completion if the substrate is composed of coarse particles (sample 1, Fig. 2). The process stops not only because of a decrease in the substrate availability, but also due to the inactivation or sink of enzymes. If a fresh portion of the enzyme is added to such a sample, the reaction of hydrolysis of natural polysaccharides is resumed and reaches a somewhat higher degree. Our experiments showed that intense mixing of the solution is necessary to remove the products efficiently from the reaction area. However, it was found that the activity of the enzyme is very sensitive to mixing. Thus, mixing solution at medium rates causes a noticeable inactivation of the enzymes. The change of activity on mixing the solution of cellulase enzymes at a frequency of 400 r.p.m. is

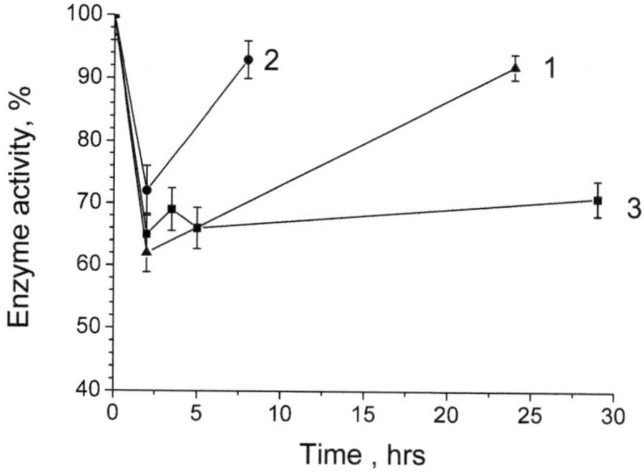

Fig. 3 Fall of enzyme activities resulting from stirring during 2 hours and renewals of the activity with storage at 15 °C. 1, 2 and 3 are independent experiments with equal times of stirring and different times of relaxation.

shown in Fig. 3. The activity of the initial solution before mixing was taken as 100%. After mixing the solution for 2 hours at a temperature of 55 °C the activity of the enzyme accounted for 60–70% of the activity of the initial enzymes. The activity of the enzyme recovered after storage at a temperature of 12–15 °C. If the activity of the enzyme decreased to 70%, it recovered almost completely within 6 hours, as in the case of experiment 1. A decrease in activity to 60% occurred either reversibly (as in case 2), or irreversibly (as in case 3). The decrease in activity was observed not only during turbulent mixing (as in our case), but also in the laminar mixing mode.[2,3] This behavior can be explained considering that the enzyme exhibits its catalytic activity in a certain conformation only. Stretching strain imposed on the enzyme during mixing causes small conformational changes. If deviations from the active conformation are small, this conformation can be recovered. If the conformation changes, however, exceed a certain critical value, the enzyme partially loses its catalytic activity, and its inactivation is observed. An alternative method of mechanical treatment *in situ* is using ultrasound. On the one hand, the action of ultrasound accelerates the enzymatic process due to the acceleration of enzyme diffusion and the removal of products. On the other hand, however, the ultrasonic action (similarly to mixing) can cause enzyme inactivation. It this happens, the best treatment mode will be the batch mode, when the ultrasonic treatment is stopped for a certain time interval, in order to allow the enzymes to relax and recover to the active conformation.

The correctness of this approach was demonstrated by experiments, the results of which are shown in Fig. 4. The treatment with ultrasound was carried out at a temperature below 10 °C, to prevent hydrolysis. The ultrasonic treatment was carried out (a) continuously for 45 minutes before the start of hydrolysis (curve 2), and (b) discretely, 3 times for 15 minutes (curve 3). In case of the discrete treatment, the sample was first treated before the hydrolysis. After that, the substrate was placed into the reactor, and the enzymatic hydrolysis continued at a temperature of 55 °C for 1 hour. The short treatment with ultrasound was repeated after hydrolysis for 1 hour and after hydrolysis for 2 hours. The discrete

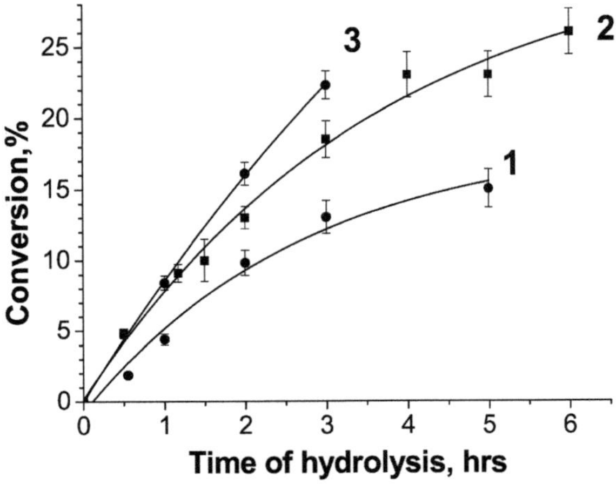

Fig. 4 Ultrasonic assistance of enzymatic hydrolysis of cellulose: 1 – without treatment, 2 – continuous treatment during 45 min, 3 – batch treatments (15 + 15 + 15) min.

(batch mode) treatment of the reaction mixture was shown to be more efficient than a continuous one; with the same consumption of mechanical energy, the efficiency of hydrolysis after 3 hours is 1.4 times higher; after 6 hours the efficiency is 1.8 times higher. We relate an increase in the energy yield of this process to the higher activity of the enzyme, which is preserved, if mechanical treatment is interrupted and the enzyme conformation can relax. Ultrasonic treatment of the reaction mixture causes an efficient increase in hydrolysis rate (curve 1 in Fig. 4). Hydrolysis without ultrasonic treatment over 3 hours is 1.4 times less efficient than that in the case of simultaneous mechanical treatment. For hydrolysis over 6 hours, the extrapolation of kinetic curves gives an estimated increase in the efficiency of ultrasonic assistance by a factor of 2.5. We propose to term the enzymatic hydrolysis proceeding simultaneously with mechanical treatment as mechano-enzymatic hydrolysis.

The case study of the heterogeneous enzymatic hydrolysis of natural oligosaccharides has shown that the effect of mechanical treatment on a process can be not simple. Understanding the mechanisms of the complicated processes involved in the heterogeneous solid-state hydrolysis allowed one to increase the efficiency of the procedure of mechanical action on the reaction.

The efficiency of the mechanical treatment could be improved also for some other reactions which are important for practical applications (Table 3). The

Table 3 Increasing the energy yield of mechanochemical transformations G_M

Substance	Products of mechanical treatment	G_M^b before treatment	Type of pretreatment or *in situ* treatment	G_M^a after treatment	G_M^a/G_M^b
Titanite $CaTiSiO_5$	$CaTiSiO_5$ amorphous	G_M^b	CO_2 medium	$4\ G_M^b$	4
Anthracene	Dianthracene	G_M^b	hν + shear deformation	$500\ G_M^b$	500
$K_2S_2O_8$	SO_5^-, SO_4^- radicals	6×10^{-4}	e^- irradiation, dose is 0.1 Mrad g^{-1}	6×10^{-1}	1000

improvements became possible after the elementary mechanisms of the reactions, including the nature of the short-lived states arising during mechanical treatment, became clear.

Increasing the efficiency of the formation of stable radicals on mechanical treatment

Another example when the energy yield of a mechanochemical reaction could be increased efficiently is provided by the decomposition of persulfates giving radicals (SO_4^- and SO_5^-) as reaction products. Persulfates, for example, are widely used in advanced oxidation processes (AOPs) for water purification and for the treatment of waste. This reaction has attracted much attention as a pure and efficient method permitting a decrease in the amount of organic pollutants. Persulfates dissociate in water to form persulfate anions with a high oxidation potential ($E_o = 2.01$ V). However, this anion reacts only slowly with many organic compounds. The sulfate radical SO_4^- is a stronger oxidizer ($E_o = 2.6$ V), than the initial persulfate anion. Similarly to hydroxyl radicals, sulfate radicals act as strong oxidizers with respect to many organic compounds. SO_4^- was shown to be a more efficient oxidant than $OH^{\cdot}$ is. There are many methods to activate a persulfate anion, for example, with the help of high temperature, UV or using heterogeneous systems, such as photocatalysis with TiO_2.[7] We demonstrated that SO_4^- and SO_5^- radicals are formed on mechanical treatment of potassium, sodium and ammonium persulfates, and can be preserved in the sample at room temperature for an indefinitely long time. This stability of the radicals allows one to carry out preliminary activation of persulfates at one place and then, even at another place and after a long time, to perform a reaction in which these preliminarily produced radicals would participate. The solid matrix acts as a specific "storage container for the radicals", since the mobility of the fragments of persulfate molecule at room temperature is low. The amount of radicals that can be obtained during the mechanical treatment of persulfate with a mechanical energy input of 1 W g^{-1} is 10^{17} spin g^{-1}. So, the energy yield of the formation of mechanical radicals G_M is about 10^{-4} mol MJ^{-1}. The value of G_M does not change within the measurement error limit, if different types of mills with different mechanical energy input are used.

The radical mechanism was proposed for the mechanochemical transformations in potassium persulfate.[8] Its limiting stage is the homolytic breaking of the S–O bond in the persulfate anion. Mechanical treatment involves the formation of the SO_4^- radical pairs, then of the isolated SO_4^- and SO_5^- radicals. Heating causes the transformation of the radicals according to scheme 2–4, to give potassium pyrosulfate and molecular oxygen, which are also the products of thermal decomposition of potassium persulfate.[8] The two types of radicals – SO_4^- and SO_5^- could be detected at a ratio of 2 : 1 during mechanical treatment, while the expected ratio of primary radicals according to the scheme of the chain radical decomposition of a persulfate ion should be 1 : 1. To explain this discrepancy, mechanical treatment was hypothesised to be accompanied, along with the primary reaction of homolytic breakage of O–O and S–O bonds, by electron transfer from one persulfate ion to another, with the formation of two anion radicals – $S_2O_8^-$ and $S_2O_8^{3-}$. These radicals can be preserved at room temperature in the dark without any noticeable changes for several years. However, under even slight mechanical action, the anion radicals decompose with the formation of SO_4^- and SO_5^- radicals. The $S_2O_8^-$ and $S_2O_8^{3-}$ anion radicals formed during preliminary irradiation are transformed quantitatively into the SO_4^- and SO_5^- radicals under mechanical treatment with the specific power of 1 W g^{-1} and the input energy dose of 300 J g^{-1}.

Enhancement of the efficiency of silicate amorphization on mechanical treatment

Mechanical treatment of isle silicates and chain silicates in laboratory mills was shown previously to result in the amorphization of their crystal structures.[9,10] When treated in the planetary mills with the specific mechanical power input up to 10 W g^{-1}, they become X-ray amorphous after 25–60 minutes, depending on the mineral. So, depending on the mineral, the energy-related yield of the amorphous phase is 0.3–0.6 mol MJ^{-1}. According to the theory of non-equilibrium short-lived active states, mechanical treatment results in the disordering of the crystal structure and deformation of covalent bonds at the surface of solids. These excited non-equilibrium molecules relax to less deformed and more equilibrated stated within short time intervals 10^{-7}–10^{-3} s. These states can be preserved at low temperatures. The second method to prevent the dissipation of excited states into heat is to use the energy accumulated in the deformed bonds in a chemical reaction, for example with carbon dioxide. During mechanical activation, the gas can react with the solid surface within 10^{-8}–10^{-7} s at atmospheric pressure, when fast relaxation processes are not yet complete. As a result, carbonates will be accumulated, which will promote the amorphization of silicates. Actually, the rate of amorphization during the mechanical treatment of diopside and titanite in the atmosphere of carbon dioxide is about 4 times higher than that for the activation of these silicates in the air (for equal energy input).[11] So, treatment in the environment of a chemically active gas causes the formation of carbonates, increases the energy-related yield of the amorphous phase, and finally causes an increase in the reactivity of the silicates.

Dimerization of solid anthracene under pressure and shear

9-10-Dimerization of anthracene is observed in its ethanol solutions under irradiation with the wavelength of 365 nm (3.4 eV). The absorption of an energy quantum with an energy of 3.4 eV causes the formation of an eximer which can, during decomposition, convert into a dimer, or relax to restore the initial molecules. To make the dimerization possible, it is important that anthracene molecules should be parallel to each other, and the distance between them should be about 4 Å (Cohen–Schmidt rule).[12]

These conditions are not fulfilled in the anthracene crystal structure, in which the planes of the nearest molecules are turned with respect to each other nearly at a right angle. Because of this, photodimerization in this solid should be impossible. Nevertheless, M. Thomas and J. Williams[13] did observe the formation of dimers on the irradiation of anthracene crystals in the absorption band at the sites of dislocations. It is necessary to combine irradiation of solid anthracene with shear deformation in order to carry out anthracene dimerization. These conditions can be created in diamond anvils, which are transparent for UV radiation. Indeed, almost complete transformation of anthracene into dianthracene was observed under shear and irradiation.[14] The reaction does not proceed at all without shear. However, unexpectedly, shear in the dark (without irradiation) also caused the formation of the dimer.[15–17] No simple explanation of this phenomenon could be suggested, though quantum chemical calculations showed that this reaction should be possible.[15] Dimerization in the dark was followed, measuring quantitatively the luminescence and analyzing the products by means of HPLC. The transformation degree was proportional to the shear angle. During dimerization in the dark, a comparable transformation degree was achieved for a shear angle about 500 times larger than that in the case of light-induced dimerization. The examples of the three reactions considered in this contribution, are very different from each other; some of them are shown in Table 3.

In all cases, in order to suggest an optimum procedure of mechanical treatment and to increase the energy-related reaction yield, it was necessary to study first the mechanisms of the processes. For mechano-enzymatic hydrolysis, it was important to interrupt mechanical treatment and let the system relax. For the formation of radicals, the generation of precursors by means of preliminary irradiation with electrons or UV light was efficient. We think that this procedure will also be applicable to similar reactions that follow the radical mechanism. These compounds include, first of all, peroxides and ionic compounds with covalent bonds, such as dithionates, dithionites, oxalates, *etc.* As far as amorphization on mechanical treatment is concerned, it should proceed in the atmosphere of gaseous carbon dioxide not only for silicates, but also for aluminosilicates and zeolites. The proposed procedures are not universal. On the contrary, each reaction requires an individual approach. This holds for mechanochemistry, as it does for reactivity of solids in general.[18] We expect that the investigation of the mechanisms of mechanochemical reactions will allow one to increase the efficiency of mechanical treatment, and to achieve numerous practical applications of mechanochemistry.[19]

References

1 P. Yu. Butyagin and I. K. Pavlichev, *React. Solids*, 1986, **1**, 361–372.
2 S. E. Charm and B. L. Wong, *Biotechnol. Bioeng.*, 1970, **12**, 1103–1109.
3 Daisuke Taneda, Yoshiki Ueno, Makoto Ikeo and Shohei Okino, *Bioresour. Technol.*, 2012, **121**, 154–160.
4 O. V. Golyazimova, A. A. Politov and O. I. Lomovsky, *Chem. Plant Raw Mater.*, 2009, **2**, 53–57 (*Khimiya Rastitelnogo Syrya*, in Russian).
5 O. V. Golyazimova and A. A. Politov, *Chem. Sus. Dev.*, 2008, **16**, 589–592, (*Chimiya v Interesach Ustoichivogo Razvitiya*, in Russian).
6 T. A. Gorshkova, *Plant Cell Wall as a Dynamic System*, Nauka, Moskow, 2007 (*Rastitelnaya Kletochnaya Stenka kak Dinamicheskaya Sistema*, in Russian).
7 B. Roshani and N. Vel Leitner, *J. Hazard. Mater.*, 2011, **190**(1–3), 403–408.
8 V. F. Radtsig and A. A. Politov, *Kinet. Catal.*, 1985, **26**(issue 1), 33–40.

9 A. M. Kalinkin, A. A. Politov, V. V. Boldyrev, E. V. Kalinkina, V. N. Makarov and V. T. Kalinnikov, *J. Mater. Synth. Process.*, 2002, **10**(1), 61–67.
10 A. M. Kalinkin, E. V. Kalinkina, A. A. Politov, V. N. Makarov and V. V. Boldyrev, *J. Mater. Sci.*, 2004, **39**, 5393–5398.
11 A. A. Politov, A. M. Kalinkin, E. V. Kalinkina, V. N. Makarov, V. V. Boldyrev, *Proceedings of the Fourth International Conference on Mechanochemistry and Mechanical Alloying*, Braunschweig, Germany, 2003.
12 A. R. West, *Solid State Chemistry and its Applications*, John Wiley & Sons, 1985, p. 742.
13 M. Thomas and J. O. Williams, *Prog. Solid State Chem.*, 1971, **6**, 119–154.
14 A. A. Politov, B. A. Fursenko and V. V. Boldyrev, *Dokl. Phys. Chem.*, 2000, **371**(issue 1–3), 28–30.
15 V. M. Tapilin, N. N. Bulgakov, A. P. Chupakhin and A. A. Politov, *J. Struct. Chem.*, 2008, **49**(4), 581–586.
16 V. M. Tapilin, N. N. Bulgakov, A. P. Chupakhin, A. A. Politov and A. G. Druganov, *J. Struct. Chem.*, 2010, **51**(4), 635–641.
17 A. A. Politov, A. P. Chupakhin, V. M. Tapilin, N. N. Bulgakov and A. G. Druganov, *J. Struct. Chem.*, 2011, **51**(6), 1064–1069.
18 V. V. Boldyrev, M. Bulens, B. Delmon, *Control of the Reactivity of Solids*, Elsevier, 1979.
19 E. V. Boldyreva, *Chem. Soc. Rev.*, 2013, **42**(18), 7719–7738.

Faraday Discussions

PAPER

Mechanically induced silyl ester cleavage under acidic conditions investigated by AFM-based single-molecule force spectroscopy in the force-ramp mode

Sebastian W. Schmidt,[abc] Michael F. Pill,[abc] Alfred Kersch,[a] Hauke Clausen-Schaumann[ac] and Martin K. Beyer*[bd]

Received 4th December 2013, Accepted 17th January 2014
DOI: 10.1039/c3fd00119a

AFM-based dynamic single-molecule force spectroscopy was used to stretch carboxymethylated amylose (CMA) polymers, which have been covalently tethered between a silanized glass substrate and a silanized AFM tip *via* acid-catalyzed ester condensation at pH 2.0. Rupture forces were measured as a function of temperature and force loading rate in the force-ramp mode. The data exhibit significant statistical scattering, which is fitted with a maximum likelihood estimation (MLE) algorithm. Bond rupture is described with a Morse potential based Arrhenius kinetics model. The fit yields a bond dissociation energy $D_e = 35$ kJ mol^{-1} and an Arrhenius pre-factor $A = 6.6 \times 10^4$ s^{-1}. The bond dissociation energy is consistent with previous experiments under identical conditions, where the force-clamp mode was employed. However, the bi-exponential decay kinetics, which the force-clamp results unambiguously revealed, are not evident in the force-ramp data. While it is possible to fit the force-ramp data with a bi-exponential model, the fit parameters differ from the force-clamp experiments. Overall, single-molecule force spectroscopy in the force-ramp mode yields data whose information content is more limited than force-clamp data. It may, however, still be necessary and advantageous to perform force-ramp experiments. The number of successful events is often higher in the force-ramp mode, and competing reaction pathways may make force-clamp experiments impossible.

[a]*Munich University of Applied Sciences, Department of Applied Natural Sciences and Mechatronics, Lothstr. 34, 80335 Munich, Germany. E-mail: clausen-schaumann@hm.edu; alfred.kersch@hm.edu; Fax: +49 89 1265 1603; Tel: +49 89 1265 1615*

[b]*Institut für Physikalische Chemie, Christian-Albrechts-Universität zu Kiel, Olshausenstrasse 40, 24098 Kiel, Germany*

[c]*Center for NanoScience (CeNS), Geschwister-Scholl-Platz 1, 80539 Munich, Germany*

[d]*Institut für Ionenphysik und Angewandte Physik, Leopold-Franzens-Universität Innsbruck, Technikerstrasse 25, 6020 Innsbruck, Austria. E-mail: martin.beyer@uibk.ac.at; Fax: +43 512 507 2932; Tel: +43 512 507 52680*

1 Introduction

Covalent mechanochemistry has been rapidly developing over the last decade.[1–3] Quantitative theory has matured, with a variety of approaches available for a wide range of problems, like COnstrained Geometry simulates External Force (COGEF),[4] External Force Explicitly Included (EFEI),[5] Force Modified Potential Energy Surfaces (FMPES)[6] including approximate models based on numerical expansion of the FMPES.[7] For quantitative experiments, the instrument of choice is the atomic force microscope (AFM). Since its introduction in 1994,[8] AFM based single-molecule force spectroscopy (SMFS) has been successfully applied in a versatile manner from the investigation of receptor ligand interactions,[8–10] protein folding,[11–15] adhesion forces and polymer elasticity,[16–23] to studies addressing the mechanical properties of covalent bonds.[1,24–29]

In early SMFS surveys, intermolecular and intramolecular forces were extracted independently of the involved mechanical transient by approximating a constant force-loading rate.[8,24] By expanding this strategy to the dynamic SMFS approach, where the force-loading rate is varied over an appropriate range, a typically nearly linear relationship between binding strength and the logarithm of the force-loading rate can be found. Thus, the structural and kinetic parameters of the underlying process become accessible.[13,30–32]

In order to obtain the kinetic parameters from such dynamic SMFS experiments, different theoretical models have been developed,[4,33–36] all considering bond rupture as a thermally activated process, where the activation barrier is lowered by the mechanical energy stored in the deflected cantilever. A general description of such a thermally activated rupture process is provided by Arrhenius type kinetics, with a force-dependent activation energy E_a, where the bond lifetime $\tau = 1/k^{\text{off}}$ is given by $k^{\text{off}} = A\exp[-E_a(f)/k_BT]$.

For the theoretical description of covalent bonds under mechanical load, an Arrhenius kinetics model combined with a Morse potential has been widely used as an approximation.[4,35,37,38] With the force-dependent deformation of the Morse potential, which enters the force-dependent activation energy in the Arrhenius equation, bond rupture probabilities can be numerically calculated and the dynamics, as well as the structural parameters of these single bond rupture events, can be extracted from the experimental data.[26]

A more recent approach in the realm of AFM–based SFMS is the force–clamp SMFS method, where an individually coupled molecule is stretched with a defined force, and retained at this force until bond scission occurs. The recorded force *versus* time plot yields the bond survival time as a function of the clamp force and temperature. A large number of statistically distributed bond survival times provides the complete unimolecular kinetics of the bond dissociation event, and allows for distinguishing multiexponential decay kinetics.[39] In contrast to dynamic SMFS, the force–clamp SMFS technique provides direct access to the reaction kinetics of mechanically activated processes on the molecular level.[12,14,15,40–43] By analyzing the recorded data with an Arrhenius kinetics model with a force-dependent activation barrier, the measured reaction rate constants can be used directly to calculate force and temperature-independent parameters, like the activation energy, the Arrhenius pre-factor and the width of the binding potential.

In a recent study,[39] we took advantage of force–clamp SMFS to record the force and temperature-dependent bond survival times of carboxymethylated amylose (CMA) covalently tethered to silane functionalized silicon oxide surfaces *via* acid-catalyzed ester condensation at pH 2.0.[44] The conducted experiments revealed bi-exponential rupture kinetics, which have been rationalized with the silyl ester hydrolysis depending on the moieties at the silicon atom under acidic conditions.

Here, we have performed temperature-dependent dynamic SMFS to stretch individual CMA polymers under similar conditions in order to determine the structural parameters of the binding potential, *i.e.* the bond dissociation energy D_e, the parameter β^{-1}, which is proportional to the potential width, and the Arrhenius pre-factor A. Comparing the obtained data with the results from force–clamp SMFS[39] reaffirm the assumption of a silyl ester hydrolysis mechanism. The advantages and disadvantages of force–clamp SMFS and the more widely used dynamic SMFS method are discussed.

2. Experimental procedure and data analysis

2.1 Materials

Carboxymethylated amylose (CMA), N^1-[3-(trimethoxysilyl)-propyl]diethylene-triamine (DETA) and phosphate buffered saline (PBS; buffer composed of 0.137 M NaCl, 0.010 M Na_2HPO_4, 0.003 M KCl, and 0.002 M KH_2PO_4, pH 7.4 at T = 25 °C) were purchased from Sigma-Aldrich (Deisenhofen, Germany). Hydrochloric acid (32% GR for analysis), acetic acid (99–100% for synthesis) and ethanol (absolute GR for analysis) were obtained from Merck (Darmstadt, Germany). All experiments were performed with silicon nitride AFM cantilevers with a nominal force constant between 10 and 20 mNm^{-1} (MLCT-AU, Veeco Instruments GmbH, Mannheim, Germany). Glass microscope slides were obtained from Menzel (Braunschweig, Germany).

2.2 Experimental setup

The sample preparation was performed as previously described.[39,44] Glass microscope slides were immersed in diluted hydrochloric acid for 120 min, sonicated in a cleaning solution for 60 min, and rinsed with water three times. The silicon nitride AFM cantilevers were irradiated with UV light for 60 min and immersed in ethanol. The glass slides and cantilever surfaces were functionalized with DETA by immersion for 60 min in a 10 : 1 ethanol : water mixture, acidified with acetic acid to pH 4.5–5.5, with a DETA content of 2 vol %. After rinsing with ethanol, the slides and cantilevers were cured at 110 °C for 20 min. Prior to individual SMFS experiments, CMA was suspended in PBS titrated to pH 2.0 with diluted hydrochloric acid and transferred to the functionalized glass substrate, followed by thorough rinsing. The slide was then mounted on the AFM stage and covered with PBS buffer, which was also acidified to pH 2.0. Previous experiments with DETA and three other organosilanes under identical experimental conditions showed that CMA is linked to DETA *via* an acid-catalyzed ester condensation.[44]

In order to perform temperature-dependent SMFS experiments, a thermostat (CF30 Kryo-Kompakt-Thermostat, Julabo Labortechnik GmbH, Seelbach, Germany) was connected with a fluid cooler (FLKU 140 G 200, Fischer Elektronik

GmbH, Lüdenscheid, Germany) serving as slide holder, as described elsewhere.[26] After a constant temperature of the solution at the stage was obtained, the SMFS experiments were conducted using an atomic force microscope (NanoWizard, JPK Instruments, Berlin, Germany) in the force spectroscopy mode. The AFM tip was approached to the glass substrate with covalently coupled CMA, and force *vs.* distance curves were recorded. The maximum contact force between the AFM tip and the substrate prior to recording a force *vs.* distance curve was kept below 0.3 nN, and the contact time between the AFM tip and the substrate was kept between 0.5 s and 1.0 s.

2.3 Data analysis

The experimental results were analyzed in a parallel fit procedure using a Morse potential based Arrhenius rate equation as previously described in detail.[26] Force-loading rate-dependent bond rupture distributions were numerically calculated and optimized by varying the parameters D_e, f_{max} and A. In order to account for the exact shape of the bond rupture distributions, parameter optimization was performed with the maximum likelihood estimator method.[45] For reasons of better comparability with the results obtained from the force-clamp SMFS experiments, the Arrhenius pre-factor A was treated as a temperature-independent parameter and the potential width β^{-1} was calculated *via* $\beta = 2f_{max}/D_e$.[4] The parameter errors were calculated as previously described[39] with the MINOS algorithm, which is included in the MINUIT analysis tool.[46] The optimum parameters obtained from the MLE procedure were varied until the maximum of the log-likelihood function was reduced by the value 0.5, which defines the confidence interval of 68.3%.[47]

3. Results and discussion

Scheme 1 displays a single CMA polymer tethered between a glass substrate and AFM tip *via* silyl ester bonds. These bonds are formed under acidic conditions, where a proton catalyzed condensation reaction takes place between the carboxyl groups in the CMA and unreacted hydroxyl groups on the glass surface or on the silane surface anchor.[38]

When the cantilever retracts at a constant velocity from the surface, the individually picked up CMA molecule is stretched until the connection between the AFM tip and substrate is lost. One can obtain the bond rupture force as well as the force-loading rate from the resulting force-extension curve.[27] Varying the retract velocity allows the coverage of a wide range of force-loading rates $\dot{f}$. If the force-loading rate is systematically varied, scatter plots can be obtained as shown in Figure 1, where the bond rupture force is plotted *versus* the force-loading rate.

In order to extract the kinetic parameters from the measured data recorded at five temperatures between 282 K and 320 K, the five data sets were analyzed with a parallel fit procedure using an Arrhenius kinetics model in combination with a Morse potential, as previously described.[27] In brief, the force-loading rate-dependent rupture force probability distributions were optimized for all five temperatures using the maximum likelihood estimation (MLE) method.

The rupture force distributions (green shaded areas) shown in Figure 1 were calculated with the obtained fit parameters $D_e = 35$ kJ mol^{-1}, $\beta^{-1} = 0.061$ Å, and

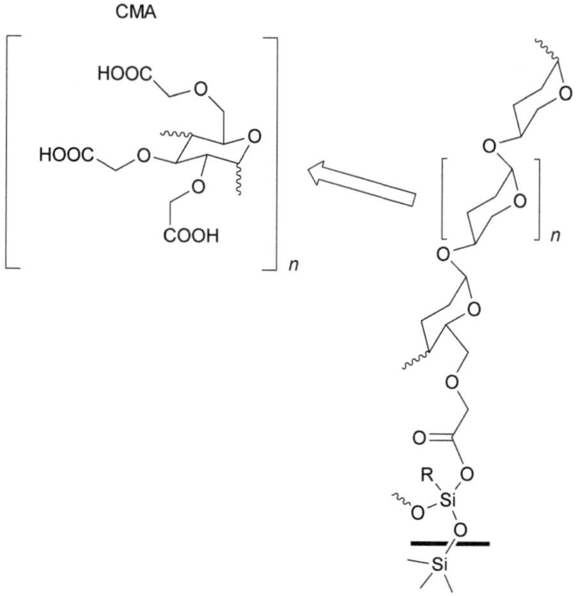

Scheme 1 An individual CMA polymer coupled between a silanized AFM tip and a silanized substrate. Under acidic conditions, the carboxyl groups of the CMA and free hydroxyl groups, which are present on free silanol groups in the silane anchor, undergo proton-catalyzed ester condensation.

the corresponding Arrhenius pre-factor $A = 6.6 \times 10^4\ s^{-1}$. The extracted parameters are summarized in Table 1 and compared to values as determined from experiments conducted under similar conditions using force-clamp SMFS.

The bond dissociation energy $D_e = 34.7$ kJ mol^{-1} as well as the Arrhenius pre-factor $A = 6.6 \times 10^4\ s^{-1}$ determined from dynamic SMFS are almost identical with the ones obtained for the slow process from the force-clamp SMFS experiments, pointing towards the observation of an identical bond rupture processes in both the dynamic and the force-clamp SMFS approach. This serves as a first hint, that the two AFM-based techniques provide consistent kinetic parameters on the single–molecule level, which clearly demonstrates the reliability of the two SMFS strategies. However, the parameters from the dynamic SMFS correspond to the parameters for the slow process and slightly differ from the values for the fast process, which become evident under force-clamp conditions. Together with $\beta^{-1} = 0.061$ Å, which is found to be twice as large as the values extracted from the force-clamp SMFS experiments, the observed variations may be explained by systemic differences concerning data extraction and analysis.

The most obvious difference between the two experimental approaches is given by the fact that experiments conducted by means of dynamic SMFS do not allow for the distinction of multi-exponential kinetics, *i.e.* different subsets of decays typically remain undetected. This can be attributed to the force-loading rate dependence in the case of dynamic SMFS: while the force-clamp SMFS experiments directly yield reaction rate constant(s) $k(f,T)$, which depend on the clamp force and temperature, the reaction rate constant, which determines the data obtained from dynamic SMFS experiments, additionally depends on the

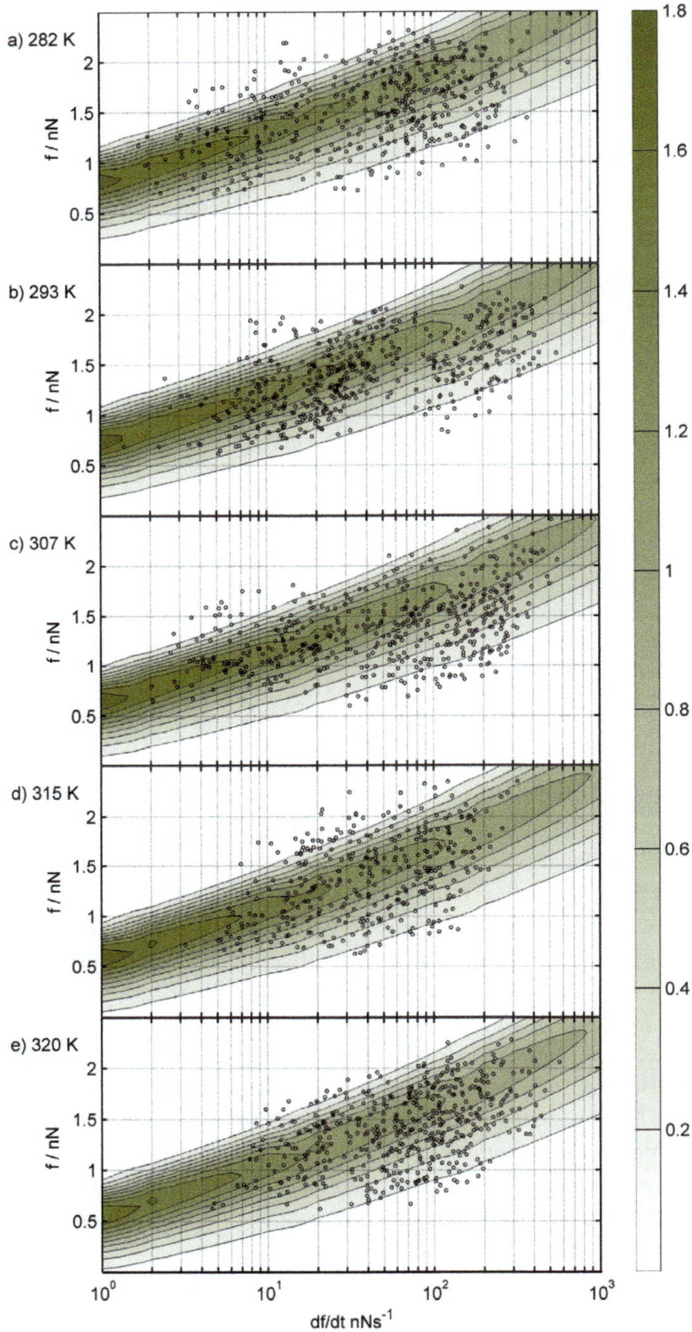

Fig. 1 The bond rupture forces f *vs.* the force loading rates df/dt of more than 2150 single molecule rupture events at 282 K (a), 293 K (b), 307 K (c), 315 K (d) and 320 K (e). Every data point corresponds to one individual rupture event. The green shaded areas show the expected distribution calculated with the fit parameters $D_e = 34.7$ kJ mol^{-1}, $\beta^{-1} = 0.061$ Å, and an Arrhenius pre-factor of $A = 6.6 \times 10^4$ s^{-1}.

Table 1 The parameters extracted from experimental data using an Arrhenius kinetics model in combination with a Morse potential

Parameter[a]	Dynamic SMFS[bd]	Dynamic SMFS[cd]		Force-clamp SMFS[ef]	
		slow	fast	slow	fast
D_e/kJ mol^{-1}	34.7 ± 1.2	54.9 ± 2.3	36.9 ± 1.7	39.2 ± 5.7	39.5 ± 5.1
β^{-1}/Å	0.061 ± 0.023	0.117 ± 0.005	0.098 ± 0.011	0.033 ± 0.0035	0.033 ± 0.020
A/s^{-1}	6.6 ± 0.2 × 10^4	2.0 ± 1.0 × 10^5	2.5 ± 1.1 × 10^4	6.0 ± 1.9 × 10^4	2.0 ± 0.5 × 10^6

[a] Free parameters from the Morse potential-based Arrhenius kinetics model. [b] Values extracted from the global fit to the experimental data shown in Fig. 1a–e. [c] Experimental data shown in Fig. 2a–e with a bi-exponential global fit where the slow process is contributing 32% and the fast one 68%. [d] Errors were obtained by calculating the standard-deviation from a numerically calculated hessian matrix. [e] Results from force-clamp SMFS experiments under similar conditions,[39] where a bi-exponential behaviour was found. According to the bi-exponential fit the overall process is subdivided into two fractions, where the slow process contributes 28% and the fast one 72%. [f] The parameters were calculated with the MLE procedure as previously described.[26] The parameter errors correspond to one-standard-deviation errors as calculated according to the MINOS algorithm as part of the MINUIT analysis tool.[47]

force-loading rate, *i.e.* $k(f,\dot{f},T)$. As a consequence, with the force-loading rate as an additional variable, the experimental data from dynamic SMFS investigations can be described by an arbitrary number of reaction rate constants, which makes a clear distinction of multiple decays practically impossible. Consequently, if the values from the dynamic SMFS summarized in Table 1 actually result from a bi-exponential decay, one has to work with the assumption that the dynamic SMFS parameters represent the optimum compromise between the parameters extracted from the two processes using force-clamp SMFS. However, fitting a data set, which apparently consists of two superimposed data entities, with one single parameter set, might be expected to be accompanied by higher parameter errors due to the influence of larger data scattering. As can be seen from Table 1, the contrary is the case here.

The fact that the parameter errors are even larger for the values extracted from force-clamp SMFS, can be ascribed to the assumption of an inconsistent bond rupture distribution in order to analyze the data recorded with dynamic SMFS. Generally, the MLE-based parameter optimization aims for a bond rupture distribution that makes the observed data most probable, *i.e.* the most likely probability for the ensemble of data point is calculated by varying the unknown fit parameters.[45] The latter is crucial in order to obtain consistent parameter estimates for the observed data. By analyzing two sub-processes with one single set of bond rupture distributions, parameters corresponding to the best fit are indeed found, but the underlying model assumption is inconsistent with the observations from force-clamp SMFS.

As a consequence, in order to be able to directly compare the parameters from dynamic SMFS and force-clamp SMFS, modelling the data shown in Figure 1 with two bond rupture distribution sets is inevitable. Figure 2 shows the results of such a fit, where the branching ratio between the two processes has been treated as a fit parameter. As expected, with the higher flexibility of the fit, the likelihood

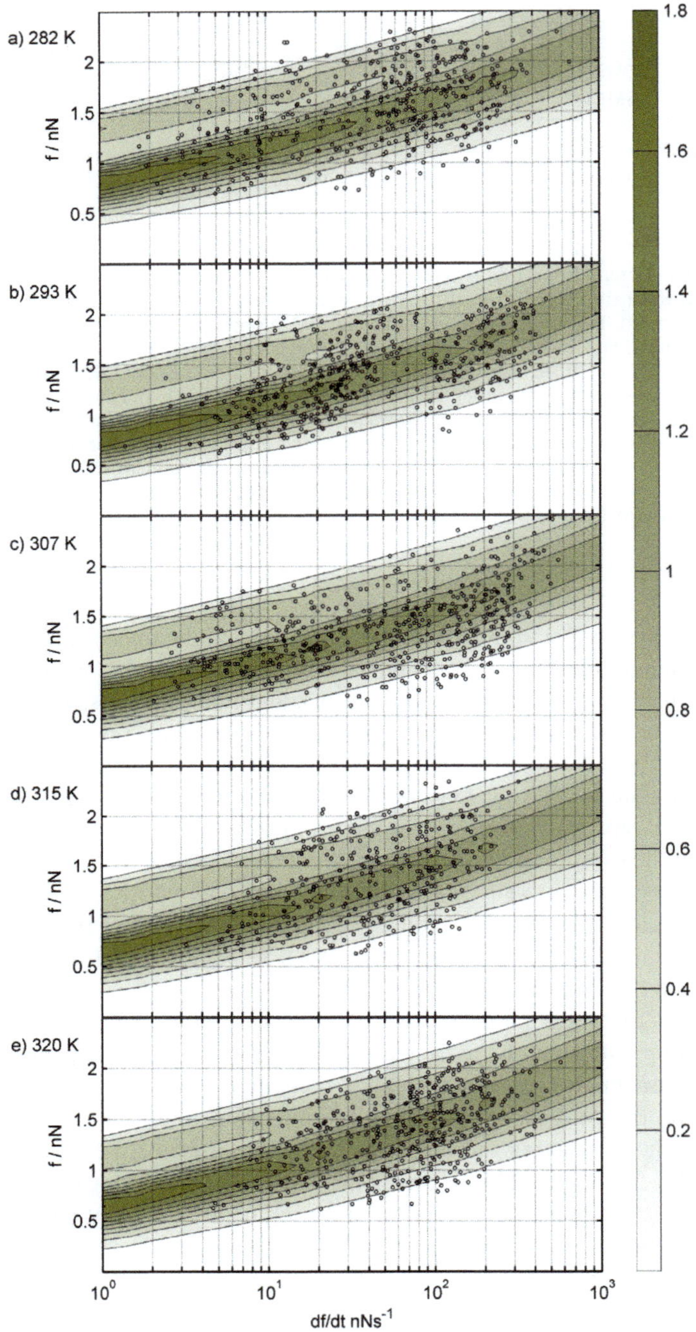

Fig. 2 Here the green shaded areas show the expected distribution calculated with the bi-exponential fit parameters for the slow process of $D_{e,s} = 54.9$ kJ mol^{-1}, $\beta^{-1}{}_s = 0.117$ Å, and an Arrhenius pre-factor of $A_s = 2.0 \times 10^5$ s^{-1}, for the fast process of $D_{e,f} = 36.9$ kJ mol^{-1}, $\beta^{-1}{}_f = 0.098$ Å, and an Arrhenius pre-factor of $A_f = 2.5 \times 10^4$ s^{-1}. The two processes are combined with a weight factor of 0.32 for the slow and 0.68 for the fast process.

function improves significantly. This alone is not unambiguous proof that the bi-exponential fit is more correct. The fact, however, that the branching ratio between the slow and the fast process is similar to the fit of the force-clamp results indicates that the two data sets are consistent, and that the key idea of a bi-exponential behaviour is correct. The fit parameters for the slow process are $D_{e,s} =$ 54.9 kJ mol^{-1}, $\beta^{-1}{}_{s} = 0.117$ Å, and an Arrhenius pre-factor of $A_s = 2.0 \times 10^5$ s^{-1}, for the fast process they are $D_{e,f} = 36.9$ kJ mol^{-1}, $\beta^{-1}{}_{f} = 0.098$ Å, and an Arrhenius pre-factor of $A_f = 2.5 \times 10^4$ s^{-1}. The two processes contribute with a weight factor of 0.32 for the slow and 0.68 for the fast process. The results of the force-ramp and force-clamp fits are summarized in Table 1 to facilitate comparison.

The parameters of the dominant fast process, in particular the dissociation energy D_e, compare favorably with the single-exponential fit. The force-clamp results deviate slightly for the fast process, where a higher dissociation energy is compensated by a higher pre-factor A. Significantly different, however, is the dissociation energy for the slow process. With 54.9 kJ mol^{-1}, it is 15.7 kJ mol^{-1} higher than the corresponding value from the force-clamp fit, and lies well outside the numerical error limits. This may in part be due to the smaller contribution of the slow process. More likely, however, is a systematic problem. The Morse potential based Arrhenius kinetics model, initially developed for the homolytic rupture of a covalent bond in the polymer backbone, is not a very good approximation for the potential energy surface of bond hydrolysis.[48] In the absence of a better model, it is still worthwhile to conduct these fits. A fully consistent picture, with quantitatively reliable results, however, will only emerge when a realistic description of the potential energy surface of the reaction is available and implemented into the fit procedure.

4. Conclusions

In the present study, we have identified structural and kinetic parameters for the mechanically induced rupture of individual covalent bonds using temperature-dependent dynamic SMFS at pH 2.0. Except for the slow process of the bi-exponential fit, the kinetic parameters agree well with the ones extracted from force-clamp SMFS experiments performed under acidic conditions. This clearly indicates the observation of the analogous bond rupture mechanism. The results corroborate that silyl ester hydrolysis is observed at low pH. The consistent results first of all demonstrate the functional equivalence of the two AFM-based single-molecule approaches. However, unambiguous identification of a bi-exponential behaviour, which has been clearly evident in force-clamp SMFS, appeared to be impossible with the dynamic SMFS technique. As a consequence, analyzing the data presented here on the basis of one single bond rupture mechanism yields parameters, which are directly comparable to the parameters from force-clamp SMFS with reservations only. When the analysis model is extended to account for two isomeric decays, dynamic SMFS experiments yield results that deviate from the force-clamp experiments. This may in part be attributed to the low number of recorded bond rupture events per force-loading rate representing a statistically weak data basis for the calculation of two weighted bond rupture distributions. Another limitation is the use of a Morse potential to describe the potential energy surface of a hydrolysis reaction.

Experiments conducted in the force-clamp SMFS mode with the clamp force as the controllable experimental parameter, allow for the direct observation of multiple isomeric decays. This substantial advantage considerably extends the capabilities of AFM-based SMFS methods to the extraction of kinetic parameters from multi-exponential processes on the single–molecule level, which are hardly accessible with the widely used dynamic SMFS approach. Force-ramp experiments will still be useful in cases where the number of successful events is too low in the force-clamp mode, and where the force–distance curve yields additional information.

Acknowledgements

S.W.S., M.F.P. and H.C.-S. gratefully acknowledge financial support of the German Excellence Initiative *via* the "Nanosystems Initiative Munich (NIM)". M.K.B. gratefully acknowledges funding from the Sonderforschungsbereich 677 "Function by Switching" supported by the Deutsche Forschungsgemeinschaft (DFG).

References

1 M. K. Beyer and H. Clausen-Schaumann, *Chem. Rev.*, 2005, **105**, 2921.
2 M. M. Caruso, D. A. Davis, Q. Shen, S. A. Odom, N. R. Sottos, S. R. White and J. S. Moore, *Chem. Rev.*, 2009, **109**, 5755.
3 J. Ribas-Arino and D. Marx, *Chem. Rev.*, 2012, **112**, 5412.
4 M. K. Beyer, *J. Chem. Phys.*, 2000, **112**, 7307.
5 J. Ribas-Arino, M. Shiga and D. Marx, *Angew. Chem., Int. Ed.*, 2009, **48**, 4190.
6 M. T. Ong, J. Leiding, H. Tao, A. M. Virshup and T. J. Martínez, *J. Am. Chem. Soc.*, 2009, **131**, 6377.
7 A. Bailey and N. J. Mosey, *J. Chem. Phys.*, 2012, **136**, 44102.
8 E. L. Florin, V. T. Moy and H. E. Gaub, *Science*, 1994, **264**, 415.
9 P. Hinterdorfer, W. Baumgartner, H. J. Gruber, K. Schilcher and H. Schindler, *Proc. Natl. Acad. Sci. U. S. A.*, 1996, **93**, 3477.
10 F. Schwesinger, R. Ros, T. Strunz, D. Anselmetti, H.-J. Guntherodt, A. Honegger, L. Jermutus, L. Tiefenauer and A. Pluckthun, *Proc. Natl. Acad. Sci. U. S. A.*, 2000, **97**, 9972.
11 M. Rief, M. Gautel, F. Oesterhelt, J. M. Fernandez and H. E. Gaub, *Science*, 1997, **276**, 1109.
12 A. F. Oberhauser, P. K. Hansma, M. Carrion-Vazquez and J. M. Fernandez, *Proc. Natl. Acad. Sci. U. S. A.*, 2001, **98**, 468.
13 M. Rief and H. Grubmüller, *ChemPhysChem*, 2002, **3**, 255.
14 J. M. Fernandez and H. B. Li, *Science*, 2004, **303**, 1674.
15 B. Bullard, T. Garcia, V. Benes, M. C. Leake, W. A. Linke and A. F. Oberhauser, *Proc. Natl. Acad. Sci. U. S. A.*, 2006, **103**, 4451.
16 M. Rief, F. Oesterhelt, B. Heymann and H. E. Gaub, *Science*, 1997, **275**, 1295.
17 P. E. Marszalek, A. F. Oberhauser, Y. P. Pang and J. M. Fernandez, *Nature*, 1998, **396**, 661.
18 T. Hugel, M. Grosholz, H. Clausen-Schaumann, A. Pfau, H. Gaub and M. Seitz, *Macromolecules*, 2001, **34**, 1039.

19 Z. Lu, W. Nowak, G. Lee, P. E. Marszalek and W. Yang, *J. Am. Chem. Soc.*, 2004, **126**, 9033.

20 E. Thormann, D. R. Evans and V. S. J. Craig, *Macromolecules*, 2006, **39**, 6180.

21 Q. Zhang and P. E. Marszalek, *Polymer*, 2006, **47**, 2526.

22 M. I. Giannotti and G. J. Vancso, *ChemPhysChem*, 2007, **8**, 2290.

23 K. Liu, Y. Song, W. Feng, N. Liu, W. Zhang and X. Zhang, *J. Am. Chem. Soc.*, 2011, **133**, 3226.

24 M. Grandbois, M. Beyer, M. Rief, H. Clausen-Schaumann and H. E. Gaub, *Science*, 1999, **283**, 1727.

25 P. Schwaderer, E. Funk, F. Achenbach, J. Weis, C. Bräuchle and J. Michaelis, *Langmuir*, 2008, **24**, 1343.

26 S. W. Schmidt, A. Kersch, M. K. Beyer and H. Clausen-Schaumann, *Phys. Chem. Chem. Phys.*, 2011, **13**, 5994.

27 S. W. Schmidt, M. K. Beyer and H. Clausen-Schaumann, *J. Am. Chem. Soc.*, 2008, **130**, 3664.

28 J. Liang and J. M. Fernández, *J. Am. Chem. Soc.*, 2011, **133**, 3528.

29 H. M. Klukovich, T. B. Kouznetsova, Z. S. Kean, J. M. Lenhardt and S. L. Craig, *Nat. Chem.*, 2012, **5**, 110.

30 R. Merkel, P. Nassoy, A. Leung, K. Ritchie and E. Evans, *Nature*, 1999, **397**, 50.

31 B. Heymann and H. Grubmüller, *Phys. Rev. Lett.*, 2000, **84**, 6126.

32 T. Hugel and M. Seitz, *Macromol. Rapid Commun.*, 2001, **22**, 989.

33 G. I. Bell, *Science*, 1978, **200**, 618.

34 E. Evans and K. Ritchie, *Biophys. J.*, 1997, **72**, 1541.

35 F. Hanke and H. J. Kreuzer, *Phys. Rev. E: Stat., Nonlinear, Soft Matter Phys.*, 2006, **74**, 31909.

36 O. Dudko, G. Hummer and A. Szabo, *Phys. Rev. Lett.*, 2006, **96**, 108101.

37 W. Kauzmann and H. Eyring, *J. Am. Chem. Soc.*, 1940, **62**, 3113.

38 M. F. Iozzi, T. Helgaker and E. Uggerud, *Mol. Phys.*, 2009, **107**, 2537.

39 S. W. Schmidt, P. Filippov, A. Kersch, M. K. Beyer and H. Clausen-Schaumann, *ACS Nano*, 2012, **6**, 1314.

40 A. P. Wiita, S. R. K. Ainavarapu, H. H. Huang and J. M. Fernandez, *Proc. Natl. Acad. Sci. U. S. A.*, 2006, **103**, 7222.

41 S. R. K. Ainavarapu, A. P. Wiita, L. Dougan, E. Uggerud and J. M. Fernandez, *J. Am. Chem. Soc.*, 2008, **130**, 6479.

42 R. Szoszkiewicz, S. R. K. Ainavarapu, A. P. Wiita, R. Perez-Jimenez, J. M. Sanchez-Ruiz and J. M. Fernandez, *Langmuir*, 2008, **24**, 1356.

43 J. Liang and J. M. Fernández, *ACS Nano*, 2009, **3**, 1628.

44 S. W. Schmidt, T. Christ, C. Glockner, M. K. Beyer and H. Clausen-Schaumann, *Langmuir*, 2010, **26**, 15333.

45 S. Brandt, *Data analysis. Statistical and computational methods for scientists and engineers*, Springer, New York, 3rd edn., 1999.

46 F. James and M. Roos, *Comp. Phys. Comm.*, 1975, **10**, 343.

47 CN/ASD Group, *MINUIT - Users Guide. nProgram Library D506*, CERN, 1993.

48 M. F. Pill, S. W. Schmidt, M. K. Beyer, H. Clausen-Schaumann and A. Kersch, *J. Chem. Phys.*, 2014, **140**, 044321.

Faraday Discussions

PAPER

Triboelectricity in insulating polymers: evidence for a mechanochemical mechanism

Lia Beraldo da Silveira Balestrin,[a] Douglas Del Duque,[a] Douglas Soares da Silva[a] and Fernando Galembeck[ab]

Received 4th December 2013, Accepted 17th February 2014
DOI: 10.1039/c3fd00118k

Transfer of reaction products formed on the surfaces of two mutually rubbed dielectric solids makes an important if not dominating contribution to triboelectricity. New evidence in support of this statement is presented in this report, based on analytical electron microscopy coupled to electrostatic potential mapping techniques. Mechanical action on contacting surface asperities transforms them into hot-spots for free-radical formation, followed by electron transfer producing cationic and anionic polymer fragments, according to their electronegativity. Polymer ions accumulate creating domains with excess charge because they are formed at fracture surfaces of pulled-out asperities. Another factor for charge segregation is the low polymer mixing entropy, following Flory and Huggins. The formation of fractal charge patterns that was previously described is thus the result of polymer fragment fractal scatter on both contacting surfaces. The present results contribute to the explanation of the centuries-old difficulties for understanding the "triboelectric series" and triboelectricity in general, as well as the dissipative nature of friction, and they may lead to better control of friction and its consequences.

1 Introduction

Triboelectricity formed on contacting surfaces has been known for centuries but the charge-formation mechanisms have never been well understood,[1–3] and they have been heavily debated in recent years.[4–7] A major issue is the identity of charge carriers and conflicting views have been presented by many authors, based on a great amount of experimental data. Disagreement on the nature of the charge carriers prevents scientists and engineers from developing a consensus on the

[a]*Institute of Chemistry, University of Campinas, Campinas, SP, Brazil 13083-970. E-mail: lia.balestrin@iqm.unicamp.br; douglas.duque@iqm.unicamp.br; dsoares@iqm.unicamp.br; fernagal@iqm.unicamp.br; Fax: +55 19 3521 2906; Tel: +55 19 3521 3014*

[b]*National Nanotechnology Laboratory, National Center for Energy and Materials Research, Campinas, SP, Brazil 13083-970. E-mail: fernando.galembeck@lnnano.cnpem.br; Fax: +55 19 3212 1004; Tel: +55 19 3518 3103*

mechanisms for charge build-up and dissipation. Since tribocharging takes place easily while handling small and large amounts of common commodities like sugar, wheat flour, polyethylene and coal, a practical consequence of the current state of knowledge on this topic is a long series of tragic events, explosions and fires, with losses of lives and property, recorded from many centuries ago to the present.[5,8–10]

1.1 Charge patterns on polymer surfaces

Since Kelvin force microscopy (KFM) and analogous techniques became available in the late nineties, they revealed[11] previously unexpected positive and negative charge domains patterned on the surfaces of any polymer sample that was examined in this and other groups.[12–15] Complex charge distribution patterns were also observed at the macroscopic scale by using scanning Kelvin electrodes,[16] suggesting a fractal character for the potential maps. Another unexpected finding was made by mapping the electrostatic potential on the surface of polytetrafluoroethylene (PTFE) rubbed with polyethylene (PE) foam and other polymer pairs: instead of the often-presumed uniform charge separation among the contacting solids that would make one object positive and the other negative, charged polymer surfaces display complex patterns and large electrostatic potential gradients are thus found along every insulating surface that has been handled or otherwise exposed to the environment. On the other hand, spatial control of surface charging is opening the way to a new kind of electrostatic lithography.[17] Thus, the widespread belief on the formation of separate positive and negative surfaces when two polymer (and other dielectric) surfaces are sheared, rubbed or rolled against each other[18] is not correct. Recognizing this now allows us to understand why there has never been full agreement on setting up a triboelectric series and also why the related quantitative data are not usually presented or available.[19]

1.2 Charge carriers on polymer surfaces

The identification of separate macroscopic domains with opposite charges in the same surface allowed the identification of the charged species formed on PTFE rubbed with PE by using a range of analytical techniques (infrared microspectrophotometry and Raman confocal microscopy, electron-energy loss spectroscopy and controlled pyrolysis), showing that the positive charge derives from the PE hydrocarbon chains while the negative charge is associated with PTFE fluorocarbon fragments, as expected considering the relative positions of PE and PTFE in most triboelectric series. In recent work from this group, charge pattern formation was interpreted as the result of polymer chain breakdown forming free radicals and ions, followed by electron transfer according to the electronegativity of each of these high-energy species formed.[17] Another puzzling observation is charge separation forming adjacent surface electrets instead of the more intuitive ion-pair formation. However, the general tendency of polymers to segregate into immiscible domains instead of mixing, albeit to a limited extent, as most other substances do easily explains charge separation at the micro- and macroscales, when charges are associated with polymer fragments. We recall that the main limitation to polymer miscibility is the positive enthalpy of mixing coupled to the very small mixing entropy of macromolecular substances, as first explained by Flory[20] and Huggins.[21]

1.3 Mechanochemical reactions, triboelectricity and friction coefficients

Mechanochemical reactions triggered by friction are thus the starting point for understanding a number of complex, challenging problems in tribology, from triboelectricity to the dependence of friction coefficients on surface modification during friction.[22]

Finding that tribocharging is the result of mechanochemical polymer reactions allows us to understand the appearance of surface charge on two rubbed samples of the same material (polisiloxane) that was presented by Baytekin *et al.*, as the result of fluctuations of chemical and micromechanical properties in the solids.[14]

One important question that was not yet fully addressed is this: is charge conserved during polymer (and other dielectrics) contact or tribocharging; this means, is the total amount of charge found in the two involved solids equal to the initial charge? Charge conservation is consistent with models and hypotheses based on charge transfer from one solid to another, either ions or electrons, due to any of the various properties of the two solids involved that have been considered by various authors. These properties are, for instance: correlations with the dielectric constant, the basic and acidic nature of the materials, the polymer work function and the surface chemistry determined by measuring the residence time of probe molecules using inverse gas chromatography.[2] On the other hand, recent work from this group[23,24] showed that the atmosphere is a charge reservoir. Water ions were identified as the carriers transferring charge between the solid (and also liquid) surfaces and the atmosphere. Thus, they participate in tribocharge formation: net positive or negative charge on a polymer surface includes contributions from polymer ions formed mechanochemically and distributed on the two involved surfaces added to excess H^+ and OH^- ions adsorbed from the atmosphere together with water vapour.

The observation of macro- and nano-scale charge patterns is consistent with the fractal nature of charge distribution that was shown experimentally earlier.[22] These findings led to additional work showing the effect of tribocharge formation on the rolling and sliding friction coefficients and also to direct experimental evidence for the modification of surface morphology at the nano-scale concurrent with friction.[22]

The identification of the role of free-radicals in tribocharge formation in polymers has already had an important outcome with a great potential for practical application: the demonstration that α-tocopherol, the important vitamin E that is effective as a free-radical suppressor in living bodies, contributes to diminish harmful charge from semiconductor encapsulants.[25]

The effect of mass transfer on the friction properties of solids has been well established in the literature for many years.[26,27] Excellent examples of direct microscopic examination of the sheared, rolled or rubbed surfaces are now found in the literature.[28,29,30] Beyond this, thin films of PTFE produced by friction deposition were studied using grazing incidence X-ray diffraction as the principal tool, showing structural surprises.[31] However, the association between all these phenomena and electrostatic charging of solid surfaces is very recent.[17]

This work presents new results on polymer fragment transfer between rubbed surfaces and their association with tribocharging, using electron microscopy and microanalysis tools that were not previously used in this context and provided previously unavailable information. A main result is the demonstration of a broad size distribution for the transferred fragments that further helps to understand the observed separation of large positive and negative charge domains.

2 Experimental

2.1 Materials

The following polymer materials were used: LDPE foam disks (∅ = 15.0 mm, 3.3 mm thick, Nalgene 6283-1850), technical grade HDPE stubs (∅ = 13.5 mm, 11.2 mm height), PTFE stubs (∅ = 13.5 mm, 11.2 mm height), HDPE films (1 mm thick), PTFE films (1 mm thick), PTFE spheres (precision, ∅ = 3.969 mm), polystyrene Petri dish (∅ = 64 mm, 1.0 mm thick body). Pellets of nylon 6, nylon 6,9, nylon 12, polyacetal and HDPE (*ca.* 2 mm × 2 mm × 2 mm) were from Aldrich. Ethanol (99.5%, Synth) was used for sample cleaning. Materials identity was verified by IR spectra or DSC.

2.2 Methods

2.2.1 Sample cleaning and drying. Polymer films and stubs were immersed in ethanol for 10 min and dried in air for 2 h to minimize pre-existing static charge.

2.2.2 Tribocharging. Two methods for tribocharging were used:

2.2.2.1 Stub spinning. Square polymer films or disks (5 × 5 cm^2) were placed on an aluminum holder mounted on a table-top balance (AM 5500 Automarte, 10 mg resolution) and were rubbed with an LDPE foam disk or polymer stubs fitted on the chuck of a drilling tool spun at 5000 rpm for 3 s. Force applied on the sample was measured using the balance, to produce 12 kPa when using the stubs and 1.5 kPa using PE foam. Fig. 1 shows a picture and schematic description of the experiment.

2.2.2.2 Shaking. The cover of a Petri dish was mounted on top of an aluminum plate that was laid on the horizontal table of a reciprocating shaker. Two grams of pellets were spread on top of the dish and the setup was shaken for 40 min, with a 2.5 Hz reciprocating frequency and 20 mm amplitude. The HDPE film was shaken with 3 g PTFE spheres, for 120 min, 5 Hz and 10 mm amplitude.

2.2.3 Potential mapping. Charge patterns on the polymer surface were determined using a Kelvin electrode mounted on a computer-controlled moving

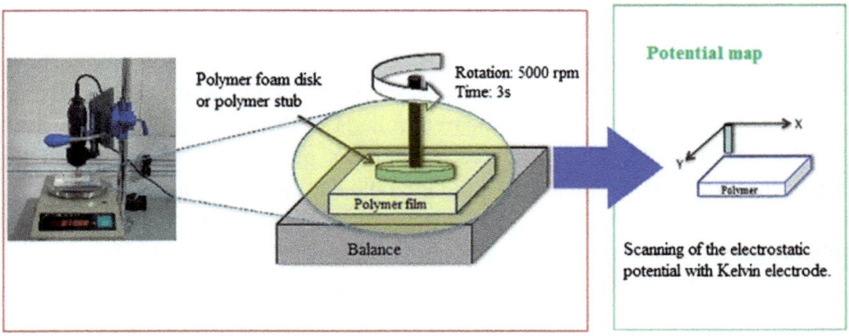

Fig. 1 Experimental sketch of a tribocharging experiment.

arm scanning the x–y plane (Optron) and connected to a voltmeter (model 347, Trek, 25 mm^2 area).

2.2.4 SEM. The samples were placed on a metallic sample holder and coated with a thin gold or carbon layer, using a MCS 010 Sputter (Bal-Tec). Morphological and chemical composition analysis of tribocharged samples were done on a JEOL JSM-6360 LV scanning electron microscope. The acceleration voltage in the reported experiments was 15 kV. Energy dispersive X-ray analysis (EDX) was done in the same microscope with a Noran System Six microanalysis system.

2.2.5 Infrared spectra. ATR/IR spectra were acquired with an IlluminatIR II instrument (Smiths) coupled to an Olympus BX51 microscopic using a ZnSe window in the spectral range between 650 and 4000 cm^{-1}, with 64 scans and 4 cm^{-1} resolution.

3 Results

The results reported here give new evidence for mass transfer between rubbed polymer surfaces producing wear and showing that mass transfer is always concurrent with charge deposition and patterning at the polymer surfaces. On the other hand, transfer is possible thanks to mechanochemical polymer chain scission that takes place either by rupturing separate chains or in a concerted way, when many chains are disconnected at once, allowing a polymer particle to be pulled out of one rubbing surface, landing on another spot on either surface.

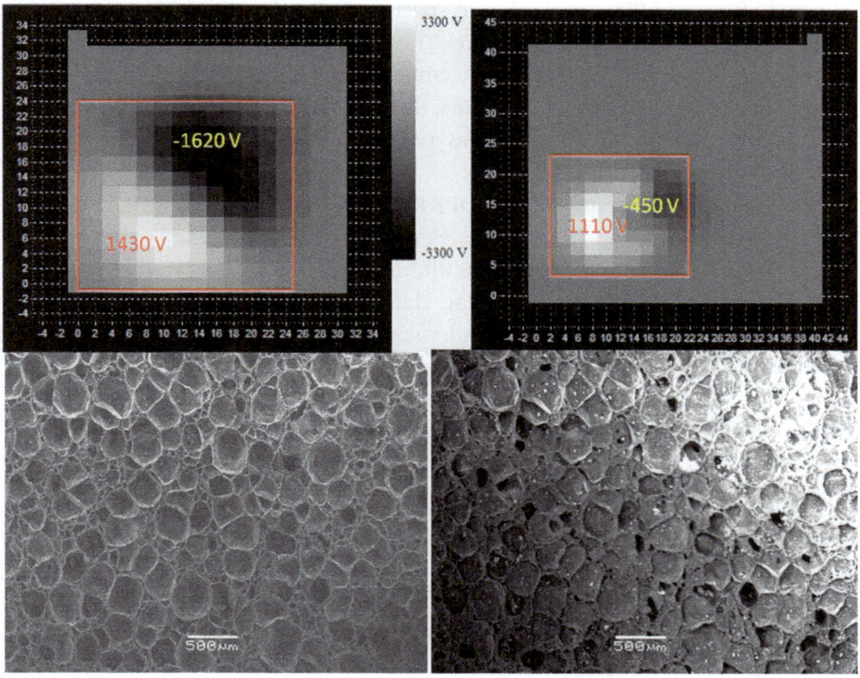

Fig. 2 (Top) Electrostatic potential maps of PTFE film (left) sheared with LDPE foam slabs (right). The average potential of the positive and negative region of each material is printed in the respective area. (Bottom left) SEI micrograph of a negative area of sheared LDPE foam. (Bottom right) BEI micrograph of the same area.

Scanning electron micrographs of a negative area of LDPE foam surface sheared with PTFE are shown in Fig. 2. The SEI micrographs show the expected morphology but BEI reveals that this surface contains many brighter or darker domains contrasting with the dominating gray tone. Brighter domains in BEI composition images are areas with a higher average atom number as compared to others. In a sample containing only hydrocarbon and fluorocarbon, the brighter regions are fluorine-rich.

EDX analysis shows the presence of 2.4 atom% fluorine in one of the bright spots while other areas contained also Ca and Mg, released by the foamed LDPE. It is important to keep in mind that the analytical information conveyed by BEI and EDX do not necessarily match, because the sampling depth is different in the two cases: usually, backscattered electrons sample a thinner layer more adjacent to the surface than EDX.

Elemental distribution maps of C, F and O for another related area are in Fig. 3, showing the superposition of these three elements on the LDPE surface and evidencing that oxygenated species were also formed during tribocharging.

Similar examination was done in an electrically positive spot on the foam surface. Again, contrast assigned to changing chemical composition was observed but point analysis did not detect F in any tested spot while the dots in the F map are barely above noise. Thus, fluorocarbon transfer to the positive areas cannot be excluded but evidence in its favor is weak.

Analogous experiments were done with the same or closely related polymer samples, but using different experimental arrangements. In one case, the HDPE film was sheared with a spinning PTFE stub. Electrostatic potential maps of the sheared film are in Fig. 4 and infrared spectra of spots in the negative area and in pristine HDPE are also in this Figure. The presence of peaks assigned to a C–F stretching vibration at 1230 and 1160 cm^{-1} show that fluorocarbon material was transferred to PTFE in the negative area; this was also described in the literature.[32] The positive area on PTFE film does not contain an amount of fluorocarbon detectable by this technique.

BEI and SEI micrographs of an HDPE film charged by shearing with a spinning PTFE stub are shown in Fig. 5, while elemental distribution maps are shown in Fig. 6. Mechanical damage on HDPE film is easily observed as well as the deposition of bits of fluorinated material, that appear as bright spots in the BEI micrographs. Fig. 5 and 6 show also some debris that are not fluorinated, as previously observed in the experiments done with PTFE film and LDPE foam.

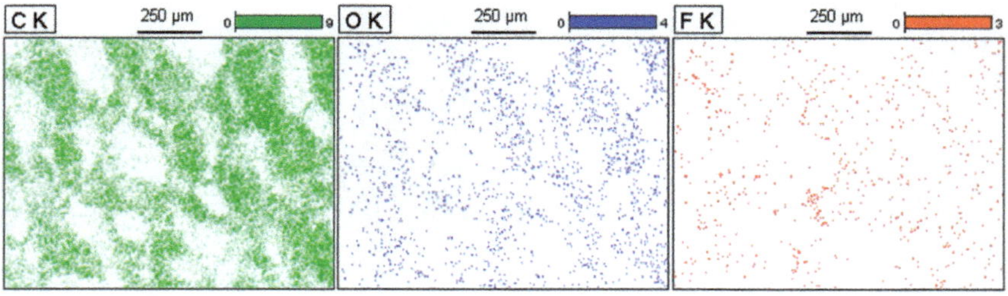

Fig. 3 EDX elemental distribution maps of a negative area of LDPE foam. Note the superposition of the dotted areas in the C, O and F maps and the variability of dot density.

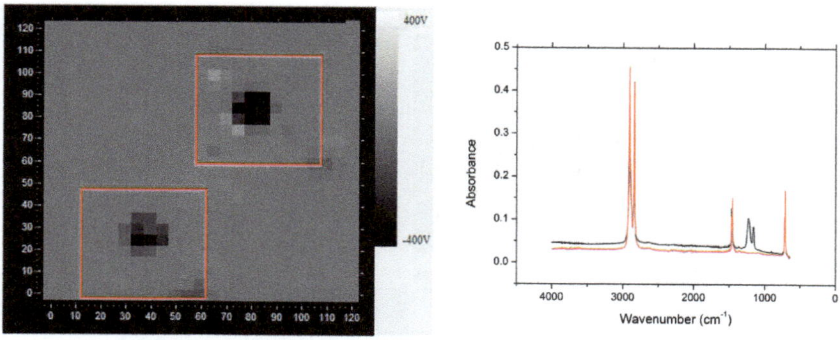

Fig. 4 (Left) Electrostatic potential maps of HDPE film sheared with a PTFE stub. (Right) Infrared spectra of a sheared area carrying negative charge (black) and a pristine area (red) of HDPE film tribocharged with PTFE.

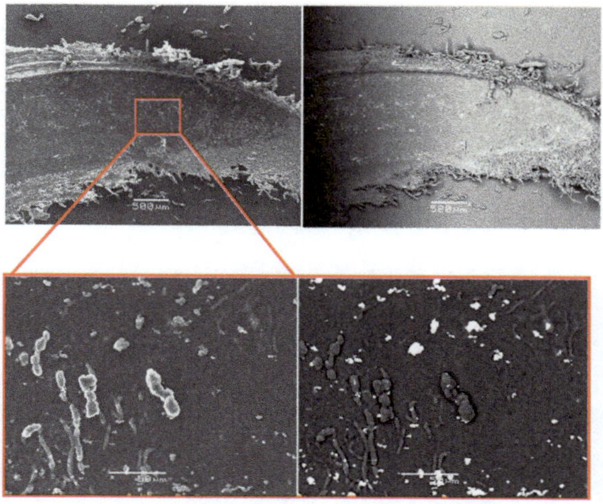

Fig. 5 SEI (left) and BEI (right) micrographs of an area of HDPE film sheared with a PTFE stub.

In other experiments, PTFE films were rubbed with HDPE stubs and they also acquired charge. The C : F atom ratio determined by EDX point analysis in the PTFE film was 1 : 0.65 instead of the expected 1 : 2 for PTFE, showing that PTFE coverage with HDPE fragments is partial.

A HDPE stub was imaged after rubbing PTFE film, SEI and BEI views are in Fig. 7, showing large debris laying on the stub surface. Microanalysis of this sample (Table 1) shows the presence of fluorine unevenly scattered in the debris surface, especially at the side which was likely formed during the tribocharging process. An intriguing feature of this sample is the presence of significant amounts of N in apparently undamaged spots of the stub surface, while O is found in all the six areas analysed. This may be understood by assuming that an atmospheric- or tribo-plasma[33] is formed in this system and its high-energy species react on the HDPE surface.

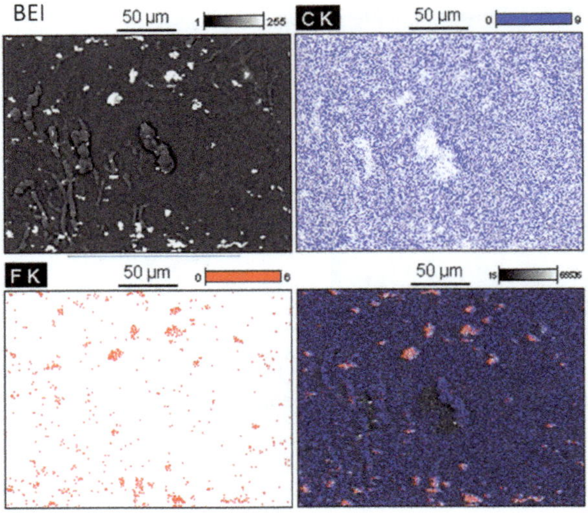

Fig. 6 Carbon and fluorine elemental distribution maps for the magnified area shown in Fig. 5. Note the correspondence between the bright spots in the BEI image and fluorine-rich areas in the elemental map.

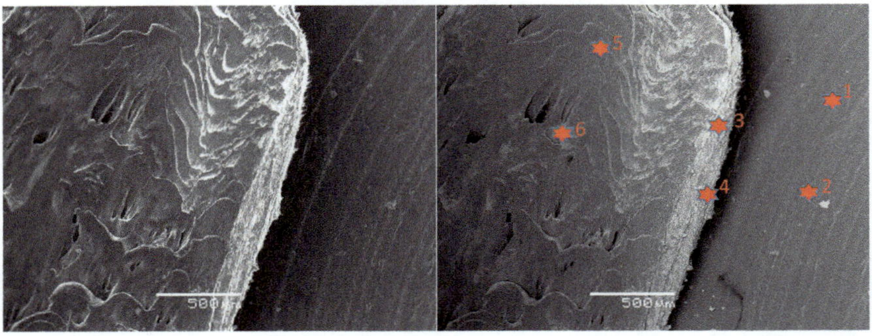

Fig. 7 (Left) SEI and (right) BEI micrograph of an area of a HDPE stub after rubbing PTFE film. The original stub surface is identified by the regular lines formed by the machining tool. Microanalysis data for the six points identified in the BEI micrograph are in the Table 1.

Table 1 Detected atom % determined by EDX point analysis in six spots of the area shown in Fig. 7

Spot	C (%)	N (%)	O (%)	F (%)
1	76.9	17.5	5.6	—
2	79.1	15.5	5.5	—
3	74.4	—	1.2	24.4
4	85.4	—	3.5	11.1
5	94.1	—	2.8	3.1
6	89.5	—	2.8	7.6

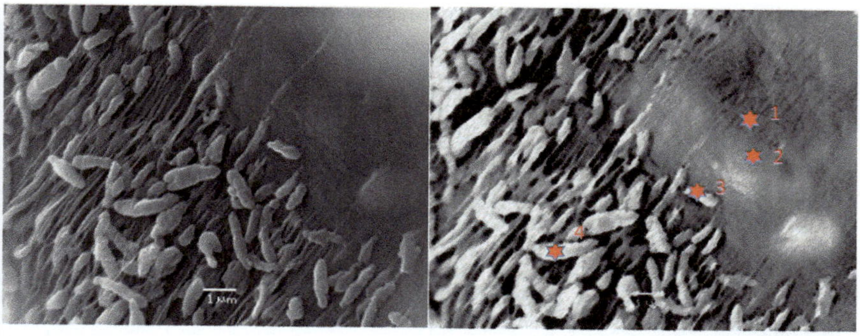

Fig. 8 SEI (left) and BEI (right) micrographs of a HDPE stub after rubbing PTFE film: left, SEI; right, BEI. The microanalysis data for the four points identified in the BEI micrograph are in Table 2.

Fig. 8 shows micrographs from another field of the surface of the HDPE stub shown in Fig. 7, with many interesting features concerning both the surface morphology and distribution of chemical constituents. First, the smooth surface in the first quadrant (upper right of the BEI image) shows flat contrasting domains with marked contrast between features as small as 100 nm and as large as 1 micron. Fractal dimensions were calculated for the flat areas in both images, yielding $D = 1.55 \pm 0.01$ for the SEI micrograph and 1.85 ± 0.02 for BEI and showing that chemical composition fluctuations follow a more complex pattern than the surface topography. Strained HDPE forming thin stripes connected to the flat matrix is also seen in the third quadrant (lower left) and adjacent areas. The surfaces of the seemingly particulate material are brighter than most other regions and they contain significant amounts of F, as shown in Table 2. This confirms that small domains containing PTFE fragments (and probably also other contaminants) are dispersed throughout this area.

A more comprehensive view of mass transfer and constituent distribution is obtained by elemental mapping at low magnification in an area of the HDPE stub that sheared PTFE, shown in Fig. 9. Fluorine-rich debris lies on top of the stub surface, while N and O are scattered unevenly throughout the surface.

Fig. 10 shows potential maps of films of PTFE and HDPE that were sheared with stubs made out of the same polymer: PTFE films sheared with PTFE stubs and HDPE films sheared with HDPE stubs. The formation of tribocharged domains is also observed but with some curious features. It is possible to observe the formation of positive domains on the PTFE film together with negative domains, even though this polymer usually tends to form anionic fragments only when rubbed with

Table 2 Detected atom % determined by EDX point analysis in spots of the area shown in Fig. 8

Spot	C	N	F
1	93.9	—	6.1
2	92.2	—	7.8
3	89.5	—	10.5
4	72.4	11.2	16.4

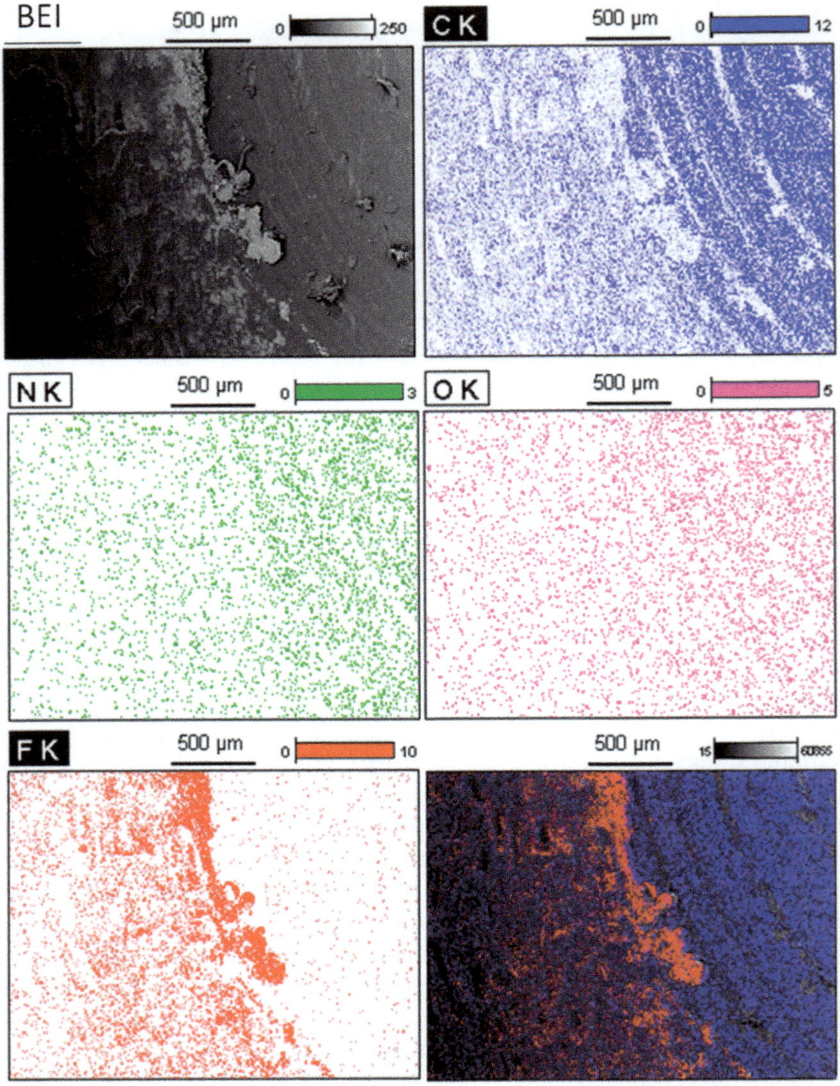

Fig. 9 Elemental maps of an area of the HDPE stub surface that sheared a PTFE film.

polyethylene. However, overall negative charge formation is observed in the stub, where a −200 V potential is measured. On the other hand, HDPE films are positively charged and the overall potential on the rubbing HDPE stub is low.

Maximum, minimum and mean square potential averages obtained for potential maps acquired in repeated independent runs are given in Fig. 10, showing the same pattern but with large quantitative differences. The average potential in the films is positive in both cases, the maximum potential is 150 V in the HDPE film and 640 V in PTFE, while minimum potentials are, respectively, 20 V and −1235 V.

The results presented so far in this report refer to PTFE and PE samples under low pressure and high speed friction but charge patterns have also been obtained

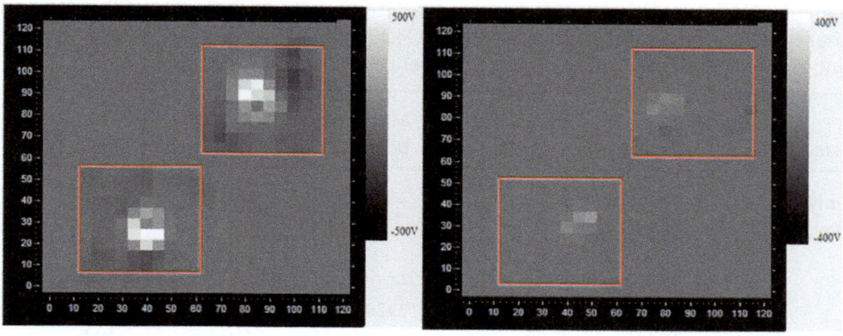

Fig. 10 (Left) Potential maps of PTFE films sheared with PTFE stubs and (right) of HDPE films sheared with HDPE stubs.

under other conditions. For instance, Fig. 11 shows micrographs of a PS slab sheared with PTFE stub, in a region where PTFE was detected. EDX point analysis on the bright fragments detects up to 20 atom percent F.

Electrostatic potential patterns have already been obtained for other polymer pairs. Pellets of different polyamides, HDPE and polyacetal were placed within PS Petri dishes and shaken in a reciprocating table, where the pellets undergo sliding motion. Table 3 shows the broad range of electrostatic potentials measured on a square in the central area of Petri dishes with the rubbing pellets. Data for HDPE film treated in the same way but with rolling PTFE spheres are also in this Table.

4 Discussion

The results described in this report confirm that the formation of charge by rubbing polymer surfaces is concurrent to mass transfer between the surfaces, without any exception among the systems examined so far. Spatial distribution of the transferred material is consistent with the formation of fractal charge patterns extending from the nano- to macroscopic scale, showing in most cases the coexistence of domains with positive and negative excess charge. Thus, the analytical electron micrographs and spectra presented here lead to a more complete explanation for the counter-intuitive charge segregation into separate

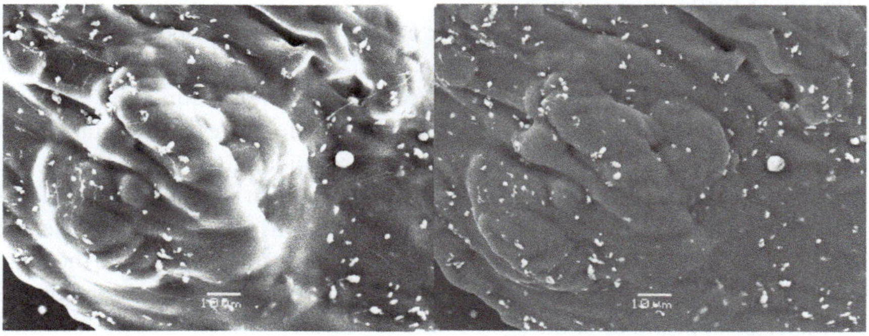

Fig. 11 Micrographs of a PS slab sheared with a PTFE stub. Left, SEI; right, BEI.

Table 3 Maximum, minimum and average potential of the central areas of PS Petri dishes shaken with polymer pellets and of a square HDPE film shaken with PTFE spheres. Pellets and spheres were not removed prior to potential measurements

Base surface	Moving pellets	Maximum potential/V	Minimum potential/V	Average potential/V	Standard deviation/V
Polystyrene (Petri dish)	Nylon 6	1185	−1780	160	685
	Nylon 6,9	830	−1010	45	390
	Nylon 12	770	−1170	−90	500
	HDPE	245	−705	−190	260
	Polyacetal	2231	2320	−657	1155
HDPE film	PTFE sphere	1550	−855	480	490

domains instead of the expected ion pairing that would contribute to abate the overall charge on the sample. On the other hand, this derives from the polymer structure and is different from findings made in other types of materials. Plastic deformation produced by indentation is the source of electric potential differences in pure and doped NaCl crystals[34] and it probably also plays a role in polymer friction, but the methods used in the present work do not allow an assessment of its relative importance.[35,36]

The backscattered-electron micrographs as well as the elemental maps agree in showing that lumps of one polymer material are pulled out and transferred to the other surface, carrying charge to it. Thus, they create large charge density fluctuations across the surface, making a strong contribution to the complex and irregular potential patterns that have been reported in recent work. On the other hand, the larger lumps shown in this work are not easily detected by other techniques such as Kelvin force microscopy (KFM), due their height which introduces significant imaging artefacts and also due to the amount of localized charge, which provokes strong deformation even on stiff KFM cantilevers.

The BEI and EDX micrographs frequently convey the same information, even though they are not expected to be always in full agreement, since the sampling depth is not the same in both cases (typically in the micron and sub-micron range). Moreover, they depend on the beam energy used in each imaging experiment and this is not kept constant, to increase the amount of information from each image. For this reason, a thin fluorine-rich surface layer may appear brighter in the BEI picture but point analysis may not show a large fluorine content, since the X-rays map a thicker surface layer.

There are two additional factors for charge patterning in polymer surfaces, beyond the formation of polymer ions. First, the charge on solid surfaces is exchanged with atmospheric or adsorbed water,[16,23,24] attracting OH^- and H^+ ions[37] to, respectively, the positive and negative surface sites. Incidentally, neat water surfaces are basic due to preferential adsorption of OH^- ions.[38] Also, the rapid formation and deposition of charged material on both rubbing surfaces should create large electrostatic potential fluctuations thus producing electric fields exceeding the thresholds for electrostatic discharge throughout the adjacent atmosphere. At this time, this is the best tentative explanation for the unexpected appearance of nitrogen bound to polymer surfaces that was detected in some samples.

Otherwise, the arguments for charge formation that were put forward earlier still prevail: charge formation is due to homolytic or heterolytic polymer chain breakdown under shearing. In the case of homolytic scission, electron transfer to and from neighbouring polymer free-radicals or molecules, as well as to atmospheric molecules and eventual polymer contaminants, leads to the formation of polymer ions.

Since chain breakdown is most likely at the surfaces of polymer particles pulled out of the surface during surface rubbing, a large fraction of the charges remain accumulated rather than dispersed throughout the tribocharged surface. Particle removal by washing surfaces with suitable liquids is usually a simple task and this explains why a large fraction of the tribocharge is easily removed from various polymer surfaces by simple immersion in both polar and apolar liquids.[39]

The fractal dimension obtained from linescans drawn on SEI micrographs in this paper agrees with previous results obtained by atomic force microscopy, as expected considering that both are determined mainly by surface topography. On the other hand, the fractal dimension for the chemical composition contrast in the BEI images (1.85) is significantly higher than figures previously obtained from Kelvin potential maps (1.64–1.72).[40] This is explained by considering the existence of a smoothing factor for potential gradients that is discharged across the surrounding atmosphere. This levels-out potential features that exceed the threshold for discharge but without transfer of the charge carrier polymer fragments.

The interdependence of friction and wear is well acknowledged in the literature, not only in polymer systems but also in metals.[41] Coefficients of friction and wear are parameters describing the state of contact of bodies and they are not constants of the contacting materials.[42,43] Wear and friction have been explained in terms of roughness, hardness, ductility, oxide film formation, reaction layers and transfer[42] but the formation of strongly charged tribolayers seems to be more conspicuous in polymers than in other types of materials. The recent recognition of the effect of tribocharges on friction coefficients[22] followed the work of nanotribologists[44–47] who paid due attention to the effect of surface charge on friction coefficients. This adds one more point to be considered in the sequence of events triggered by mutual mechanical action of contacting surfaces, from the mechanochemical reactions to the formation of triboelectricity, wear, adhesion and changes in the friction coefficients. On the other hand, it shows that tribometer and related experiments could benefit from concurrent electrostatic potential measurements that not often done.

One additional point that should be brought to this discussion is the time-temperature superposition as introduced by Williams, Landel and Ferry.[48] This implies that the mechanical behavior of two polymers mutually acting at high speed or high frequency resemble that of the same polymer but at some lower temperature, following a non-Arrhenius pattern. In practice, this means that polyethylene behavior at high frequency or high speeds resembles its behavior at a lower temperature, when the polymer is harder and less ductile. For this reason, simply contacting polymer surfaces or applying different types of mechanical action at different speeds should be expected to produce different chemical, structural, morphological and electrostatic results analogous to those described by Distler in ionic solids.[49]

To sum up, mechanochemical reactions triggered by friction produce ionic polymer chain fragments, especially at the surface of particles pulled away from the rubbing surfaces that can be transferred from one rubbing body surface to another. Amphiphilic species are thus formed, containing charged end-groups that are excluded from the polymer–air interface, since their contribution to the material surface tension is more positive than the contribution of other, less polar groups. Charges are thus occluded and partly protected from water adsorbed at the polymer surface and from water vapor or aerosols, explaining the tribocharge stability that was previously reported.

5 Conclusion

Tribolayer formation in polymers is initiated by mechanochemical reactions during friction producing free-radicals that are at least partly transformed into ionic species with half-lives exceeding tens of hours and have a major if not dominating role in the formation of triboelectricity. This explains the ubiquity of charge patterns on insulating polymer surfaces following contact with any other materials, which has been evidenced for the past fifteen years, including patterns obtained by rubbing two identical solids. Moreover, it supports a new explanation for the dissipative character of friction, based on the large enthalpy requirements for the scission of strong carbon–carbon covalent bonds.

Acknowledgements

This is a contribution from the INCT Inomat, a project supported by Brazilian agencies MCTI/CNPq and Fapesp. L. B. S. B. and D. D. D. acknowledge graduate fellowships from Fapesp and CNPq.

References

1 G. S. P. Castle and L. B. Schein, *J. Electrost.*, 1995, **36**, 165.
2 L. B. Schein, *Science*, 2007, **316**, 1572.
3 A. G. Bailey, *J. Electrost.*, 2001, **51–52**, 82.
4 M. W. Williams, *IEEE Trans. Ind. Appl.*, 2011, **47**, 1093.
5 M. W. Williams, *Am. Sci.*, 2012, **100**, 316.
6 M. W. Williams, *AIP Adv.*, 2012, **2**, 010701.
7 D. J. Lacks and R. M. Sankaran, *J. Phys. D: Appl. Phys.*, 2011, **44**, 453001.
8 M. Glor, *J. Electrost.*, 1985, **16**, 175.
9 N. Gibson, *J. Electrost.*, 1997, **40–41**, 21.
10 L. Perrin, A. Laurent, V. Falk, O. Dufaud and M. Traoré, *J. Loss Prev. Process Ind.*, 2007, **20**, 207.
11 A. Galembeck, C. A. R. Costa, M. C. V. M. da Silva, E. F. Souza and F. Galembeck, *Polymer*, 2001, **42**, 4845.
12 E. Teixeira Neto and F. Galembeck, *Colloids Surf., A*, 2002, **207**, 147.
13 A. L. H. Cardoso, C. A. P. Leite and F. Galembeck, *Langmuir*, 1999, **15**, 4447.
14 H. T. Baytekin, A. Z. Patashinski, M. Branicki, B. Baytekin, S. Soh and B. A. Grzybowski, *Science*, 2011, **333**, 308.
15 Y. Martin, C. C. Williams and H. K. Wickramasinghe, *J. Appl. Phys.*, 1987, **61**, 4723.

16 C. A. Rezende, R. F. Gouveia, M. A. da Silva and F. Galembeck, *J. Phys.: Condens. Matter*, 2009, **21**, 263002.
17 T. A. L. Burgo, T. R. D. Ducati, K. R. Francisco, K. J. Clinckspoor, F. Galembeck and S. Galembeck, *Langmuir*, 2012, **28**, 7407.
18 R. P. Feynman, R. B. Leighton and M. Sands, *The Feynman Lectures on Physics*, 1, 2nd edn, Addison-Wesley Publishing Company, Reading MA, 2005, pp. 4–7.
19 A. F. Diaz and R. M. Felix-Navarro, *J. Electrost.*, 2004, **62**, 277.
20 P. J. Flory, *J. Chem. Phys.*, 1942, **10**, 51.
21 M. J. Huggins, *The Journal of Physical Chemistry*, 1942, **46**, 151.
22 T. A. L. Burgo, C. S. Souza, L. B. S. Balestrin and F. Galembeck, *Sci. Rep.*, 2013, **3**, 2384.
23 R. F. Gouveia and F. Galembeck, *J. Am. Chem. Soc.*, 2009, **131**, 11381.
24 J. S. Bernardes, C. A. Rezende and F. Galembeck, *J. Phys. Chem. C*, 2010, **114**, 19016.
25 H. T. Baytekin, B. Baytekin, T. M. Hermans, B. Kowalczyk and B. A. Grzybowski, *Science*, 2013, **341**, 1368.
26 C. M. Pooley and D. Tabor, *Proc. R. Soc. London, Ser. A*, 1972, **329**, 251.
27 V. Quaglini and P. Dubini, *Adv. Tribol.*, 2011, 178943.
28 J. Ye, H. S. Khare and D. L. Burris, *Wear*, 2013, **297**, 1095.
29 T. A. Blanchet and F. E. Kennedy, *Wear*, 1992, **153**, 229.
30 C. Lhymn, *Wear*, 1986, **107**, 95.
31 D. W. Breiby, T. I. Sølling, O. Bunk, R. B. Nyberg, K. Norrman and M. M. Nielsen, *Macromolecules*, 2005, **38**, 2383.
32 A. I. Sviridyonok, V. A. Bely, V. A. Smurugov and V. G. Savkin, *Wear*, 1973, **25**, 301.
33 G. Heinicke, *Tribochemistry*, Hanser, Munich, 1984.
34 F. Frohlich and P. Seifert, *Phys. Status Solidi B*, 1968, **25**, 303.
35 W. M. Rainforth, P. Zeng, L. Ma, A. N. Valdez and T. Stewart, *Faraday Discuss.*, 2012, **156**, 41.
36 A. Vernes, S. Eder, G. Vorlaufer and G. Betz, *Faraday Discuss.*, 2012, **156**, 173.
37 L. P. Santos, T. R. D. Ducati, L. S. Balestrin and F. Galembeck, *J. Phys. Chem. C*, 2011, **115**, 11226.
38 J. K. Beattie, A. N. Djerdjev and G. G. Warr, *Faraday Discuss.*, 2009, **141**, 31.
39 K. R. Francisco, T. A. L. Burgo and F. Galembeck, *Chem. Lett.*, 2012, **41**, 1256.
40 J. P. Santos, P. Corpart, K. Wong and F. Galembeck, *Langmuir*, 2004, **20**, 10576.
41 F. P. Bowden and D. Tabor, *The Friction and Lubrication of Solids*, Oxford, Clarendon, 2001, p. 341.
42 K. Kato, *Wear*, 2000, **241**, 151.
43 T. Polcar, F. Gustavsson, T. Thersleff, S. Jacobson and A. Cavaleiro, *Faraday Discuss.*, 2012, **156**, 383.
44 C. M. Mate, G. M. MacClelland, R. Erlandsson and S. Chiang, *Phys. Rev. Lett.*, 1987, **59**, 1942.
45 D. F. Ogletree, R. W. Carpick and M. Salmeron, *Rev. Sci. Instrum.*, 1996, **67**, 3298.
46 G. Marti and N. D. Spencer, *Langmuir*, 1995, **11**, 4632.
47 R. W. Carpick, N. Agrait, D. F. Ogletree and M. Salmeron, *Langmuir*, 1996, **12**, 3334.
48 M. L. Williams, R. F. Landel and D. Ferry, *J. Am. Chem. Soc.*, 1955, **77**, 3701.
49 G. I. Distler, *Krist. Tech.*, 1970, **5**, 73.

Faraday Discussions

PAPER

The mechanochemical production of phenyl cations through heterolytic bond scission†

Tomohiro Shiraki,[ab] Charles E. Diesendruck[ab] and Jeffrey S. Moore*[ab]

Received 28th February 2014, Accepted 26th March 2014
DOI: 10.1039/c4fd00027g

High mechanical forces applied to polymeric materials typically induce unselective chain scission. For the last decade, mechanoresponsive molecules, mechanophores, have been designed to harness the mechanical energy applied to polymers and provide a productive chemical response. The selective homolysis of chemical bonds was achieved by incorporating peroxide and azo mechanophores into polymer backbones. However, selective heterolysis in polymer mechanochemistry is still mostly unachieved. We hypothesized that highly polarized bonds in ionic species are likely to undergo heterolytic bond scission. To test this, we examined a triarylsulfonium salt (TAS) as a mechanophore. Poly(methyl acrylate) possessing TAS at the center of the chain (PMA-TAS) is synthesized by a single electron transfer living radical polymerization (SET-LRP) method. Computational and experimental studies in solution reveal the mechanochemical production of phenyl cations from PMA-TAS. Interestingly, the generated phenyl cation reacts with its counter-anion (trifluoromethanesulfonate) to produce a terminal trifluoromethyl benzene structure that, to the best of our knowledge, is not observed in the photolysis of TAS. Moreover, the phenyl cation can be trapped by the addition of a nucleophile. These findings emphasize the interesting reaction pathways that become available by mechanical activation.

Introduction

Biological systems are able to detect mechanical forces and harness mechanical energy through mechanochemical transduction that utilizes complex functional networks.[1] In terms of synthetic mechanoresponsiveness, polymer mechanochemistry has recently gathered significant interest, using mechanical force as an

[a]*Beckman Institute for Advanced Science and Technology, University of Illinois at Urbana-Champaign, 05 N. Mathews Ave., Urbana, IL 61801, United States. E-mail: jsmoore@illinois.edu; Fax: +1(217) 244-0181; Tel: +1(217) 244-5289*

[b]*Department of Chemistry, University of Illinois at Urbana-Champaign, United States*

† Electronic supplementary information (ESI) available: the NMR data, CoGEF energies and GPC data are included. See DOI: 10.1039/c4fd00027g

external stimulus to induce chemical reactions leading to new types of smart, functional materials.[2]

Achieving a selective mechanochemical response requires the judicious design of force-responsive molecules, called mechanophores. Mechanophores are tethered to polymer chains through a specific chemical bond which transmits the external force to the mechanophore through the polymer chain stretching from elongational flow or mechanical deformation. Useful mechanochemical responses such as deformation indicators,[3] mechanocatalysis[4] and small molecule release[5] have been reported. The activation processes usually involve chemical bond scission in mechanophores; thus, the design of the resulting products after bond cleavage is an important factor to exploit in terms of function and response.

The mechanical scission of chemical bonds in homopolymers is mostly homolytic, producing radical species.[6] This response has been amplified synthetically by introducing strained rings[7] or weak bonds such as peroxide,[8a] azo[8b] and disulfide[8c] linkages in polymer backbones. On the other hand, reports regarding heterolytic bond scission are very limited. For example, Sakaguchi *et al.* demonstrated that the heterolytic scission of polymers is partially observed when ball-milling bacteria cellulose, poly(vinylidene fluoride) and poly(tetrafluoroethylene) under careful experimental conditions (vacuum, dark, 77 K).[9] Therefore, we designed a new mechanophore with a highly polarized chemical bond that would favor heterolytic bond scission. This new mechanophore could lead to novel reactivity since the generated ionic species (such as phenyl cations and carbocations) may be useful as highly reactive-intermediates in various chemical reactions.

Given our interests in mechanochemical acid production,[5a] we investigated a triarylsulfonium salt (TAS, Fig. 1) as a mechanophore candidate. TASs are well-known photoacid generators commonly used in photolithography applications.[10] Different mechanisms for photochemical acid production have been proposed and debated in the literature, but they always involve heterolysis or homolysis of the C–S bond.[10] Here, we investigate the mechanochemical scission of TAS using both computational simulations and sonication experiments in solution.

Experimental

Materials

Unless otherwise stated, all chemical reagents were obtained from commercial suppliers and used without purification. α-Bromoisobutyryl bromide, 4,4′-thiodiphenol, methyl acrylate (MA), Cu(0) powder and tris[2-(dimethylamino)ethyl]

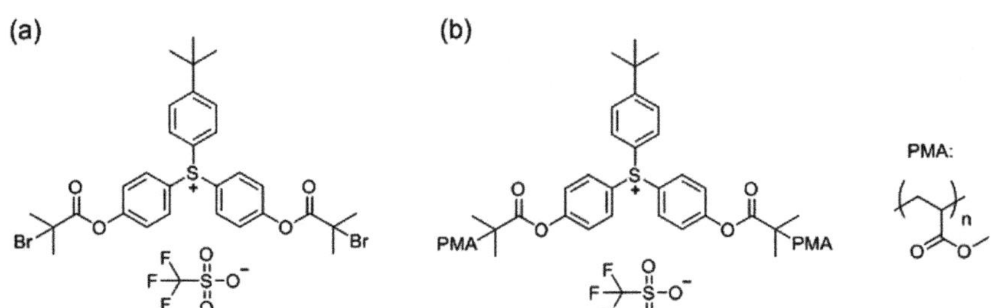

Fig. 1 The chemical structures of (a) the TAS initiator and (b) PMA-TAS.

amine (Me_6TREN) were purchased from Sigma-Aldrich. Bis(4-*tert*-butylphenyl) iodonium trifluoromethanesulfonate was purchased from TCI. Silica gel 60 (230–400 mesh) was obtained from Silicycle.

Instruments

The ^1H, ^{13}C and ^{19}F NMR spectra were collected by using a Varian 500 MHz spectrometer in the VOICE NMR laboratory at the University of Illinois; the residual solvent proton was used to reference the chemical shifts.[11] The coupling constants (*J*) are reported in Hertz (Hz) and the splitting patterns are designated as s (singlet), d (doublet), m (multiplet) and br (broad). The electrospray ionization (ESI) and high resolution ESI mass spectra were obtained through the Mass Spectrometry Facility, SCS, University of Illinois. The elemental analysis data were obtained through the Microanalysis Facility, SCS, University of Illinois (CHN Analysis - Exeter Analytical CE 440 and Perkin Elmer 2440, Series II). The gel permeation chromatography (GPC) analyses were carried out in a Waters 1515 Isocratic HPLC pump, with a Waters (2707) 96-well autosampler and a series of 4 Waters HR Styragel columns (7.8 × 300 mm, HR1, HR3, HR4, and HR5) in THF at 30 °C, coupled to a Waters (2998) Photodiode Array Detector and a Waters (2414) Refractive Index Detector. Polystyrene standard samples (Showa Denko K.K.) were used for the standard calibration. Ultrasound sonication experiments were performed on a Vibra Cell 505 liquid processor with a ½″ diameter solid horn from Sonics and Materials. The distance between the titanium tip and the bottom of the Suslick cell was 1 cm. The Suslick cells were made by the Glass Shop, SCS, University of Illinois. Novacure 2100 (EXFO America, 4000 mW cm^{-2}, 15 min) was used for the UV light irradiation experiments. CoGEF (constrained geometries simulate external force) calculations were conducted in Spartan '10 at the B3LYP/6-31G* level of theory.

Synthesis

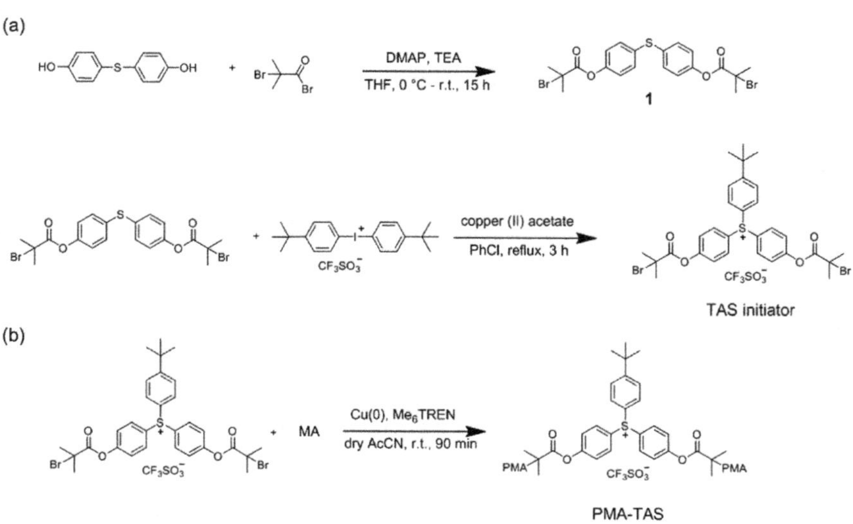

Scheme 1 The synthetic routes of (a) TAS initiator and (b) PMA-TAS *via* SET-LRP.

Thiobis(4,1-phenylene) bis(2-bromo-2-methylpropanoate) (1). In a 100 mL three-neck round-bottom flask equipped with an addition funnel, 4,4′-thiodiphenol (1.00 g, 4.58 mmol) and 4-dimethylaminopyridine (DMAP, 14.0 mg, 0.115 mmol) were dissolved in 20 mL THF and the solution was cooled in an ice bath. Triethylamine (TEA, 1.60 mL, 11.5 mmol) was added and then α-bromoisobutyryl bromide (1.42 mL, 11.5 mmol) in 15 mL THF was added dropwise. After stirring at 0 °C for 1 h, the mixture was allowed to warm to room temperature and was stirred overnight. The solids were filtered out and the filtrate evaporated. The obtained solid was dissolved in 30 mL CH_2Cl_2 and washed with saturated $NaHCO_3$ (twice), water and brine (30 mL each). The organic layer was dried over anhydrous magnesium sulfate, filtered and evaporated to provide **1** as an off-white solid (2.35 g, yield 99%); ^{1}H NMR (500 MHz, $CDCl_3$, Figure S1†): δ/ppm 7.37 (d, J = 8.7 Hz, 4H), 7.09 (d, J = 8.7 Hz, 4H), 2.06 (s, 12H), ^{13}C NMR (500 MHz, $CDCl_3$): δ/ppm 170.2, 150.2, 133.5, 132.4, 122.2, 55.3, 30.8, HRMS-ESI (m/z): calculated for $C_{20}H_{21}O_4SBr_2$ $[M + H]^+$: 514.9527; found: 514.9531.

Bis(4-((2-bromo-2-methylpropanoyl)oxy)phenyl)(4-(tert-butyl)phenyl)sulfonium trifluoromethanesulfonate (TAS initiator).[12] Compound **1** (1.03 g, 2.00 mmol), bis(4-*tert*-butylphenyl)iodonium trifluoromethanesulfonate (1.08 g, 2.00 mmol) and Cu(II) acetate (8.7 mg, 0.024 mmol) were weighed in a 50 mL two-neck round-bottom flask equipped with a condenser. Chlorobenzene (20 mL) was added and the mixture was refluxed for 3 h. After cooling to room temperature, the mixture was filtered and the solvent was evaporated to dryness. The product was recrystallized from a 2 : 8 (v/v) CH_2Cl_2 : diethyl ether mixture to provide colorless crystals (346 mg, yield 22%); ^{1}H NMR (500 MHz, $CDCl_3$, Figure S2†): δ/ppm 7.85 (d, J = 9.0 Hz, 4H), 7.70 (d, J = 4.63 Hz, 4H), 7.51 (d, J = 9.0 Hz, 4H), 2.06 (s, 12H), 1.35 (s, 9H). ^{13}C NMR (500 MHz, $CDCl_3$): δ/ppm 169.4, 159.6, 155.6, 133.0, 131.2, 129.2, 124.8, 121.9, 120.8, 119.7, 54.9, 35.7, 31.0, 30.5. MS-ESI (m/z): calculated for $C_{30}H_{33}Br_2O_4S+$ $[M]^+$: 649.0 and for CF_3O_3S- $[M]^-$: 149.0; found: 649.1 and 149.0, respectively. Elemental analysis: $C_{31}H_{33}Br_2F_3O_7S_2$: calc.% C 46.63, H 4.17: found C 46.38, H 4.05.

PMA-TAS. Poly(methyl acrylate) having a TAS group at the center position (PMA-TAS) was synthesized through a single electron transfer living radical polymerization (SET-LRP)[13] using the functionalized TAS initiator. MA was filtered through basic alumina to remove inhibitor. To an N_2-purged 10 mL Schlenk flask equipped with a Teflon stir bar, Me_6TREN (4.01 μL, 0.0150 mmol), TAS initiator (3.99 mg, 0.00500 mmol) and MA (1.00 mL, 11.1 mmol) were added. Cu(0) powder (0.95 mg, 0.0150 mmol) dispersed in 1.0 mL of dry acetonitrile was injected to the reaction mixture. Three freeze-pump-thaw cycles were applied. The flask was backfilled with N_2 and allowed to stir in a water bath for 90 min at room temperature. The polymerization was quenched by immersing the flask in liquid N_2. The resulting solution was diluted with 10 mL THF and filtered through a pad of silica gel. After concentrating the solution in vacuo, the polymer was precipitated by the dropwise addition to stirring methanol. The resulting polymer was collected and dried under vacuum at room temperature. The product was obtained as a colorless solid (269 mg); GPC (THF eluent, polystyrene standard) M_p = 94 kDa, M_w/M_n = 1.10. ^{1}H NMR (500 MHz, $CDCl_3$, Figure S3†): δ/ppm 7.80 (d br, 4H of TAS), 7.69 (s br, 4H of TAS), 7.46 (d br, 4H of TAS), 3.64 (s br, 3H of PMA), 2.39–2.23 (br, 1H of PMA), 1.98–1.39 (m br, 2H of PMA). Low molecular weight PMA-TAS was synthesized in the similar procedure ($M_p = 19$ kDa, $M_w/M_n = 1.17$).

The procedure for the sonication experiments

The general apparatus for the sonication experiments has been described in a previous publication.[14] The polymer was dissolved in acetonitrile (10 mg in 10 mL) and transferred to a Suslick cell, which was placed into the collar and screwed to the horn. An N_2 line was introduced into the cell and N_2 was sparged through the system for 30 min prior to any sonication runs, as well as during the run itself. The Suslick cell was cooled with an ice bath throughout the entire sonication in order to maintain a constant temperature of 6 °C. Pulsed ultrasound (1.0 s on, 2.0 s off, 8.7 W cm^{-2}) was applied to the system.

Results and discussion

As an initial step, we used CoGEF simulations to predict the mechanochemical reaction.[7c,15] Therein, the simulation is employed to examine the effects of the induced molecular scale mechanical deformation by artificially constraining the distance between two points (*i.e.*, two atoms at each end of the mechanophore). The CoGEF calculation was applied to a model TAS and revealed that selective scission occurred at the C–S bond in a constrained geometry (Fig. 2a and S4†). Moreover, the LUMO of the resulting mechanically-cleaved structure was calculated and found to be consistent with the phenyl cation and not with the phenyl radical (Fig. 2b).[16] Thus, the DFT calculations support our hypothesis that the polarized bond in TAS undergoes heterolytic C–S scission by mechanical activation to generate a phenyl cation, even given that the phenyl cation is 530 kJ mol^{-1} more stable than the phenyl radical. Homolytic bond scission would produce a diaryl sulfide radical cation, which is 526 kJ mol^{-1} less stable than the diaryl sulfide; *i.e.*, the energy difference between the homolytic and heterolytic bond scission is negligible (energies calculated with the addition of acetonitrile solvation using the SM8 solvation model).[17]

We investigated the mechanoresponsiveness of PMA-TAS experimentally by applying solvodynamic shear forces to dilute polymer solutions using an

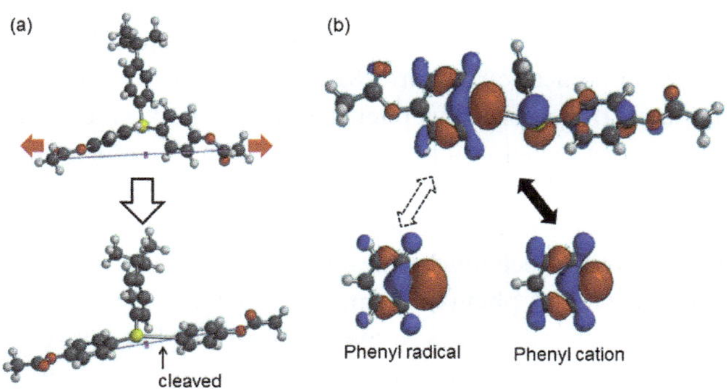

Fig. 2 (a) The calculated TAS structure through a CoGEF simulation and (b) the calculated LUMO of the cleaved structure compared to the LUMO of the phenyl radical and cation. All calculations were conducted in Spartan' 10 at the B3LYP/6-31G* level of theory.

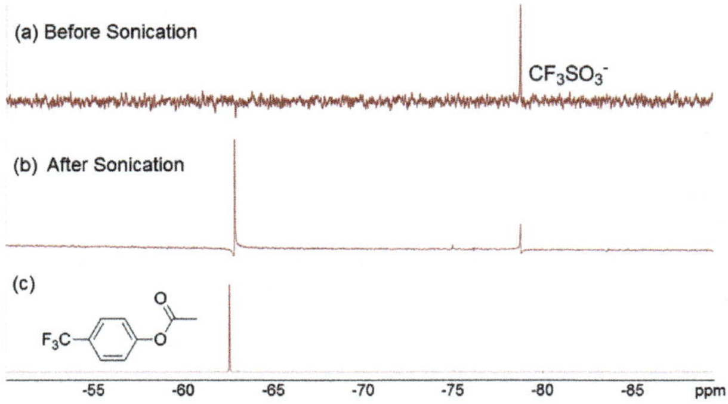

Fig. 3 The ^{19}F NMR spectra of PMA-TAS (a) before and (b) after sonication and (c) 4-(trifluoromethyl)phenyl acetate as a control compound; solvent: $CDCl_3$, sonication: 60 min in acetonitrile under an N_2 atmosphere at 6 °C.

ultrasound horn.[2a,b] The GPC analyses of the sonicated PMA-TAS indicates that the polymer was cleaved approximately in half (Figure S5†) and the NMR signals pertaining to the aromatic groups of TAS were significantly shifted (Figure S6†), indicating that chain scission occurs at the TAS center. For the control experiments, the TAS initiator itself and its mixture with homo PMA were sonicated under the same conditions. The NMR of the sonicated products in both cases showed no significant changes, indicating that the reaction is not thermal (Figure S7†).

The ^{19}F NMR measurements revealed the structural changes of the resulting product which were unexpected (Fig. 3). Before sonication, a single peak at −78.9 ppm was observed, which is ascribable to the trifluoromethanesulfonate counter anion of PMA-TAS. Interestingly, sonication gave rise to a single new peak at −62.5 ppm. After consulting the ^{19}F NMR tables, the only plausible assignment would be to a trifluoromethylphenyl group. We confirmed this assignment by preparing 4-(trifluoromethyl)phenyl acetate through the simple acetylation of *p*-trifluoromethylphenol with acetic anhydride. The ^{19}F NMR analysis showed a peak at −62.3 ppm, very close to the one observed in PMA-TAS after sonication. In contrast, the TAS initiator sonicated or photo-irradiated by itself or in the presence of homo PMA exhibited no change in the ^{19}F NMR spectra (Figure S8†). Moreover, low-molecular weight PMA-TAS (19 kDa), which did not fragment under sonication conditions (Figure S9†), showed no spectral shifts in the ^{19}F and ^{1}H NMR data after sonication (Figures S10 and S11,† respectively). Thus, these results point to a trifluoromethyl phenyl terminated product, caused by the reaction of the mechanochemical product of TAS with its counter anion, which is the only source of fluoride atoms in the mixture. Furthermore, this product is not observed in the photochemical reaction.

We propose that the mechanochemical heterolytic scission of the C–S bond in the TAS moiety creates a phenyl cation, as anticipated by the computational simulation. The generated phenyl cation is an extremely reactive intermediate and is attacked by the trifluoromethanesulfonate to provide the trifluoromethylphenyl structure (Scheme 2). Importantly, the photochemical activation of TAS is proposed to produce phenyl cations, which rapidly react with diphenyl sulfide to give (phenylthio)biphenyls. These biphenyls form through a

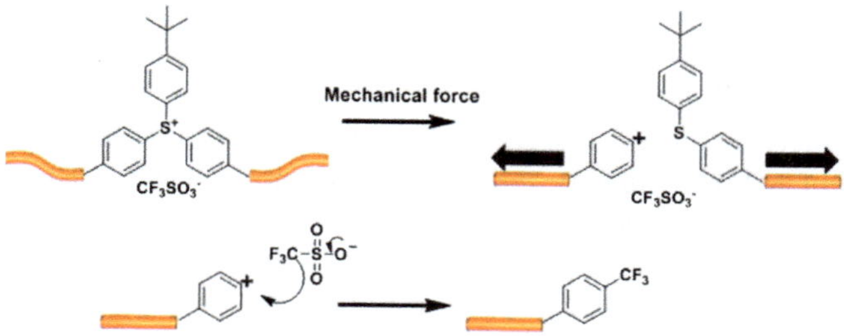

Scheme 2 The proposed reaction mechanism of TAS in the mechanochemically-induced pathway to generate the trifluorotoluene group.

direct reaction with the generated diphenyl sulfide or *via* the formation of a phenyl radical and diphenylsulfinyl radical cation pair.[10] Subsequently, the following proton extraction and transfer reactions result in acid production. While we cannot disprove this mechanism for our system, this reaction would leave the ^{19}F NMR unchanged and, as observed in Fig. 3, a significant portion of the fluoride atoms are shifted. A possible explanation for this significant change in the products is that the pulling force employed during sonication causes the generated species to move rapidly apart from each other. Consequently, the produced phenyl cation is not near the diphenyl sulfide by-product for an electron or hydrogen transfer reaction. Therefore, the cation may react with other components present in solution. Thus, in dilute solutions, the mechanochemical scission of TAS not only generates phenyl cations through heterolytic bond scission, it also creates a unique product for this reaction system.

Our hypothesis was also supported by a trapping experiment. The phenyl cation is a high energy intermediate compound in organic chemistry that rapidly reacts with nucleophiles such as amines and alcohols.[18] Based on previous research experiments,[14] we chose 9-(methylaminomethyl) anthracene (MAMA,

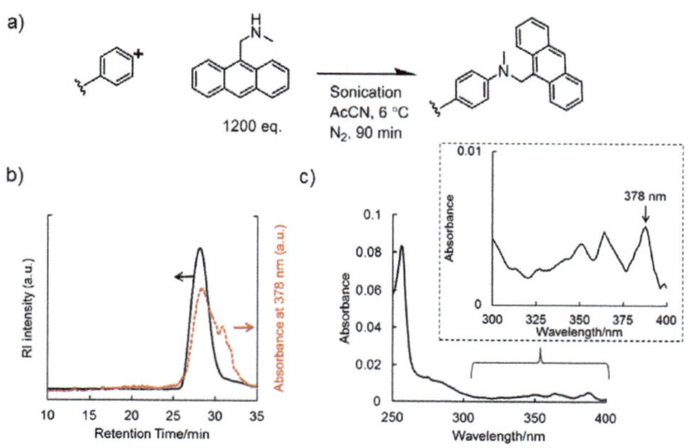

Fig. 4 (a) The reaction scheme of a phenyl cation with MAMA, the GPC analysis results monitoring (b) RI and UV (378 nm) signals (black solid line and red dash line, respectively) and (c) the absorption spectrum of the sonicated PMA-TAS with MAMA (retention time 27.5 min). The UV absorption data were collected by a PDA detector equipped in the GPC system.

Fig. 4a) as a trapping nucleophile to confirm our reaction mechanism. The sonication of PMA-TAS was repeated under similar conditions, but in the presence of an excess amount of the trapping MAMA (1200 eq. molar with respect to the TAS concentration). Fig. 4b shows the GPC traces of sonicated PMA-TAS in the presence of MAMA using RI and photo diode array (PDA) detectors. The UV spectrum of the polymer can be seen in Fig. 4c. The absorption band of MAMA (Figure S12†) was clearly observed in the sonicated PMA-TAS in the presence of MAMA. In addition, the mixing (no sonication) of PMA-TAS and MAMA results in no absorbance increase, indicating no nucleophilic attack of MAMA to the TAS (Figure S13†). Thus, the mechanochemical scission of TAS is necessary to produce a stronger electrophile. These results strongly support our proposed mechanism, and suggest that the mechanochemically generated phenyl cations can be utilized as reactants for chemical reactions with electrophiles. The production of a highly electrophilic reagent will be considerably useful in the design of new mechanochemically triggered responses.

Conclusion

We designed and prepared a new mechanophore that selectively undergoes heterolytic bond scission. CoGEF DFT calculations supported our hypothesis and we prepared PMA-TAS to experimentally study the mechanochemical reaction. Surprisingly, the generated phenyl cations reacted with the counter anion (trifluoromethanesulfonate) to produce a trifluormethylphenyl structure. Although heterolytic scission of C–S bonds in TAS is expected in photochemistry, we observed unique products derived from the polymer mechanochemistry process; herein, the reactive phenyl cation product could react with different solution components and not be limited to reaction with the diphenyl sulfide byproduct. Further studies on this novel mechanochemical reaction are underway in order to elucidate with more detail the mechanism and to direct the reactivity of the phenyl cation. Our efforts will contribute in the development of new mechanochemical self-healing applications utilizing mechanical damage-induced cross-linking/polymerization and mechanical deformation-induced site selective reactions of polymeric materials.

This work was financially supported by the National Science Foundation (CHE-1300313). T.S. thanks the Japan Society for the Promotion of Science for a JAPS postdoctoral fellowship for research abroad.

References

1 (*a*) C. C. DuFort, M. J. Paszek and V. M. Weaver, *Nat. Rev. Mol. Cell Biol.*, 2011, **12**, 308–319; (*b*) P. Delmas, J. Hao and L. Rodat-Despoix, *Nat. Rev. Neurosci.*, 2011, **12**, 139–153; (*c*) D. Jones, G. Leivseth and J. Tenbosch, *Biochem. Cell Biol.*, 1995, **73**, 525–534.

2 For recent reviews, see: (*a*) P. A. May and J. S. Moore, *Chem. Soc. Rev.*, 2013, **42**, 7497–7506; (*b*) K. M. Wiggins, J. N. Brantley and C. W. Bielawski, *Chem. Soc. Rev.*, 2013, **42**, 7130–7147; (*c*) R. Groote, R. T. M. Jakobs and R. P. Sijbesma, *Polym. Chem.*, 2013, **4**, 4846–4859; (*d*) Z. S. Kean and S. L. Craig, *Polymer*, 2012, **53**, 1035–1048.

3 (*a*) D. A. Davis, A. Hamilton, J. Yang, L. D. Cremar, D. V. Gough, S. L. Potisek, M. T. Ong, P. V. Braun, T. J. Martínez, S. R. White, J. S. Moore and N. R. Sottos, *Nature*, 2009, **459**, 68–72; (*b*) Y. Chen, A. J. H. Spiering, S. Karthikeyan, G. W. M. Peters, E. W. Meijer and R. P. Sijbesma, *Nat. Chem.*, 2012, **4**, 559–562; (*c*) Y.-K. Song, K.-H. Lee, W.-S. Hong, S.-Y. Cho, H.-C. Yu and C.-M. Chung, *J. Mater. Chem.*, 2012, **22**, 1380–1386.

4 (*a*) A. Piermattei, S. Karthikeyan and R. P. Sijbesma, *Nat. Chem.*, 2009, **1**, 133–137; (*b*) K. M. Wiggins, T. W. Hudnall, A. G. Tennyson and C. W. Bielawski, *J. Mater. Chem.*, 2011, **21**, 8355–8359; (*c*) R. T. M. Jakobs, S. Ma and R. P. Sijbesma, *ACS Macro Lett.*, 2013, **2**, 613–616.

5 (*a*) C. E. Diesendruck, B. D. Steinberg, N. Sugai, M. N. Silberstein, N. R. Sottos, S. R. White, P. V. Braun and J. S. Moore, *J. Am. Chem. Soc.*, 2012, **134**, 12446–12449; (*b*) M. B. Larsen and A. J. Boydston, *J. Am. Chem. Soc.*, 2014, **136**, 1276–1279.

6 (*a*) W. Kauzmann and H. Eyring, *J. Am. Chem. Soc.*, 1940, **62**, 3113–3125; (*b*) A. M. Basedow and K. H. Ebert, *Adv. Polym. Sci.*, 1977, **22**, 83–148.

7 (*a*) Z. S. Kean, Z. Niu, G. B. Hewage, A. L. Rheingold and S. L. Craig, *J. Am. Chem. Soc.*, 2013, **135**, 13598–13604; (*b*) J. M. Lenhardt, M. T. Ong, R. Choe, C. R. Evenhuis, T. J. Martinez and S. L. Craig, *Science*, 2010, **329**, 1057–1060; (*c*) M. J. Kryger, A. M. Munaretto and J. S. Moore, *J. Am. Chem. Soc.*, 2011, **133**, 18992–18998.

8 (*a*) M. V. Encina, E. Lissi, M. Sarasúa, L. Gargallo and D. J. Radic, *J. Polym. Sci., Polym. Lett. Ed.*, 1980, **18**, 757–760; (*b*) K. L. Berkowski, S. L. Potisek, C. R. Hickenboth and J. S. Moore, *Macromolecules*, 2005, **38**, 8975–8978; (*c*) I. Park, S. S. Sheiko, A. Nese and K. Matyjaszewski, *Macromolecules*, 2009, **42**, 1805–1807.

9 (*a*) M. Sakaguchi, M. Makino, T. Ohura and T. Iwata, *J. Phys. Chem. A*, 2012, **116**, 9872–9877; (*b*) M. Sakaguchi, H. Kinpara, Y. Hori, S. Shimada and H. Kashiwabara, *J. Polym. Sci., Part B: Polym. Phys.*, 1988, **26**, 1307–1312; (*c*) M. Sakaguchi, H. Kinpara, Y. Hori, S. Shimada and H. Kashiwabara, *J. Polym. Sci., Part B: Polym. Phys.*, 1987, **25**, 1431–1437.

10 (*a*) M. Shirai and M. Tsunooka, *Prog. Polym. Sci.*, 1996, **21**, 1–45; (*b*) J. L. Dektar and N. P. Hacker, *J. Am. Chem. Soc.*, 1990, **112**, 6004–6015; (*c*) N. P. Hackerm, D. C. Hofer and K. M. Weksh, *J. Photopolym. Sci. Technol.*, 1992, **5**, 35–46; (*d*) S. Oikawa, N. Fujii, M. Hata and M. Tsuda, *J. Photopolym. Sci. Technol.*, 1994, **7**, 483–486; (*e*) J. V. Crivello, J. L. Lee and D. A. Conlon, *Makromol. Chem., Macromol. Symp.*, 1988, **13–14**, 145–160.

11 H. E. Gottlieb, V. Kotlyar and A. Nudelman, *J. Org. Chem.*, 1997, **62**, 7512–7515.

12 J. V. Crivello and J. H. W. Lam, *J. Org. Chem.*, 1978, **43**, 3055–3058.

13 (*a*) V. Percec, T. Guliashvili, J. S. Ladislaw, A. Wistrand, A. Stjerndahl, M. J. Sienkowska, M. J. Monteiro and S. Sahoo, *J. Am. Chem. Soc.*, 2006, **128**, 14156–14165; (*b*) N. H. Nguyen and V. Percec, *Polym. Chem.*, 2011, **49**, 4227–4240; (*c*) D. Konkolewicz, Y. Wang, M. Zhong, P. Krys, A. A. Isse, A. Gennaro and K. Matyjaszewski, *Macromolecules*, 2013, **46**, 8749–8772.

14 M. J. Kryger, M. T. Ong, S. A. Odom, N. R. Sottos, S. R. White, T. J. Martinez and J. S. Moore, *J. Am. Chem. Soc.*, 2010, **132**, 4558–4559.

15 M. K. Beyer, *J. Chem. Phys.*, 2000, **112**, 7307–7312.

16 D. Aktah and I. Frank, *J. Am. Chem. Soc.*, 2002, **124**, 3402–3406.

17 A. V. Marenich, R. M. Olson, C. P. Kelly, C. J. Cramer and D. G. Truhlar, *J. Chem. Theory Comput.*, 2007, **3**, 2011–2033.

18 (*a*) K. Hori, T. Sonoda, M. Harada and S. Yamazaki-Nishida, *Tetrahedron*, 2000, **56**, 1429–1436; (*b*) V. Dichiarante, M. Fagnoni and A. Albini, *J. Org. Chem.*, 2008, **73**, 1282–1289; (*c*) T. A. Kochina, D. V. Vrazhnov, I. S. Ignatyev, E. O. Kalinin and E. N. Sinotova, *J. Radioanal. Nucl. Chem.*, 2002, **253**, 435–437.

Faraday Discussions

DISCUSSIONS

Sonication and macromolecular mechanochemistry: General discussion

DOI: 10.1039/C4FD90008A

Mr Stauch opened the discussion of the paper by Professor Stephen L. Craig: In your talk you have mentioned that if a polymer is stretched, stress is distributed heterogeneously in the polymer. In particular, you have mentioned that most stress is localized in a specific polymer strand. How can we be sure that this is the case rather than a more homogeneous distribution of mechanical stress among a large number of polymer strands?

Professor Craig responded: Finite elements modeling and strain mapping (for example, using embedded particles) both bear out the idea of stress/strain concentration. Modeling of stress in polymer melts under shear shows that the same is true on molecular-length scales, but the details of how heterogeneous the distribution is on what sort of length scale are not known and represent, I think, an exciting potential opportunity for the field.

Professor Beyer asked: Only a small fraction of cyclopropane rings are opened. Does this mean that a small fraction of polymer strands undergoes complete ring-opening, or are the ring-opened cyclopropane units randomly distributed over all polymer strands?

Professor Craig replied: We know in a previous system (ref. 6 in our paper), but not here, that a ring-opened cyclopropane is more likely to be adjacent to another opened ring than not (average correlation length is about 4 repeats). So it is not completely random, but we do not know if it is a lot of random and a few very long runs, or if it is a smooth distribution that peaks around a few cyclopropanes.

Dr Nagapudi asked: Based on your paper I am wondering if you did fiber spinning with your polymers, thereby giving an orientation to the chains, if you would see increased conversion?

Professor Craig responded: I think this is likely. Orientation is almost certainly necessary, and anything that can be done to pre-organize the mechanophores and their chains in that way should help. This was part of the thinking with the electrospinning, but melt spinning might give even better orientation and perhaps an improved response.

Professor Suslick asked: There is a mechanochemical reaction that you may not see from NMR of dissolved polymer and that is homolytic cleavage of the majority backbone bonds (*e.g.*, simple C–C bonds). Although these bonds are stronger and therefore less likely to break during mechanical activation, they are much more abundant than the single mechanophore. Can one do EPR *in situ* to estimate the number of macroradicals formed and thereby estimate the number of regular chains that fail mechanochemically compared to the number of mechanophores that break during mechanochemical activation of polymer fibers?

Professor Craig answered: Indeed, this experiment could be done. We have not done it. In cases of substantial bond breaking, one can characterize the scission by changes in molecular weight of the dissolved polymers, and we have done that (ref. 25 and 26 in our paper). Many more mechanophores are opened than are simple backbone bonds broken. In cases where there is less damage, one would need something more like EPR as you suggest, but I do not think that the qualitative picture would change.

Professor Suslick responded: It would be interesting to look at a series of different mechanophores competing within single fibers. Could one correlate relative yields of mechanophore breakage with their bond strengths (*i.e.*, enthalpies of ring opening, *etc.*) or their "stretchability"?

Professor Craig replied: This is a great idea. In principle, one could do this to answer the very interesting question (in my mind) of the relationship between isolated mechanical activity (as we characterize by force spectroscopy) and mechanical activity in the context of a bulk system.

Dr Diesendruck noted: This work is very impressive, a nice way to improve the "grip" between the chains and therefore improve the force transmittance to the chains containing the mechanophores. Since "grip" is related to T_g, did you measure the T_g for the block co-polymers, specifically the polymer that was fiber-spun and annealed, which presented no phase separation?

Professor Craig answered: We do not have good thermal data, but interestingly spanning a comparable range of T relative to T_g in another system (ref. 6 in our paper) did not lead to measurable activation. We believe an effect beyond only a shift in effective T_g is at play, although it certainly must contribute to some extent.

Dr Diesendruck replied: Given that you observed no phase separation by electrospinning, do you think you need block co-polymers to get the effect you observed or would solutions of polymer mixtures get the same effect?

Professor Craig responded: I think that mixtures very well could have an effect. What is interesting here is that even when we varied temperature in relation to T_g in the homopolymer case, we were unable to achieve measurable activation (ref. 6 in our paper). So there is something more at play than simply shifting T_g in this case, but that could be due to alignment in the electrospinning.

Professor Beyer addressed Professor Craig and Professor Suslick: The answer to the question whether ring-opening of *gem*-dihalogenated cyclopropane mechanophores competes with breaking of any other bond in the polymer may lie in the AFM data shown. The brominated cyclopropane seems completely stretched out, and failure of the anchoring of the polymer requires an increase in the force, as evident from the force curves.

Professor Suslick answered in reply: This is certainly true for single-molecule AFM mechanochemical rupture of bonds most of the time. My question, however, is for the more general case where a mechanochemical breakage occurs in sonicated solutions or in stretched strands of polymers. It is important to realize that every polymer chain has "mechanophores" built into it: *e.g.*, the carbon–carbon bonds that make up the vast majority of most polymer backbones. While it is true that these bonds are stronger than synthetically designed mechanophores and therefore less likely to break upon mechanical force, there are 1000s of the standard C–C backbone bonds compared to a single mechanophore in a long polymer chain.

Professor Craig added: I agree. When tension is constant along a chain length, the cyclopropanes will essentially always open before the chain breaks.

Professor Beyer again addressed Professor Craig: The kinetics of rupture of competing mechanophores has been investigated by AFM experiments of single biopolymers for non-covalent interactions by the Gaub group.

Professor Craig responded: And we have put different covalent mechanophores along the same polymer backbone, although that work has not yet been published.

Dr Khripin asked: My question is, what would mechanophore activation be like in fatigue experiments? By "fatigue experiment" I mean one that subjects the material to moderate strain (10–100%) over thousands of cycles, keeping strain constant while creep is observed. I wonder if stress redistribution during creep could result in higher levels of mechanophore activation?

Professor Craig explained: We make one such comparison in our paper, and do not see a substantial difference when we slowly increase strain from one cycle to the next. I think maintaining the same strain would not have much effect in subsequent cycles, because creep would limit the stress that accumulates. But pre-straining is known to enhance subsequent activation (see, for example, ref. 35 in our paper).

Dr Irikura commented: It appears to be challenging to establish that a polymer chain has broken at the mechanophore, and not elsewhere along the chain. Are there mechanophores that signal their activation by releasing a small molecule, such as N_2 or CO_2, or by creating a chromophore?

Professor Craig answered: Remember that in our case an activated mechanophore does not break, but rather it opens. We can see that by conventional spectroscopy, but there are definitely cases where more sensitive indicators are desirable. The most effective that I know of to date are colorimetric (ref. 21 in our paper) and luminescent (ref. 22 in our paper) probes developed by others.

Mr Julien added: In the activation of your polymers, you first apply mechanochemical force to break the polymer, have the diffusion of reagents to the "broken" sites, and finally the subsequent reaction. Do you have a feeling for the extent to which each of these steps is determining the rate and extent of polymer activation, and if so which you are most focused on improving?

Professor Craig responded: When thinking about this activation in the context of self-strengthening (as in ref. 26 in our paper), the rate-determining step is generally the application of the force – nothing happens until the mechanophores experience the necessary load. Activation itself is very dependent on the mechanical environment. In our previous work, we saw substantial cross-linking reactivity in tens of minutes, but subsequent reactivity occurred on the time scale of hours/days. I think there is great room for designing systems that react at the encounter-controlled limit once a latent reactant is activated mechanically.

Mr Julien continued: After you break and activate your polymer fibers by unidirectional stretching and reaction, to what extent are the resulting effects on the polymer unidirectional? Do you know: to what extent do the properties of the new activated polymer change in direction perpendicular to the original stretching?

Professor Craig replied: This is a great question, and we do not have data to address it. One would suppose that activation requires alignment with the macroscopic stretching, and previous work by Moore, Sottos, and White at UIUC has shown this to be the case in another system using a spiropyran indicator.

Ms Nielsen asked: Do you know of mechanophores that can be reliably activated with a variety of linker placements?

Professor Craig concluded by saying: The best examples are with differing stereochemistry. The dihalocyclopropanes discussed in this paper can be activated with either *cis* or *trans* stereochemistry, but the activities differ. The same has been reported for benzocyclobutanes, epoxides, and perfluorocyclobutyl aryl ethers. Regiochemical examples have been reported for triazoles and spiropyrans. Nice examples of how different attachments might lead to very different activation modes can be found in Boulatov's force probe work on phosphodiesters.

Dr Diesendruck opened the discussion of the paper by Professor Martin Beyer by communicating: In tensile testing, the force distribution is heterogeneous, concentrating at the middle of the material. Is AFM similar in that regard?

Professor Beyer answered: AFM is different. Since the force is applied from endpoint to endpoint, the molecule experiences the same force in any position between the anchoring points at the cantilever and the substrate. I am not very familiar with mechanical properties of bulk polymers, but it seems to me that in stretching a bulk polymer block, deformation precedes bond rupture, with the polymer block thinning in the center. When the cross-section in the middle of the material sample is reduced, the force per unit area is increasing.

Dr Diesendruck then asked: Could you explain the chemistry of the binding – in the paper you describe surface activation using DETA (a triamine), while the amylose has carboxylic acid groups; however, you present the activation of silyl ester bonds, when I would expect you obtain an amide bond.

Professor Beyer explained: The amide linkage is obtained at pH 7.4, with the EDC/NHS coupling chemistry in PBS buffer. Going to lower pH, we found that coupling was more efficient, and that EDC/NHS was not required. We systematically changed the silane used for surface functionalization, and verified that the amine group was not needed to get covalent attachment. We therefore concluded that at pH 2.0, a silyl ester is formed.

Professor Craig then asked: You fit your kinetic data with a modified Morse potential. How confident are you that this potential is appropriate for the mechanism of the force-induced reaction?

Professor Beyer responded: The Morse potential is surely not ideal to model a hydrolysis. We have tried the Bell potential, and found it yields very similar results. It seems that the exact shape of the potential has only a small effect. The probably best way would be to do EFEI calculations at a series of forces, and describe the force-dependent activation energy with an analytical fit function.

Dr Irikura communicated: The experiments vary in their speed. Are they all slow enough to allow proper thermalization by the heat bath?

Professor Beyer confirmed: Yes. The heat capacity of the single stretched molecule is small. Thermalization therefore should occur within picoseconds. Even the fastest experiments take several milliseconds, so there is sufficient time for equilibration.

Dr Diesendruck remarked: You did your AFM measurements in a complicated system having multiple bonds. How do you know that the bond you are activating is the silyl ester, and not, say, the sugar acetal bond, considering you are doing the experiment in water at pH 2?

Professor Beyer explained in reply: The biexponential decay kinetics in the force-ramp experiments at pH 2.0 show that there are two chemically distinct subsets of molecules rupturing. Therefore, we conclude that the rupture occurs in the surface anchor, where these two different anchoring moieties can be realized. We also know from quantum chemical calculations, CoGEF as well as EFEI, that the silyl ester is the mechanically weakest bond in the system.

Professor Craig added: I want to call attention to Professor Beyer's 2000 paper on mechanical bond strength, that shows nicely how the relative strength of covalent bonds is force-dependent. Some bonds are stronger (that is, they survive longer) than another bond at low forces, whereas the relative survival probability swaps at higher forces. I think this was a quite important paper, and one still deserving of experimental characterization.

Dr Friščić: I would like to ask about using mechanochemical action to build molecules onto a surface rather than to break them apart. Has any work been done in that direction?

Professor Beyer replied: There are several examples of mechanochemical nanolithography, see *e.g.* Okmar A. Nafday, "Scanning Probe Lithography", in *Nanofabrication Handbook*, ed. Stefano Cabrini & Satoshi Kawata, 2012, CRC Press.

Professor Craig: I agree with Professor Beyer's point that it will be very difficult to achieve the level of precision in applying force under compression that one can under tension with scanning probe techniques. I would point out that Adam Braunschweig (not attending) is doing nice work involving the topic asked about by Professor Friščić, and Diels–Alder reactions are in fact effective in this regard.

Professor Bolm: What is the size of the polymer used in your investigation, and what is the smallest molecule that can be studied with the technique you presented?

Professor Beyer replied: Carboxymethylated amylose (CMA) is around 1000 nm long. With the technique used here, we have studied poly(ethylene glycol) (PEG) with less than 50 nm length. The minimum chain length is probably around 10–20 nm. Long polymers make it possible to pull the cantilever sufficiently far from the surface so that the cantilever–surface interaction becomes negligible. The force–extension curve of the stretched molecules is important to verify that indeed only one molecule is attached to the cantilever. In force-clamp experiments, the elasticity of the molecule helps to maintain a constant force, since small changes in the cantilever position do not affect the force too much. With other techniques, like a combined conductance/force measurement as performed by the Venkataraman group, individual small molecules like bipyridine or octanedithiol are accessible.

Professor Mack asked: Do outside environmental forces (*i.e.* solvent polarity) aid in the rupturing of a bond under mechanochemical force?

Professor Beyer explained in reply: The interaction of the stressed molecule with the solvent certainly modifies the potential energy surface of the bond-rupture event. Solvent effects may promote heterolytic bond scissions, *e.g.* bond hydrolysis or ion pair formation. If the solvent, however, destabilizes the transition state, it is also possible that the environment hinders bond scission. For a correct description of mechanochemical bond rupture, the environment has to be taken into account.

Professor Mack opened the discussion of the paper by Professor Fernando Galembeck by communicating: In one of the triboelectric reactions nitrogen was incorporated into the polymer from nitrogen gas, have you tried to insert other atoms into the polymer using other gases?

Professor Galembeck responded: Not yet, but this is an excellent idea. I believe that not only gases but also sorbed or wetting liquids could be used as precursors of the desired atoms.

Professor Suslick asked: In principle, with the high-resolution Kelvin force microscope, may one be able to see the charged ends of the polymer chains after triboelectrification?

Professor Galembeck confirmed: Yes, since KFM high resolution is now sub-atomic. The main difficulty that I can foresee is that the sample surfaces should also be very smooth.

Dr Diesendruck then asked: From your analytical techniques, can you calculate the amount of electrons transferred and from that find out the number of chain scission events that occurred?

Professor Galembeck replied: Charge measurements tell us the final number of charges acquired by the rubbing samples, following a complex chain of fast reactions including competitive ones leading to uncharged products. To obtain the total number of chain-scission events we need to consider concurrent products as described in some references within Martin K. Beyer and Hauke Clausen-Schaumann.[1] We are currently examining techniques that could provide complete information, such as XPS (X-ray photoelectron spectroscopy).

1 M. K. Beyer and H. Clausen-Schaumann, *Chem. Rev.*, 2005, **105**, 2921.

Dr Diesendruck continued: What are the molecular weights of the polymers you are using? Did you consider using polymers with lower molecular weights, say 1 kDa, below entanglement limit? It would be a nice control, since no mechanochemical bond scissions would occur.

Professor Galembeck clarified in his reply: Polymers used in this work are grades for technical applications, thus their molecular weights are above the entanglement limit. We plan to look at lower molecular weights as well as to small molecules susceptible to mechanochemical reactions.

Dr Blair remarked: Your high content of surface nitrogen is quite surprising. Can you comment on the nature of this nitrogen? Is it physisorbed or chemisorbed with a reduction in the N_2 bond order? Have you though about testing under a 1 : 3 mix of nitrogen and hydrogen? You may observe the formation of ammonia.

Professor Galembeck communicated in reply: I agree, surface nitrogen was not expected and it was only observed because the EDS survey scans quickly show all the elements present in the sample. It is chemisorbed, because it persists while the sample is handled and examined, under high vacuum, in the electron microscope. I did not yet test it under mixed nitrogen and hydrogen but I plan to do so.

Dr Blair added: If nitrogen is chemisorbed this may be the initial steps toward a process that could replace the Haber process for the production of ammonia.

Professor Galembeck verified in his reply: I agree. We are currently looking at tribochemical modification of biomass residues such as sugarcane bagasse under air. If this also leads to nitrogen chemisorption, it can become an interesting alternative to petrochemical ammonia.

Mr Julien then commented: You describe this process as hot spots generating free radicals which undergo electron transfer to cationic and anionic polymer fragments which subsequently arrange into domains. Is this arrangement of domains localized to the "hot spot" or does it propagate through the material?

Professor Galembeck responded: Cationic and anionic polymer fragments propagate through the material but with finite mobility. This happens around micron-size "hot-spots" but also around the overall rubbed area, since charges are detected many millimetres outside the rubbing disk limits.

Mr Huskic noted: After milling of powders made of small molecules we often observe that the product is statically charged. Would you say that there is a common, underlying mechanism?

Professor Galembeck replied: I believe that in the case of small molecules there is an effect of charge created in the mill walls and transferred to the molecules. However, this is an open question.

Professor Boldyreva communicated: This is a very interesting study. I have several comments on behalf of myself and Dr. A. Politov. The title is very general, but in fact the paper deals with one class of dielectrics only, namely with polymers.

I want to mention, that many years ago similar studies were carried out for other dielectrics – crystals of alkali metal halides (see, *e.g.*, F. Frohlich and P. Seifert, *Phys. Status Solidi*, 1968, **25**, 303; G. I. Distler *et al.*, Decoration of solid surfaces, *Nauka*, Moscow, 1976). The radical formation at the surfaces of quartz and polymers on mechanical treatment was also studied in detail by several research groups in the USSR in the 1970s–1990s. It would be very interesting, if the authors could compare the previously obtained results and proposed models with their own findings. I give a few examples only.

The authors of this paper interpret the charge pattern formation as the "result of polymer chain breakdown forming free radicals and ions, followed by electron transfer according to the electronegativity of each of these high-energy species formed". The role of local melting and "hot spot" formation is discussed. I would like to mention that a real situation may be much more complex than considered in this work. Thus, P. Yu. Butyagin and A. M. Dubinskaya have shown that on mechanical treatment of polymers the radicals can be formed not at hot spots but, on the contrary, following a non-thermal mechanism, which is more efficient at low temperatures (A. M. Dubinskaya, Transformations of organic solids under mechanical stress, *Russ. Chem. Rev.*,

1999, **68**(8), 708–724; P. Yu. Butyagin, *Russ. Chem. Rev.*, 1971, **40**, 1935). A detailed mechanism of bond-breaking in polymers at constant and at variable mechanical stress was described in the papers by Zhurkov and his research group (V. R. Regel, A. I. Slutcker, E. E. Tomashevskii, Kinetic nature of mechanical strength of solids, *Nauka*, Moscow, 1974). V. A. Zakrevskii has shown that low-temperature deformation of polymers (such as the one described in this work) often results in electron emission and formation of radical cations (V. A. Zakrevskii, V. A. Pakhotin, Autoionization and the bond breaking in macromolecules, *Polym. Sci.*, 1981, **23**(3), 658–662). A. I. Gubanov has developed a theoretical model in which he considered a tunneling mechanism of electron emission in the deformed polymer matrices, in order to interpret the charge generation accompanying the deformation of polymers (A. I. Gubanov, V. A. Kosobukin, *Mechanics of Polymers*, 1975, **1**, 33). This mechanism is significantly different from that proposed in the ref. 17 cited in the present work and accepted by the authors. Numerous examples of mass-transfer between the two surfaces during their friction resulting from co-melting can be found in the book by F. P. Bowden and D. Tabor, *The Friction and Lubrication of Solids*, Oxford, Clarendon, 2001, p. 341, which is cited in this paper as ref. 40. After the melt is cooled down, the two polymers are welded to each other, and mechanical destruction results in the transfer of a fragment of one polymer to the other. In the papers by L. S. Zarkhin *et al.* (Mechanical destruction of polymers. Molecular dynamics, *Russ. Chem. Rev.*, 1989, LVII, issue 4, 644–662) it was theoretically predicted – and in the works by V. R. Regel *et al.* experimentally confirmed – that mechanical treatment of a polymer can result in the emission of small molecular fragments with a distribution of molecular masses (A. V. Amelin, T. M. Muinov, O. F. Pozdnyakov, V. R. Regel, *Mechanics of Polymers*, 1967, **1**, 80]. Thus, the charged fragments not only of another polymer, but also of the original polymer can be present at the surface after mechanical treatment.

Professor Galembeck communicated: Yes, polymer chain breaking can take place under lower temperatures, especially because fragile fracture can then take place easier, especially under fast mechanical action following the Williams–Landell–Ferry model. Our statement "transforms them into hot-spots for free-radical formation" indicates that this is where action takes place, not necessarily that these are warmed-up sites, although local heating may also be part of the picture. Polymer melting allows viscous transfer of whole chains between surfaces, while breaking protrusions actually requires covalent bond breaking, in agreement with our model. Our main contribution is the identification of charged species as ionic polymer fragments formed by mechanical action using their spectral signatures coupled to imaging and charge mapping and this is unprecedented, for polymers. Ionic polymer fragments are not any molecular fragments and their identification on separated positive or negative domains allows us to bridge successfully the abundant literature on mass transfer and surface wear and the also abundant literature on tribocharging. Fröhlich's work is already cited as reference 34 and the authors thank Professor Boldyreva for the other references that will be used for future pertinent referencing.

Professor Craig opened the discussion of the paper by Dr Tomohiro Shiraki by communicating: Could you comment on the efficiency of your MAMA trapping and what it means? Is inefficient trapping a consequence of competing reactions or of inefficient activation?

Dr Diesendruck responding on behalf of Dr Shiraki said: We calculated the efficiency of the trapping from the UV intensity of the MAMA peaks. We got a rough estimate of about 23% trapping. We believe that is a consequence of competing reactions, given the high reactivity of the phenyl cations.

Professor Craig then asked: Is it possible that the triflate dissociates in sonication, but recombines in the absence of a phenyl cation? I am not sure of the relative rates of the putative reaction of trifluoromethyl anion with phenyl cation *vs.* those of the trapping reactions related to triflate that you employed.

Dr Diesendruck replied: We are still not sure about the mechanism of the reaction with the triflate. The reaction with phenyl cation occurs only when there is no other nucleophile in solution; however, we are still investigating reactions with other nucleophiles.

Mr Stauch commented: You have carried out CoGEF calculations of organic cationic species, in which you investigate the radical character of the systems. For this, you have used the DFT hybrid functional B3LYP. Density Functional Theory, however, has been shown to deliver large errors when applied to organic radical cations. Why did you choose DFT? Have you compared your results to other levels of theory, *e.g.* wavefunction-based methods?

Dr Diesendruck explained: We have used B3LYP since it has been the level of theory we have always used, and it gives us an idea of how "hard" it is to activate the mechanophore, as compared to other mechanophores we developed. We did not run other levels of theory, since we do not give much value to the energies found. We use CoGEF only for initial assessment of our hypothesis by pointing out what bond is going to be cleaved.

Professor Mack then added: How do you know the bond cleavage is heterocleavage and not homocleavage? Did you check for radical formation?

Dr Diesendruck responded: We expected to have radicals, since there are always radicals as a consequence of sonochemistry. One of our controls was a homo-PMA which undergoes homolytic bond scission, and it did not incorporate the amine trap. That indicates to us that the mechanophore is undergoing heterolytic bond scission in order to react with the trap.

Professor Mack asked: In the trapping experiment with 9-(methylaminomethyl)anthracene how do you know the phenyl cation is on the nitrogen and not on the aromatic ring?

Dr Diesendruck said in reply: A Friedel–Crafts reaction mechanism was indeed possible. However, addition of phenyl cation to the anthracene system would change the UV-Vis spectrum of the π-system, but the spectrum remains unchanged after the mechanochemical reaction.

Dr Baláž then asked: In the experiments, 1200 equivalents of trapping agent are being used. This seems like a very large amount. Is it necessary to use such high concentration of the substance?

Dr Diesendruck explained: Sonication is done at a concentration of 1 mg polymer per millilitre of solvent, and since each chain has a single mechanophore, the concentration of actives produced is extremely low. 1200 equiv. is only around 2.5 mg mL^{-1}, so still not a high concentration. In this case, since we believe there is reaction of the phenyl cation with the solvent, I believe higher concentrations would give us better results.

Dr Lamaty remarked: You present a mechanism of aryl cation trapping by the amine? Could it be a nucleophilic aromatic substitution? Did you perform sonication in the absence of the polymers?

Dr Diesendruck replied: A Friedel–Crafts reaction mechanism was indeed possible. However, addition of phenyl cation to the anthracene system would change the UV-Vis spectrum of the π-system, but the spectrum remains unchanged after the mechanochemical reaction. We did control experiments using small molecules, or low molecular weight polymers, and we did not observe significant thermal reactions.

Professor Bolm communicated: I am wondering if your CF_3 incorporation (through $F_3CSO_3^-$) is mechanistically related to trifluoromethylations with Langlois's reagent ($F_3CSO_2^-$). In an early paper[1] he suggested the intermediacy of radical species. Assuming an electron transfer mechanism in your case, you might reach analogous intermediates.

What is mechanochemistry? A very broad area. Use of mechanical force to enable assembly or dissociation of objects at different lengths scales, from molecules to microscopic sizes. Agitation is also relevant.

What are the underlying mechanisms? It all happens at the same time (K Suslick), there is more than one effect and there is not only one answer.

How do we proceed? As it is a very complex field, we need to work together (W Jones), through enhanced interactions at national and regional meetings

Where it will go? Microscopy and nanoindentation studies, exploration of the materials at very small length scales.
Understanding of mechanochemical activation and defects in terms of supramolecular chemistry and intermolecular interactions

Observation of mechanochemical reactions as they take place, bulk → assemblies → individual events

Understanding of mechanochemical processes, as presented at this meeting, shows a clear and inherent connection to the fundamental understanding of materials; Green Chemistry.

Fig. 1

1 B. R. Langlois, E. Laurent and N. Roidot, *Tetrahedron Lett.*, 1991, **32**, 7525.

Dr Diesendruck responded: We initially proposed this mechanism imagining a Sandmeyer-like reaction. However, our trials to do a Schiemann reaction, which would be related to the mechanism we proposed, did not provide any phenyl fluoride. We are currently studying other nucleophiles, but we had not considered the possibility of a radical mechanism related to Langlois reagent. Thanks for your suggestion, we will definitely look into it.

Professor Beyer added: Have you tried different pulling points in the CoGEF calculations? It looks to me that pulling in the *meta* position from the C–S bond should lead to lower rupture forces.

Dr Diesendruck remarked: We have not. That is a good idea! We are having difficulties activating this mechanophore in the solid state and anything that would facilitate activation is helpful and worth testing, thanks.

Professor Mack posed the question: In the formation of phenyl cation have you looked at other counter ions aside from triflate?

Dr Diesendruck answered: We did hexafluorophosphate and tetrafluoroborate, and they did not react with the cation; we were expecting a Schiemann reaction but we did not observe it in the ^{19}F NMR. We are currently using other counter ions to study reaction with other nucleophiles.

Professor Mack then asked: Once you form the phenyl cation you suggested the triflate decomposed to liberate sulfur trioxide, do you observed a change in the weight of the polymer after the reaction?

Dr Diesendruck replied: We never tried. For each sonication we use around 10 mg of ~100 kDa polymer, each chain containing a single triflate group. The change in sample weight is too small.

Dr Blair addressed Dr Diesendruck and Professor Mack: As a follow up to James Mack's comment, the polymer weight may be measurable by incorporating a QCM into the apparatus.

Mr Kochhar also communicated: In you CoGEF-based simulations, did you consider different pulling rates? Do you believe this would alter the outcome of the reaction?

Dr Diesendruck repsonded: When running CoGEF, we typically use initially a "fast-pulling rate" to find the distance where scission occurs. Afterwards we re-run the calculation using a much slower pulling rate to get more points around the scission and a more precise peak energy. In our experience, fast pulling rates lead sometimes to scission in the wrong places, so we try always to use low rates.

Mr Kochhar commented: There does not appear to be a smooth transition after bond scission in your CoGEF potentials. Is this significant?

Dr Diesendruck replied in conclusion: We do not get an energy plateau after scission, which is what is typically observed. However, given that the product is ionic, we think the energy continues to increase due to an electrostatic interaction that is being lost with increasing the distance between the two parts.

Dr Friščić opened the discussion of the concluding remarks by Kenneth Suslick by communicating: The attached slide (Fig. 1) is a tentative summary of our meeting. I have attempted to address the questions suggested at the beginning of the meeting. Overall, it appears very clear that mechanochemistry is a very broad area and we are aware that mechanochemical effects can take place over a variety of length scales (from single molecules to microns). Mechanisms are not clearly understood and, to make things more complicated and, as Professor Suslick has noted, everything is taking place at the same (often very short) time. A proper understanding and the development of this area is certainly a job for a cohesive community, rather than individual, isolated researchers, as noted by Professor Jones. In that context, we should be looking forward to further high-impact meetings at regional and international levels.

In the future, we can expect a growing importance of nanoindentation studies and exploration of mechanochemical effects at very small length scales. It seems as if supramolecular chemistry and understanding of intermolecular interactions could provide important insight into mechanochemical activation and associated structural defects in solids. An important contribution will be made by the development of new diffraction or spectroscopic techniques for direct, real-time and *in-situ* monitoring of milling reactions and transformations.

The papers and discussions presented at this meeting show that mechanochemistry has the potential to become a central element in the contexts of green and sustainable chemistry, as well as solid-materials science.

Further comments and views of all delegates are very welcome!

After the meeting had concluded, **Dr Baláž** opened the discussion of the paper by Dr Adam Michalchuk and Dr Elena Boldryeva (which could not be presented at the meeting) by communicating: (1) Congratulations on the great paper, which significantly contributes to the aspect of the mechanisms of the mechanochemical reactions. My question deals with the novel phase you observed. Do you have any idea about its exact chemical composition, or what kind of compound that might be?

(2) My other question deals with the Section 3.1.3.1 ('Reactive regions in a ball mill'). Do you know any other systems, in which such different products were observed on the walls and at the ends of milling jars? Are there such examples also in the case of inorganic reactions?

Professor Boldyreva replied to these questions: (1) At present the exact identity of the novel phase remains unknown. However, considering the minute quantities obtainable through mechanical treatment (discussed in the manuscript *via* tableting mechanisms), current thinking is that the novel phase is a multi-component crystal, a salt or co-crystal of the two reactant components. Should a single phase crystal be suggested, one would expect much larger portions of the tablet to undergo this transformation. It has, unfortunately, proven itself rather elusive due to the current inability to obtain single crystals of this phase. Phase identification remains an ongoing segment of this project.

(2) As far as we are aware this is the first system to have been identified as having this character in a mill, very possibly the result of very simple variation in experimental protocol in which the entire milling jar is emptied for phase analysis. Some remote hint on the possibility to have different products at different parts of a milling jar can be found in a review on inorganic oxides by Zyryanov.[1]

However, we do not imagine this is unique to the system mentioned in our study. In fact, a number of years ago, a lecture given by Mark Descampes at the University of Lille presented results (unpublished) in which a similar concept of varying shear and impact suggested similar effects. Different products formed on shearing and on impact have been reported by Tumanov *et al.*[2] It therefore offers a fruitful direction to further search for systems that may share this phenomenon.

1 V. V. Zyryanov, Mechanochemical synthesis of complex oxides *Russ. Chem. Rev.*, 2008, **77**, 105–135.
2 I. A. Tumanov, A. F. Achkasov, S. A. Myz, E. V. Boldyreva and V. V. Boldyrev, Different effect of shear and impact on mechanochemical co-crystallization of piroxicam and succinic acid, *Dokl. Phys. Chem.*, 2014, in press.

In addition **Dr Baláž** opened the discussion of the paper by Dr Anatoly Politov (which also could not be presented at the meeting) by communicating: (1) In your paper you write that by performing the mechanical treatment *in situ*, the products which accumulate in the reaction zone are removed. How is it achieved in the contact reaction zone?

(2) In the part dealing with radicals you wrote that the ratio between SO_4^- and SO_5^- radicals was 1 : 3 during mechanical treatment, although according to the scheme it should be 1 : 1. In further text, this discrepancy is explained by effect of mechanical action; however, I was unable to find any sentence stating why the creation of SO_5^- is preferred. Maybe I have overlooked something basic, but anyway, could you please clarify this issue?

Dr Politov responded to these questions: (1) In the cited place, we mean accumulation of intermediate products of enzymatic hydrolysis that cover the surface of the solid substrate, in our case a surface of natural polysaccharides. Hydrolysis of carbohydrates is the result of a number of parallel and consequent reactions and due to action of several enzymes. Hydrolysis begins on a surface of a solid substrate. Some enzymes facilitate solid-state reactions and produce long-chain and short-chain polysaccharides; other enzymes react with obtained saccharides in solution near the surface and produce mono- and di-saccharides. Taking into account that only short-chain sugars have a good solubility in water one can expect the formation of a viscous layer on the reactive surface that drastically increases the coefficient of diffusion and, therefore, decelerates the hydrolysis of native polysaccharides. Thus, the reaction zone in our example is the high-viscous layer between the solid substrate and the low-viscous aqueous solution. On experiments, the enzyme hydrolysis of the solid substrate stops soon after beginning. The reaction recommences again after mechanical treatment of the reaction mixture *in situ*, *i.e.* in the same reactor where hydrolysis takes place. Since reactions decelerate on a solid–liquid interface, we applied ultrasound as a way of mechanical treatment and the contact reaction zone is achieved by ultrasonic lengthwise waves.

(2) Unfortunately, there is a misprint in paper. The ratio between SO_4^- and SO_5^- is 2 : 1 rather than 1 : 3.

Nevertheless, there is no good explanation also in the case when the ratio of primary radicals SO_4^- and SO_5^- is 2 : 1 because we had to assume that some of the primary radicals are more reactive than other primary radicals are. The suggestion has only indirect evidence and occurs in the literature as well.

It should be specified that the above-mentioned ratio is valid only for primary radicals SO_4^- and SO_5^-, *i.e.* radicals that result from mechanical treatment at the temperature of liquid nitrogen, excluding secondary heat transformations. The subsequent heating of the treated substance changes this ratio.

The radicals SO_4^- and SO_5^- can be supposed to form as a result from from the decay of anion radicals $S_2O_8^-$ and $S_2O_8^{3-}$, following the scheme:

$$6\ S_2O_8^{2-} = 3\ S_2O_8^{3-} + 3\ S_2O_8^- \quad (1)$$

$$3\ S_2O_8^{3-} = 3\ SO_4^{2-} + 3\ SO_4^- \quad (2)$$

$$2\ S_2O_8^- = 2\ SO_3 + 2\ SO_5^- \quad (3)$$

$$S_2O_8^- + SO_4^{2-} = S_2O_8^{2-} + SO_4^- \quad (4)$$

The resulting ratio of SO_4^- and SO_5^- formed according to the equations (1)–(4) is thus 2 : 1.

The transformations (1)–(3) were observed experimentally, and the reaction described by eqn (4) is our suggestion.

It is well known that the electron affinity and ionization potential of an anion can be changed when the anion is deformed in solids. Thus, it seems likely that the persulfate anion takes away electrons from the sulfate anion.

One should note that ratio of primary radicals SO_4^- and SO_5^- is approximately 2 : 1. Sometimes it is less than the indicated value; sometimes it is more. However, the amount of prime SO_4^- radicals is always higher than amount of primary SO_5^- radicals.

Faraday Discussions

PAPER

Mechanochemistry and sonochemistry: concluding remarks

Kenneth S. Suslick*

Received 8th July 2014, Accepted 9th July 2014
DOI: 10.1039/c4fd00148f

This paper offers a perspective on mechanochemistry and offers summarizing commentary on the *Faraday Discussion* **170**, "Mechanochemistry: From Functional Solids to Single Molecules". The connection between the mechanical and the chemical worlds dates back to our earliest written records and beyond, but its renaissance over the past decade or so has had an impact on a huge swathe of modern science and engineering: from metallurgists to polymer scientists to synthetic organic and inorganic chemists to cellular biologists. Connections among the different subfields of mechanochemistry (tribochemistry, trituration, macromolecular, and sonochemistry) are drawn out and the common themes and open questions are considered.

1 Introduction

The use of friction to make fire takes our connection between the mechanical and the chemical worlds back to the dawn of our species. The written records on mechanochemistry date back to the ancient Greeks, when Theophrastus (one of Aristotle's students) wrote "On Stones" in *ca.* 315 BCE and referred to the reduction of cinnabar to mercury by grinding in a copper mortar and pestle.[1] 2100 years later, Faraday described the reduction of silver chloride by grinding with various metals, his "dry way" of inducing chemical reactions.[2] The full history of mechanochemistry is fascinating and has been delightfully examined by Laszlo Takacs elsewhere.[3]

While the chemical consequences of mechanical actions have a long heritage, it was Ostwald (the ninth chemistry Nobel Laureate) who systemized chemistry in 1919 into the branches of thermochemistry, electrochemistry, photochemistry and mechanochemistry.[4] It is only in the past decade or so, however, that mechanochemistry has gained notable attention in the chemical community with the diverse confluence of related but quite different specialties, ranging from metallurgists to polymer scientists to synthetic organic and inorganic chemists to cellular biologists.[5–18]

School of Chemical Sciences, University of Illinois at Urbana-Champaign, 600 S. Mathews Av., Urbana, IL 61801, USA. E-mail: ksuslick@illinois.edu; Tel: +1-217-333-2794

This very recent renaissance, moreover, makes this symposium (FD170, "Mechanochemistry: From Functional Solids to Single Molecules") the first significant appearance of mechanochemistry in the *Faraday Discussions*. The 26 papers and 48 posters presented at FD170 spanned most of the modern topics of mechanochemistry, with the notable exception of bio-mechanochemical topics (*i.e.*, cellular motion, biomolecular motors, and intracellular active transport[16,19,20]). While the relationship between mechanochemical and the bio-mechanical is clear, the latter is sufficiently complex to require its own, future *Faraday Discussion*.

The diversity of mechanochemistry reflects the multitude of ways in which mechanical phenomena can bring about chemical consequences; I provide one useful organizational chart in Fig. 1. One may break mechanochemistry into four areas: tribochemistry (the chemistry of surfaces in contact), trituration (chemistry induced by grinding and milling), macromolecular mechanochemistry (from breakage of polymer chains to molecular motors and biological motion), and sonochemistry (the chemistry generated from the mechanical consequences of sound). One might also add shock-induced solid-state chemistry as a fifth subfield, but this was not included in these discussions. These separate categories have each developed largely independently of one another in different

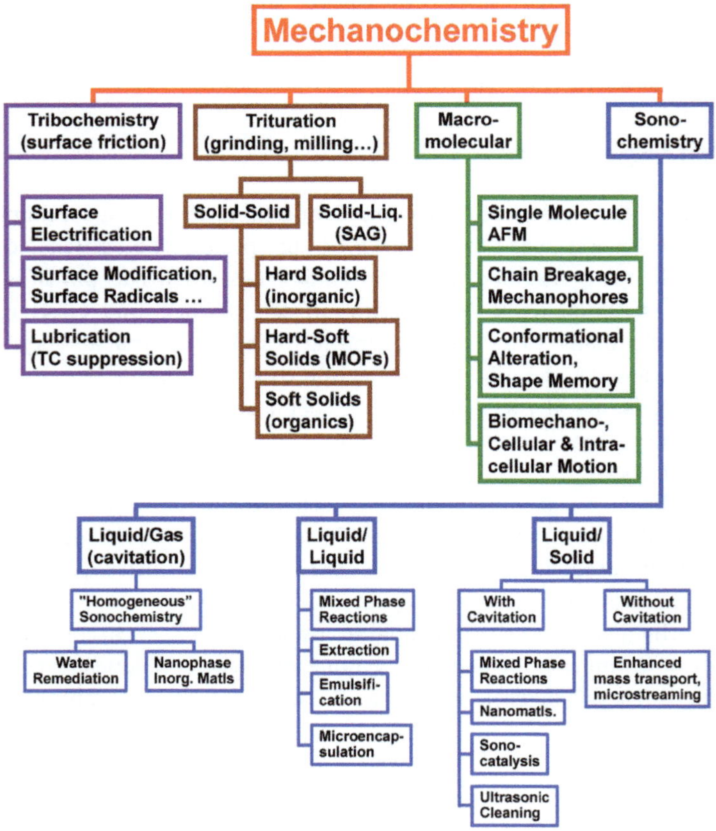

Fig. 1 Mechanochemistry may be divided into four distinct, but related areas.

communities and often with different goals and concerns. One may hope that FD170 will have contributed substantially to a convergence of these diverse areas under the more general rubric of mechanochemistry.

My role in FD170 was two-fold: I was given the honour of presenting the summary lecture, but I was also the principal representative of a related branch of non-traditional chemistry dealing with the chemical effects of ultrasound. With apologies to Churchill, sonochemists and mechanochemists are one people, separated by a common phenomenon. That common phenomenon is the conversion of mechanical energy into chemical consequences. As such, my commentary below will first draw out the similarities among all forms of mechanochemistry (with some focus on my own sonochemical interests) and second will provide an overview of the presentations and discussions of FD170. I do not intend to provide a blow-by-blow summary of all the presentations and the accompanying discussions: obviously, this can be far better gained by reading the Discussion volume. I will, however, attempt to bring out common themes and unresolved questions that the Discussions developed.

2 The sonochemical-mechanochemical connection

Sonochemistry is a consequence of the mechanical effects of sound on liquids and originates from acoustic cavitation (the formation, growth, and implosive collapse of bubbles in liquids, Fig. 2).[21–24] The bubble collapse creates intense compressional heating within the bubble and generates extreme transient conditions in the resultant hot spots that can achieve temperatures above 5000 K, pressures exceeding 1000 atmospheres, and heating and cooling rates in excess of 10^{10} K s^{-1}.[22,25–28] These conditions are distinct from other conventional synthetic techniques such as photochemistry, wet chemistry, or hydrothermal synthesis (Fig. 3).

Sonochemical reactions can be categorized as primary sonochemistry (gas-phase chemistry occurring inside collapsing bubbles), secondary sonochemistry

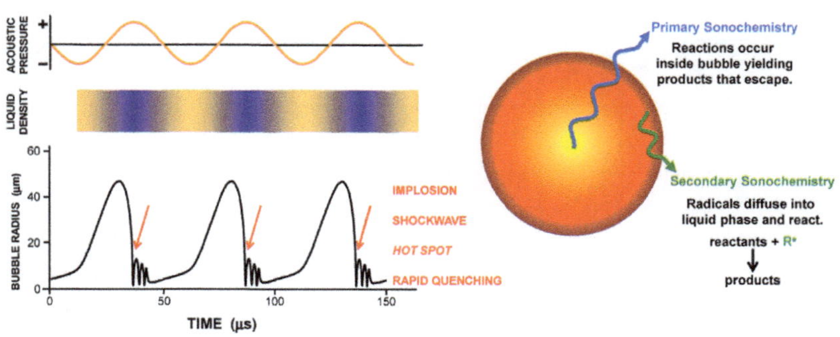

Fig. 2 Schematic illustration of the process of acoustic cavitation: the formation, growth and implosive collapse of bubbles in a liquid irradiated with high intensity ultrasound. Homogeneous sonochemistry derives from the reactions that occur within the hot spot generated by the collapsing bubble. In liquid–solid systems, acoustic cavitation can have other mechanochemical consequences due to shock waves, interparticle collisions in slurries, and microjetting near surfaces.

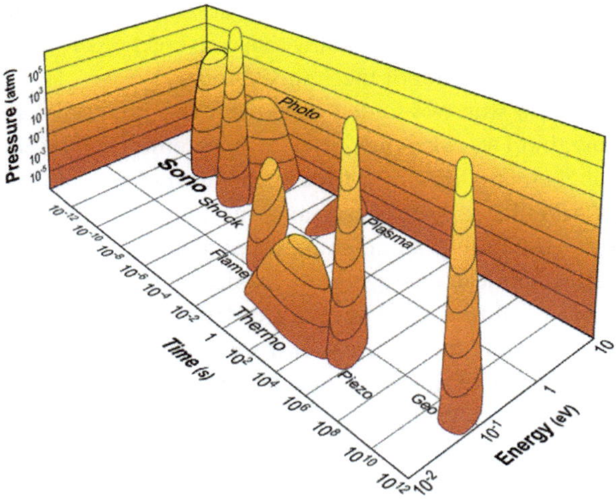

Fig. 3 Islands of chemistry as a function of time, pressure, and energy. Sonochemistry occupies a unique short-time, high-energy, and high-pressure space. Reprinted with permission from ref. 22. Copyright 2008 Annual Reviews.

(solution-phase chemistry occurring outside the bubbles from species produced in or near the bubble interface), and physical modifications (caused by high-speed jets of liquid driven at extended surfaces (Fig. 4a) or by shock waves in the liquid created by bubble collapse, Fig. 4b).[21,29,30]

It is the physical effects of sonochemistry that are most reminiscent of other forms of mechanochemistry. For example, the shock waves created by bubble rebound can accelerate solid particles suspended in the liquid. Interparticle collisions in slurries can reach velocities of hundreds of meters per second, causing changes in particle size distributions, particle morphologies, and surface compositions.[32–34] Particle agglomeration (for malleable materials, Fig. 5), particle

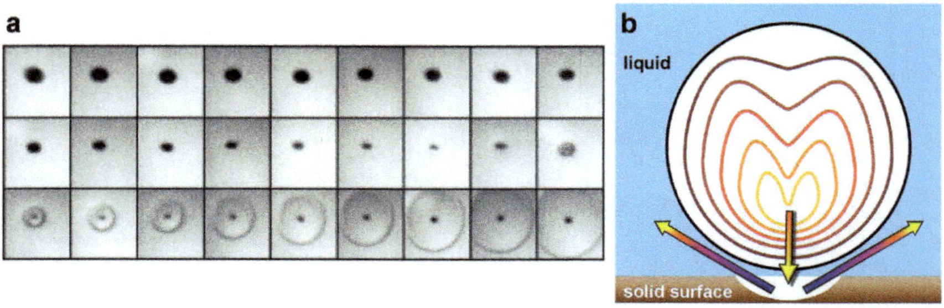

Fig. 4 (a) A spherical shock wave launched from laser-induced cavitation. The rebound of the collapsing bubble generates a shock wave in the surrounding liquid that can reach 60 kBar with velocities of 4 km s^{-1} (frame size 1.5 × 1.8 mm^2, 5 ns exposure, 20.8 million fps). Reprinted with permission from ref. 31. Copyright 1999 The Royal Society. (b) Cavitation near an extended solid surface is highly asymmetric and generates liquid jets with velocities of hundreds of m s^{-1}. The jets cause pitting, erosion, and removal of passivating surface coatings and can thus enhance surface reactions on reactive surfaces, such as Grignards and lithiations.

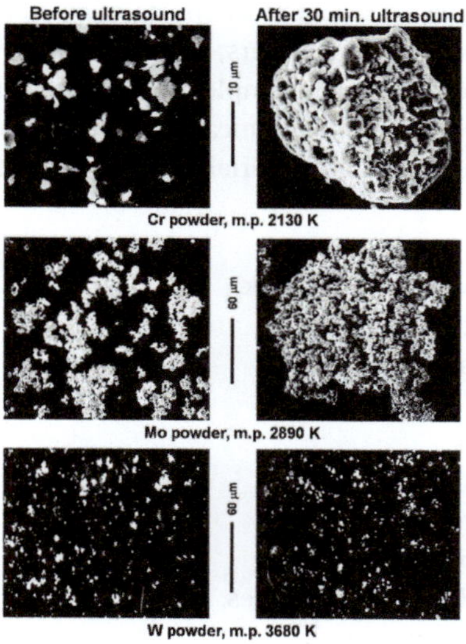

Fig. 5 Effects of ultrasonic irradiation on slurries of fine metal powders (~5 μm diameter). Sonicated in decane slurries of 2.5% metal v/v. Reprinted with permission from ref. 32. Copyright 1990 AAAS.

fragmentation (for brittle materials, Fig. 6), and exfoliation of layered materials into 2D layers have all been observed.[21,29,35,36] The consequences of cavitation lend themselves readily to the production of novel nanomaterials (especially since every bubble is an isolated nanoreactor) and provide an unusual route to novel materials without bulk high temperatures, high pressures, or long reaction times.[21,29]

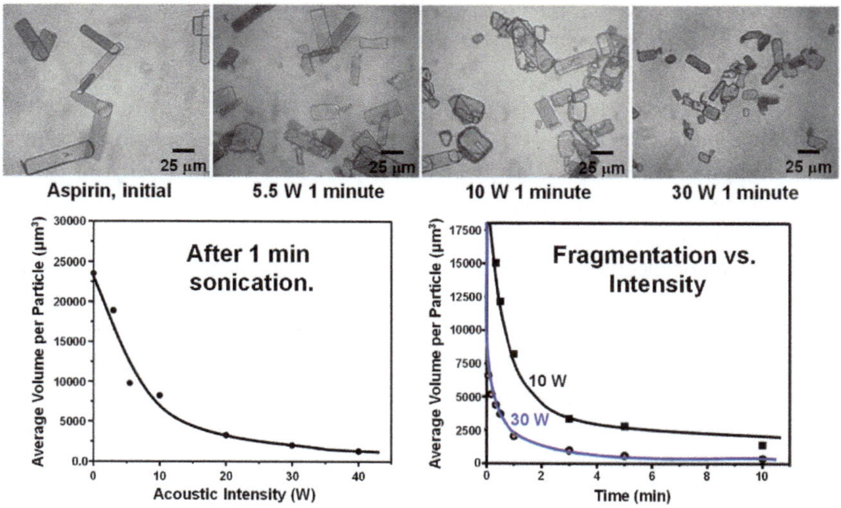

Fig. 6 Effects of ultrasonic irradiation on slurries of fine metal powders (~5 μm diameter). Sonicated in decane slurries of 2.5% metal v/v. Reprinted with permission from ref. 35. Copyright 2011 ACS.

Given the similarities in mechanical effects of cavitation of slurries and trituration (grinding, milling) of powders, there is an inevitable convergence of chemical consequences. As an example, if one examines the chemical and physical consequences of sonochemical and mechanochemical (specifically grinding) processes of inorganic materials, one arises at exactly the same list for both, as given below.

- Bulk heating.
- Local heating and pressure/shock.
- Crystal deformation.
- Formation of many defects.
- Shear stresses.
- Phase transitions.
- Amorphization.
- Reduction of crystallite sizes.
- Aggregation of crystallites.
- Acceleration of diffusion processes.
- Emission of light (common).
- Emission of electrons (uncommon).
- Chemical bond breaking.

3 Summarization of mechanochemistry: from functional solids to single molecules

As at any *Faraday Discussion*, the contributed papers formed a complementary web of topics with an associated wealth of specific questions and comments, among which several recurring themes and issues can be identified. The Introductory Lecturer (Professor Bill Jones, University of Cambridge) provided a masterly overview of the current state-of-the-art of mechanochemistry, touching on its history, areas of application, and mechanistic aspects.[37]

As was clear from the start of the Discussions, the technological and industrial applications of mechanochemistry are diverse. The earliest studies of mechanochemistry dealt with the effect of grinding on minerals and have already resulted in real applications. Mechanical activation of minerals and mechanical alloying (for example, oxide dispersions) have been produced commercially on the ton scale.[3] Fine milling of lignocellulosic biomass is already coming into use for biofuel production[38] and higher value added materials, including graphene.[39] Applications of mechanochemical activation to inorganic and ceramic synthesis are increasingly abundant at the laboratory scale, where ball milling becomes a common primary step. The use of mechanochemical methods for organic and organometallic synthesis remains a relatively new, but extremely active research area. Mechanochemistry is often viewed as a "greener" alternative because it often avoids the use of solvents.

One may break the applications discussed in FD170 into perhaps five categories: (1) inorganic compounds, metal coordination and organometallic complexes and extended coordination polymers (*e.g.*, MOFs);[40–43] (2) organic synthesis with the making and breaking of covalent bonds;[41,44–47] (3) supramolecular and pharmaceutical reactions and the making of co-crystals;[37,42,48] (4) inorganic and organic nanomaterials;[39,40,49–51] and (5) polymer chain scission, intra- and inter-chain reactions.[52–55]

Mechanochemical preparation of co-crystals of pharmaceuticals became a prominent theme that emerged in multiple papers during FD170.[37,42,48] Co-crystals are one example of the broader issues of supramolecular solid state chemistry, but one that has important biomedical applications for drug delivery, pharmaceutical efficacy, and admittedly pharma intellectual property extension. The use of twin screw extruders for this purpose was particularly well received.[48]

The production of nanomaterials (*i.e.*, particles no more than a few nm in size) and nanoscaled materials (*i.e.*, bulk materials with features or crystallites at the few nm scale) through mechanochemical means was another major theme of the Discussions. There are two general approaches to such nanomaterials: top down or bottom up, as illustrated in Fig. 7. Inherently, sufficient application of trituration (*i.e.*, comminution, grinding, or milling) will lead to amorphization. One may expect that such processes will take crystalline materials and induce defects and fragmentation, followed by agglomeration and formation of mosaic particles, to finally amorphous products, but the meanings of these terms implies a length scale: from mm to μm to nm to Å. One must realize that the ultimate defect of any crystal is its outer surface. The meaning of "amorphous" is inherently unclear unless one states a length scale. In discussions of mechanochemistry and crystallinity of products, the community must take more care to specify the relevant length scales.

The continued expansion of the use of mechanophores in polymer mechanochemistry was another significant aspect of FD170.[53,54] The use of well-defined "weak-links" in polymer chains as a means of controlling bond scission has developed well over the past decade.[8,56] Progress towards coupling mechanical forces into more general chemical reactions will be an important aspect of mechanochemistry over the next decade.

Surprisingly, mechanistic aspects of mechanochemical reactions induced by trituration, grinding, or milling were not heavily discussed in FD170. For both experimentalists and theorists, understanding the fundamental nature of mechanochemical reactions remains an important but largely unsolved problem.[57] Mechanochemical processes are complex; they involve very wide

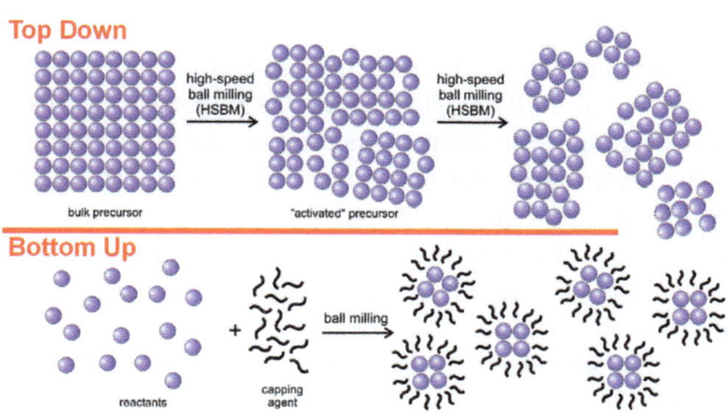

Fig. 7 Mechanochemical synthesis of nanomaterials and nanoscale materials through amorphization, comminution, and grinding can occur either from bulk (*i.e.*, mm scale) materials down or from molecular or quantum dot precursors.[51]

length and time scales; they are diverse and often system specific. Among these we have surface modification, crack propagation, surface electrification and even plasma-like transients, dislocations and amorphization, formation of metastable polymorphs, and localized impulse heating at points of impact, all occurring simultaneously with timescales ranging from ns to minutes.[57] Clear definition of the relative importance of these phenomena has not yet emerged and indeed is likely to vary substantially from system to system.

Some mechanistic aspects of mechanochemistry, however, were part of our Discussions. While macroscopic experimental investigations, empirical models, and the characterization of mechanically activated materials have provided the lion's share of our empirical knowledge of mechanochemistry, single molecule studies using atomic force microscopy[54] is a very promising tool to address fundamental questions on the atomic scale. *In situ* XRD characterization of milling processes has received particular attention using new techniques available through the high intensity X-ray beams available at the European Synchrotron Radiation Facility.[43] The sudden onset of solid state phase transitions during milling was particularly striking, and especially so in light of Takacs' extremely interesting paper on exothermic mechanochemical reactions, where again, sudden and reproducible onset of reaction (*e.g.*, between Sn and Se powders) was observed.[58] The electrification of surfaces was also discussed, and Galembeck's paper introduced the participants to the current state of the art of the closely related field of tribochemistry;[59] comparisons between high resolution electrostatic potential maps (using a non-contact Kelvin scanning electrode) were compared with SEM images of the same regions of polymer films after tribocharging (*i.e.*, rubbing of surface), and it was argued that surface electrification occurs through transfer of tribochemical reaction products.

Within the context of mechanistic concerns, it is obvious that computer simulations have become an increasingly important component of all theoretical investigations, from modelling the macroscopic operation of milling devices to molecular dynamics and quantum chemical calculations.[13,60,61] The difficulties in such approaches, particularly for MD-QM calculations under mechanochemical conditions, reveals the underlying complexity of such systems, and it may be some time before such results prove predictive.

The issues of mechanochemical equipment and scale-up were a plentiful source of discussion. A significant practical problem remains in the field of mechanochemistry: how does one compare results that use different apparatus, how does one interpret changes in procedures (*e.g.*, different grinding balls), how does one compare ball mills to jet impact mills to planetary mills, and so on? There are no well-defined mechanochemical equivalents of actinometry. It seems to me that the sharp onset that is observed for highly exothermic solid–solid reactions,[58] however, provides a real possibility of using such self-sustaining reactions as a means of calibrating different mechanochemical apparatus with respect to one another. This would be a major advance to mechanochemical experimentalists.

In terms of equipment itself, a wide range of apparatus studies were discussed, from the small to the very large. A clever and simple small-scale photo-mechanochemical apparatus was described by MacGillivray based on a modification of a common vortex mixer.[45] Improvements in efficiency and scale-up were examined for a variety of different larger scale laboratory equipment for both organic and

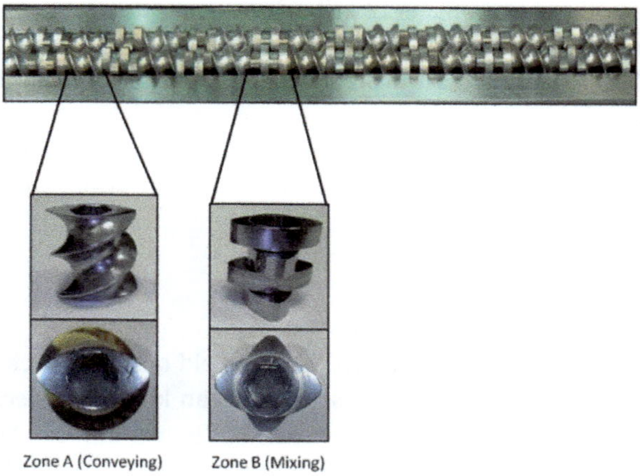

Fig. 8 A Prism PharmaLab twin screw extruder showing conveying and mixing elements.[48]

inorganic systems.[39,47,55,62] Blair's use of finite element modelling of ball milling apparatus provided some visual insights into more effective operation of commercially available devices.[39] Excitingly, the work of Daurio *et al.*[48] on twin-screw extruders (Fig. 8) for solid state manufacturing of pharmaceutical co-crystals provides a new route to large scale, continuous flow mechanochemical processing; while such extruders are common for polymer blending and food processing, their chemical consequences have only rarely been examined.[63] The development of such extruders for a wider range of other mechanochemical reactions will be very interesting.

Finally, let me end this discussion with a list of a dozen open issues and questions that I see, at least within the context of FD170, as worthy of further investigation in the field of mechanochemistry.

• The ability to predict mechanochemical reactions is still mostly lacking.

• Our fundamental QM understanding of breakage of chemical bonds during mechanical stress remains to be further developed.

• Initial conditions at point of impact during milling or grinding (*i.e.*, temperature and pressure resolved at ns timescale) are generally unknown.

• Role of materials hardness remains to be clarified: fracture *vs.* defects *vs.* plastic deformation *vs.* agglomeration *vs.* local eutectic/glassy solid/solid solution.

• Why do supramolecular rearrangements dominate with organics *vs.* covalent/ionic rearrangements with inorganics?

• How wide is the range of molecular chemistry and synthesis that can be done mechanochemically, especially without the use of organic solvents?

• Can endothermic mechanochemical reactions dissipate energy from shocks?

• Control of selectivity of macromolecular chemistry using mechanophores: how generally can one change the specificity of macromolecular chemistry mechanochemically?

• What are the connections between chemistry and luminescence in mechanoluminescence? Can mechanoluminescence serve as a probe of conditions during mechanochemical reactions?

• What are the mechanisms of triboelectrification (*e.g.*, surface chemistry, electron transfer, proton transfer, ion transfer…)?

• Can self-propagating reactions be used for calibration of mechanochemical apparatuses?

• Scale-up issues require further investigation: energy efficiency *vs.* mechanochemical apparatuses, improvements beyond the kg scale, and flow *vs.* batch processing.

4 Conclusions

It became clear during the Discussion that the field of mechanochemistry is still in active development. In a very real sense, human knowledge recapitulates our anthropological evolution: we start as hunter-gatherers and ask "Where?" as in "Where's the food?" We develop into naturalists and ask "What?" as in "What's that long-toothed creature and is it good to eat?". We progress further and become engineers asking "How?" as in "How do we keep that long-toothed creature from eating us?", and finally only as scientists do we turn to "Why?" as in "Why did that spear work so well to puncture that long-toothed creature?". In mechanochemistry, the progression from descriptive questions to more mechanistic concerns is still at an early stage. We do not yet have a firm grasp of the underlying conditions created during most mechanochemical events; indeed we do not even have a fully resolved theory of the underlying quantum mechanics that connects chemical and mechanical phenomena.

The fundamental importance of the connection between the chemical and the mechanical makes it certain that, although this was the first Faraday Discussion dealing with mechanochemistry, it is unlikely to be the last. So let me conclude by congratulating, and thanking, the organizers and scientific committee of FD170—Tomislav Friscic, Elena Boldyreva, James Mack, Stuart James, Carsten Bolm, and Jonathan Steed—for proposing and coordinating such an interesting, diverse, and thought-provoking meeting.

Acknowledgements

This work was supported by the NSF (CHE-1011972 and DMR-1206355) and in part by the US DoD (Navy N00014-11-1).

References

1 Theophrastus, *On Stones, ca. 315 BCE, trans., E. R. Caley and J. F. C. Richards*, Ohio State University Press, Columbus, OH, 1956.

2 M. Faraday, *Q. J. Sci., Lit., Arts*, 1820, **8**, 374–376.

3 L. Takacs, *Chem. Soc. Rev.*, 2013, **42**, 7649–7659.

4 W. Ostwald, Die chemische Literatur und die Organisation der Wissenschaft, in *Handbuch der allgemeinen Chemie*, Akademische Verlagsgesellschaft, Leipzig, 1919.

5 K. M. Wiggins, J. N. Brantley and C. W. Bielawski, *Chem. Soc. Rev.*, 2013, **42**, 7130–7147.

6 G.-W. Wang, *Chem. Soc. Rev.*, 2013, **42**, 7668–7700.

7 K. Ralphs, C. Hardacre and S. L. James, *Chem. Soc. Rev.*, 2013, **42**, 7701–7718.

8 P. A. May and J. S. Moore, *Chem. Soc. Rev.*, 2013, **42**, 7497–7506.
9 G. K. Kole and J. J. Vittal, *Chem. Soc. Rev.*, 2013, **42**, 1755–1775.
10 R. Groote, R. T. M. Jakobs and R. P. Sijbesma, *Polym. Chem.*, 2013, **4**, 4846–4859.
11 D. Braga, L. Maini and F. Grepioni, *Chem. Soc. Rev.*, 2013, **42**, 7638–7648.
12 E. Boldyreva, *Chem. Soc. Rev.*, 2013, **42**, 7719–7738.
13 J. Ribas-Arino and D. Marx, *Chem. Rev.*, 2012, **112**, 5412–5487.
14 S. L. James, C. J. Adams, C. Bolm, D. Braga, P. Collier, T. Friscic, F. Grepioni, K. D. M. Harris, G. Hyett, W. Jones, A. Krebs, J. Mack, L. Maini, A. G. Orpen, I. P. Parkin, W. C. Shearouse, J. W. Steed and D. C. Waddell, *Chem. Soc. Rev.*, 2012, **41**, 413–447.
15 T. Friscic, *Chem. Soc. Rev.*, 2012, **41**, 3493–3510.
16 H. Noji, D. Okuno and T. Ikeda, *Chem. Sci.*, 2011, **2**, 2086–2093.
17 J. Liang and J. M. Fernandez, *ACS Nano*, 2009, **3**, 1628–1645.
18 M. K. Beyer and H. Clausen-Schaumann, *Chem. Rev.*, 2005, **105**, 2921–2948.
19 C. Bustamante, D. Keller and G. Oster, *Acc. Chem. Res.*, 2001, **34**, 412–420.
20 D. E. Ingber, N. Wang and D. Stamenovic, *Rep. Prog. Phys.*, 2014, **77**, 046603.
21 H. Xu, B. W. Zeiger and K. S. Suslick, *Chem. Soc. Rev.*, 2013, **42**, 2555–2567.
22 K. S. Suslick and D. J. Flannigan, *Annu. Rev. Phys. Chem.*, 2008, **59**, 659–683.
23 K. S. Suslick, *Science*, 1990, **247**, 1439–1445.
24 K. S. Suslick, *Ultrasound: Its Chemical, Physical, and Biological Effects*, 1988.
25 D. J. Flannigan and K. S. Suslick, *Nature*, 2005, **434**, 52–55.
26 D. J. Flannigan and K. S. Suslick, *Nat. Phys.*, 2010, **6**, 598–601.
27 W. B. McNamara, Y. T. Didenko and K. S. Suslick, *Nature*, 1999, **401**, 772–775.
28 Y. T. Didenko and K. S. Suslick, *Nature*, 2002, **418**, 394–397.
29 K. S. Suslick and G. J. Price, *Annu. Rev. Mater. Sci.*, 1999, **29**, 295–326.
30 J. H. Bang and K. S. Suslick, *Adv. Mater.*, 2010, **22**, 1039–1059.
31 C. D. Ohl, T. Kurz, R. Geisler, O. Lindau and W. Lauterborn, *Philos. Trans. R. Soc., A*, 1999, **357**, 269.
32 S. J. Doktycz and K. S. Suslick, *Science*, 1990, **247**, 1067–1069.
33 K. S. Suslick and S. J. Doktycz, *J. Am. Chem. Soc.*, 1989, **111**, 2342–2344.
34 T. Prozorov, R. Prozorov and K. S. Suslick, *J. Am. Chem. Soc.*, 2004, **126**, 13890–13891.
35 B. W. Zeiger and K. S. Suslick, *J. Am. Chem. Soc.*, 2011, **133**, 14530–14533.
36 H. Xu and K. S. Suslick, *J. Am. Chem. Soc.*, 2011, **133**, 9148–9151.
37 W. Jones, *Faraday Discuss.*, 2014, **170**, in press.
38 S. Fujimoto, H. Inoue, S. Yano, T. Sakaki, T. Minowa, T. Endo, S. Sawayama and K. Sakanishi, *J. Jpn. Pet. Inst.*, 2008, **51**, 264–273.
39 R. G. Blair, K. Chagoya, S. Biltek, S. Jackson, A. Sinclair, A. Taraboletti and D. T. Restrepo, *Faraday Discuss.*, 2014, **170**, DOI: 10.1039/c4fd00007b.
40 V. Šepelák, M. Myndyk, R. Witte, J. Röder, D. Menzel, R. H. Schuster, H. Hahn, P. Heitjans and K.-D. Becker, *Faraday Discuss.*, 2014, **170**, DOI: 10.1039/c3fd00137g.
41 I. Z. Hlova, S. Gupta, J. F. Goldston, T. Kobayashi, M. Pruski and V. K. Pecharsky, *Faraday Discuss.*, 2014, **170**, DOI: 10.1039/c3fd00161j.
42 L. Tröbs and F. Emmerling, *Faraday Discuss.*, 2014, **170**, DOI: 10.1039/c3fd00163f.

43 I. Halasz, T. Friščić, S. A. J. Kimber, K. Užarević, A. Puškarić, C. Mottillo, P. Julien, V. Štrukil, V. Honkimäki and R. E. Dinnebier, *Faraday Discuss.*, 2014, **170**, DOI: 10.1039/c4fd00013g.
44 A. A. L. Michalchuk, I. A. Tumanov, V. A. Drebushchak and E. V. Boldyreva, *Faraday Discuss.*, 2014, **170**, DOI: 10.1039/c3fd00150d.
45 J. Stojaković, B. S. Farris and L. R. MacGillivray, *Faraday Discuss.*, 2014, **170**, DOI: 10.1039/c4fd00006d.
46 H. Shy, P. Mackin, A. S. Orvieto, D. Gharbharan, G. R. Peterson, N. Bampos and T. D. Hamilton, *Faraday Discuss.*, 2014, **170**, DOI: 10.1039/c3fd00140g.
47 A. Stolle, R. Schmidt and K. Jacob, *Faraday Discuss.*, 2014, **170**, DOI: 10.1039/c3fd00144j.
48 D. Daurio, K. Nagapudi, L. Li, P. Quan and F.-A. Nunez, *Faraday Discuss.*, 2014, **170**, DOI: 10.1039/c3fd00153a.
49 P. Baláž, M. Baláž, M. Čaplovičová, A. Zorkovská, Ľ. Čaplovič and M. Psotka, *Faraday Discuss.*, 2014, **170**, DOI: 10.1039/c3fd00117b.
50 L. Maini, P. P. Mazzeo, F. Farinella, V. Fattori and D. Braga, *Faraday Discuss.*, 2014, **170**, DOI: 10.1039/c3fd00164d.
51 M. J. Rak, T. Friščić and A. Moores, *Faraday Discuss.*, 2014, DOI: 10.1039/c4fd00053f.
52 A. L. B. Ramirez, A. K. Schmitt, M. K. Mahanthappa and S. L. Craig, *Faraday Discuss.*, 2014, **170**, DOI: 10.1039/c4fd00001c.
53 T. Shiraki, C. E. Diesendruck and J. S. Moore, *Faraday Discuss.*, 2014, **170**, DOI: 10.1039/c4fd00027g.
54 S. W. Schmidt, M. F. Pill, A. Kersch, H. Clausen-Schaumann and M. K. Beyer, *Faraday Discuss.*, 2014, **170**, DOI: 10.1039/c3fd00119a.
55 A. Politov and O. Golyazimova, *Faraday Discuss.*, 2014, **170**, DOI: 10.1039/c3fd00143a.
56 J. N. Brantley, K. M. Wiggins and C. W. Bielawski, *Polym. Int.*, 2013, **62**, 2–12.
57 P. Baláž, *Mechanochemistry in Nanoscience and Minerals Engineering*, Springer-Verlag, Berlin, 2008.
58 L. Takacs, *Faraday Discuss.*, 2014, **170**, DOI: 10.1039/c3fd00133d.
59 L. B. d. S. Balestrin, D. D. Duque, D. S. d. Silva and F. Galembeck, *Faraday Discuss.*, 2014, **170**, DOI: 10.1039/c3fd00118k.
60 P. J. Bygrave, D. H. Case and G. M. Day, *Faraday Discuss.*, 2014, **170**, DOI: 10.1039/c3fd00162h.
61 Z. Huang and R. Boulatov, *Chem. Soc. Rev.*, 2011, **40**, 2359–2384.
62 D. W. Peters and R. G. Blair, *Faraday Discuss.*, 2014, **170**, DOI: 10.1039/c3fd00157a.
63 A. L. Black, J. A. Orlicki and S. L. Craig, *J. Mater. Chem.*, 2011, **21**, 8460–8465.

Poster titles

Solvent-directed polymorph formation by solvent-free DCC reaction, **A. Belenguer, G. Lampronti, D. Wales and J. Sanders,** *University of Cambridge, UK*

Solvent-free mechanochemical synthesis of gold nanoparticles stabilized with amines or biomass matrix, **A. Moores, M. J. Rak and T. Friščić,** *McGill University, Canada*

Molecular switches for AFM measurements, **I. Koehl, U. Luening and M. Beyer,** *Christian-Albrechts-Universität zu Kiel, Germany*

1,2,3-Triazoles for mechanical induced 1,3-dipolar cycloreversion, **K. Holz and U. Lüning,** *Otto Diels-Institut für Organische Chemie, Germany*

Energetic considerations of organic reactions in a planetary ball mill, **R. Schmidt, K. Jacob, C. Burmeister, S. Breitung-Faes, A. Kwade and A. Stolle,** *Friedrich-Schiller-University Jena, Germany*

Good or no good - that's the question! Assessment of organic syntheses in planetary ball mills using auxiliaries, **K. Jacob, R. Schmidt and A. Stolle,** *Friedrich-Schiller-University Jena, Germany*

Mechano-chemical synthesis of $CoFe_2O_4$ magnetic nanoparticles, **A. Ataie and S. Rashidi,** *University of Tehran, Iran*

Component discrimination in the preparation of ternary cocrystals, **F. Topic and K. Rissanen,** *University of Jyväskylä, Finland*

Electrochemical quantification of solution-phase mechanochemistry: insulating polymers, **T. Varley and K. Holt,** *University College London, UK*

Synthesis and evaluation α,β-(*S*)-proline-containing dipeptides as organocatalysts in solvent-free asymmetric aldol reactions under ball-milling conditions, **E. Machuca, E. Juaristi and Y. Rojas,** *CINVESTAV, Zacatenco, Mexico*

Stretching molecular force probes: Where is the force stored?, **Tim S. and A. Dreuw,** *University of Heidelberg, Germany*

Sonochemical *versus* mechanochemical activation of polyaluminium chloride for the low temperature synthesis of α-Al_2O_3, **P. Billik, M. Novotný and M. Čaplovičová,** *Slovak Academy of Sciences, Slovakia*

Identification of siloxane mechanochemistry with single molecule force spectroscopy and *ab initio* simulations, **M. Pill, A. Kersch, M. Beyer and H. Clausen-Schaumann,** *Munich University of Applied Sciences, Germany*

Organocatalytic solvent-free asymmetric Michael additions under ball milling conditions, **M. Jörres and C. Bolm,** *RWTH Aachen University, Germany*

Mechanochemical reactions of sulfathiazole and sulfamerazine with carboxylic acids: Formation of salts, co-crystals and co-amorphous solids, **A. Erxleben, K. Gniado and P. McArdle,** *National University of Ireland, Galway, Republic of Ireland*

Destruction of decabromodiphenyl ether into visible light photocatalyst BiOBr by ball milling, **K. Zhang, J. Huang, H. Wang, G. Yu and Q. Zhang,** *Tsinghua University, China*

Tetra-aryl and tetra-alkyl-substituted porphyrins synthesized by mechanochemistry, **T. Hamilton, Q. Su, P. Mackin and D. Gharbharan,** *Barry University, USA*

Purification of porphyrins from mechanochemistry using entrainment sublimation, **T. Hamilton and V. Hoelscher,** *Barry University, USA*

Mechanochemical assembly of IRMOFs enabled by pre-designed oxo-zinc amidate precursors, **D. Prochowicz, K. Sokołowski, A. Kornowicz, I. Justyniak, T. Friščić and J. Lewiński,** *Warsaw University of Technology, Poland*

Towards chitin depolymerization using mechanochemical methods, **G. Margoutidis and F. Kerton,** *Memorial University of Newfoundland, Canada*

Modelling mechanochemical processes under constant force and constant extension conditions, **G. Kochhar and N. Mosey,** *Queen's University, Canada*

Mechanochemistry and its applications: from organic and peptide synthesis to catalysis, **F. Lamaty, E. Colacino, T.-X. Métro, X. Bantreil and J. Martinez,** *Institut des Biomolécules Max Mousseron, France*

Solvent vapour effects on the kinetics of solid state phase transitions – an aid for the understanding liquid assisted mechanochemical reactions, **V. Stilinovic,** *University of Zagreb, Croatia*

Conductance and force in angstrom-scale molecular junctions, **A. Nielsen, K. Bharat, L. Venkataraman and C. Nuckolls,** *Columbia University, USA*

The black polymorph of TTF-CA: TTF polymorphism and solvent effects in mechanochemical and vapor digestion syntheses, crystal structure and properties, **S. Pagola, S. H. Lapidus, A. Naik, Alex W., N. E. Massa, V. T. Phuoc, L. del Campo, S. Lebegue, J. G. Angyan, T. Abdel-Fattah and S. Pagola,** *College of William and Mary, USA*

Two faces of mechanochemistry: From mechanosynthesis to new compounds and from new compounds to mechanical motion, **J. G. Hernández, C. Mottillo, I. S. Butler, O. Magydszuk, R. E. Dinnebier and T. Friščić,** *McGill University, Canada*

Medicine balls: Mechanochemical catalytic synthesis of pharmaceutically relevant sulfonyl-ureas, -thioureas and –guanidines, **D. Tan, V. Štrukil, C. Mottillo, A. D. Katsenis and T. Friščić,** *McGill University, Canada*

Photomechanical movement of azobenzene crystals, **O. Bushuyev, A. Tomberg, C. Corkery, T. Friščić and C. Barrett,** *McGill University, Canada*

Investigating the solid-state supramolecular chemistry of steroids using mechanochemistry, **K. Ardila Fierro and T. Friščić,** *McGill University, Canada*

Olefin metathesis by mechanochemical milling, **J.-L. Do, V. Štrukil and T. Friščić,** *McGill University, Canada*

Solvothermal and mechanochemical approach to the synthesis of homochiral coordination polymers based on CuX and *Cinchona* alkaloids, **P. Krupinski, A. Kornowicz, D. Prochowicz, I. Justyniak and J. Lewiński,** *Polish Academy of Sciences, Poland*

A short story about breaking peroxide bonds by mechanical force, **J. Lewiński, K. Zelga, M. Wolska, I. Justyniak, M. Kubisiak, Z. Kaszkur and B. Grzybowski,** *Polish Academy of Sciences, Poland*

Biomimetic mineral neogenesis as a simple, low-energy method for solid-state separation of minerals and synthesis of metal–organic frameworks from metal oxides, **I. Huskic, F. Qi and T. Friščić,** *McGill University, Canada*

Solid state synthesis of ZIFs: Templation effects and novel ZIF structures, **I. Brekalo, C. Kane, J. R. Ramirez and K. T. Holman,** *Georgetown University,*

Exploring the solid-state synthesis of CO-releasing molecules (CORMs) based on group VI metal carbonyls, **N. Macdonald, J. Wong, C. Mottillo, I. Butler and T. Friščić,** *McGill University, Canada*

Synthesis of porous functional materials by solid-state mechanochemical methods, **C. Mottillo, Y. Lu, M. Frenette, V. Glembockyte, G. Cosa and T. Friščić,** *McGill University, Canada*

Characterization of mechanochemically synthesized CdS@ZnS quantum dots: Focus on surface properties and potential interaction with capping agent, **M. Baláž, P. Baláž, A. Zorkovská, M. Čaplovičová and L. Čaplovič,** *Slovak Academy of Sciences, Slovakia*

Mechanochemical synthesis of coordination polymers *via* acid-base reactions, **V. K. Gramm and U. Ruschewitz,** *University of Cologne, Germany*

Controlling the outcome of mechanochemical synthesis, **N. Judas, T. Friščić and W. Jones,** *University of Zagreb, Croatia*

Effect of the binary and ternary exchanges on crystallinity and textural properties of X zeolites, modification, characterisation and application, **H. Habib, B. Souhila, M.-K. Kheira, R. Marouf, S. Jacques and K. Amine,** *University of Mostaganem, Algeria*

Enantiomer-specific oriented attachment in conglomerate crystals of guanidine carbonate, **T. P. Thao Nguyen, P. Ramrup, J. Kwiatoszynski, P. S. M. Cheung, L. Cuccia, P. Cintas and C. Viedma,** *Concordia University, Canada*

Monitoring the mechanochemical transformations of organic pharmaceutical solids *via* solid-state fluorescence emission spectroscopy, **J. Wong, M. Frenette, G. Cosa and T. Friščić,** *McGill University, Canada*

Controlled assembly of semi-quinone radical based materials in the solid state, **M. Glavinovic, F. Qi, A. D. Katsenis, C. Mottillo, J.-P. Lumb and T. Friščić**, *McGill University, Canada*

The Skinner Prize for the best poster was awarded to Mr Gurpaul Kochhar for his poster on modelling mechanochemical processes under constant force and constant extension conditions.

List of participants

Mr Bassim Abdulwahaab, *University of Hull, United Kingdom*
Dr Lea Anderson-Smith, *SPEX SamplePrep*
Ms Karen Ardila, *Mcgill University, Canada*
Professor Abolghasem Ataie, *University of Tehran, Iran*
Professor Peter Baláž, *Slovak Academy of Sciences, Slovakia*
Dr Matej Baláž, *Slovak Academy of Sciences, Slovakia*
Dr Ana Belenguer, *University of Cambridge, United Kingdom*
Professor Martin K. Beyer, *Leopold-Franzens-Universität Innsbruck, Germany*
Mr Peter Billik, *Slovak Academy of Sciences, Slovakia*
Dr Richard Blair, *University of Central Florida, U. S. A.*
Professor Carsten Bolm, *RWTH Aachen University, Germany*
Ms Ivana Brekalo, *Georgetown University, U. S. A.*
Mr Cameron Brown, *Duke University, U. S. A.*
Mr Oleksandr Bushuyev, *McGill University, Canada*
Dr Peter Bygrave, *University of Southampton, United Kingdom*
Dr Andres Cardozo Perez, *L'Oreal*
Mr Batsho Chite, *IHHSM*
Mr Jan Christopherson, *McGill University, Canada*
Professor Stephen Craig, *Duke University, U. S. A.*
Professor Louis Cuccia, *Concordia University, Canada*
Dr Graeme Day, *University of Southampton, United Kingdom*
Dr Charles Diesendruck, *University of Illinois, U. S. A.*
Mr Jean-Louis Do, *McGill University, Canada*
Dr Andrea Erxleben, *National University of Ireland, Galway, Ireland*
Dr Daniel Frankel, *Bruker AXS*
Dr Tomislav Friščić, *McGill University, Canada*
Professor Fernando Galembeck, *Brazilian National Nanotechnology Laboratory, Brazil*
Mr Martin Glavinovic, *McGill University, Canada*
Ms Verena Katharina Gramm, *University of Cologne, Germany*
Dr Shalabh Gupta, *The Ames Laboratory, U. S. A.*
Dr Hammoudi Habib, *University of Mostaganem, Algeria*
Dr Ivan Halasz, *Rudjer Boskovic Institute, Croatia*
Dr Tamara Hamilton, *Barry University, U. S. A.*
Dr José G. Hernández, *McGill University, Canada*
Professor Saeed Heshmati-Manesh, *University of Tehran, Iran*
Miss Victoria Hoelscher, *Barry University, U. S. A.*
Miss Katharina Holz, *Otto Diels-Institut für Organische Chemie, Germany*
Mr Igor Huskic, *McGill University, Canada*
Dr Karl Irikura, *National Institute of Standards & Technology, U. S. A.*
Mrs Katharina Jacob, *Friedrich-Schiller University Jena, Germany*

Professor Stuart James, *Queen's University Belfast, MOF Technologies, United Kingdom*
Dr Phil Jones, *Imerys*
Professor Bill Jones, *University of Cambridge, United Kingdom*
Mr Manuel Jörres, *RWTH Aachen University, Germany*
Professor Nenad Judas, *University of Zagreb, Croatia*
Mr Patrick Julien, *McGill University, Canada*
Mr Pritam Kadam, *Ruhr University of Bochum, Germany*
Dr Constantine Khripin, *Michelin*
Mr Ross Koby, *McGill University, Canada*
Mr Gurpaul Kochhar, *Queen's University, United Kingdom*
Miss Isabel Koehl, *Christian-Albrechts-Universität zu Kiel, Germany*
Mr Piotr Krupinski, *Polish Academy of Sciences, Poland*
Dr Frédéric Lamaty, *Institut des Biomolécules Max Mousseron, France*
Professor Janusz Lewiński, *Polish Academy of Sciences, Poland*
Mr Neil Macdonald, *McGill University, Canada*
Professor Len MacGillivray, *University of Iowa, U. S. A.*
Miss Elizabeth Machuca, *CINVESTAV, Mexico*
Professor James Mack, *University of Cincinnati, U. S. A.*
Miss Paula Mackin, *Barry University, U. S. A.*
Dr Lucia Maini, *Università di Bologna, Italy*
Mr George Margoutidis, *Memorial University of Newfoundland, Canada*
Dr Heather Montgomery, *Royal Society of Chemistry, United Kingdom*
Dr Audrey Moores, *McGill University, Canada*
Miss Cristina Mottillo, *McGill University, Canada*
Mr Christopher Nickels, *McGill University, Canada*
Ms Aileen Nielsen, *Columbia University, U. S. A.*
Dr Heiko Oertling, *Nestec Ltd.*
Mr Yussuf Omotayo, *University of Leeds, United Kingdom*
Mr Adebare Matthew Oyebode, *ICMA Limited*
Dr Silvina Pagola, *College of Wiliam and Mary, U. S. A.*
Dr Vitalij K. Pecharsky, *The Ames Laboratory, U. S. A.*
Dr David Peters, *ATMI*
Mr Michael Pill, *Munich University of Applied Sciences, Germany*
Dr Daniel Prochowicz, *Warsaw University of Technology, Poland*
Ms Monika Rak, *McGill University, Canada*
Mr Xavier Saintigny, *Michelin*
Mr Robert Schmidt, *Friedrich-Schiller University Jena, Germany*
Professor Vladimir Šepélak, *Karlsruhe Institute of Technology, Germany*
Professor Peter Skabara, *University of Strathclyde, United Kingdom*
Mr Kamil Sokolowski, *Polish Academy of Sciences, Poland*
Dr Michael Spencelayh, *Royal Society of Chemistry, United Kingdom*
Mr Tim Stauch, *University of Heidelberg, Germany*
Dr Vladimir Stilinovic, *University of Zagreb, Croatia*
Dr Achim Stolle, *Friedrich-Schiller University Jena, Germany*
Ms Qiwen Su, *Barry University, U. S. A.*
Professor Kenneth Suslick, *University of Illinois, U. S. A.*
Professor Laszlo Takacs, *University of Maryland Baltimore County, U. S. A.*
Mr Davin Tan, *McGill University, Canada*
Mrs Rachel Thompson, *Royal Society of Chemistry, United Kingdom*
Mr Filip Topic, *University of Jyväskylä, Finland*
Ms Lisa Tröbs, *BAM Federal Institute for Materials Research and Testing, Germany*
Dr Thomas Varley, *University College London, United Kingdom*

Dr Florian Viton, *Nestec Ltd.*
Miss Jessica Wong, *McGill University*, *Canada*
Miss Elizabeth Woodhouse, *Royal Society of Chemistry*, *United Kingdom*
Professor Varoujan Yaylayan, *McGill University*, *Canada*
Mr Kunlun Zhang, *Tsinghua University*, *PR China*